Instructor's Solutions Manual

Tamas Wiandt
Rochester Institute of Technology

to accompany

Elementary Differential Equations, Tenth Edition
and
Elementary Differential Equations and Boundary Value Problems, Tenth Edition

William E. Boyce
Edward P. Hamilton Professor Emeritus

Richard C. DiPrima
formerly Eliza Ricketts Foundation Professor

Department of Mathematical Sciences
Rensselaer Polytechnic Institute

WILEY

PUBLISHER	Laurie Rosatone
ACQUISITIONS EDITOR	David Dietz
FREELANCE DEVELOPMENTAL EDITOR	Anne Scanlon-Rohrer
ASSOCIATE CONTENT EDITOR	Beth Pearson
SENIOR EDITORIAL ASSISTANT	Jacqueline Sinacori
SENIOR CONTENT MANAGER	Karoline Luciano
SENIOR PRODUCTION EDITOR	Kerry Weinstein

Founded in 1807, John Wiley & Sons, Inc. has been a valued source of knowledge and understanding for more than 200 years, helping people around the world meet their needs and fulfill their aspirations. Our company is built on a foundation of principles that include responsibility to the communities we serve and where we live and work. In 2008, we launched a Corporate Citizenship Initiative, a global effort to address the environmental, social, economic, and ethical challenges we face in our business. Among the issues we are addressing are carbon impact, paper specifications and procurement, ethical conduct within our business and among our vendors, and community and charitable support. For more information, please visit our website: www.wiley.com/go/citizenship.

Copyright © 2013 John Wiley & Sons, Inc. All rights reserved. No part of this publication may be reproduced, stored in a retrieval system, or transmitted in any form or by any means, electronic, mechanical, photocopying, recording, scanning or otherwise, except as permitted under Sections 107 or 108 of the 1976 United States Copyright Act, without either the prior written permission of the Publisher, or authorization through payment of the appropriate per-copy fee to the Copyright Clearance Center, Inc., 222 Rosewood Drive, Danvers, MA 01923 (Web site: www.copyright.com). Requests to the Publisher for permission should be addressed to the Permissions Department, John Wiley & Sons, Inc., 111 River Street, Hoboken, NJ 07030-5774, (201) 748-6011, fax (201) 748-6008, or online at: www.wiley.com/go/permissions.

ISBN 978-0-470-45834-1

10 9 8 7 6 5 4 3 2 1

CONTENTS

1 Introduction **1**

    1.1 1
    1.2 6
    1.3 9

2 First Order Differential Equations **13**

    2.1 13
    2.2 19
    2.3 26
    2.4 32
    2.5 35
    2.6 44
    2.7 47
    2.8 50
    2.9 57
    PR. 59

3 Second Order Linear Equations **65**

    3.1 65
    3.2 69
    3.3 72
    3.4 78
    3.5 83
    3.6 88
    3.7 95
    3.8 100

4 Higher Order Linear Equations **109**

    4.1 109
    4.2 112
    4.3 117
    4.4 122

5 Series Solutions of Second Order Linear Equations **127**

    5.1 127
    5.2 132
    5.3 146
    5.4 154
    5.5 163
    5.6 173
    5.7 183

6 The Laplace Transform 191

- 6.1 . . . 191
- 6.2 . . . 197
- 6.3 . . . 205
- 6.4 . . . 212
- 6.5 . . . 229
- 6.6 . . . 243

7 Systems of First Order Linear Equations 253

- 7.1 . . . 253
- 7.2 . . . 260
- 7.3 . . . 265
- 7.4 . . . 273
- 7.5 . . . 276
- 7.6 . . . 289
- 7.7 . . . 307
- 7.8 . . . 315
- 7.9 . . . 324

8 Numerical Methods 333

- 8.1 . . . 333
- 8.2 . . . 342
- 8.3 . . . 351
- 8.4 . . . 358
- 8.5 . . . 369
- 8.6 . . . 375

9 Nonlinear Differential Equations and Stability 379

- 9.1 . . . 379
- 9.2 . . . 385
- 9.3 . . . 396
- 9.4 . . . 414
- 9.5 . . . 433
- 9.6 . . . 445
- 9.7 . . . 449
- 9.8 . . . 459

10 Partial Differential Equations and Fourier Series 467

- 10.1 . . . 467
- 10.2 . . . 472
- 10.3 . . . 483
- 10.4 . . . 491
- 10.5 . . . 506
- 10.6 . . . 516

Contents

    10.7 527
    10.8 540

11 Boundary Value Problems and Sturm-Liouville Theory 549

    11.1 549
    11.2 558
    11.3 566
    11.4 578
    11.5 581
    11.6 583

CHAPTER 1

Introduction

1.1

1.

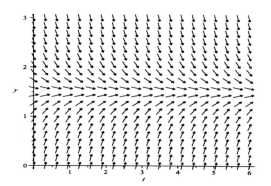

For $y > 3/2$, the slopes are negative, therefore the solutions are decreasing. For $y < 3/2$, the slopes are positive, hence the solutions are increasing. The equilibrium solution appears to be $y(t) = 3/2$, to which all other solutions converge.

3.

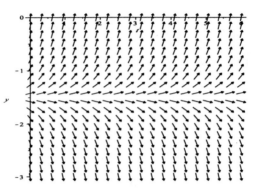

For $y > -3/2$, the slopes are positive, therefore the solutions increase. For $y < -3/2$, the slopes are negative, and hence the solutions decrease. All solutions appear to diverge away from the equilibrium solution $y(t) = -3/2$.

5.

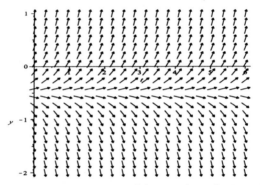

For $y > -1/2$, the slopes are positive, and hence the solutions increase. For $y < -1/2$, the slopes are negative, and hence the solutions decrease. All solutions diverge away from the equilibrium solution $y(t) = -1/2$.

6.

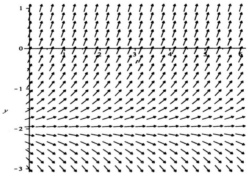

For $y > -2$, the slopes are positive, and hence the solutions increase. For $y < -2$, the slopes are negative, and hence the solutions decrease. All solutions diverge away from the equilibrium solution $y(t) = -2$.

8. For all solutions to approach the equilibrium solution $y(t) = 2/3$, we must have $y' < 0$ for $y > 2/3$, and $y' > 0$ for $y < 2/3$. The required rates are satisfied by the differential equation $y' = 2 - 3y$.

10. For solutions other than $y(t) = 1/3$ to diverge from $y = 1/3$, we must have $y' < 0$ for $y < 1/3$, and $y' > 0$ for $y > 1/3$. The required rates are satisfied by the differential equation $y' = 3y - 1$.

12.

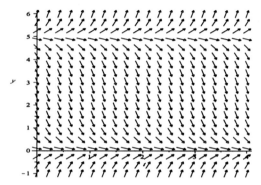

Note that $y' = 0$ for $y = 0$ and $y = 5$. The two equilibrium solutions are $y(t) = 0$ and $y(t) = 5$. Based on the direction field, $y' > 0$ for $y > 5$; thus solutions with initial values greater than 5 diverge from the solution $y(t) = 5$. For $0 < y < 5$, the slopes are negative, and hence solutions with initial values between 0 and 5 all decrease toward the solution $y(t) = 0$. For $y < 0$, the slopes are all positive; thus solutions with initial values less than 0 approach the solution $y(t) = 0$.

14.

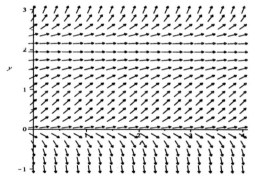

Observe that $y' = 0$ for $y = 0$ and $y = 2$. The two equilibrium solutions are $y(t) = 0$ and $y(t) = 2$. Based on the direction field, $y' > 0$ for $y > 2$; thus solutions with initial values greater than 2 diverge from $y(t) = 2$. For $0 < y < 2$, the slopes are also positive, and hence solutions with initial values between 0 and 2 all increase toward the solution $y(t) = 2$. For $y < 0$, the slopes are all negative; thus solutions with initial values less than 0 diverge from the solution $y(t) = 0$.

15. -(j) $y' = 2 - y$.

17. -(g) $y' = -2 - y$.

18. -(b) $y' = 2 + y$.

20. -(e) $y' = y(y - 3)$.

23. The difference between the temperature of the object and the ambient temperature is $u - 70$ (u in °F). Since the object is cooling when $u > 70$, and the rate constant is $k = 0.05$ min^{-1}, the governing differential equation for the temperature of the object is $du/dt = -.05(u - 70)$.

24.(a) Let $M(t)$ be the total amount of the drug (in milligrams) in the patient's body at any given time t (hr). The drug is administered into the body at a constant rate of 500 mg/hr. The rate at which the drug leaves the bloodstream is given by $0.4M(t)$. Hence the accumulation rate of the drug is described by the differential equation $dM/dt = 500 - 0.4\,M$ (mg/hr).

(b)

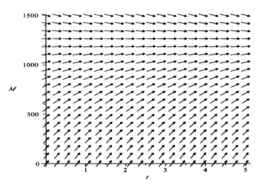

Based on the direction field, the amount of drug in the bloodstream approaches the equilibrium level of 1250 mg (within a few hours).

26.

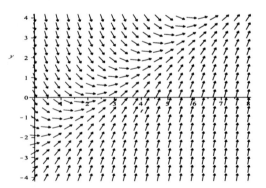

All solutions appear to approach a linear asymptote (with slope equal to 1). It is easy to verify that $y(t) = t - 3$ is a solution.

27.

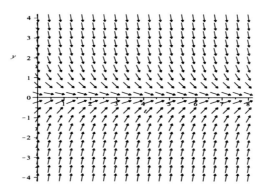

All solutions appear to approach $y = 0$.

30.

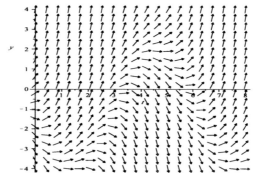

All solutions (except $y(0) = -5/2$) appear to diverge from the sinusoid $y(t) = -3\sin(t + \pi/4)/\sqrt{2} - 1$, which is also a solution corresponding to the initial value $y(0) = -5/2$.

32.

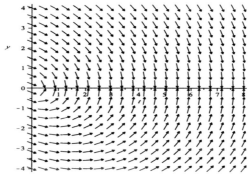

All solutions appear to converge to $y(t) = 0$. Solutions above the line $y = -2t$ (but below the t-axis) have positive slope and increase rapidly to meet the t axis.

Solutions that begin below the line $y = -2t$ eventually cross it and have positive slope.

33.

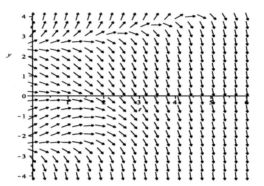

The direction field is rather complicated. Nevertheless, the collection of points at which the slope field is zero, is given by the implicit equation $y^3 - 6y = 2t^2$. The graph of these points is shown below:

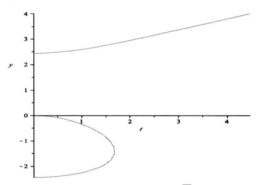

The y-intercepts of these curves are at $y = 0, \pm\sqrt{6}$. It follows that for solutions with initial values $y > \sqrt{6}$, all solutions increase without bound. For solutions with initial values in the range $y < -\sqrt{6}$ or $0 < y < \sqrt{6}$, the slopes remain negative, and hence these solutions decrease without bound. Solutions with initial conditions in the range $-\sqrt{6} < y < 0$ initially increase. Once the solutions reach the critical value, given by the equation $y^3 - 6y = 2t^2$, the slopes become negative and remain negative. These solutions eventually decrease without bound.

1.2

4.(a) The equilibrium solution satisfies the differential equation $dy_e/dt = 0$. Setting $ay_e - b = 0$, we obtain $y_e(t) = b/a$.

(b) Since $dY/dt = dy/dt$, it follows that $dY/dt = a(Y + y_e) - b = aY$.

6.(a) Consider the simpler equation $dy_1/dt = -ay_1$. As in the previous solutions, rewrite the equation as $(1/y_1)dy_1 = -a\,dt$. Integrating both sides results in $y_1(t) = ce^{-at}$.

(b) Now set $y(t) = y_1(t) + k$, and substitute into the original differential equation. We find that $-ay_1 + 0 = -a(y_1 + k) + b$. That is, $-ak + b = 0$, and hence $k = b/a$.

(c) The general solution of the differential equation is $y(t) = ce^{-at} + b/a$. This is exactly the form given by Eq.(17) in the text. Invoking an initial condition $y(0) = y_0$, the solution may also be expressed as $y(t) = b/a + (y_0 - b/a)e^{-at}$.

8.(a) The general solution is $p(t) = p_0 e^{rt}$. Based on the discussion in the text, time t is measured in months. Assuming 1 month= 30 days, the hypothesis can be expressed as $p_0 e^{r \cdot 1} = 2p_0$. Solving for the rate constant, $r = \ln(2)$, with units of per month.

(b) N days= $N/30$ months. The hypothesis is stated mathematically as $p_0 e^{rN/30} = 2p_0$. It follows that $rN/30 = \ln(2)$, and hence the rate constant is given by $r = 30\ln(2)/N$. The units are understood to be per month.

10.(a) Assuming no air resistance, with the positive direction taken as downward, Newton's Second Law can be expressed as $mdv/dt = mg$, in which g is the gravitational constant measured in appropriate units. The equation can be written as $dv/dt = g$, with solution $v(t) = gt + v_0$. The object is released with an initial velocity v_0.

(b) Suppose that the object is released from a height of h units above the ground. Using the fact that $v = dx/dt$, in which x is the downward displacement of the object, we obtain the differential equation for the displacement as $dx/dt = gt + v_0$. With the origin placed at the point of release, direct integration results in $x(t) = gt^2/2 + v_0 t$. Based on the chosen coordinate system, the object reaches the ground when $x(t) = h$. Let $t = T$ be the time that it takes the object to reach the ground. Then $gT^2/2 + v_0 T = h$. Using the quadratic formula to solve for T, we obtain $T = (-v_0 \pm \sqrt{v_0 + 2gh})/g$. The positive answer corresponds to the time it takes for the object to fall to the ground. The negative answer represents a previous instant at which the object could have been launched upward (with the same impact speed), only to ultimately fall downward with speed v_0, from a height of h units above the ground. The numerical value is $T = \sqrt{2 \cdot 9.8 \cdot 300}/9.8 \approx 7.82$ s.

(c) The impact speed is calculated by substituting $t = T$ into $v(t)$ in part (a). That is, $v(T) = \sqrt{v_0 + 2gh}$. The numerical value is $v = \sqrt{2 \cdot 9.8 \cdot 300} \approx 76.68$ m/s.

12. The general solution of the differential equation $dQ/dt = -rQ$ is $Q(t) = Q_0 e^{-rt}$, in which $Q_0 = Q(0)$ is the initial amount of the substance. Let τ be the time that it takes the substance to decay to one-half of its original amount, Q_0. Setting $t = \tau$ in the solution, we have $0.5 Q_0 = Q_0 e^{-r\tau}$. Taking the natural logarithm of both sides, it follows that $-r\tau = \ln(0.5)$ or $r\tau = \ln 2$.

14. The differential equation governing the amount of radium-226 is $dQ/dt = -rQ$, with solution $Q(t) = Q(0)e^{-rt}$. Using the result in Problem 13, and the fact that the half-life $\tau = 1620$ years, the decay rate is given by $r = \ln(2)/1620$ per year. The amount of radium-226, after t years, is therefore $Q(t) = Q(0)e^{-0.00042786t}$. Let T be the time that it takes the isotope to decay to 3/4 of its original amount. Then setting $t = T$, and $Q(T) = (3/4)Q(0)$, we obtain $(3/4)Q(0) = Q(0)e^{-0.00042786T}$. Solving for the decay time, it follows that $-0.00042786\,T = \ln(3/4)$ or $T \approx 672.36$ years.

16. Based on Problem 15, the governing differential equation for the temperature in the room is $du/dt = -0.15\,(u-10)$. Setting $t = 0$ at the instant that the heating system fail, the initial condition is $u(0) = 70\,°F$. Using separation of variables, the general solution of the differential equation is $u(t) = 10 + Ce^{-0.15t}$. Invoking the given initial condition, the temperature in the room is given by $u(t) = 10 + 60\,e^{-0.15t}$. Setting $u(t) = 32$, we obtain $t = 6.69$ hr.

18.(a) The accumulation rate of the chemical is $(0.01)(300)$ grams per hour. At any given time t, the concentration of the chemical in the pond is $Q(t)/10^6$ grams per gallon. Consequently, the chemical leaves the pond at a rate of $(3 \times 10^{-4})Q(t)$ grams per hour. Hence, the rate of change of the chemical is given by

$$\frac{dQ}{dt} = 3 - 0.0003\,Q(t)\ \text{g/hr}.$$

Since the pond is initially free of the chemical, $Q(0) = 0$.

(b) The differential equation can be rewritten as $dQ/(10000-Q) = 0.0003\,dt$. Integrating both sides of the equation results in $-\ln|10000-Q| = 0.0003t + C$. Taking the exponential of both sides gives $10000 - Q = ce^{-0.0003t}$. Since $Q(0) = 0$, the value of the constant is $c = 10000$. Hence the amount of chemical in the pond at any time is $Q(t) = 10000(1 - e^{-0.0003t})$ grams. Note that 1 year= 8760 hours. Setting $t = 8760$, the amount of chemical present after one year is $Q(8760) \approx 9277.77$ grams, that is, 9.27777 kilograms.

(c) With the accumulation rate now equal to zero, the governing equation becomes $dQ/dt = -0.0003\,Q(t)$ g/hr. Resetting the time variable, we now assign the new initial value as $Q(0) = 9277.77$ grams.

(d) The solution of the differential equation in part (c) is $Q(t) = 9277.77\,e^{-0.0003t}$. Hence, one year after the source is removed, the amount of chemical in the pond is $Q(8760) \approx 670.1$ grams.

(e) Letting t be the amount of time after the source is removed, we obtain the equation $10 = 9277.77\,e^{-0.0003t}$. Taking the natural logarithm of both sides, $-0.0003\,t = \ln(10/9277.77)$ or $t \approx 22,776$ hours≈ 2.6 years.

(f)

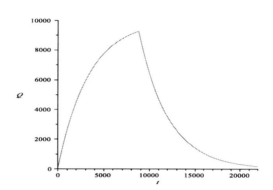

19.(a) It is assumed that dye is no longer entering the pool. In fact, the rate at which the dye leaves the pool is $200 \cdot [q(t)/60000]$ g/min. Hence the equation that governs the amount of dye in the pool is $dq/dt = -q/300$ (g/min). The initial amount of dye in the pool is $q(0) = 5000$ grams.

(b) The solution of the governing differential equation, with the specified initial value, is $q(t) = 5000\, e^{-t/300}$.

(c) The amount of dye in the pool after four hours is obtained by setting $t = 240$. That is, $q(4) = 5000\, e^{-0.8} = 2246.64$ grams. Since the size of the pool is $60,000$ gallons, the concentration of the dye is 0.0374 grams/gallon, and the answer is no.

(d) Let T be the time that it takes to reduce the concentration level of the dye to 0.02 grams/gallon. At that time, the amount of dye in the pool is $1,200$ grams. Using the answer in part (b), we have $5000\, e^{-T/300} = 1200$. Taking the natural logarithm of both sides of the equation results in the required time $T \approx 7.14$ hours.

(e) Consider the differential equation $dq/dt = -(r/60,000)\, q$. Here the parameter r corresponds to the flow rate, measured in gallons per minute. Using the same initial value, the solution is given by $q(t) = 5000\, e^{-rt/60,000}$. In order to determine the appropriate flow rate, set $t = 240$ and $q = 1200$. (Recall that 1200 grams of dye has a concentration of 0.02 g/gal). We obtain the equation $1200 = 5000\, e^{-r/250}$. Taking the natural logarithm of both sides of the equation results in the required flow rate $r \approx 357$ gallons per minute.

1.3

1. The differential equation is second order, since the highest derivative in the equation is of order two. The equation is linear, since the left hand side is a linear function of y and its derivatives.

3. The differential equation is fourth order, since the highest derivative of the function y is of order four. The equation is also linear, since the terms containing

the dependent variable is linear in y and its derivatives.

4. The differential equation is first order, since the only derivative is of order one. The dependent variable is squared, hence the equation is nonlinear.

5. The differential equation is second order. Furthermore, the equation is nonlinear, since the dependent variable y is an argument of the sine function, which is not a linear function.

7. $y_1(t) = e^t$, so $y_1'(t) = y_1''(t) = e^t$. Hence $y_1'' - y_1 = 0$. Also, $y_2(t) = \cosh t$, so $y_2'(t) = \sinh t$ and $y_2''(t) = \cosh t$. Thus $y_2'' - y_2 = 0$.

9. $y(t) = 3t + t^2$, so $y'(t) = 3 + 2t$. Substituting into the differential equation, we have $t(3 + 2t) - (3t + t^2) = 3t + 2t^2 - 3t - t^2 = t^2$. Hence the given function is a solution.

10. $y_1(t) = t/3$, so $y_1'(t) = 1/3$ and $y_1''(t) = y_1'''(t) = y_1''''(t) = 0$. Clearly, $y_1(t)$ is a solution. Likewise, $y_2(t) = e^{-t} + t/3$, so $y_2'(t) = -e^{-t} + 1/3$, $y_2''(t) = e^{-t}$, $y_2'''(t) = -e^{-t}$, $y_2''''(t) = e^{-t}$. Substituting into the left hand side of the equation, we find that $e^{-t} + 4(-e^{-t}) + 3(e^{-t} + t/3) = e^{-t} - 4e^{-t} + 3e^{-t} + t = t$. Hence both functions are solutions of the differential equation.

12. $y_1(t) = t^{-2}$, so $y_1'(t) = -2t^{-3}$ and $y_1''(t) = 6t^{-4}$. Substituting into the left hand side of the differential equation, we have $t^2(6t^{-4}) + 5t(-2t^{-3}) + 4t^{-2} = 6t^{-2} - 10t^{-2} + 4t^{-2} = 0$. Likewise, $y_2(t) = t^{-2} \ln t$, so $y_2'(t) = t^{-3} - 2t^{-3} \ln t$ and $y_2''(t) = -5t^{-4} + 6t^{-4} \ln t$. Substituting into the left hand side of the equation, we have

$$t^2(-5t^{-4} + 6t^{-4} \ln t) + 5t(t^{-3} - 2t^{-3} \ln t) + 4(t^{-2} \ln t) =$$

$$= -5t^{-2} + 6t^{-2} \ln t + 5t^{-2} - 10t^{-2} \ln t + 4t^{-2} \ln t = 0.$$

Hence both functions are solutions of the differential equation.

13. $y(t) = (\cos t) \ln \cos t + t \sin t$, so $y'(t) = -(\sin t) \ln \cos t + t \cos t$ and $y''(t) = -(\cos t) \ln \cos t - t \sin t + \sec t$. Substituting into the left hand side of the differential equation, we have $(-(\cos t) \ln \cos t - t \sin t + \sec t) + (\cos t) \ln \cos t + t \sin t = -(\cos t) \ln \cos t - t \sin t + \sec t + (\cos t) \ln \cos t + t \sin t = \sec t$. Hence the function $y(t)$ is a solution of the differential equation.

15. Let $y(t) = e^{rt}$. Then $y'(t) = re^{rt}$, and substitution into the differential equation results in $re^{rt} + 2e^{rt} = 0$. Since $e^{rt} \neq 0$, we obtain the algebraic equation $r + 2 = 0$. The root of this equation is $r = -2$.

17. $y(t) = e^{rt}$, so $y'(t) = re^{rt}$ and $y''(t) = r^2 e^{rt}$. Substituting into the differential equation, we have $r^2 e^{rt} + re^{rt} - 6e^{rt} = 0$. Since $e^{rt} \neq 0$, we obtain the algebraic equation $r^2 + r - 6 = 0$, that is, $(r-2)(r+3) = 0$. The roots are $r_1 = 2$ and $r_2 = -3$.

18. Let $y(t) = e^{rt}$. Then $y'(t) = re^{rt}$, $y''(t) = r^2 e^{rt}$ and $y'''(t) = r^3 e^{rt}$. Substituting the derivatives into the differential equation, we have $r^3 e^{rt} - 3r^2 e^{rt} + 2re^{rt} = 0$. Since $e^{rt} \neq 0$, we obtain the algebraic equation $r^3 - 3r^2 + 2r = 0$. By inspection, it follows that $r(r-1)(r-2) = 0$. Clearly, the roots are $r_1 = 0, r_2 = 1$ and $r_3 = 2$.

20. $y(t) = t^r$, so $y'(t) = rt^{r-1}$ and $y''(t) = r(r-1)t^{r-2}$. Substituting the derivatives into the differential equation, we have $t^2 \left[r(r-1)t^{r-2} \right] - 4t(rt^{r-1}) + 4t^r = 0$. After some algebra, it follows that $r(r-1)t^r - 4rt^r + 4t^r = 0$. For $t \neq 0$, we obtain the algebraic equation $r^2 - 5r + 4 = 0$. The roots of this equation are $r_1 = 1$ and $r_2 = 4$.

21. The order of the partial differential equation is two, since the highest derivative, in fact each one of the derivatives, is of second order. The equation is linear, since the left hand side is a linear function of the partial derivatives.

23. The partial differential equation is fourth order, since the highest derivative, and in fact each of the derivatives, is of order four. The equation is linear, since the left hand side is a linear function of the partial derivatives.

24. The partial differential equation is second order, since the highest derivative of the function $u(x, y)$ is of order two. The equation is nonlinear, due to the product $u \cdot u_x$ on the left hand side of the equation.

25. If $u_1(x, y) = \cos x \cosh y$, then $\partial^2 u_1 / \partial x^2 = -\cos x \cosh y$ and $\partial^2 u_1 / \partial y^2 = \cos x \cosh y$. It is evident that $\partial^2 u_1 / \partial x^2 + \partial^2 u_1 / \partial y^2 = 0$. Also, when $u_2(x, y) = \ln(x^2 + y^2)$, the second derivatives are

$$\frac{\partial^2 u_2}{\partial x^2} = \frac{2}{x^2 + y^2} - \frac{4x^2}{(x^2 + y^2)^2} \quad \text{and} \quad \frac{\partial^2 u_2}{\partial y^2} = \frac{2}{x^2 + y^2} - \frac{4y^2}{(x^2 + y^2)^2}.$$

Adding the partial derivatives,

$$\frac{\partial^2 u_2}{\partial x^2} + \frac{\partial^2 u_2}{\partial y^2} = \frac{2}{x^2 + y^2} - \frac{4x^2}{(x^2 + y^2)^2} + \frac{2}{x^2 + y^2} - \frac{4y^2}{(x^2 + y^2)^2} =$$

$$= \frac{4}{x^2 + y^2} - \frac{4(x^2 + y^2)}{(x^2 + y^2)^2} = 0.$$

Hence $u_2(x, y)$ is also a solution of the differential equation.

27. Let $u_1(x, t) = \sin(\lambda x) \sin(\lambda a t)$. Then the second derivatives are

$$\frac{\partial^2 u_1}{\partial x^2} = -\lambda^2 \sin \lambda x \sin \lambda a t \quad \text{and} \quad \frac{\partial^2 u_1}{\partial t^2} = -\lambda^2 a^2 \sin \lambda x \sin \lambda a t.$$

It is easy to see that $a^2 \partial^2 u_1 / \partial x^2 = \partial^2 u_1 / \partial t^2$. Likewise, given $u_2(x, t) = \sin(x - at)$, we have

$$\frac{\partial^2 u_2}{\partial x^2} = -\sin(x - at) \quad \text{and} \quad \frac{\partial^2 u_2}{\partial t^2} = -a^2 \sin(x - at).$$

Clearly, $u_2(x, t)$ is also a solution of the partial differential equation.

28. Given the function $u(x,t) = \sqrt{\pi/t}\, e^{-x^2/4\alpha^2 t}$, the partial derivatives are

$$u_{xx} = -\frac{\sqrt{\pi/t}\, e^{-x^2/4\alpha^2 t}}{2\alpha^2 t} + \frac{\sqrt{\pi/t}\, x^2 e^{-x^2/4\alpha^2 t}}{4\alpha^4 t^2}$$

$$u_t = -\frac{\sqrt{\pi t}\, e^{-x^2/4\alpha^2 t}}{2t^2} + \frac{\sqrt{\pi}\, x^2 e^{-x^2/4\alpha^2 t}}{4\alpha^2 t^2 \sqrt{t}}$$

It follows that

$$\alpha^2 u_{xx} = u_t = -\frac{\sqrt{\pi}\,(2\alpha^2 t - x^2)e^{-x^2/4\alpha^2 t}}{4\alpha^2 t^2 \sqrt{t}}.$$

Hence $u(x,t)$ is a solution of the partial differential equation.

30.(a) The kinetic energy of a particle of mass m is given by $T = mv^2/2$, in which v is its speed. A particle in motion on a circle of radius L has speed $L\,(d\theta/dt)$, where θ is its angular position and $d\theta/dt$ is its angular speed.

(b) Gravitational potential energy is given by $V = mgh$, where h is the height above a certain datum. Choosing the lowest point of the swing as the datum, it follows from trigonometry that $h = L(1 - \cos\theta)$.

(c) From parts (a) and (b),

$$E = \frac{1}{2}mL^2(\frac{d\theta}{dt})^2 + mgL(1 - \cos\theta).$$

Applying the chain rule for differentiation,

$$\frac{dE}{dt} = mL^2 \frac{d\theta}{dt}\frac{d^2\theta}{dt^2} + mgL\sin\theta\,\frac{d\theta}{dt}.$$

Setting $dE/dt = 0$ and dividing both sides of the equation by $d\theta/dt$ results in

$$mL^2\frac{d^2\theta}{dt^2} + mgL\sin\theta = 0,$$

which leads to Equation (12).

31.(a) The angular momentum is the moment of the linear momentum about a given point. The linear momentum is given by $mv = mLd\theta/dt$. Taking the moment about the point of support, the angular momentum is

$$M = mvL = mL^2\frac{d\theta}{dt}.$$

(b) The moment of the gravitational force is $-mgL\sin\theta$. The negative sign is included since positive moments are counterclockwise. Setting dM/dt equal to the moment of the gravitational force gives

$$\frac{dM}{dt} = mL^2\frac{d^2\theta}{dt^2} = -mgL\sin\theta,$$

which leads to Equation (12).

CHAPTER 2

First Order Differential Equations

2.1

5.(a)

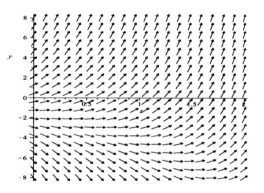

(b) If $y(0) > -3$, solutions eventually have positive slopes, and hence increase without bound. If $y(0) \leq -3$, solutions have negative slopes and decrease without bound.

(c) The integrating factor is $\mu(t) = e^{-\int 2dt} = e^{-2t}$. The differential equation can be written as $e^{-2t}y' - 2e^{-2t}y = 3e^{-t}$, that is, $(e^{-2t}y)' = 3e^{-t}$. Integration of both sides of the equation results in the general solution $y(t) = -3e^t + ce^{2t}$. It follows that all solutions will increase exponentially if $c > 0$ and will decrease exponentially

13

if $c \leq 0$. Letting $c = 0$ and then $t = 0$, we see that the boundary of these behaviors is at $y(0) = -3$.

9.(a)

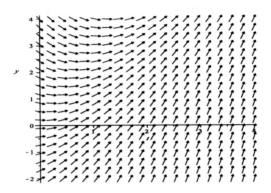

(b) All solutions eventually have positive slopes, and hence increase without bound.

(c) The integrating factor is $\mu(t) = e^{\int (1/2)\, dt} = e^{t/2}$. The differential equation can be written as $e^{t/2}y' + e^{t/2}y/2 = 3t\, e^{t/2}/2$, that is, $(e^{t/2}\, y/2)' = 3t\, e^{t/2}/2$. Integration of both sides of the equation results in the general solution $y(t) = 3t - 6 + c\, e^{-t/2}$. All solutions approach the specific solution $y_0(t) = 3t - 6$.

10.(a)

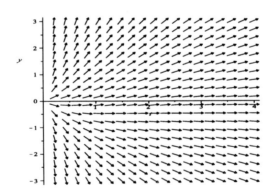

(b) For $y > 0$, the slopes are all positive, and hence the corresponding solutions increase without bound. For $y < 0$, almost all solutions have negative slopes, and hence solutions tend to decrease without bound.

(c) First divide both sides of the equation by t ($t > 0$). From the resulting standard form, the integrating factor is $\mu(t) = e^{-\int (1/t)\, dt} = 1/t$. The differential equation can be written as $y'/t - y/t^2 = t\, e^{-t}$, that is, $(y/t)' = t\, e^{-t}$. Integration leads to the general solution $y(t) = -t e^{-t} + ct$. For $c \neq 0$, solutions diverge, as implied by the direction field. For the case $c = 0$, the specific solution is $y(t) = -t e^{-t}$, which evidently approaches zero as $t \to \infty$.

12.(a)

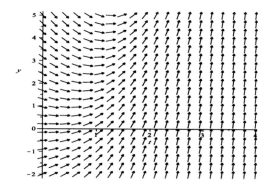

(b) All solutions eventually have positive slopes, and hence increase without bound.

(c) The integrating factor is $\mu(t) = e^{t/2}$. The differential equation can be written as $e^{t/2} y' + e^{t/2} y/2 = 3t^2/2$, that is, $(e^{t/2} y/2)' = 3t^2/2$. Integration of both sides of the equation results in the general solution $y(t) = 3t^2 - 12t + 24 + c e^{-t/2}$. It follows that all solutions converge to the specific solution $3t^2 - 12t + 24$.

14. The integrating factor is $\mu(t) = e^{2t}$. After multiplying both sides by $\mu(t)$, the equation can be written as $(e^{2t} y)' = t$. Integrating both sides of the equation results in the general solution $y(t) = t^2 e^{-2t}/2 + c e^{-2t}$. Invoking the specified condition, we require that $e^{-2}/2 + c e^{-2} = 0$. Hence $c = -1/2$, and the solution to the initial value problem is $y(t) = (t^2 - 1)e^{-2t}/2$.

16. The integrating factor is $\mu(t) = e^{\int (2/t)\,dt} = t^2$. Multiplying both sides by $\mu(t)$, the equation can be written as $(t^2 y)' = \cos t$. Integrating both sides of the equation results in the general solution $y(t) = \sin t/t^2 + c t^{-2}$. Substituting $t = \pi$ and setting the value equal to zero gives $c = 0$. Hence the specific solution is $y(t) = \sin t/t^2$.

17. The integrating factor is $\mu(t) = e^{-2t}$, and the differential equation can be written as $(e^{2t} y)' = 1$. Integrating, we obtain $e^{2t} y(t) = t + c$. Invoking the specified initial condition results in the solution $y(t) = (t + 2)e^{2t}$.

19. After writing the equation in standard form, we find that the integrating factor is $\mu(t) = e^{\int (4/t)\,dt} = t^4$. Multiplying both sides by $\mu(t)$, the equation can be written as $(t^4 y)' = t e^{-t}$. Integrating both sides results in $t^4 y(t) = -(t+1)e^{-t} + c$. Letting $t = -1$ and setting the value equal to zero gives $c = 0$. Hence the specific solution of the initial value problem is $y(t) = -(t^{-3} + t^{-4})e^{-t}$.

22.(a)

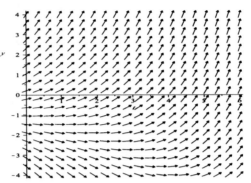

The solutions eventually increase or decrease, depending on the initial value a. The critical value seems to be $a_0 = -2$.

(b) The integrating factor is $\mu(t) = e^{-t/2}$, and the general solution of the differential equation is $y(t) = -3e^{t/3} + ce^{t/2}$. Invoking the initial condition $y(0) = a$, the solution may also be expressed as $y(t) = -3e^{t/3} + (a+3)e^{t/2}$. The critical value is $a_0 = -3$.

(c) For $a_0 = -3$, the solution is $y(t) = -3e^{t/3}$, which diverges to $-\infty$ as $t \to \infty$.

23.(a)

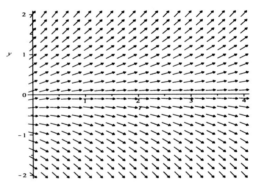

Solutions appear to grow infinitely large in absolute value, with signs depending on the initial value $y(0) = a_0$. The direction field appears horizontal for $a_0 \approx -1/8$.

(b) Dividing both sides of the given equation by 3, the integrating factor is $\mu(t) = e^{-2t/3}$. Multiplying both sides of the original differential equation by $\mu(t)$ and integrating results in $y(t) = (2e^{2t/3} - 2e^{-\pi t/2} + a(4+3\pi)e^{2t/3})/(4+3\pi)$. The qualitative behavior of the solution is determined by the terms containing $e^{2t/3}$: $2e^{2t/3} + a(4+3\pi)e^{2t/3}$. The nature of the solutions will change when $2 + a(4+3\pi) = 0$. Thus the critical initial value is $a_0 = -2/(4+3\pi)$.

(c) In addition to the behavior described in part (a), when $y(0) = -2/(4+3\pi)$, the solution is $y(t) = (-2e^{-\pi t/2})/(4+3\pi)$, and that specific solution will converge to $y = 0$.

24.(a)

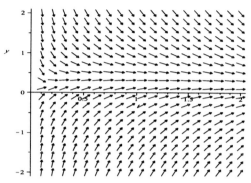

As $t \to 0$, solutions increase without bound if $y(1) = a > 0.4$, and solutions decrease without bound if $y(1) = a < 0.4$.

(b) The integrating factor is $\mu(t) = e^{\int (t+1)/t \, dt} = t e^t$. The general solution of the differential equation is $y(t) = t e^{-t} + c e^{-t}/t$. Since $y(1) = a$, we have that $1 + c = ae$. That is, $c = ae - 1$. Hence the solution can also be expressed as $y(t) = t e^{-t} + (ae - 1) e^{-t}/t$. For small values of t, the second term is dominant. Setting $ae - 1 = 0$, the critical value of the parameter is $a_0 = 1/e$.

(c) When $a = 1/e$, the solution is $y(t) = t e^{-t}$, which approaches 0 as $t \to 0$.

27. The integrating factor is $\mu(t) = e^{\int (1/2) \, dt} = e^{t/2}$. Therefore the general solution is $y(t) = (4 \cos t + 8 \sin t)/5 + c e^{-t/2}$. Invoking the initial condition, the specific solution is $y(t) = (4 \cos t + 8 \sin t - 9 e^{-t/2})/5$. Differentiating, it follows that $y'(t) = (-4 \sin t + 8 \cos t + 4.5 e^{-t/2})/5$ and $y''(t) = (-4 \cos t - 8 \sin t - 2.25 e^{-t/2})/5$. Setting $y'(t) = 0$, the first solution is $t_1 = 1.3643$, which gives the location of the first stationary point. Since $y''(t_1) < 0$, the first stationary point in a local maximum. The coordinates of the point are $(1.3643, 0.82008)$.

28. The integrating factor is $\mu(t) = e^{\int (2/3) \, dt} = e^{2t/3}$, and the differential equation can be written as $(e^{2t/3} y)' = e^{2t/3} - t e^{2t/3}/2$. The general solution is $y(t) = (21 - 6t)/8 + c e^{-2t/3}$. Imposing the initial condition, we have $y(t) = (21 - 6t)/8 + (y_0 - 21/8) e^{-2t/3}$. Since the solution is smooth, the desired intersection will be a point of tangency. Taking the derivative, $y'(t) = -3/4 - (2y_0 - 21/4) e^{-2t/3}/3$. Setting $y'(t) = 0$, the solution is $t_1 = (3/2) \ln[(21 - 8y_0)/9]$. Substituting into the solution, the respective value at the stationary point is $y(t_1) = 3/2 + (9/4) \ln 3 - (9/8) \ln(21 - 8y_0)$. Setting this result equal to zero, we obtain the required initial value $y_0 = (21 - 9 e^{4/3})/8 \approx -1.643$.

29.(a) The integrating factor is $\mu(t) = e^{t/4}$, and the differential equation can be written as $(e^{t/4} y)' = 3 e^{t/4} + 2 e^{t/4} \cos 2t$. After integration, we get that the general solution is $y(t) = 12 + (8 \cos 2t + 64 \sin 2t)/65 + c e^{-t/4}$. Invoking the initial condition, $y(0) = 0$, the specific solution is $y(t) = 12 + (8 \cos 2t + 64 \sin 2t - 788 e^{-t/4})/65$. As $t \to \infty$, the exponential term will decay, and the solution will oscillate about an average value of 12, with an amplitude of $8/\sqrt{65}$.

(b) Solving $y(t) = 12$, we obtain the desired value $t \approx 10.0658$.

31. The integrating factor is $\mu(t) = e^{-3t/2}$, and the differential equation can be written as $(e^{-3t/2} y)' = 3t\, e^{-3t/2} + 2\, e^{-t/2}$. The general solution is $y(t) = -2t - 4/3 - 4\,e^t + c\,e^{3t/2}$. Imposing the initial condition, $y(t) = -2t - 4/3 - 4\,e^t + (y_0 + 16/3)\,e^{3t/2}$. Now as $t \to \infty$, the term containing $e^{3t/2}$ will dominate the solution. Its sign will determine the divergence properties. Hence the critical value of the initial condition is $y_0 = -16/3$. The corresponding solution, $y(t) = -2t - 4/3 - 4\,e^t$, will also decrease without bound.

Note on Problems 34-37:

Let $g(t)$ be given, and consider the function $y(t) = y_1(t) + g(t)$, in which $y_1(t) \to 0$ as $t \to \infty$. Differentiating, $y'(t) = y_1'(t) + g'(t)$. Letting a be a constant, it follows that $y'(t) + ay(t) = y_1'(t) + ay_1(t) + g'(t) + ag(t)$. Note that the hypothesis on the function $y_1(t)$ will be satisfied, if $y_1'(t) + ay_1(t) = 0$. That is, $y_1(t) = c\,e^{-at}$. Hence $y(t) = c\,e^{-at} + g(t)$, which is a solution of the equation $y' + ay = g'(t) + ag(t)$. For convenience, choose $a = 1$.

34. Here $g(t) = 3$, and we consider the linear equation $y' + y = 3$. The integrating factor is $\mu(t) = e^t$, and the differential equation can be written as $(e^t y)' = 3e^t$. The general solution is $y(t) = 3 + c\,e^{-t}$.

36. Here $g(t) = 2t - 5$. Consider the linear equation $y' + y = 2 + 2t - 5$. The integrating factor is $\mu(t) = e^t$, and the differential equation can be written as $(e^t y)' = (2t - 3)e^t$. The general solution is $y(t) = 2t - 5 + c\,e^{-t}$.

37. $g(t) = 4 - t^2$. Consider the linear equation $y' + y = 4 - 2t - t^2$. The integrating factor is $\mu(t) = e^t$, and the equation can be written as $(e^t y)' = (4 - 2t - t^2)e^t$. The general solution is $y(t) = 4 - t^2 + c\,e^{-t}$.

38.(a) Differentiating y and using the fundamental theorem of calculus we obtain that $y' = Ae^{-\int p(t)dt} \cdot (-p(t))$, and then $y' + p(t)y = 0$.

(b) Differentiating y we obtain that

$$y' = A'(t)e^{-\int p(t)dt} + A(t)e^{-\int p(t)dt} \cdot (-p(t)).$$

If this satisfies the differential equation then

$$y' + p(t)y = A'(t)e^{-\int p(t)dt} = g(t)$$

and the required condition follows.

(c) Let us denote $\mu(t) = e^{\int p(t)dt}$. Then clearly $A(t) = \int \mu(t)g(t)dt$, and after substitution $y = \int \mu(t)g(t)dt \cdot (1/\mu(t))$, which is just Eq. (33).

40. We assume a solution of the form $y = A(t)e^{-\int (1/t)\,dt} = A(t)e^{-\ln t} = A(t)t^{-1}$, where $A(t)$ satisfies $A'(t) = 3t \cos 2t$. This implies that

$$A(t) = \frac{3 \cos 2t}{4} + \frac{3t \sin 2t}{2} + c$$

and the solution is

$$y = \frac{3 \cos 2t}{4t} + \frac{3 \sin 2t}{2} + \frac{c}{t}.$$

41. First rewrite the differential equation as

$$y' + \frac{2}{t} y = \frac{\sin t}{t}.$$

Assume a solution of the form $y = A(t)e^{-\int (2/t)\,dt} = A(t)t^{-2}$, where $A(t)$ satisfies the ODE $A'(t) = t \sin t$. It follows that $A(t) = \sin t - t \cos t + c$ and thus $y = (\sin t - t \cos t + c)/t^2$.

2.2

Problems 1 through 20 follow the pattern of the examples worked in this section. The first eight problems, however, do not have an initial condition, so the integration constant c cannot be found.

2. For $x \neq -1$, the differential equation may be written as $y\,dy = \left[x^2/(1+x^3)\right] dx$. Integrating both sides, with respect to the appropriate variables, we obtain the relation $y^2/2 = (1/3) \ln |1 + x^3| + c$. That is, $y(x) = \pm\sqrt{(2/3) \ln |1 + x^3| + c}$.

3. The differential equation may be written as $y^{-2} dy = -\sin x\,dx$. Integrating both sides of the equation, with respect to the appropriate variables, we obtain the relation $-y^{-1} = \cos x + c$. That is, $(c - \cos x)y = 1$, in which c is an arbitrary constant. Solving for the dependent variable, explicitly, $y(x) = 1/(c - \cos x)$.

5. Write the differential equation as $\cos^{-2} 2y\,dy = \cos^2 x\,dx$, which also can be written as $\sec^2 2y\,dy = \cos^2 x\,dx$. Integrating both sides of the equation, with respect to the appropriate variables, we obtain the relation $\tan 2y = \sin x \cos x + x + c$.

7. The differential equation may be written as $(y + e^y)dy = (x - e^{-x})dx$. Integrating both sides of the equation, with respect to the appropriate variables, we obtain the relation $y^2 + 2e^y = x^2 + 2e^{-x} + c$.

8. Write the differential equation as $(1 + y^2)dy = x^2\,dx$. Integrating both sides of the equation, we obtain the relation $y + y^3/3 = x^3/3 + c$.

9.(a) The differential equation is separable, with $y^{-2} dy = (1 - 2x)dx$. Integration yields $-y^{-1} = x - x^2 + c$. Substituting $x = 0$ and $y = -1/6$, we find that $c = 6$. Hence the specific solution is $y = 1/(x^2 - x - 6)$.

(b)

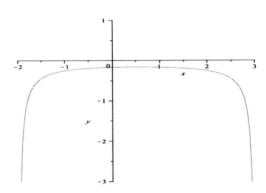

(c) Note that $x^2 - x - 6 = (x+2)(x-3)$. Hence the solution becomes singular at $x = -2$ and $x = 3$, so the interval of existence is $(-2, 3)$.

11.(a) Rewrite the differential equation as $x e^x dx = -y\, dy$. Integrating both sides of the equation results in $x e^x - e^x = -y^2/2 + c$. Invoking the initial condition, we obtain $c = -1/2$. Hence $y^2 = 2e^x - 2x e^x - 1$. The explicit form of the solution is $y(x) = \sqrt{2e^x - 2x e^x - 1}$. The positive sign is chosen, since $y(0) = 1$.

(b)

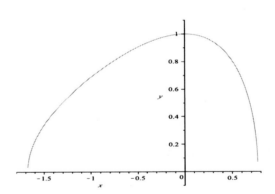

(c) The function under the radical becomes negative near $x \approx -1.7$ and $x \approx 0.77$.

12.(a) Write the differential equation as $r^{-2} dr = \theta^{-1} d\theta$. Integrating both sides of the equation results in the relation $-r^{-1} = \ln \theta + c$. Imposing the condition $r(1) = 2$, we obtain $c = -1/2$. The explicit form of the solution is $r = 2/(1 - 2 \ln \theta)$.

(b)

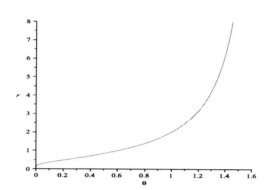

(c) Clearly, the solution makes sense only if $\theta > 0$. Furthermore, the solution becomes singular when $\ln \theta = 1/2$, that is, $\theta = \sqrt{e}$.

14.(a) Write the differential equation as $y^{-3}dy = x(1+x^2)^{-1/2}\,dx$. Integrating both sides of the equation, with respect to the appropriate variables, we obtain the relation $-y^{-2}/2 = \sqrt{1+x^2} + c$. Imposing the initial condition, we obtain $c = -3/2$. Hence the specific solution can be expressed as $y^{-2} = 3 - 2\sqrt{1+x^2}$. The explicit form of the solution is $y(x) = 1/\sqrt{3 - 2\sqrt{1+x^2}}$. The positive sign is chosen to satisfy the initial condition.

(b)

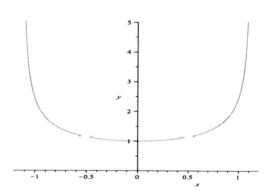

(c) The solution becomes singular when $2\sqrt{1+x^2} = 3$. That is, at $x = \pm\sqrt{5}/2$.

16.(a) Rewrite the differential equation as $4y^3 dy = x(x^2+1)dx$. Integrating both sides of the equation results in $y^4 = (x^2+1)^2/4 + c$. Imposing the initial condition, we obtain $c = 0$. Hence the solution may be expressed as $(x^2+1)^2 - 4y^4 = 0$. The explicit form of the solution is $y(x) = -\sqrt{(x^2+1)/2}$. The sign is chosen based on $y(0) = -1/\sqrt{2}$.

(b)

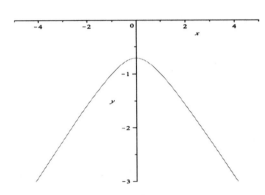

(c) The solution is valid for all $x \in \mathbb{R}$.

18.(a) Write the differential equation as $(3 + 4y)dy = (e^{-x} - e^x)dx$. Integrating both sides of the equation, with respect to the appropriate variables, we obtain the relation $3y + 2y^2 = -(e^x + e^{-x}) + c$. Imposing the initial condition, $y(0) = 1$, we obtain $c = 7$. Thus, the solution can be expressed as $3y + 2y^2 = -(e^x + e^{-x}) + 7$. Now by completing the square on the left hand side, $2(y + 3/4)^2 = -(e^x + e^{-x}) + 65/8$. Hence the explicit form of the solution is $y(x) = -3/4 + \sqrt{65/16 - \cosh x}$.

(b)

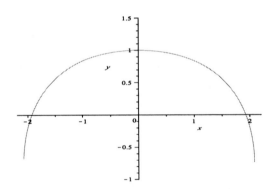

(c) Note the $65 - 16 \cosh x \geq 0$ as long as $|x| > 2.1$ (approximately). Hence the solution is valid on the interval $-2.1 < x < 2.1$.

20.(a) Rewrite the differential equation as $y^2 dy = \arcsin x/\sqrt{1-x^2}\, dx$. Integrating both sides of the equation results in $y^3/3 = (\arcsin x)^2/2 + c$. Imposing the condition $y(0) = 1$, we obtain $c = 1/3$. The explicit form of the solution is $y(x) = (3(\arcsin x)^2/2 + 1)^{1/3}$.

(b)

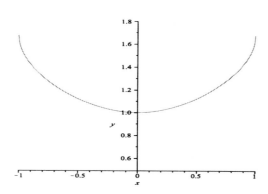

(c) Since $\arcsin x$ is defined for $-1 \leq x \leq 1$, this is the interval of existence.

22. The differential equation can be written as $(3y^2 - 4)dy = 3x^2 dx$. Integrating both sides, we obtain $y^3 - 4y = x^3 + c$. Imposing the initial condition, the specific solution is $y^3 - 4y = x^3 - 1$. Referring back to the differential equation, we find that $y' \to \infty$ as $y \to \pm 2/\sqrt{3}$. The respective values of the abscissas are $x \approx -1.276$, 1.598. Hence the solution is valid for $-1.276 < x < 1.598$.

24. Write the differential equation as $(3 + 2y)dy = (2 - e^x)dx$. Integrating both sides, we obtain $3y + y^2 = 2x - e^x + c$. Based on the specified initial condition, the solution can be written as $3y + y^2 = 2x - e^x + 1$. Completing the square, it follows that $y(x) = -3/2 + \sqrt{2x - e^x + 13/4}$. The solution is defined if $2x - e^x + 13/4 \geq 0$, that is, $-1.5 \leq x \leq 2$ (approximately). In that interval, $y' = 0$ for $x = \ln 2$. It can be verified that $y''(\ln 2) < 0$. In fact, $y''(x) < 0$ on the interval of definition. Hence the solution attains a global maximum at $x = \ln 2$.

26. The differential equation can be written as $(1 + y^2)^{-1}dy = 2(1 + x)dx$. Integrating both sides of the equation, we obtain $\arctan y = 2x + x^2 + c$. Imposing the given initial condition, the specific solution is $\arctan y = 2x + x^2$. Therefore, $y = \tan(2x + x^2)$. Observe that the solution is defined as long as $-\pi/2 < 2x + x^2 < \pi/2$. It is easy to see that $2x + x^2 \geq -1$. Furthermore, $2x + x^2 = \pi/2$ for $x \approx -2.6$ and 0.6. Hence the solution is valid on the interval $-2.6 < x < 0.6$. Referring back to the differential equation, the solution is stationary at $x = -1$. Since $y''(-1) > 0$, the solution attains a global minimum at $x = -1$.

28.(a) Write the differential equation as $y^{-1}(4 - y)^{-1}dy = t(1 + t)^{-1}dt$. Integrating both sides of the equation, we obtain $\ln|y| - \ln|y - 4| = 4t - 4\ln|1 + t| + c$. Taking the exponential of both sides $|y/(y - 4)| = c e^{4t}/(1 + t)^4$. It follows that as $t \to \infty$, $|y/(y - 4)| = |1 + 4/(y - 4)| \to \infty$. That is, $y(t) \to 4$.

(b) Setting $y(0) = 2$, we obtain that $c = 1$. Based on the initial condition, the solution may be expressed as $y/(y - 4) = -e^{4t}/(1 + t)^4$. Note that $y/(y - 4) < 0$, for all $t \geq 0$. Hence $y < 4$ for all $t \geq 0$. Referring back to the differential equation, it follows that y' is always positive. This means that the solution is monotone increasing. We find that the root of the equation $e^{4t}/(1+t)^4 = 399$ is near $t = 2.844$.

(c) Note the $y(t) = 4$ is an equilibrium solution. Examining the local direction field we see that if $y(0) > 0$, then the corresponding solutions converge to $y = 4$. Referring back to part (a), we have $y/(y-4) = [y_0/(y_0-4)]\, e^{4t}/(1+t)^4$, for $y_0 \neq 4$. Setting $t = 2$, we obtain $y_0/(y_0-4) = (3/e^2)^4 y(2)/(y(2)-4)$. Now since the function $f(y) = y/(y-4)$ is monotone for $y < 4$ and $y > 4$, we need only solve the equations $y_0/(y_0-4) = -399(3/e^2)^4$ and $y_0/(y_0-4) = 401(3/e^2)^4$. The respective solutions are $y_0 = 3.6622$ and $y_0 = 4.4042$.

32.(a) Observe that $(x^2 + 3y^2)/2xy = (1/2)(y/x)^{-1} + (3/2)(y/x)$. Hence the differential equation is homogeneous.

(b) The substitution $y = x v$ results in $v + x v' = (x^2 + 3x^2 v^2)/2x^2 v$. The transformed equation is $v' = (1 + v^2)/2xv$. This equation is separable, with general solution $v^2 + 1 = c x$. In terms of the original dependent variable, the solution is $x^2 + y^2 = c x^3$.

(c) The integral curves are symmetric with respect to the origin.

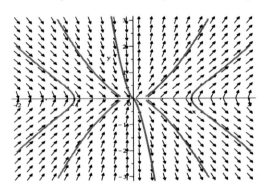

34.(a) Observe that $-(4x + 3y)/(2x + y) = -2 - (y/x)\,[2 + (y/x)]^{-1}$. Hence the differential equation is homogeneous.

(b) The substitution $y = x v$ results in $v + x v' = -2 - v/(2 + v)$. The transformed equation is $v' = -(v^2 + 5v + 4)/(2 + v)x$. This equation is separable, with general solution $(v + 4)^2 |v + 1| = c/x^3$. In terms of the original dependent variable, the solution is $(4x + y)^2 |x + y| = c$.

(c) The integral curves are symmetric with respect to the origin.

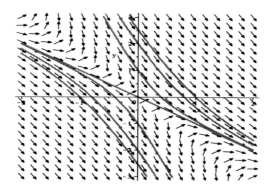

36.(a) Divide by x^2 to see that the equation is homogeneous. Substituting $y = xv$, we obtain $xv' = (1+v)^2$. The resulting differential equation is separable.

(b) Write the equation as $(1+v)^{-2}dv = x^{-1}dx$. Integrating both sides of the equation, we obtain the general solution $-1/(1+v) = \ln|x| + c$. In terms of the original dependent variable, the solution is $y = x(c - \ln|x|)^{-1} - x$.

(c) The integral curves are symmetric with respect to the origin.

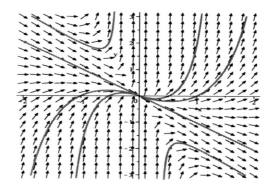

37.(a) The differential equation can be expressed as $y' = (1/2)(y/x)^{-1} - (3/2)(y/x)$. Hence the equation is homogeneous. The substitution $y = xv$ results in $xv' = (1 - 5v^2)/2v$. Separating variables, we have $2vdv/(1 - 5v^2) = dx/x$.

(b) Integrating both sides of the transformed equation yields $-(\ln|1 - 5v^2|)/5 = \ln|x| + c$, that is, $1 - 5v^2 = c/|x|^5$. In terms of the original dependent variable, the general solution is $5y^2 = x^2 - c/|x|^3$.

(c) The integral curves are symmetric with respect to the origin.

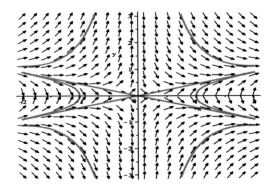

38.(a) The differential equation can be expressed as $y' = (3/2)(y/x) - (1/2)(y/x)^{-1}$. Hence the equation is homogeneous. The substitution $y = xv$ results in $xv' = (v^2 - 1)/2v$, that is, $2v\,dv/(v^2 - 1) = dx/x$.

(b) Integrating both sides of the transformed equation yields $\ln|v^2 - 1| = \ln|x| + c$, that is, $v^2 - 1 = c|x|$. In terms of the original dependent variable, the general solution is $y^2 = cx^2|x| + x^2$.

(c) The integral curves are symmetric with respect to the origin.

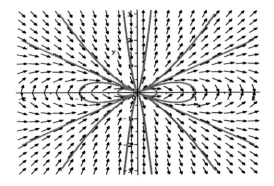

2.3

1. Let $Q(t)$ be the amount of dye in the tank at time t. Clearly, $Q(0) = 200$ g. The differential equation governing the amount of dye is $Q'(t) = -2Q(t)/200$. The solution of this separable equation is $Q(t) = Q(0)e^{-t/100} = 200e^{-t/100}$. We need the time T such that $Q(T) = 2$ g. This means we have to solve $2 = 200e^{-T/100}$ and we obtain that $T = -100\ln(1/100) = 100\ln 100 \approx 460.5$ min.

5.(a) Let Q be the amount of salt in the tank. Salt enters the tank of water at a rate of $2(1/4)(1 + (1/2)\sin t) = 1/2 + (1/4)\sin t$ oz/min. It leaves the tank at a rate of $2\,Q/100$ oz/min. Hence the differential equation governing the amount of salt at any time is

$$\frac{dQ}{dt} = \frac{1}{2} + \frac{1}{4}\sin t - \frac{Q}{50}\,.$$

The initial amount of salt is $Q_0 = 50$ oz. The governing differential equation is linear, with integrating factor $\mu(t) = e^{t/50}$. Write the equation as $(e^{t/50}Q)' = e^{t/50}(1/2 + (1/4)\sin t)$. The specific solution is $Q(t) = 25 + (12.5\sin t - 625\cos t + 63150\,e^{-t/50})/2501$ oz.

(b)

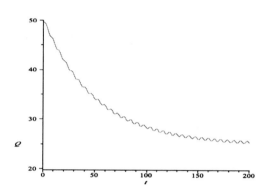

(c) The amount of salt approaches a steady state, which is an oscillation of approximate amplitude $1/4$ about a level of 25 oz.

6.(a) Using the Principle of Conservation of Energy, the speed v of a particle falling from a height h is given by

$$\frac{1}{2}mv^2 = mgh\,.$$

(b) The outflow rate is (outflow cross-section area)$\times$(outflow velocity): $\alpha a\sqrt{2gh}$. At any instant, the volume of water in the tank is $V(h) = \int_0^h A(u)du$. The time rate of change of the volume is given by $dV/dt = (dV/dh)(dh/dt) = A(h)dh/dt$. Since the volume is decreasing, $dV/dt = -\alpha a\sqrt{2gh}$.

(c) With $A(h) = \pi$, $a = 0.01\,\pi$, $\alpha = 0.6$, the differential equation for the water level h is $\pi(dh/dt) = -0.006\,\pi\sqrt{2gh}$, with solution $h(t) = 0.000018gt^2 - 0.006\sqrt{2gh(0)}\,t + h(0)$. Setting $h(0) = 3$ and $g = 9.8$, $h(t) = 0.0001764\,t^2 - 0.046\,t + 3$, resulting in $h(t) = 0$ for $t \approx 130.4$ s.

7.(a) The equation governing the value of the investment is $dS/dt = r\,S$. The value of the investment, at any time, is given by $S(t) = S_0 e^{rt}$. Setting $S(T) = 2S_0$, the required time is $T = \ln(2)/r$.

(b) For the case $r = .07$, $T \approx 9.9$ yr.

(c) Referring to part (a), $r = \ln(2)/T$. Setting $T = 8$, the required interest rate is to be approximately $r = 8.66\%$.

12.(a) Using Eq.(15) we have $dS/dt - 0.005S = -(800 + 10t)$, $S(0) = 150,000$. Using an integrating factor and integration by parts we obtain that $S(t) = 560,000 - 410,000 e^{0.005t} + 2000t$. Setting $S(t) = 0$ and solving numerically for t yields $t = 146.54$ months.

(b) The solution we obtained in part (a) with a general initial condition $S(0) = S_0$ is $S(t) = 560,000 - 560,000 e^{0.005t} + S_0 e^{0.005t} + 2000t$. Solving the equation $S(240) = 0$ yields $S_0 = 246,758$.

13.(a) Let $Q' = -rQ$. The general solution is $Q(t) = Q_0 e^{-rt}$. Based on the definition of half-life, consider the equation $Q_0/2 = Q_0 e^{-5730\, r}$. It follows that $-5730\, r = \ln(1/2)$, that is, $r = 1.2097 \times 10^{-4}$ per year.

(b) The amount of carbon-14 is given by $Q(t) = Q_0\, e^{-1.2097 \times 10^{-4} t}$.

(c) Given that $Q(T) = Q_0/5$, we have the equation $1/5 = e^{-1.2097 \times 10^{-4} T}$. Solving for the decay time, the apparent age of the remains is approximately $T = 13,305$ years.

15.(a) The differential equation $dy/dt = r(t)\, y - k$ is linear, with integrating factor $\mu(t) = e^{-\int r(t) dt}$. Write the equation as $(\mu\, y)' = -k\, \mu(t)$. Integration of both sides yields the general solution $y = \left[-k \int \mu(\tau) d\tau + y_0\, \mu(0) \right] /\mu(t)$. In this problem, the integrating factor is $\mu(t) = e^{(\cos t - t)/5}$.

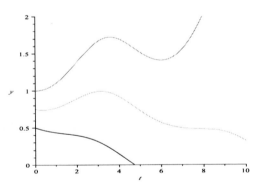

(b) The population becomes extinct, if $y(t^*) = 0$, for some $t = t^*$. Referring to part (a), we find that $y(t^*) = 0$ when

$$\int_0^{t^*} e^{(\cos \tau - \tau)/5} d\tau = 5\, e^{1/5} y_c.$$

It can be shown that the integral on the left hand side increases monotonically, from zero to a limiting value of approximately 5.0893. Hence extinction can happen only if $5\, e^{1/5} y_0 < 5.0893$. Solving $5 e^{1/5} y_c = 5.0893$ yields $y_c = 0.8333$.

(c) Repeating the argument in part (b), it follows that $y(t^*) = 0$ when

$$\int_0^{t^*} e^{(\cos \tau - \tau)/5} d\tau = \frac{1}{k} e^{1/5} y_c.$$

Hence extinction can happen only if $e^{1/5} y_0/k < 5.0893$, so $y_c = 4.1667\, k$.

(d) Evidently, y_c is a linear function of the parameter k.

17.(a) The solution of the governing equation satisfies $u^3 = u_0^3/(3\alpha u_0^3 t + 1)$. With the given data, it follows that $u(t) = 2000/\sqrt[3]{6t/125 + 1}$.

(b)

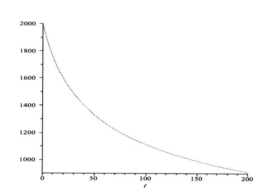

(c) Numerical evaluation results in $u(t) = 600$ for $t \approx 750.77$ s.

22.(a) The differential equation for the upward motion is $mdv/dt = -\mu v^2 - mg$, in which $\mu = 1/1325$. This equation is separable, with $m/(\mu v^2 + mg)\, dv = -dt$. Integrating both sides and invoking the initial condition, $v(t) = 44.133 \tan(0.425 - 0.222\, t)$. Setting $v(t_1) = 0$, the ball reaches the maximum height at $t_1 = 1.916$ s. Integrating $v(t)$, the position is given by $x(t) = 198.75 \ln\left[\cos(0.222\, t - 0.425)\right] + 48.57$. Therefore the maximum height is $x(t_1) = 48.56$ m.

(b) The differential equation for the downward motion is $m\, dv/dt = +\mu v^2 - mg$. This equation is also separable, with $m/(mg - \mu v^2)\, dv = -dt$. For convenience, set $t = 0$ at the top of the trajectory. The new initial condition becomes $v(0) = 0$. Integrating both sides and invoking the initial condition, we obtain $\ln((44.13 - v)/(44.13 + v)) = t/2.25$. Solving for the velocity, $v(t) = 44.13(1 - e^{t/2.25})/(1 + e^{t/2.25})$. Integrating $v(t)$, we obtain $x(t) = 99.29 \ln(e^{t/2.25}/(1 + e^{t/2.25})^2) + 186.2$. To estimate the duration of the downward motion, set $x(t_2) = 0$, resulting in $t_2 = 3.276$ s. Hence the total time that the ball spends in the air is $t_1 + t_2 = 5.192$ s.

(c)

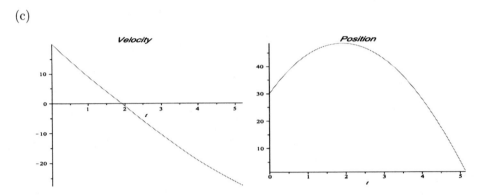

24.(a) Setting $-\mu v^2 = v(dv/dx)$, we obtain $dv/dx = -\mu v$.

(b) The speed v of the sled satisfies $\ln(v/v_0) = -\mu x$. Noting that the unit conversion factors cancel, solution of $\ln(15/150) = -2000\,\mu$ results in $\mu = \ln(10)/2000$ ft$^{-1} \approx$ 0.00115 ft$^{-1} \approx 6.0788$ mi^{-1}.

(c) Solution of $dv/dt = -\mu v^2$ can be expressed as $1/v - 1/v_0 = \mu t$. Noting that 1 mi/hr $= 5280/3600$ ft/s, the elapsed time is

$$t = (1/15 - 1/150)/((5280/3600)(\ln(10)/2000)) \approx 35.53\,\text{s}.$$

25.(a) Measure the positive direction of motion upward. The equation of motion is given by $m\,dv/dt = -kv - mg$. The initial value problem is $dv/dt = -kv/m - g$, with $v(0) = v_0$. The solution is $v(t) = -mg/k + (v_0 + mg/k)e^{-kt/m}$. Setting $v(t_m) = 0$, the maximum height is reached at time $t_m = (m/k)\ln\left[(mg + k v_0)/mg\right]$. Integrating the velocity, the position of the body is

$$x(t) = -mg\,t/k + \left[(\frac{m}{k})^2 g + \frac{m v_0}{k}\right](1 - e^{-kt/m}).$$

Hence the maximum height reached is

$$x_m = x(t_m) = \frac{m v_0}{k} - g(\frac{m}{k})^2 \ln\left[\frac{mg + k v_0}{mg}\right].$$

(b) Recall that for $\delta \ll 1$, $\ln(1 + \delta) = \delta - \delta^2/2 + \delta^3/3 - \delta^4/4 + \ldots$.

(c) The dimensions of the quantities involved are $[k] = MT^{-1}$, $[v_0] = LT^{-1}$, $[m] = M$ and $[g] = LT^{-2}$. This implies that kv_0/mg is dimensionless.

31.(a) Both equations are linear and separable. Initial conditions: $v(0) = u\cos A$ and $w(0) = u\sin A$. We obtain the solutions $v(t) = (u\cos A)e^{-rt}$ and $w(t) = -g/r + (u\sin A + g/r)e^{-rt}$.

(b) Integrating the solutions in part (a), and invoking the initial conditions, the coordinates are $x(t) = u\cos A(1 - e^{-rt})/r$ and

$$y(t) = -\frac{gt}{r} + \frac{g + ur\sin A + hr^2}{r^2} - (\frac{u}{r}\sin A + \frac{g}{r^2})e^{-rt}.$$

(c)

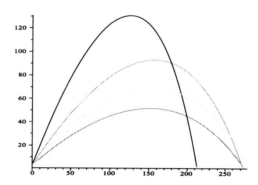

(d) Let T be the time that it takes the ball to go 350 ft horizontally. Then from above, $e^{-T/5} = (u \cos A - 70)/u \cos A$. At the same time, the height of the ball is given by

$$y(T) = -160T + 803 + 5u \sin A - \frac{(800 + 5u \sin A)(u \cos A - 70)}{u \cos A}.$$

Hence A and u must satisfy the equality

$$800 \ln \left[\frac{u \cos A - 70}{u \cos A}\right] + 803 + 5u \sin A - \frac{(800 + 5u \sin A)(u \cos A - 70)}{u \cos A} = 10$$

for the ball to touch the top of the wall. To find the optimal values for u and A, consider u as a function of A and use implicit differentiation in the above equation to find that

$$\frac{du}{dA} = -\frac{u(u^2 \cos A - 70u - 11200 \sin A)}{11200 \cos A}.$$

Solving this equation simultaneously with the above equation yields optimal values for u and A: $u \approx 145.3$ ft/s, $A \approx 0.644$ rad.

32.(a) Solving equation (i), $y'(x) = \left[(k^2 - y)/y\right]^{1/2}$. The positive answer is chosen, since y is an increasing function of x.

(b) Let $y = k^2 \sin^2 t$. Then $dy = 2k^2 \sin t \cos t\, dt$. Substituting into the equation in part (a), we find that

$$\frac{2k^2 \sin t \cos t\, dt}{dx} = \frac{\cos t}{\sin t}.$$

Hence $2k^2 \sin^2 t\, dt = dx$.

(c) Setting $\theta = 2t$, we further obtain $k^2 \sin^2(\theta/2)\, d\theta = dx$. Integrating both sides of the equation and noting that $t = \theta = 0$ corresponds to the origin, we obtain the solutions $x(\theta) = k^2(\theta - \sin \theta)/2$ and (from part (b)) $y(\theta) = k^2(1 - \cos \theta)/2$.

(d) Note that $y/x = (1 - \cos\theta)/(\theta - \sin\theta)$. Setting $x = 1$, $y = 2$, the solution of the equation $(1 - \cos\theta)/(\theta - \sin\theta) = 2$ is $\theta \approx 1.401$. Substitution into either of the expressions yields $k \approx 2.193$.

2.4

2. Rewrite the differential equation as $y' + 1/(t(t-4))\,y = 0$. It is evident that the coefficient $1/t(t-4)$ is continuous everywhere except at $t = 0, 4$. Since the initial condition is specified at $t = 2$, Theorem 2.4.1 assures the existence of a unique solution on the interval $0 < t < 4$.

3. The function $\tan t$ is discontinuous at odd multiples of $\pi/2$. Since $\pi/2 < \pi < 3\pi/2$, the initial value problem has a unique solution on the interval $(\pi/2, 3\pi/2)$.

5. $p(t) = 2t/(4 - t^2)$ and $g(t) = 3t^2/(4 - t^2)$. These functions are discontinuous at $x = \pm 2$. The initial value problem has a unique solution on the interval $(-2, 2)$.

6. The function $\ln t$ is defined and continuous on the interval $(0, \infty)$. At $t = 1$, $\ln t = 0$, so the normal form of the differential equation has a singularity there. Also, $\cot t$ is not defined at integer multiples of π, so the initial value problem will have a solution on the interval $(1, \pi)$.

7. The function $f(t, y)$ is continuous everywhere on the plane, except along the straight line $y = -2t/5$. The partial derivative $\partial f/\partial y = -7t/(2t + 5y)^2$ has the same region of continuity.

9. The function $f(t, y)$ is discontinuous along the coordinate axes, and on the hyperbola $t^2 - y^2 = 1$. Furthermore,

$$\frac{\partial f}{\partial y} = \frac{\pm 1}{y(1 - t^2 + y^2)} - 2\frac{y \ln|ty|}{(1 - t^2 + y^2)^2}$$

has the same points of discontinuity.

10. $f(t, y)$ is continuous everywhere on the plane. The partial derivative $\partial f/\partial y$ is also continuous everywhere.

12. The function $f(t, y)$ is discontinuous along the lines $t = \pm k\pi$ for $k = 0, 1, 2, \ldots$ and $y = -1$. The partial derivative $\partial f/\partial y = \cot t/(1 + y)^2$ has the same region of continuity.

14. The equation is separable, with $dy/y^2 = 2t\,dt$. Integrating both sides, the solution is given by $y(t) = y_0/(1 - y_0 t^2)$. For $y_0 > 0$, solutions exist as long as $t^2 < 1/y_0$. For $y_0 \leq 0$, solutions are defined for all t.

15. The equation is separable, with $dy/y^3 = -dt$. Integrating both sides and invoking the initial condition, $y(t) = y_0/\sqrt{2y_0^2 t + 1}$. Solutions exist as long as

$2y_0^2 t + 1 > 0$, that is, $2y_0^2 t > -1$. If $y_0 \neq 0$, solutions exist for $t > -1/2y_0^2$. If $y_0 = 0$, then the solution $y(t) = 0$ exists for all t.

16. The function $f(t, y)$ is discontinuous along the straight lines $t = -1$ and $y = 0$. The partial derivative $\partial f/\partial y$ is discontinuous along the same lines. The equation is separable, with $y\,dy = t^2\,dt/(1 + t^3)$. Integrating and invoking the initial condition, the solution is $y(t) = \left[(2/3)\ln\left|1 + t^3\right| + y_0^2\right]^{1/2}$. Solutions exist as long as $(2/3)\ln\left|1 + t^3\right| + y_0^2 \geq 0$, that is, $y_0^2 \geq -(2/3)\ln\left|1 + t^3\right|$. For all y_0 (it can be verified that $y_0 = 0$ yields a valid solution, even though Theorem 2.4.2 does not guarantee one), solutions exist as long as $\left|1 + t^3\right| \geq e^{-3y_0^2/2}$. From above, we must have $t > -1$. Hence the inequality may be written as $t^3 \geq e^{-3y_0^2/2} - 1$. It follows that the solutions are valid for $(e^{-3y_0^2/2} - 1)^{1/3} < t < \infty$.

18.

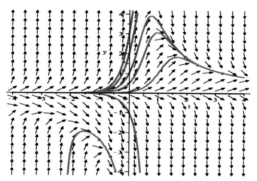

Based on the direction field, and the differential equation, for $y_0 < 0$, the slopes eventually become negative, and hence solutions tend to $-\infty$. For $y_0 > 0$, solutions increase without bound if $t_0 < 0$. Otherwise, the slopes eventually become negative, and solutions tend to zero. Furthermore, $y_0 = 0$ is an equilibrium solution. Note that slopes are zero along the curves $y = 0$ and $ty = 3$.

19.

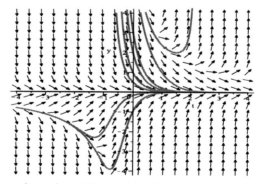

For initial conditions (t_0, y_0) satisfying $ty < 3$, the respective solutions all tend to zero. For $y_0 \leq 9$, the solutions tend to 0; for $y_0 > 9$, the solutions tend to ∞. Also, $y_0 = 0$ is an equilibrium solution.

20.

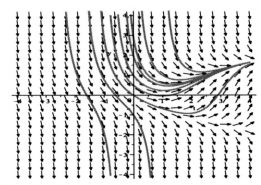

Solutions with $t_0 < 0$ all tend to $-\infty$. Solutions with initial conditions (t_0, y_0) to the right of the parabola $t = 1 + y^2$ asymptotically approach the parabola as $t \to \infty$. Integral curves with initial conditions above the parabola (and $y_0 > 0$) also approach the curve. The slopes for solutions with initial conditions below the parabola (and $y_0 < 0$) are all negative. These solutions tend to $-\infty$.

21.(a) No. There is no value of $t_0 \geq 0$ for which $(2/3)(t - t_0)^{2/3}$ satisfies the condition $y(1) = 1$.

(b) Yes. Let $t_0 = 1/2$ in Eq.(19).

(c) For $t_0 > 0$, $|y(2)| \leq (4/3)^{3/2} \approx 1.54$.

24. The assumption is $\phi'(t) + p(t)\phi(t) = 0$. But then $c\phi'(t) + p(t)c\phi(t) = 0$ as well.

26.(a) Recalling Eq.(33) in Section 2.1,

$$y = \frac{1}{\mu(t)} \int_{t_0}^{t} \mu(s)g(s)\,ds + \frac{c}{\mu(t)}.$$

It is evident that $y_1(t) = 1/\mu(t)$ and $y_2(t) = (1/\mu(t)) \int_{t_0}^{t} \mu(s)g(s)\,ds$.

(b) By definition, $1/\mu(t) = e^{-\int p(t)dt}$. Hence $y_1' = -p(t)/\mu(t) = -p(t)y_1$. That is, $y_1' + p(t)y_1 = 0$.

(c) $y_2' = (-p(t)/\mu(t)) \int_0^t \mu(s)g(s)\,ds + \mu(t)g(t)/\mu(t) = -p(t)y_2 + g(t)$. This implies that $y_2' + p(t)y_2 = g(t)$.

30. Since $n = 3$, set $v = y^{-2}$. It follows that $v' = -2y^{-3}y'$ and $y' = -(y^3/2)v'$. Substitution into the differential equation yields $-(y^3/2)v' - \varepsilon y = -\sigma y^3$, which further results in $v' + 2\varepsilon v = 2\sigma$. The latter differential equation is linear, and can be written as $(ve^{2\varepsilon t})' = 2\sigma e^{2\varepsilon t}$. The solution is given by $v(t) = \sigma/\varepsilon + ce^{-2\varepsilon t}$. Converting back to the original dependent variable, $y = \pm v^{-1/2} = \pm(\sigma/\varepsilon + ce^{-2\varepsilon t})^{-1/2}$.

31. Since $n = 3$, set $v = y^{-2}$. It follows that $v' = -2y^{-3}y'$ and $y' = -(y^3/2)v'$. The differential equation is written as $-(y^3/2)v' - (\Gamma \cos t + T)y = \sigma y^3$, which upon

further substitution is $v' + 2(\Gamma\cos t + T)v = 2$. This ODE is linear, with integrating factor $\mu(t) = e^{2\int(\Gamma\cos t+T)dt} = e^{2\Gamma\sin t+2Tt}$. The solution is

$$v(t) = 2e^{-(2\Gamma\sin t+2Tt)}\int_0^t e^{2\Gamma\sin \tau+2T\tau}d\tau + ce^{-(2\Gamma\sin t+2Tt)}.$$

Converting back to the original dependent variable, $y = \pm v^{-1/2}$.

33. The solution of the initial value problem $y_1' + 2y_1 = 0$, $y_1(0) = 1$ is $y_1(t) = e^{-2t}$. Therefore $y(1^-) = y_1(1) = e^{-2}$. On the interval $(1,\infty)$, the differential equation is $y_2' + y_2 = 0$, with $y_2(t) = ce^{-t}$. Therefore $y(1^+) = y_2(1) = ce^{-1}$. Equating the limits $y(1^-) = y(1^+)$, we require that $c = e^{-1}$. Hence the global solution of the initial value problem is

$$y(t) = \begin{cases} e^{-2t}, & 0 \le t \le 1 \\ e^{-1-t}, & t > 1 \end{cases}.$$

Note the discontinuity of the derivative

$$y'(t) = \begin{cases} -2e^{-2t}, & 0 < t < 1 \\ -e^{-1-t}, & t > 1 \end{cases}.$$

2.5

1.

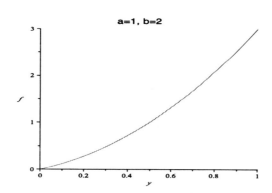

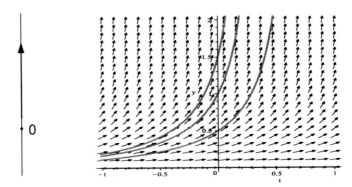

For $y_0 \geq 0$, the only equilibrium point is $y^* = 0$, and $y' = ay + by^2 > 0$ when $y > 0$, hence the equilibrium solution $y = 0$ is unstable.

2.

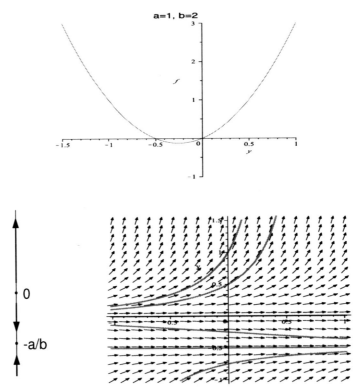

The equilibrium points are $y^* = -a/b$ and $y^* = 0$, and $y' > 0$ when $y > 0$ or $y < -a/b$, and $y' < 0$ when $-a/b < y < 0$, therefore the equilibrium solution $y = -a/b$ is asymptotically stable and the equilibrium solution $y = 0$ is unstable.

4.

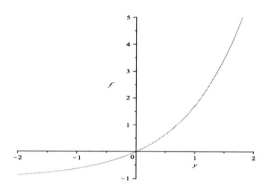

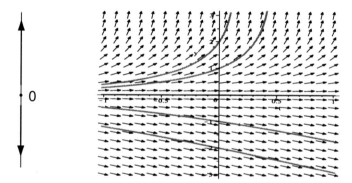

The only equilibrium point is $y^* = 0$, and $y' > 0$ when $y > 0$, $y' < 0$ when $y < 0$, hence the equilibrium solution $y = 0$ is unstable.

6.

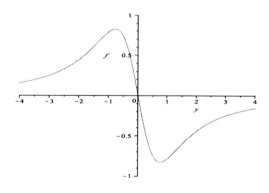

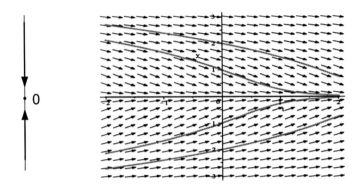

The only equilibrium point is $y^* = 0$, and $y' > 0$ when $y < 0$, $y' < 0$ when $y > 0$, hence the equilibrium solution $y = 0$ is asymptotically stable.

8.

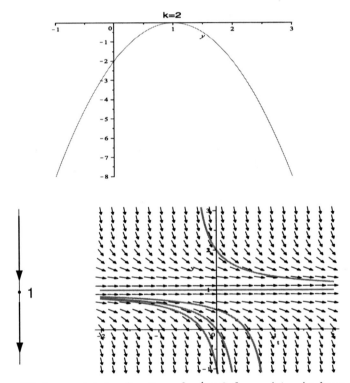

The only equilibrium point is $y^* = 1$, and $y' < 0$ for $y \neq 1$. As long as $y_0 \neq 1$, the corresponding solution is monotone decreasing. Hence the equilibrium solution $y = 1$ is semistable.

10.

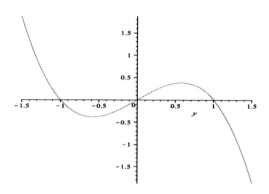

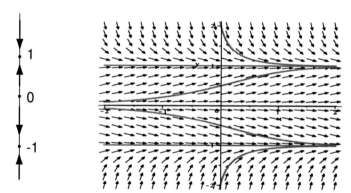

The equilibrium points are $y^* = 0, \pm 1$, and $y' > 0$ for $y < -1$ or $0 < y < 1$ and $y' < 0$ for $-1 < y < 0$ or $y > 1$. The equilibrium solution $y = 0$ is unstable, and the remaining two are asymptotically stable.

12.

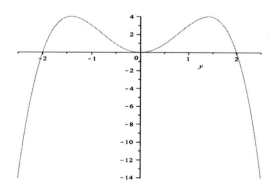

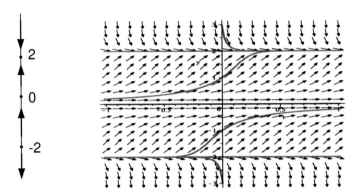

The equilibrium points are $y^* = 0, \pm 2$, and $y' < 0$ when $y < -2$ or $y > 2$, and $y' > 0$ for $-2 < y < 0$ or $0 < y < 2$. The equilibrium solutions $y = -2$ and $y = 2$ are unstable and asymptotically stable, respectively. The equilibrium solution $y = 0$ is semistable.

13.

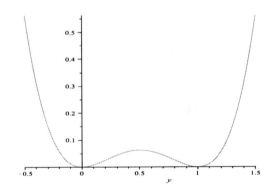

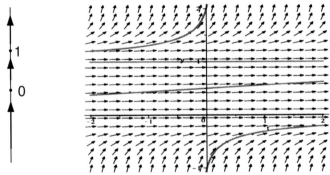

The equilibrium points are $y^* = 0, 1$. $y' > 0$ for all y except $y = 0$ and $y = 1$. Both equilibrium solutions are semistable.

15.(a) Inverting Eq.(11), Eq.(13) shows t as a function of the population y and the carrying capacity K. With $y_0 = K/3$,

$$t = -\frac{1}{r} \ln \left| \frac{(1/3)[1 - (y/K)]}{(y/K)[1 - (1/3)]} \right|.$$

Setting $y = 2y_0$,

$$\tau = -\frac{1}{r} \ln \left| \frac{(1/3)[1 - (2/3)]}{(2/3)[1 - (1/3)]} \right|.$$

That is, $\tau = (\ln 4)/r$. If $r = 0.025$ per year, $\tau \approx 55.45$ years.

(b) In Eq.(13), set $y_0/K = \alpha$ and $y/K = \beta$. As a result, we obtain

$$T = -\frac{1}{r} \ln \left| \frac{\alpha[1 - \beta]}{\beta[1 - \alpha]} \right|.$$

Given $\alpha = 0.1$, $\beta = 0.9$ and $r = 0.025$ per year, $\tau \approx 175.78$ years.

19.(a) The rate of increase of the volume is given by rate of flow in − rate of flow out. That is, $dV/dt = k - \alpha a \sqrt{2gh}$. Since the cross section is constant, $dV/dt = A dh/dt$. Hence the governing equation is $dh/dt = (k - \alpha a \sqrt{2gh})/A$.

(b) Setting $dh/dt = 0$, the equilibrium height is $h_e = (1/2g)(k/\alpha a)^2$. Furthermore, since $dh/dt < 0$ for $h > h_e$ and $dh/dt > 0$ for $h < h_e$, it follows that the equilibrium height is asymptotically stable.

22.(a) The equilibrium points are at $y^* = 0$ and $y^* = 1$. Since $f'(y) = \alpha - 2\alpha y$, the equilibrium solution $y = 0$ is unstable and the equilibrium solution $y = 1$ is asymptotically stable.

(b) The differential equation is separable, with $[y(1 - y)]^{-1} dy = \alpha \, dt$. Integrating both sides and invoking the initial condition, the solution is

$$y(t) = \frac{y_0 \, e^{\alpha t}}{1 - y_0 + y_0 \, e^{\alpha t}} = \frac{y_0}{y_0 + (1 - y_0)e^{-\alpha t}}.$$

It is evident that (independent of y_0) $\lim_{t \to -\infty} y(t) = 0$ and $\lim_{t \to \infty} y(t) = 1$.

23.(a) $y(t) = y_0 \, e^{-\beta t}$.

(b) From part (a), $dx/dt = -\alpha x y_0 e^{-\beta t}$. Separating variables, $dx/x = -\alpha y_0 e^{-\beta t} dt$. Integrating both sides, the solution is $x(t) = x_0 \, e^{-\alpha y_0 (1 - e^{-\beta t})/\beta}$.

(c) As $t \to \infty$, $y(t) \to 0$ and $x(t) \to x_0 \, e^{-\alpha y_0/\beta}$. Over a long period of time, the proportion of carriers vanishes. Therefore the proportion of the population that escapes the epidemic is the proportion of susceptibles left at that time, $x_0 \, e^{-\alpha y_0/\beta}$.

26.(a) For $a < 0$, the only critical point is at $y = 0$, which is asymptotically stable. For $a = 0$, the only critical point is at $y = 0$, which is asymptotically stable. For $a > 0$, the three critical points are at $y = 0, \pm\sqrt{a}$. The critical point at $y = 0$ is unstable, whereas the other two are asymptotically stable.

(b) Below, we graph solutions in the case $a = -1$, $a = 0$ and $a = 1$ respectively.

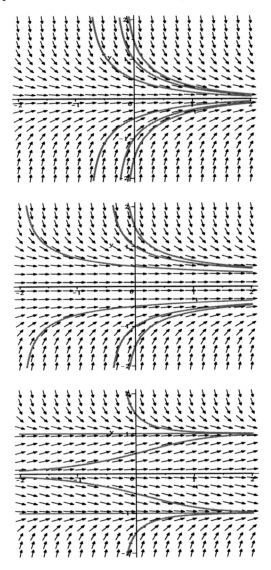

(c)

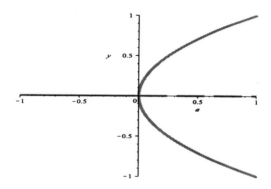

27.(a) $f(y) = y(a-y)$; $f'(y) = a - 2y$. For $a < 0$, the critical points are at $y = a$ and $y = 0$. Observe that $f'(a) > 0$ and $f'(0) < 0$. Hence $y = a$ is unstable and $y = 0$ asymptotically stable. For $a = 0$, the only critical point is at $y = 0$, which is semistable since $f(y) = -y^2$ is concave down. For $a > 0$, the critical points are at $y = 0$ and $y = a$. Observe that $f'(0) > 0$ and $f'(a) < 0$. Hence $y = 0$ is unstable and $y = a$ asymptotically stable.

(b) Below, we graph solutions in the case $a = -1$, $a = 0$ and $a = 1$ respectively.

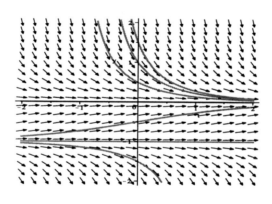

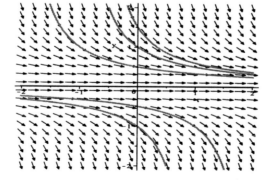

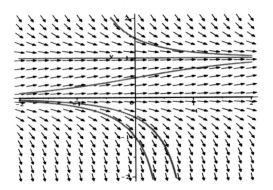

(c)

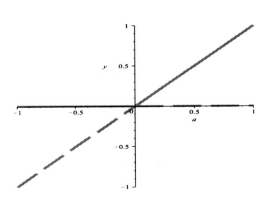

2.6

1. $M(x,y) = 2x + 3$ and $N(x,y) = 2y - 2$. Since $M_y = N_x = 0$, the equation is exact. Integrating M with respect to x, while holding y constant, yields $\psi(x,y) = x^2 + 3x + h(y)$. Now $\psi_y = h'(y)$, and equating with N results in the possible function $h(y) = y^2 - 2y$. Hence $\psi(x,y) = x^2 + 3x + y^2 - 2y$, and the solution is defined implicitly as $x^2 + 3x + y^2 - 2y = c$.

2. $M(x,y) = 2x + 4y$ and $N(x,y) = 2x - 2y$. Note that $M_y \neq N_x$, and hence the differential equation is not exact.

4. First divide both sides by $(2xy + 2)$. We now have $M(x,y) = y$ and $N(x,y) = x$. Since $M_y = N_x = 0$, the resulting equation is exact. Integrating M with respect to x, while holding y constant, results in $\psi(x,y) = xy + h(y)$. Differentiating with respect to y, $\psi_y = x + h'(y)$. Setting $\psi_y = N$, we find that $h'(y) = 0$, and hence $h(y) = 0$ is acceptable. Therefore the solution is defined implicitly as $xy = c$. Note that if $xy + 1 = 0$, the equation is trivially satisfied.

6. Write the equation as $(ax - by)dx + (bx - cy)dy = 0$. Now $M(x,y) = ax - by$ and $N(x,y) = bx - cy$. Since $M_y \neq N_x$, the differential equation is not exact.

8. $M(x,y) = e^x \sin y + 3y$ and $N(x,y) = -3x + e^x \sin y$. Note that $M_y \neq N_x$, and hence the differential equation is not exact.

10. $M(x,y) = y/x + 6x$ and $N(x,y) = \ln x - 2$. Since $M_y = N_x = 1/x$, the given equation is exact. Integrating N with respect to y, while holding x constant, results in $\psi(x,y) = y \ln x - 2y + h(x)$. Differentiating with respect to x, $\psi_x = y/x + h'(x)$. Setting $\psi_x = M$, we find that $h'(x) = 6x$, and hence $h(x) = 3x^2$. Therefore the solution is defined implicitly as $3x^2 + y \ln x - 2y = c$.

11. $M(x,y) = x \ln y + xy$ and $N(x,y) = y \ln x + xy$. Note that $M_y \neq N_x$, and hence the differential equation is not exact.

13. $M(x,y) = 2x - y$ and $N(x,y) = 2y - x$. Since $M_y = N_x = -1$, the equation is exact. Integrating M with respect to x, while holding y constant, yields $\psi(x,y) = x^2 - xy + h(y)$. Now $\psi_y = -x + h'(y)$. Equating ψ_y with N results in $h'(y) = 2y$, and hence $h(y) = y^2$. Thus $\psi(x,y) = x^2 - xy + y^2$, and the solution is given implicitly as $x^2 - xy + y^2 = c$. Invoking the initial condition $y(1) = 3$, the specific solution is $x^2 - xy + y^2 = 7$. The explicit form of the solution is $y(x) = (x + \sqrt{28 - 3x^2})/2$. Hence the solution is valid as long as $3x^2 \leq 28$.

16. $M(x,y) = y e^{2xy} + x$ and $N(x,y) = bx e^{2xy}$. Note that $M_y = e^{2xy} + 2xy e^{2xy}$, and $N_x = b e^{2xy} + 2bxy e^{2xy}$. The given equation is exact, as long as $b = 1$. Integrating N with respect to y, while holding x constant, results in $\psi(x,y) = e^{2xy}/2 + h(x)$. Now differentiating with respect to x, $\psi_x = y e^{2xy} + h'(x)$. Setting $\psi_x = M$, we find that $h'(x) = x$, and hence $h(x) = x^2/2$. We conclude that $\psi(x,y) = e^{2xy}/2 + x^2/2$. Hence the solution is given implicitly as $e^{2xy} + x^2 = c$.

17. Note that ψ is of the form $\psi(x,y) = f(x) + g(y)$, since each of the integrands is a function of a single variable. It follows that $\psi_x = f'(x)$ and $\psi_y = g'(y)$. That is, $\psi_x = M(x, y_0)$ and $\psi_y = N(x_0, y)$. Furthermore,

$$\frac{\partial^2 \psi}{\partial x \partial y}(x_0, y_0) = \frac{\partial M}{\partial y}(x_0, y_0) \text{ and } \frac{\partial^2 \psi}{\partial y \partial x}(x_0, y_0) = \frac{\partial N}{\partial x}(x_0, y_0),$$

based on the hypothesis and the fact that the point (x_0, y_0) is arbitrary, $\psi_{xy} = \psi_{yx}$ and $M_y(x,y) = N_x(x,y)$.

18. Observe that $(M(x))_y = (N(y))_x = 0$.

20. $M_y = y^{-1} \cos y - y^{-2} \sin y$ and $N_x = -2 e^{-x}(\cos x + \sin x)/y$. Multiplying both sides by the integrating factor $\mu(x,y) = y e^x$, the given equation can be written as $(e^x \sin y - 2y \sin x)dx + (e^x \cos y + 2 \cos x)dy = 0$. Let $\tilde{M} = \mu M$ and $\tilde{N} = \mu N$. Observe that $\tilde{M}_y = \tilde{N}_x$, and hence the latter ODE is exact. Integrating $\tilde{N}$ with respect to y, while holding x constant, results in $\psi(x,y) = e^x \sin y + 2y \cos x + h(x)$. Now differentiating with respect to x, $\psi_x = e^x \sin y - 2y \sin x + h'(x)$. Setting $\psi_x = \tilde{M}$, we find that $h'(x) = 0$, and hence $h(x) = 0$ is feasible. Hence the solution of the given equation is defined implicitly by $e^x \sin y + 2y \cos x = c$.

21. $M_y = 1$ and $N_x = 2$. Multiply both sides by the integrating factor $\mu(x,y) = y$ to obtain $y^2 dx + (2xy - y^2 e^y)dy = 0$. Let $\tilde{M} = yM$ and $\tilde{N} = yN$. It is easy to see that $\tilde{M}_y = \tilde{N}_x$, and hence the latter ODE is exact. Integrating $\tilde{M}$ with respect to x yields $\psi(x,y) = xy^2 + h(y)$. Equating ψ_y with $\tilde{N}$ results in $h'(y) = -y^2 e^y$, and hence $h(y) = -e^y(y^2 - 2y + 2)$. Thus $\psi(x,y) = xy^2 - e^y(y^2 - 2y + 2)$, and the solution is defined implicitly by $xy^2 - e^y(y^2 - 2y + 2) = c$.

24. The equation $\mu M + \mu N y' = 0$ has an integrating factor if $(\mu M)_y = (\mu N)_x$, that is, $\mu_y M - \mu_x N = \mu N_x - \mu M_y$. Suppose that $N_x - M_y = R(xM - yN)$, in which R is some function depending only on the quantity $z = xy$. It follows that the modified form of the equation is exact, if $\mu_y M - \mu_x N = \mu R(xM - yN) = R(\mu xM - \mu yN)$. This relation is satisfied if $\mu_y = (\mu x)R$ and $\mu_x = (\mu y)R$. Now consider $\mu = \mu(xy)$. Then the partial derivatives are $\mu_x = \mu' y$ and $\mu_y = \mu' x$. Note that $\mu' = d\mu/dz$. Thus μ must satisfy $\mu'(z) = R(z)$. The latter equation is separable, with $d\mu = R(z)dz$, and $\mu(z) = \int R(z)dz$. Therefore, given $R = R(xy)$, it is possible to determine $\mu = \mu(xy)$ which becomes an integrating factor of the differential equation.

28. The equation is not exact, since $N_x - M_y = 2y - 1$. However, $(N_x - M_y)/M = (2y - 1)/y$ is a function of y alone. Hence there exists $\mu = \mu(y)$, which is a solution of the differential equation $\mu' = (2 - 1/y)\mu$. The latter equation is separable, with $d\mu/\mu = 2 - 1/y$. One solution is $\mu(y) = e^{2y - \ln y} = e^{2y}/y$. Now rewrite the given ODE as $e^{2y}dx + (2x e^{2y} - 1/y)dy = 0$. This equation is exact, and it is easy to see that $\psi(x,y) = x e^{2y} - \ln|y|$. Therefore the solution of the given equation is defined implicitly by $x e^{2y} - \ln|y| = c$.

30. The given equation is not exact, since $N_x - M_y = 8x^3/y^3 + 6/y^2$. But note that $(N_x - M_y)/M = 2/y$ is a function of y alone, and hence there is an integrating factor $\mu = \mu(y)$. Solving the equation $\mu' = (2/y)\mu$, an integrating factor is $\mu(y) = y^2$. Now rewrite the differential equation as $(4x^3 + 3y)dx + (3x + 4y^3)dy = 0$. By inspection, $\psi(x,y) = x^4 + 3xy + y^4$, and the solution of the given equation is defined implicitly by $x^4 + 3xy + y^4 = c$.

32. Multiplying both sides of the ODE by $\mu = [xy(2x + y)]^{-1}$, the given equation is equivalent to $[(3x + y)/(2x^2 + xy)] dx + [(x + y)/(2xy + y^2)] dy = 0$. Rewrite the differential equation as

$$\left[\frac{2}{x} + \frac{2}{2x + y}\right] dx + \left[\frac{1}{y} + \frac{1}{2x + y}\right] dy = 0.$$

It is easy to see that $M_y = N_x$. Integrating M with respect to x, while keeping y constant, results in $\psi(x,y) = 2\ln|x| + \ln|2x + y| + h(y)$. Now taking the partial derivative with respect to y, $\psi_y = (2x + y)^{-1} + h'(y)$. Setting $\psi_y = N$, we find that $h'(y) = 1/y$, and hence $h(y) = \ln|y|$. Therefore $\psi(x,y) = 2\ln|x| + \ln|2x + y| + \ln|y|$, and the solution of the given equation is defined implicitly by $2x^3 y + x^2 y^2 = c$.

2.7

2. The Euler formula is given by $y_{n+1} = y_n + h(2y_n - 1) = (1 + 2h)y_n - h$.

(a) 1.1, 1.22, 1.364, 1.5368

(b) 1.105, 1.23205, 1.38578, 1.57179

(c) 1.10775, 1.23873, 1.39793, 1.59144

(d) The differential equation is linear with solution $y(t) = (1 + e^{2t})/2$. The values are 1.1107, 1.24591, 1.41106, 1.61277.

5.

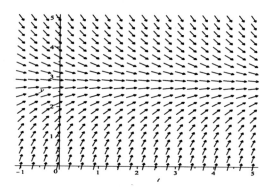

All solutions seem to converge to $y = 25/9$.

7.

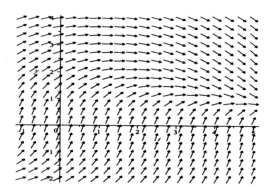

All solutions seem to converge to a specific function.

8.

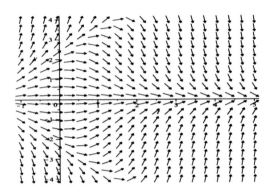

Solutions with initial conditions $|y(0)| > 2.5$ seem to diverge. On the other hand, solutions with initial conditions $|y(0)| < 2.5$ seem to converge to zero. Also, $y = 0$ is an equilibrium solution.

10.

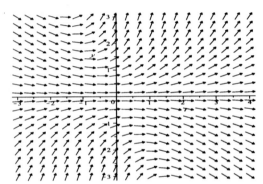

Solutions with positive initial conditions increase without bound. Solutions with negative initial conditions decrease without bound. Note that $y = 0$ is an equilibrium solution.

11. The Euler formula is $y_{n+1} = y_n - 3h\sqrt{y_n} + 5h$. The initial value is $y_0 = 2$.

(a) 2.30800, 2.49006, 2.60023, 2.66773, 2.70939, 2.73521

(b) 2.30167, 2.48263, 2.59352, 2.66227, 2.70519, 2.73209

(c) 2.29864, 2.47903, 2.59024, 2.65958, 2.70310, 2.73053

(d) 2.29686, 2.47691, 2.58830, 2.65798, 2.70185, 2.72959

12. The Euler formula is $y_{n+1} = (1 + 3h)y_n - ht_n y_n^2$. The initial value is $(t_0, y_0) = (0, 0.5)$.

(a) 1.70308, 3.06605, 2.44030, 1.77204, 1.37348, 1.11925

(b) 1.79548, 3.06051, 2.43292, 1.77807, 1.37795, 1.12191

(c) 1.84579, 3.05769, 2.42905, 1.78074, 1.38017, 1.12328

(d) 1.87734, 3.05607, 2.42672, 1.78224, 1.38150, 1.12411

14. The Euler formula is $y_{n+1} = (1 - ht_n)y_n + hy_n^3/10$, with $(t_0, y_0) = (0, 1)$.

(a) 0.950517, 0.687550, 0.369188, 0.145990, 0.0421429, 0.00872877

(b) 0.938298, 0.672145, 0.362640, 0.147659, 0.0454100, 0.0104931

(c) 0.932253, 0.664778, 0.359567, 0.148416, 0.0469514, 0.0113722

(d) 0.928649, 0.660463, 0.357783, 0.148848, 0.0478492, 0.0118978

17. The Euler formula is $y_{n+1} = y_n + h(y_n^2 + 2t_n y_n)/(3 + t_n^2)$. The initial point is $(t_0, y_0) = (1, 2)$. Using this iteration formula with the specified h values, the value of the solution at $t = 2.5$ is somewhere between 18 and 19. At $t = 3$ there is no reliable estimate.

19.(a)

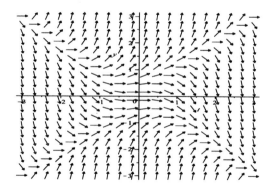

(b) The iteration formula is $y_{n+1} = y_n + h y_n^2 - h t_n^2$. The critical value α_0 appears to be between 0.67 and 0.68. For $y_0 > \alpha_0$, the iterations diverge.

20.(a) The ODE is linear, with general solution $y(t) = t + ce^t$. Invoking the specified initial condition, $y(t_0) = y_0$, we have $y_0 = t_0 + ce^{t_0}$. Hence $c = (y_0 - t_0)e^{-t_0}$. Thus the solution is given by $\phi(t) = (y_0 - t_0)e^{t-t_0} + t$.

(b) The Euler formula is $y_{n+1} = (1 + h)y_n + h - ht_n$. Now set $k = n + 1$.

(c) We have $y_1 = (1 + h)y_0 + h - ht_0 = (1 + h)y_0 + (t_1 - t_0) - ht_0$. Rearranging the terms, $y_1 = (1 + h)(y_0 - t_0) + t_1$. Now suppose that $y_k = (1 + h)^k(y_0 - t_0) + t_k$, for some $k \geq 1$. Then $y_{k+1} = (1 + h)y_k + h - ht_k$. Substituting for y_k, we find

that
$$y_{k+1} = (1+h)^{k+1}(y_0 - t_0) + (1+h)t_k + h - ht_k = (1+h)^{k+1}(y_0 - t_0) + t_k + h.$$
Noting that $t_{k+1} = t_k + h$, the result is verified.

(d) Substituting $h = (t-t_0)/n$, with $t_n = t$, $y_n = (1 + (t-t_0)/n)^n(y_0 - t_0) + t$. Taking the limit of both sides, and using the fact that $\lim_{n\to\infty}(1+a/n)^n = e^a$, pointwise convergence is proved.

21. The exact solution is $y(t) = e^t$. The Euler formula is $y_{n+1} = (1+h)y_n$. It is easy to see that $y_n = (1+h)^n y_0 = (1+h)^n$. Given $t > 0$, set $h = t/n$. Taking the limit, we find that $\lim_{n\to\infty} y_n = \lim_{n\to\infty}(1+t/n)^n = e^t$.

23. The exact solution is $y(t) = t/2 + e^{2t}$. The Euler formula is $y_{n+1} = (1+2h)y_n + h/2 - ht_n$. Since $y_0 = 1$, $y_1 = (1+2h) + h/2 = (1+2h) + t_1/2$. It is easy to show by mathematical induction, that $y_n = (1+2h)^n + t_n/2$. For $t > 0$, set $h = t/n$ and thus $t_n = t$. Taking the limit, we find that $\lim_{n\to\infty} y_n = \lim_{n\to\infty} [(1+2t/n)^n + t/2] = e^{2t} + t/2$. Hence pointwise convergence is proved.

2.8

2. Let $z = y - 3$ and $\tau = t + 1$. It follows that $dz/d\tau = (dz/dt)(dt/d\tau) = dz/dt$. Furthermore, $dz/dt = dy/dt = 1 - y^3$. Hence $dz/d\tau = 1 - (z+3)^3$. The new initial condition is $z(0) = 0$.

3.(a) The approximating functions are defined recursively by
$$\phi_{n+1}(t) = \int_0^t 2[\phi_n(s) + 1]\, ds.$$
Setting $\phi_0(t) = 0$, $\phi_1(t) = 2t$. Continuing, $\phi_2(t) = 2t^2 + 2t$, $\phi_3(t) = 4t^3/3 + 2t^2 + 2t$, $\phi_4(t) = 2t^4/3 + 4t^3/3 + 2t^2 + 2t$, ... Based upon these we conjecture that $\phi_n(t) = \sum_{k=1}^n 2^k t^k/k!$ and use mathematical induction to verify this form for $\phi_n(t)$. First, let $n = 1$, then $\phi_n(t) = 2t$, so it is certainly true for $n = 1$. Then, using Eq.(7) again we have
$$\phi_{n+1}(t) = \int_0^t 2[\phi_n(s) + 1]\, ds = \int_0^t 2\left[\sum_{k=1}^n \frac{2^k}{k!}s^k + 1\right] ds = \sum_{k=1}^{n+1} \frac{2^k}{k!}t^k,$$
and we have verified our conjecture.

(b)

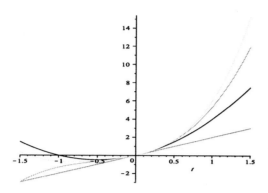

(c) Recall from calculus that $e^{at} = 1 + \sum_{k=1}^{\infty} a^k t^k/k!$. Thus
$$\phi(t) = \sum_{k=1}^{\infty} \frac{2^k}{k!} t^k = e^{2t} - 1.$$

(d)

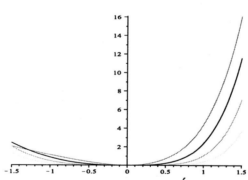

From the plot it appears that ϕ_4 is a good estimate for $|t| < 1/2$.

5.(a) The approximating functions are defined recursively by
$$\phi_{n+1}(t) = \int_0^t [-\phi_n(s)/2 + s]\, ds.$$

Setting $\phi_0(t) = 0$, $\phi_1(t) = t^2/2$. Continuing, $\phi_2(t) = t^2/2 - t^3/12$, $\phi_3(t) = t^2/2 - t^3/12 + t^4/96$, $\phi_4(t) = t^2/2 - t^3/12 + t^4/96 - t^5/960$, ... Based upon these we conjecture that $\phi_n(t) = \sum_{k=1}^n 4(-1/2)^{k+1} t^{k+1}/(k+1)!$ and use mathematical induction to verify this form for $\phi_n(t)$.

(b)

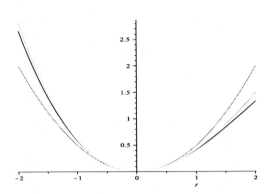

(c) Recall from calculus that $e^{at} = 1 + \sum_{k=1}^{\infty} a^k t^k/k!$. Thus

$$\phi(t) = \sum_{k=1}^{\infty} 4 \frac{(-1/2)^{k+1}}{k+1!} t^{k+1} = 4e^{-t/2} + 2t - 4.$$

(d)

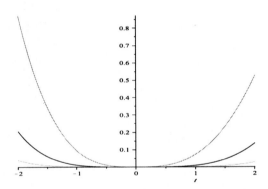

From the plot it appears that ϕ_4 is a good estimate for $|t| < 2$.

6.(a) The approximating functions are defined recursively by

$$\phi_{n+1}(t) = \int_0^t [\phi_n(s) + 1 - s] \, ds.$$

Setting $\phi_0(t) = 0$, $\phi_1(t) = t - t^2/2$, $\phi_2(t) = t - t^3/6$, $\phi_3(t) = t - t^4/24$, $\phi_4(t) = t - t^5/120$, ... Based upon these we conjecture that $\phi_n(t) = t - t^{n+1}/(n+1)!$ and use mathematical induction to verify this form for $\phi_n(t)$.

(b)

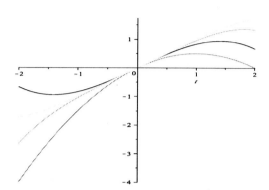

(c) Clearly $\phi(t) = t$.

(d)

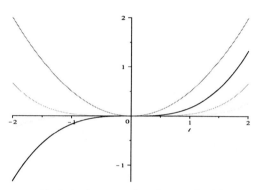

From the plot it appears that ϕ_4 is a good estimate for $|t| < 1$.

8.(a) The approximating functions are defined recursively by

$$\phi_{n+1}(t) = \int_0^t \left[s^2 \phi_n(s) - s \right] ds.$$

Set $\phi_0(t) = 0$. The iterates are given by $\phi_1(t) = -t^2/2$, $\phi_2(t) = -t^2/2 - t^5/10$, $\phi_3(t) = -t^2/2 - t^5/10 - t^8/80$, $\phi_4(t) = -t^2/2 - t^5/10 - t^8/80 - t^{11}/880$,.... Upon inspection, it becomes apparent that

$$\phi_n(t) = -t^2 \left[\frac{1}{2} + \frac{t^3}{2 \cdot 5} + \frac{t^6}{2 \cdot 5 \cdot 8} + \ldots + \frac{(t^3)^{n-1}}{2 \cdot 5 \cdot 8 \ldots [2 + 3(n-1)]} \right] =$$

$$= -t^2 \sum_{k=1}^{n} \frac{(t^3)^{k-1}}{2 \cdot 5 \cdot 8 \ldots [2 + 3(k-1)]}.$$

(b)

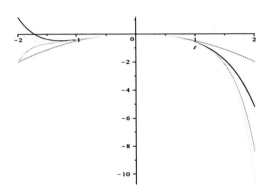

(c) Using the identity $\phi_n(t) = \phi_1(t) + [\phi_2(t) - \phi_1(t)] + [\phi_3(t) - \phi_2(t)] + \ldots + [\phi_n(t) - \phi_{n-1}(t)]$, consider the series $\phi_1(t) + \sum_{k=1}^{\infty}[\phi_{k+1}(t) - \phi_k(t)]$. Fix any t value now. We use the Ratio Test to prove the convergence of this series:

$$\left|\frac{\phi_{k+1}(t) - \phi_k(t)}{\phi_k(t) - \phi_{k-1}(t)}\right| = \left|\frac{\frac{(-t^2)(t^3)^k}{2\cdot 5 \cdots (2+3k)}}{\frac{(-t^2)(t^3)^{k-1}}{2\cdot 5 \cdots (2+3(k-1))}}\right| = \frac{|t|^3}{2+3k}.$$

The limit of this quantity is 0 for any fixed t as $k \to \infty$, and we obtain that $\phi_n(t)$ is convergent for any t.

9.(a) The approximating functions are defined recursively by

$$\phi_{n+1}(t) = \int_0^t \left[s^2 + \phi_n^2(s)\right] ds.$$

Set $\phi_0(t) = 0$. The first three iterates are given by $\phi_1(t) = t^3/3$, $\phi_2(t) = t^3/3 + t^7/63$, $\phi_3(t) = t^3/3 + t^7/63 + 2t^{11}/2079 + t^{15}/59535$.

(b)

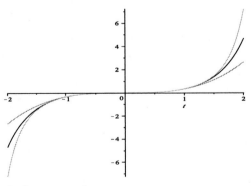

The iterates appear to be converging.

12.(a) The approximating functions are defined recursively by
$$\phi_{n+1}(t) = \int_0^t \left[\frac{3s^2+4s+2}{2(\phi_n(s)-1)}\right]ds.$$

Note that $1/(2y-2) = -(1/2)\sum_{k=0}^6 y^k + O(y^7)$. For computational purposes, use the geometric series sum to replace the above iteration formula by
$$\phi_{n+1}(t) = -\frac{1}{2}\int_0^t \left[(3s^2+4s+2)\sum_{k=0}^6 \phi_n^k(s)\right]ds.$$

Set $\phi_0(t) = 0$. The first four approximations are given by $\phi_1(t) = -t - t^2 - t^3/2$, $\phi_2(t) = -t - t^2/2 + t^3/6 + t^4/4 - t^5/5 - t^6/24 + \ldots$, $\phi_3(t) = -t - t^2/2 + t^4/12 - 3t^5/20 + 4t^6/45 + \ldots$, $\phi_4(t) = -t - t^2/2 + t^4/8 - 7t^5/60 + t^6/15 + \ldots$

(b)

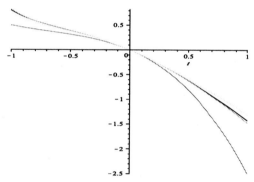

The approximations appear to be converging to the exact solution, which can be found by separating the variables: $\phi(t) = 1 - \sqrt{1+2t+2t^2+t^3}$.

14.(a) $\phi_n(0) = 0$, for every $n \geq 1$. Let $a \in (0,1]$. Then $\phi_n(a) = 2na\,e^{-na^2} = 2na/e^{na^2}$. Using l'Hospital's rule, $\lim_{z\to\infty} 2az/e^{az^2} = \lim_{z\to\infty} 1/ze^{az^2} = 0$. Hence $\lim_{n\to\infty} \phi_n(a) = 0$.

(b) $\int_0^1 2nx\,e^{-nx^2}dx = -e^{-nx^2}\big|_0^1 = 1 - e^{-n}$. Therefore,
$$\lim_{n\to\infty}\int_0^1 \phi_n(x)dx \neq \int_0^1 \lim_{n\to\infty}\phi_n(x)dx.$$

15. Let t be fixed, such that $(t,y_1), (t,y_2) \in D$. Without loss of generality, assume that $y_1 < y_2$. Since f is differentiable with respect to y, the mean value theorem asserts that there exists $\xi \in (y_1, y_2)$ such that $f(t,y_1) - f(t,y_2) = f_y(t,\xi)(y_1-y_2)$. This means that $|f(t,y_1) - f(t,y_2)| = |f_y(t,\xi)|\,|y_1-y_2|$. Since, by assumption, $\partial f/\partial y$ is continuous in D, f_y attains a maximum K on any closed and bounded subset of D. Hence $|f(t,y_1) - f(t,y_2)| \leq K\,|y_1-y_2|$.

16. For a sufficiently small interval of t, $\phi_{n-1}(t), \phi_n(t) \in D$. Since f satisfies a Lipschitz condition, $|f(t,\phi_n(t)) - f(t,\phi_{n-1}(t))| \leq K\,|\phi_n(t) - \phi_{n-1}(t)|$. Here $K = \max|f_y|$.

17.(a) $\phi_1(t) = \int_0^t f(s,0)ds$. Hence $|\phi_1(t)| \leq \int_0^{|t|} |f(s,0)|\,ds \leq \int_0^{|t|} M\,ds = M\,|t|$, in which M is the maximum value of $|f(t,y)|$ on D.

(b) By definition, $\phi_2(t) - \phi_1(t) = \int_0^t [f(s,\phi_1(s)) - f(s,0)]\,ds$. Taking the absolute value of both sides, $|\phi_2(t) - \phi_1(t)| \leq \int_0^{|t|} |[f(s,\phi_1(s)) - f(s,0)]|\,ds$. Based on the results in Problems 16 and 17,

$$|\phi_2(t) - \phi_1(t)| \leq \int_0^{|t|} K\,|\phi_1(s) - 0|\,ds \leq KM \int_0^{|t|} |s|\,ds.$$

Evaluating the last integral, we obtain that $|\phi_2(t) - \phi_1(t)| \leq MK\,|t|^2/2$.

(c) Suppose that

$$|\phi_i(t) - \phi_{i-1}(t)| \leq \frac{MK^{i-1}\,|t|^i}{i!}$$

for some $i \geq 1$. By definition,

$$\phi_{i+1}(t) - \phi_i(t) = \int_0^t [f(s,\phi_i(s)) - f(s,\phi_{i-1}(s))]\,ds.$$

It follows that

$$|\phi_{i+1}(t) - \phi_i(t)| \leq \int_0^{|t|} |f(s,\phi_i(s)) - f(s,\phi_{i-1}(s))|\,ds$$

$$\leq \int_0^{|t|} K\,|\phi_i(s) - \phi_{i-1}(s)|\,ds \leq \int_0^{|t|} K \frac{MK^{i-1}\,|s|^i}{i!}\,ds =$$

$$= \frac{MK^i\,|t|^{i+1}}{(i+1)!} \leq \frac{MK^i h^{i+1}}{(i+1)!}.$$

Hence, by mathematical induction, the assertion is true.

18.(a) Use the triangle inequality, $|a+b| \leq |a| + |b|$.

(b) For $|t| \leq h$, $|\phi_1(t)| \leq Mh$, and $|\phi_n(t) - \phi_{n-1}(t)| \leq MK^{n-1}h^n/(n!)$. Hence

$$|\phi_n(t)| \leq M \sum_{i=1}^n \frac{K^{i-1}h^i}{i!} = \frac{M}{K} \sum_{i=1}^n \frac{(Kh)^i}{i!}.$$

(c) The sequence of partial sums in (b) converges to $M(e^{Kh} - 1)/K$. By the comparison test, the sums in (a) also converge. Since individual terms of a convergent series must tend to zero, $|\phi_n(t) - \phi_{n-1}(t)| \to 0$, and it follows that the sequence $|\phi_n(t)|$ is convergent.

19.(a) Let $\phi(t) = \int_0^t f(s,\phi(s))ds$ and $\psi(t) = \int_0^t f(s,\psi(s))ds$. Then by linearity of the integral, $\phi(t) - \psi(t) = \int_0^t [f(s,\phi(s)) - f(s,\psi(s))]\,ds$.

(b) It follows that $|\phi(t) - \psi(t)| \leq \int_0^t |f(s,\phi(s)) - f(s,\psi(s))|\,ds$.

(c) We know that f satisfies a Lipschitz condition, $|f(t,y_1) - f(t,y_2)| \leq K\,|y_1 - y_2|$, based on $|\partial f/\partial y| \leq K$ in D. Therefore,

$$|\phi(t) - \psi(t)| \leq \int_0^t |f(s,\phi(s)) - f(s,\psi(s))|\,ds \leq \int_0^t K\,|\phi(s) - \psi(s)|\,ds.$$

2.9

1. Writing the equation for each $n \geq 0$, $y_1 = -0.9\,y_0$, $y_2 = -0.9\,y_1 = (-0.9^2)y_0$, $y_3 = -0.9\,y_2 = (-0.9)^3 y_0$ and so on, it is apparent that $y_n = (-0.9)^n\,y_0$. The terms constitute an alternating series, which converge to zero, regardless of y_0.

3. Write the equation for each $n \geq 0$, $y_1 = \sqrt{3}\,y_0$, $y_2 = \sqrt{4/2}\,y_1$, $y_3 = \sqrt{5/3}\,y_2$, ... Upon substitution, we find that $y_2 = \sqrt{(4 \cdot 3)/2}\,y_1$, $y_3 = \sqrt{(5 \cdot 4 \cdot 3)/(3 \cdot 2)}\,y_0$, ... It can be proved by mathematical induction, that

$$y_n = \frac{1}{\sqrt{2}}\sqrt{\frac{(n+2)!}{n!}}\,y_0 = \frac{1}{\sqrt{2}}\sqrt{(n+1)(n+2)}\,y_0.$$

This sequence is divergent, except for $y_0 = 0$.

4. Writing the equation for each $n \geq 0$, $y_1 = -y_0$, $y_2 = y_1$, $y_3 = -y_2$, $y_4 = y_3$, and so on. It can be shown that

$$y_n = \begin{cases} y_0, & \text{for } n = 4k \text{ or } n = 4k-1 \\ -y_0, & \text{for } n = 4k-2 \text{ or } n = 4k-3 \end{cases}$$

The sequence is convergent only for $y_0 = 0$.

6. Writing the equation for each $n \geq 0$,

$y_1 = -0.5\,y_0 + 6$
$y_2 = -0.5\,y_1 + 6 = -0.5(-0.5\,y_0 + 6) + 6 = (-0.5)^2 y_0 + 6 + (-0.5)6$
$y_3 = -0.5\,y_2 + 6 = -0.5(-0.5\,y_1 + 6) + 6 = (-0.5)^3 y_0 + 6\left[1 + (-0.5) + (-0.5)^2\right]$
$\vdots$
$y_n = (-0.5)^n y_0 + 4\left[1 - (-0.5)^n\right]$

which follows from Eq.(13) and (14). The sequence is convergent for all y_0, and in fact $y_n \to 4$.

8. Let y_n be the balance at the end of the nth month. Then $y_{n+1} = (1 + r/12)y_n + 25$. We have $y_n = \rho^n[y_0 - 25/(1-\rho)] + 25/(1-\rho)$, in which $\rho = (1 + r/12)$. Here r is the annual interest rate, given as 8%. Thus $y_{36} = (1.0066)^{36}[1000 + 12 \cdot 25/r] - 12 \cdot 25/r = \$2,283.63$.

9. Let y_n be the balance due at the end of the nth month. The appropriate difference equation is $y_{n+1} = (1 + r/12)\,y_n - P$. Here r is the annual interest rate

and P is the monthly payment. The solution, in terms of the amount borrowed, is given by $y_n = \rho^n[y_0 + P/(1-\rho)] - P/(1-\rho)$, in which $\rho = (1 + r/12)$ and $y_0 = 8,000$. To figure out the monthly payment P, we require that $y_{36} = 0$. That is, $\rho^{36}[y_0 + P/(1-\rho)] = P/(1-\rho)$. After the specified amounts are substituted, we find that $P = \$258.14$.

11. Let y_n be the balance due at the end of the nth month. The appropriate difference equation is $y_{n+1} = (1 + r/12)\,y_n - P$, in which $r = .09$ and P is the monthly payment. The initial value of the mortgage is $y_0 = \$100,000$. Then the balance due at the end of the n-th month is $y_n = \rho^n[y_0 + P/(1-\rho)] - P/(1-\rho)$, where $\rho = (1 + r/12)$. In terms of the specified values, $y_n = (1.0075)^n[10^5 - 12P/r] + 12P/r$. Setting $n = 30 \cdot 12 = 360$, and $y_{360} = 0$, we find that $P = \$804.62$. For the monthly payment corresponding to a 20 year mortgage, set $n = 240$ and $y_{240} = 0$ to find that $P = \$899.73$. The total amount paid during the term of the loan is $360 \times 804.62 = \$289,663.20$ for the 30-year loan and is $240 \times 899.73 = \$215,935.20$ for the 20-year loan.

12. Let y_n be the balance due at the end of the nth month, with y_0 the initial value of the mortgage. The appropriate difference equation is $y_{n+1} = (1 + r/12)\,y_n - P$, in which $r = 0.1$ and $P = \$1000$ is the maximum monthly payment. Given that the life of the mortgage is 20 years, we require that $y_{240} = 0$. The balance due at the end of the n-th month is $y_n = \rho^n[y_0 + P/(1-\rho)] - P/(1-\rho)$. In terms of the specified values for the parameters, the solution of $(1.00833)^{240}[y_0 - 12 \cdot 1000/0.1] = -12 \cdot 1000/0.1$ is $y_0 = \$103,624.62$.

19.(a) $\delta_2 = (\rho_2 - \rho_1)/(\rho_3 - \rho_2) = (3.449 - 3)/(3.544 - 3.449) = 4.7263$.

(b) diff $= (|\delta - \delta_2|/\delta) \cdot 100 = (|4.6692 - 4.7363|/4.6692) \cdot 100 \approx 1.22\%$.

(c) Assuming $(\rho_3 - \rho_2)/(\rho_4 - \rho_3) = \delta$, $\rho_4 \approx 3.5643$

(d) A period 16 solution appears near $\rho \approx 3.565$.

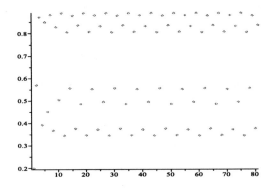

(e) Note that $(\rho_{n+1} - \rho_n) = \delta_n^{-1}(\rho_n - \rho_{n-1})$. With the assumption that $\delta_n = \delta$, we have $(\rho_{n+1} - \rho_n) = \delta^{-1}(\rho_n - \rho_{n-1})$, which is of the form $y_{n+1} = \alpha\,y_n$, $n \geq 3$. It

follows that $(\rho_k - \rho_{k-1}) = \delta^{3-k}(\rho_3 - \rho_2)$ for $k \geq 4$. Then
$$\begin{aligned}\rho_n &= \rho_1 + (\rho_2 - \rho_1) + (\rho_3 - \rho_2) + (\rho_4 - \rho_3) + \ldots + (\rho_n - \rho_{n-1}) \\ &= \rho_1 + (\rho_2 - \rho_1) + (\rho_3 - \rho_2)\left[1 + \delta^{-1} + \delta^{-2} + \ldots + \delta^{3-n}\right] \\ &= \rho_1 + (\rho_2 - \rho_1) + (\rho_3 - \rho_2)\left[\frac{1 - \delta^{4-n}}{1 - \delta^{-1}}\right].\end{aligned}$$

Hence $\lim_{n \to \infty} \rho_n = \rho_2 + (\rho_3 - \rho_2)\left[\frac{\delta}{\delta - 1}\right]$. Substitution of the appropriate values yields
$$\lim_{n \to \infty} \rho_n = 3.5699$$

PROBLEMS

1. The equation is *linear*. It can be written in the form $y' + 2y/x = x^2$, and the integrating factor is $\mu(x) = e^{\int (2/x)\,dx} = e^{2\ln x} = x^2$. Multiplication by $\mu(x)$ yields $x^2 y' + 2yx = (yx^2)' = x^4$. Integration with respect to x and division by x^2 gives that $y = x^3/5 + c/x^2$.

5. The equation is *exact*. Algebraic manipulations give the symmetric form of the equation, $(2xy + y^2 + 1)dx + (x^2 + 2xy)dy = 0$. We can check that $M_y = 2x + 2y = N_x$, so the equation is really exact. Integrating M with respect to x gives that $\psi(x, y) = x^2 y + xy^2 + x + g(y)$, then $\psi_y = x^2 + 2xy + g'(y) = x^2 + 2xy$, so we get that $g'(y) = 0$, so we obtain that $g(y) = 0$ is acceptable. Therefore the solution is defined implicitly as $x^2 y + xy^2 + x = c$.

6. The equation is *linear*. It can be written in the form $y' + (1 + (1/x))y = 1/x$ and the integrating factor is $\mu(x) = e^{\int 1+(1/x)\,dx} = e^{x+\ln x} = xe^x$. Multiplication by $\mu(x)$ yields $xe^x y' + (xe^x + e^x)y = (xe^x y)' = e^x$. Integration with respect to x and division by xe^x shows that the general solution of the equation is $y = 1/x + c/(xe^x)$. The initial condition implies that $0 = 1 + c/e$, which means that $c = -e$ and the solution is $y = 1/x - e/(xe^x) = x^{-1}(1 - e^{1-x})$.

7. The equation is *separable*. Separation of variables gives the differential equation $y(2 + 3y)dy = (4x^3 + 1)dx$, and then after integration we obtain that the solution is $x^4 + x - y^2 - y^3 = c$.

8. The equation is *linear*. It can be written in the form $y' + 2y/x = \sin x/x^2$ and the integrating factor is $\mu(x) = e^{\int (2/x)\,dx} = e^{2\ln x} = x^2$. Multiplication by $\mu(x)$ gives $x^2 y' + 2xy = (x^2 y)' = \sin x$, and after integration with respect to x and division by x^2 we obtain the general solution $y = (c - \cos x)/x^2$. The initial condition implies that $c = 4 + \cos 2$ and the solution becomes $y = (4 + \cos 2 - \cos x)/x^2$.

11. The equation is *exact*. It is easy to check that $M_y = 1 = N_x$. Integrating M with respect to x gives that $\psi(x, y) = x^3/3 + xy + g(y)$, then $\psi_y = x + g'(y) =$

$x + e^y$, which means that $g'(y) = e^y$, so we obtain that $g(y) = e^y$. Therefore the solution is defined implicitly as $x^3/3 + xy + e^y = c$.

13. The equation is *separable*. Factoring the right hand side leads to the equation $y' = (1 + y^2)(1 + 2x)$. We separate the variables to obtain $dy/(1 + y^2) = (1 + 2x)dx$, then integration gives us $\arctan y = x + x^2 + c$. The solution is $y = \tan(x + x^2 + c)$.

14. The equation is *exact*. We can check that $M_y = 1 = N_x$. Integrating M with respect to x gives that $\psi(x, y) = x^2/2 + xy + g(y)$, then $\psi_y = x + g'(y) = x + 2y$, which means that $g'(y) = 2y$, so we obtain that $g(y) = y^2$. Therefore the general solution is defined implicitly as $x^2/2 + xy + y^2 = c$. The initial condition gives us $c = 17$, so the solution is $x^2 + 2xy + 2y^2 = 34$.

15. The equation is *separable*. Separation of variables leads us to the equation
$$\frac{dy}{y} = \frac{1 - e^x}{1 + e^x} dx.$$
Note that $1 + e^x - 2e^x = 1 - e^x$. We obtain that
$$\ln|y| = \int \frac{1 - e^x}{1 + e^x} dx = \int 1 - \frac{2e^x}{1 + e^x} dx = x - 2\ln(1 + e^x) + \tilde{c}.$$
This means that $y = ce^x(1 + e^x)^{-2}$, which also can be written as $y = c/\cosh^2(x/2)$ after some algebraic manipulations.

16. The equation is *exact*. The symmetric form is $(-e^{-x}\cos y + e^{2y}\cos x)dx + (-e^{-x}\sin y + 2e^{2y}\sin x)dy = 0$. We can check that $M_y = e^{-x}\sin y + 2e^{2y}\cos x = N_x$. Integrating M with respect to x gives that $\psi(x, y) = e^{-x}\cos y + e^{2y}\sin x + g(y)$, then $\psi_y = -e^{-x}\sin y + 2e^{2y}\sin x + g'(y) = -e^{-x}\sin y + 2e^{2y}\sin x$, so we get that $g'(y) = 0$, so we obtain that $g(y) = 0$ is acceptable. Therefore the solution is defined implicitly as $e^{-x}\cos y + e^{2y}\sin x = c$.

17. The equation is *linear*. The integrating factor is $\mu(x) = e^{-\int 3 dx} = e^{-3x}$, which turns the equation into $e^{-3x}y' - 3e^{-3x}y = (e^{-3x}y)' = e^{-x}$. We integrate with respect to x to obtain $e^{-3x}y = -e^{-x} + c$, and the solution is $y = ce^{3x} - e^{2x}$ after multiplication by e^{3x}.

18. The equation is *linear*. The integrating factor is $\mu(x) = e^{\int 2 dx} = e^{2x}$, which gives us $e^{2x}y' + 2e^{2x}y = (e^{2x}y)' = e^{-x^2}$. The antiderivative of the function on the right hand side can not be expressed in a closed form using elementary functions, so we have to express the solution using integrals. Let us integrate both sides of this equation from 0 to x. We obtain that the left hand side turns into
$$\int_0^x (e^{2s}y(s))' ds = e^{2x}y(x) - e^0 y(0) = e^{2x}y - 3.$$
The right hand side gives us $\int_0^x e^{-s^2} ds$. So we found that
$$y = e^{-2x}\int_0^x e^{-s^2} ds + 3e^{-2x}.$$

19. The equation is *exact*. Algebraic manipulations give us the symmetric form $(y^3 + 2y - 3x^2)dx + (2x + 3xy^2)dy = 0$. We can check that $M_y = 3y^2 + 2 = N_x$. Integrating M with respect to x gives that $\psi(x,y) = xy^3 + 2xy - x^3 + g(y)$, then $\psi_y = 3xy^2 + 2x + g'(y) = 2x + 3xy^2$, which means that $g'(y) = 0$, so we obtain that $g(y) = 0$ is acceptable. Therefore the solution is $xy^3 + 2xy - x^3 = c$.

20. The equation is *separable*, because $y' = e^{x+y} = e^x e^y$. Separation of variables yields the equation $e^{-y}dy = e^x dx$, which turns into $-e^{-y} = e^x + c$ after integration and we obtain the implicitly defined solution $e^x + e^{-y} = c$.

22. The equation is *separable*. Separation of variables turns the equation into $(y^2 + 1)dy = (x^2 - 1)dx$, which, after integration, gives $y^3/3 + y = x^3/3 - x + c$. The initial condition yields $c = 2/3$, and the solution is $y^3 + 3y - x^3 + 3x = 2$.

23. The equation is *linear*. Division by t gives $y' + (1 + (1/t))y = e^{2t}/t$, so the integrating factor is $\mu(t) = e^{\int (1+(1/t))dt} = e^{t+\ln t} = te^t$. The equation turns into $te^t y' + (te^t + e^t)y = (te^t y)' = e^{3t}$. Integration therefore leads to $te^t y = e^{3t}/3 + c$ and the solution is $y = e^{2t}/(3t) + ce^{-t}/t$.

24. The equation is *exact*. We can check that $M_y = 2\cos y \sin x \cos x = N_x$. Integrating M with respect to x gives that $\psi(x,y) = \sin y \sin^2 x + g(y)$, then $\psi_y = \cos y \sin^2 x + g'(y) = \cos y \sin^2 x$, which means that $g'(y) = 0$, so we obtain that $g(y) = 0$ is acceptable. Therefore the solution is defined implicitly as $\sin y \sin^2 x = c$.

25. The equation is *exact*. We can check that

$$M_y = -\frac{2x}{y^2} - \frac{x^2 - y^2}{(x^2+y^2)^2} = N_x.$$

Integrating M with respect to x gives that $\psi(x,y) = x^2/y + \arctan(y/x) + g(y)$, then $\psi_y = -x^2/y^2 + x/(x^2+y^2) + g'(y) = x/(x^2+y^2) - x^2/y^2$, which means that $g'(y) = 0$, so we obtain that $g(y) = 0$ is acceptable. Therefore the solution is defined implicitly as $x^2/y + \arctan(y/x) = c$.

28. The equation can be made *exact* by choosing an appropriate integrating factor. We can check that $(M_y - N_x)/N = (2 - 1)/x = 1/x$ depends only on x, so $\mu(x) = e^{\int (1/x)dx} = e^{\ln x} = x$ is an integrating factor. After multiplication, the equation becomes $(2yx + 3x^2)dx + x^2 dy = 0$. This equation is exact now, because $M_y = 2x = N_x$. Integrating M with respect to x gives that $\psi(x,y) = yx^2 + x^3 + g(y)$, then $\psi_y = x^2 + g'(y) = x^2$, which means that $g'(y) = 0$, so we obtain that $g(y) = 0$ is acceptable. Therefore the solution is defined implicitly as $x^3 + x^2 y = c$.

29. The equation is *homogeneous*. (See Section 2.2, Problem 30) We can see that

$$y' = \frac{x+y}{x-y} = \frac{1+(y/x)}{1-(y/x)}.$$

We substitute $u = y/x$, which means also that $y = ux$ and then $y' = u'x + u =$

$(1+u)/(1-u)$, which implies that

$$u'x = \frac{1+u}{1-u} - u = \frac{1+u^2}{1-u},$$

a separable equation. Separating the variables yields

$$\frac{1-u}{1+u^2}du = \frac{dx}{x},$$

and then integration gives $\arctan u - \ln(1+u^2)/2 = \ln|x| + c$. Substituting $u = y/x$ back into this expression and using that

$$-\ln(1+(y/x)^2)/2 - \ln|x| = -\ln(|x|\sqrt{1+(y/x)^2}) = -\ln(\sqrt{x^2+y^2})$$

we obtain that the solution is $\arctan(y/x) - \ln(\sqrt{x^2+y^2}) = c$.

30. The equation is *homogeneous*. (See Section 2.2, Problem 30) Algebraic manipulations show that it can be written in the form

$$y' = \frac{3y^2 + 2xy}{2xy + x^2} = \frac{3(y/x)^2 + 2(y/x)}{2(y/x) + 1}.$$

Substituting $u = y/x$ gives that $y = ux$ and then

$$y' = u'x + u = \frac{3u^2 + 2u}{2u + 1},$$

which implies that

$$u'x = \frac{3u^2 + 2u}{2u + 1} - u = \frac{u^2 + u}{2u + 1},$$

a separable equation. We obtain that $(2u+1)du/(u^2+u) = dx/x$, which in turn means that $\ln(u^2+u) = \ln|x| + \tilde{c}$. Therefore, $u^2 + u = cx$ and then substituting $u = y/x$ gives us the solution $(y^2/x^3) + (y/x^2) = c$.

31. The equation can be made *exact* by choosing an appropriate integrating factor. We can check that $(M_y - N_x)/M = -(3x^2 + y)/(y(3x^2 + y)) = -1/y$ depends only on y, so $\mu(y) = e^{\int (1/y)dy} = e^{\ln y} = y$ is an integrating factor. After multiplication, the equation becomes $(3x^2y^2 + y^3)dx + (2x^3y + 3xy^2)dy = 0$. This equation is exact now, because $M_y = 6x^2y + 3y^2 = N_x$. Integrating M with respect to x gives that $\psi(x,y) = x^3y^2 + y^3x + g(y)$, then $\psi_y = 2x^3y + 3y^2x + g'(y) = 2x^3y + 3xy^2$, which means that $g'(y) = 0$, so we obtain that $g(y) = 0$ is acceptable. Therefore the general solution is defined implicitly as $x^3y^2 + xy^3 = c$. The initial condition gives us $4 - 8 = c = -4$, and the solution is $x^3y^2 + xy^3 = -4$.

33. Let y_1 be a solution, i.e. $y_1' = q_1 + q_2y_1 + q_3y_1^2$. Now let $y = y_1 + (1/v)$ also be a solution. Differentiating this expression with respect to t and using that y is also a solution we obtain $y' = y_1' - (1/v^2)v' = q_1 + q_2y + q_3y^2 = q_1 + q_2(y_1 + (1/v)) + q_3(y_1 + (1/v))^2$. Now using that y_1 was also a solution we get that $-(1/v^2)v' = q_2(1/v) + 2q_3(y_1/v) + q_3(1/v^2)$, which, after some simple algebraic manipulations turns into $v' = -(q_2 + 2q_3y_1)v - q_3$.

35.(a) The equation is $y' = (1-y)(x+by) = x + (b-x)y - by^2$. We set $y = 1 + (1/v)$ and differentiate: $y' = -v^{-2}v' = x + (b-x)(1+(1/v)) - b(1+(1/v))^2$, which, after simplification, turns into $v' = (b+x)v + b$.

(b) When $x = at$, the equation is $v' - (b+at)v = b$, so the integrating factor is $\mu(t) = e^{-bt - at^2/2}$. This turns the equation into $(v\mu(t))' = b\mu(t)$, so $v\mu(t) = \int b\mu(t)dt$, and then $v = (b\int \mu(t)dt)/\mu(t)$.

36. Substitute $v = y'$, then $v' = y''$. The equation turns into $t^2v' + 2tv = (t^2v)' = 1$, which yields $t^2v = t + c_1$, so $y' = v = (1/t) + (c_1/t^2)$. Integrating this expression gives us the solution $y = \ln t - (c_1/t) + c_2$.

37. Set $v = y'$, then $v' = y''$. The equation with this substitution is $tv' + v = (tv)' = 1$, which gives $tv = t + c_1$, so $y' = v = 1 + (c_1/t)$. Integrating this expression yields the solution $y = t + c_1 \ln t + c_2$.

38. Set $v = y'$, so $v' = y''$. The equation is $v' + tv^2 = 0$, which is a separable equation. Separating the variables we obtain $dv/v^2 = -tdt$, so $-1/v = -t^2/2 + c$, and then $y' = v = 2/(t^2 + c_1)$. Now depending on the value of c_1, we have the following possibilities: when $c_1 = 0$, then $y = -2/t + c_2$, when $0 < c_1 = k^2$, then $y = (2/k)\arctan(t/k) + c_2$, and when $0 > c_1 = -k^2$ then

$$y = (1/k)\ln|(t-k)/(t+k)| + c_2.$$

We also divided by $v = y'$ when we separated the variables, and $v = 0$ (which is $y = c$) is also a solution.

39. Substitute $v = y'$ and $v' = y''$. The equation is $2t^2v' + v^3 = 2tv$. This is a Bernoulli equation (See Section 2.4, Problem 27), so the substitution $z = v^{-2}$ yields $z' = -2v^{-3}v'$, and the equation turns into $2t^2v'v^3 + 1 = 2t/v^2$, i.e. into $-2t^2z'/2 + 1 = 2tz$, which in turn simplifies to $t^2z' + 2tz = (t^2z)' = 1$. Integration yields $t^2z = t + c$, which means that $z = (1/t) + (c/t^2)$. Now $y' = v = \pm\sqrt{1/z} = \pm t/\sqrt{t+c_1}$ and another integration gives

$$y = \pm\frac{2}{3}(t - 2c_1)\sqrt{t+c_1} + c_2.$$

The substitution also loses the solution $v = 0$, i.e. $y = c$.

40. Set $v = y'$, then $v' = y''$. The equation reads $v' + v = e^{-t}$, which is a linear equation with integrating factor $\mu(t) = e^t$. This turns the equation into $e^tv' + e^tv = (e^tv)' = 1$, which means that $e^tv = t + c$ and then $y' = v = te^{-t} + ce^{-t}$. Another integration yields the solution $y = -te^{-t} + c_1e^{-t} + c_2$.

41. Let $v = y'$ and $v' = y''$. The equation is $t^2v' = v^2$, which is a separable equation. Separating the variables we obtain $dv/v^2 = dt/t^2$, which gives us $-1/v = -(1/t) + c_1$, and then $y' = v = t/(1+c_1t)$. Now when $c_1 = 0$, then $y = t^2/2 + c_2$, and when $c_1 \neq 0$, then $y = t/c_1 - (\ln|1+c_1t|)/c_1^2 + c_2$. Also, at the separation we divided by $v = 0$, which also gives us the solution $y = c$.

43. Set $y' = v(y)$. Then $y'' = v'(y)(dy/dt) = v'(y)v(y)$. We obtain the equation $v'v + y = 0$, where the differentiation is with respect to y. This is a separable equation which simplifies to $vdv = -ydy$. We obtain that $v^2/2 = -y^2/2 + c$, so $y' = v(y) = \pm\sqrt{c - y^2}$. We separate the variables again to get $dy/\sqrt{c - y^2} = \pm dt$, so $\arcsin(y/\sqrt{c}) = t + d$, which means that $y = \sqrt{c}\sin(\pm t + d) = c_1\sin(t + c_2)$.

44. Set $y' = v(y)$. Then $y'' = v'(y)(dy/dt) = v'(y)v(y)$. We obtain the equation $v'v + yv^3 = 0$, where the differentiation is with respect to y. Separation of variables turns this into $dv/v^2 = -ydy$, which gives us $y' = v = 2/(c_1 + y^2)$. This implies that $(c_1 + y^2)dy = 2dt$ and then the solution is defined implicitly as $c_1 y + y^3/3 = 2t + c_2$. Also, $y = c$ is a solution which we lost when divided by $y' = v = 0$.

46. Set $y' = v(y)$. Then $y'' = v'(y)(dy/dt) = v'(y)v(y)$. We obtain the equation $yv'v - v^3 = 0$, where the differentiation is with respect to y. This separable equation gives us $dv/v^2 = dy/y$, which means that $-1/v = \ln|y| + c$, and then $y' = v = 1/(c - \ln|y|)$. We separate variables again to obtain $(c - \ln|y|)dy = dt$, and then integration yields the implicitly defined solution $cy - (y\ln|y| - y) = t + d$. Also, $y = c$ is a solution which we lost when we divided by $v = 0$.

49. Set $y' = v(y)$. Then $y'' = v'(y)(dy/dt) = v'(y)v(y)$. We obtain the equation $v'v - 3y^2 = 0$, where the differentiation is with respect to y. Separation of variables gives $vdv = 3y^2 dy$, and after integration this turns into $v^2/2 = y^3 + c$. The initial conditions imply that $c = 0$ here, so $(y')^2 = v^2 = 2y^3$. This implies that $y' = \sqrt{2}y^{3/2}$ (the sign is determined by the initial conditions again), and this separable equation now turns into $y^{-3/2}dy = \sqrt{2}dt$. Integration yields $-2y^{-1/2} = \sqrt{2}t + d$, and the initial conditions at this point give that $d = -\sqrt{2}$. Algebraic manipulations find that $y = 2(1 - t)^{-2}$.

50. Set $v = y'$, then $v' = y''$. The equation with this substitution turns into the equation $(1 + t^2)v' + 2tv = ((1 + t^2)v)' = -3t^{-2}$. Integrating this we get that $(1 + t^2)v = 3t^{-1} + c$, and $c = -5$ from the initial conditions. This means that $y' = v = 3/(t(1 + t^2)) - 5/(1 + t^2)$. The partial fraction decomposition of the first expression shows that $y' = 3/t - 3t/(1 + t^2) - 5/(1 + t^2)$ and then another integration here gives us that $y = 3\ln t - (3/2)\ln(1 + t^2) - 5\arctan t + d$. The initial conditions identify $d = 2 + (3/2)\ln 2 + 5\pi/4$, and we obtained the solution.

CHAPTER 3

Second Order Linear Equations

3.1

1. Let $y = e^{rt}$, so that $y' = r\,e^{rt}$ and $y'' = r^2\,e^{rt}$. Direct substitution into the differential equation yields $(r^2 + 2r - 3)e^{rt} = 0$. Canceling the exponential, the characteristic equation is $r^2 + 2r - 3 = 0$. The roots of the equation are $r = -3, 1$. Hence the general solution is $y = c_1 e^t + c_2 e^{-3t}$.

2. Let $y = e^{rt}$. Substitution of the assumed solution results in the characteristic equation $r^2 + 3r + 2 = 0$. The roots of the equation are $r = -2, -1$. Hence the general solution is $y = c_1 e^{-t} + c_2 e^{-2t}$.

4. Substitution of the assumed solution $y = e^{rt}$ results in the characteristic equation $2r^2 - 3r + 1 = 0$. The roots of the equation are $r = 1/2, 1$. Hence the general solution is $y = c_1 e^{t/2} + c_2 e^t$.

6. The characteristic equation is $4r^2 - 9 = 0$, with roots $r = \pm 3/2$. Therefore the general solution is $y = c_1 e^{-3t/2} + c_2 e^{3t/2}$.

8. The characteristic equation is $r^2 - 2r - 2 = 0$, with roots $r = 1 \pm \sqrt{3}$. Hence the general solution is $y = c_1 e^{(1-\sqrt{3})t} + c_2 e^{(1+\sqrt{3})t}$.

9. Substitution of the assumed solution $y = e^{rt}$ results in the characteristic equation $r^2 + r - 2 = 0$. The roots of the equation are $r = -2, 1$. Hence the general solution is $y = c_1 e^{-2t} + c_2 e^t$. Its derivative is $y' = -2c_1 e^{-2t} + c_2 e^t$. Based on the

65

first condition, $y(0) = 1$, we require that $c_1 + c_2 = 1$. In order to satisfy $y'(0) = 1$, we find that $-2c_1 + c_2 = 1$. Solving for the constants, $c_1 = 0$ and $c_2 = 1$. Hence the specific solution is $y(t) = e^t$. It clearly increases without bound as $t \to \infty$.

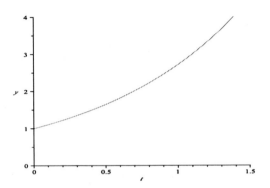

11. Substitution of the assumed solution $y = e^{rt}$ results in the characteristic equation $6r^2 - 5r + 1 = 0$. The roots of the equation are $r = 1/3, 1/2$. Hence the general solution is $y = c_1 e^{t/3} + c_2 e^{t/2}$. Its derivative is $y' = c_1 e^{t/3}/3 + c_2 e^{t/2}/2$. Based on the first condition, $y(0) = 1$, we require that $c_1 + c_2 = 4$. In order to satisfy the condition $y'(0) = 1$, we find that $c_1/3 + c_2/2 = 0$. Solving for the constants, $c_1 = 12$ and $c_2 = -8$. Hence the specific solution is $y(t) = 12\, e^{t/3} - 8\, e^{t/2}$. It clearly decreases without bound as $t \to \infty$.

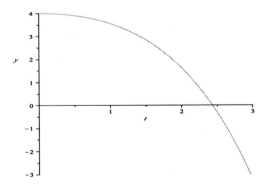

12. The characteristic equation is $r^2 + 3r = 0$, with roots $r = -3,\, 0$. Therefore the general solution is $y = c_1 + c_2 e^{-3t}$, with derivative $y' = -3 c_2 e^{-3t}$. In order to satisfy the initial conditions, we find that $c_1 + c_2 = -2$, and $-3 c_2 = 3$. Hence the specific solution is $y(t) = -1 - e^{-3t}$. This converges to -1 as $t \to \infty$.

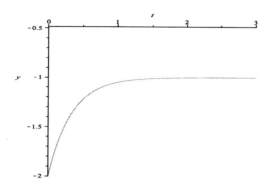

13. The characteristic equation is $r^2 + 5r + 3 = 0$, with roots $r = (-5 \pm \sqrt{13})/2$. The general solution is $y = c_1 e^{(-5-\sqrt{13})t/2} + c_2 e^{(-5+\sqrt{13})t/2}$, with derivative

$$y' = \frac{-5-\sqrt{13}}{2} c_1 e^{(-5-\sqrt{13})t/2} + \frac{-5+\sqrt{13}}{2} c_2 e^{(-5+\sqrt{13})t/2}.$$

In order to satisfy the initial conditions, we require that

$$c_1 + c_2 = 1 \quad \text{and} \quad \frac{-5-\sqrt{13}}{2} c_1 + \frac{-5+\sqrt{13}}{2} c_2 = 0.$$

Solving for the coefficients, $c_1 = (1 - 5/\sqrt{13})/2$ and $c_2 = (1 + 5/\sqrt{13})/2$. The solution clearly converges to 0 as $t \to \infty$.

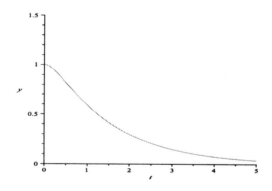

14. The characteristic equation is $2r^2 + r - 4 = 0$, with roots $r = (-1 \pm \sqrt{33})/4$. The general solution is $y = c_1 e^{(-1-\sqrt{33})t/4} + c_2 e^{(-1+\sqrt{33})t/4}$, with derivative

$$y' = \frac{-1-\sqrt{33}}{4} c_1 e^{(-1-\sqrt{33})t/4} + \frac{-1+\sqrt{33}}{4} c_2 e^{(-1+\sqrt{33})t/4}.$$

In order to satisfy the initial conditions, we require that

$$c_1 + c_2 = 0 \quad \text{and} \quad \frac{-1-\sqrt{33}}{4} c_1 + \frac{-1+\sqrt{33}}{4} c_2 = 1.$$

Solving for the coefficients, $c_1 = -2/\sqrt{33}$ and $c_2 = 2/\sqrt{33}$. The specific solution is

$$y(t) = -2 \left[e^{(-1-\sqrt{33})t/4} - e^{(-1+\sqrt{33})t/4} \right] / \sqrt{33}.$$

It clearly increases without bound as $t \to \infty$.

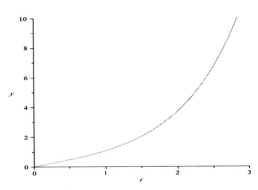

16. The characteristic equation is $4r^2 - 1 = 0$, with roots $r = \pm 1/2$. Therefore the general solution is $y = c_1 e^{-t/2} + c_2 e^{t/2}$. Since the initial conditions are specified at $t = -2$, is more convenient to write $y = d_1 e^{-(t+2)/2} + d_2 e^{(t+2)/2}$. The derivative is given by $y' = - \left[d_1 e^{-(t+2)/2}\right]/2 + \left[d_2 e^{(t+2)/2}\right]/2$. In order to satisfy the initial conditions, we find that $d_1 + d_2 = 1$, and $-d_1/2 + d_2/2 = -1$. Solving for the coefficients, $d_1 = 3/2$, and $d_2 = -1/2$. The specific solution is

$$y(t) = \frac{3}{2} e^{-(t+2)/2} - \frac{1}{2} e^{(t+2)/2} = \frac{3}{2e} e^{-t/2} - \frac{e}{2} e^{t/2}.$$

It clearly decreases without bound as $t \to \infty$.

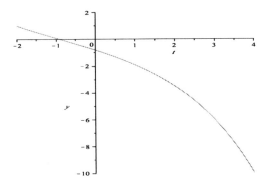

18. An algebraic equation with roots -2 and $-1/2$ is $2r^2 + 5r + 2 = 0$. This is the characteristic equation for the differential equation $2y'' + 5y' + 2y = 0$.

20. The characteristic equation is $2r^2 - 3r + 1 = 0$, with roots $r = 1/2, 1$. Therefore the general solution is $y = c_1 e^{t/2} + c_2 e^t$, with derivative $y' = c_1 e^{t/2}/2 + c_2 e^t$. In order to satisfy the initial conditions, we require $c_1 + c_2 = 2$ and $c_1/2 + c_2 = 1/2$. Solving for the coefficients, $c_1 = 3$, and $c_2 = -1$. The specific solution is $y(t) = 3e^{t/2} - e^t$. To find the stationary point, set $y' = 3e^{t/2}/2 - e^t = 0$. There is a unique solution, with $t_1 = \ln(9/4)$. The maximum value is then $y(t_1) = 9/4$. To find the x-intercept, solve the equation $3e^{t/2} - e^t = 0$. The solution is readily found to be $t_2 = \ln 9 \approx 2.1972$.

22. The characteristic equation is $4r^2 - 1 = 0$, with roots $r = \pm 1/2$. Hence the general solution is $y = c_1 e^{-t/2} + c_2 e^{t/2}$ and $y' = -c_1 e^{-t/2}/2 + c_2 e^{t/2}/2$. Invoking the initial conditions, we require that $c_1 + c_2 = 2$ and $-c_1 + c_2 = 2\beta$. The specific solution is $y(t) = (1 - \beta)e^{-t/2} + (1 + \beta)e^{t/2}$. Based on the form of the solution, it is evident that as $t \to \infty$, $y(t) \to 0$ as long as $\beta = -1$.

23. The characteristic equation is $r^2 - (2\alpha - 1)r + \alpha(\alpha - 1) = 0$. Examining the coefficients, the roots are $r = \alpha$, $\alpha - 1$. Hence the general solution of the differential equation is $y(t) = c_1 e^{\alpha t} + c_2 e^{(\alpha-1)t}$. Assuming $\alpha \in \mathbb{R}$, all solutions will tend to zero as long as $\alpha < 0$. On the other hand, all solutions will become unbounded as long as $\alpha - 1 > 0$, that is, $\alpha > 1$.

26.(a) The characteristic roots are $r = -3, -2$. The solution of the initial value problem is $y(t) = (6 + \beta)e^{-2t} - (4 + \beta)e^{-3t}$.

(b) The maximum point has coordinates $t_0 = \ln[(3(4 + \beta))/(2(6 + \beta))]$, $y_0 = 4(6 + \beta)^3/(27(4 + \beta)^2)$.

(c) $y_0 = 4(6 + \beta)^3/(27(4 + \beta)^2) \geq 4$, as long as $\beta \geq 6 + 6\sqrt{3}$.

(d) $\lim_{\beta \to \infty} t_0 = \ln(3/2)$, $\lim_{\beta \to \infty} y_0 = \infty$.

27.(a) Assuming that y is a constant, the differential equation reduces to $cy = d$. Hence the only equilibrium solution is $y = d/c$.

(b) Setting $y = Y + d/c$, substitution into the differential equation results in the equation $aY'' + bY' + c(Y + d/c) = d$. The equation satisfied by Y is $aY'' + bY' + cY = 0$.

3.2

1.
$$W(e^{2t}, e^{-3t/2}) = \begin{vmatrix} e^{2t} & e^{-3t/2} \\ 2e^{2t} & -\frac{3}{2}e^{-3t/2} \end{vmatrix} = -\frac{7}{2}e^{t/2}.$$

3.
$$W(e^{-2t}, te^{-2t}) = \begin{vmatrix} e^{-2t} & te^{-2t} \\ -2e^{-2t} & (1 - 2t)e^{-2t} \end{vmatrix} = e^{-4t}.$$

5.
$$W(e^t \sin t, e^t \cos t) = \begin{vmatrix} e^t \sin t & e^t \cos t \\ e^t(\sin t + \cos t) & e^t(\cos t - \sin t) \end{vmatrix} = -e^{2t}.$$

6.
$$W(\cos^2 \theta, 1 + \cos 2\theta) = \begin{vmatrix} \cos^2 \theta & 1 + \cos 2\theta \\ -2\sin \theta \cos \theta & -2\sin 2\theta \end{vmatrix} = 0.$$

7. Write the equation as $y'' + (3/t)y' = 1$. $p(t) = 3/t$ is continuous for all $t > 0$. Since $t_0 > 0$, the IVP has a unique solution for all $t > 0$.

9. Write the equation as $y'' + (3/(t-4))y' + (4/t(t-4))y = 2/t(t-4)$. The coefficients are not continuous at $t = 0$ and $t = 4$. Since $t_0 \in (0,4)$, the largest interval is $0 < t < 4$.

10. The coefficient $3\ln|t|$ is discontinuous at $t = 0$. Since $t_0 > 0$, the largest interval of existence is $0 < t < \infty$.

11. Write the equation as $y'' + (x/(x-3))y' + (\ln|x|/(x-3))y = 0$. The coefficients are discontinuous at $x = 0$ and $x = 3$. Since $x_0 \in (0,3)$, the largest interval is $0 < x < 3$.

13. $y_1'' = 2$. We see that $t^2(2) - 2(t^2) = 0$. $y_2'' = 2t^{-3}$, with $t^2(y_2'') - 2(y_2) = 0$. Let $y_3 = c_1 t^2 + c_2 t^{-1}$, then $y_3'' = 2c_1 + 2c_2 t^{-3}$. It is evident that y_3 is also a solution.

16. No. Substituting $y = \sin(t^2)$ into the differential equation,
$$-4t^2 \sin(t^2) + 2\cos(t^2) + 2t\cos(t^2)p(t) + \sin(t^2)q(t) = 0.$$
At $t = 0$, this equation becomes $2 = 0$ (if we suppose that $p(t)$ and $q(t)$ are continuous), which is impossible.

17. $W(e^{2t}, g(t)) = e^{2t} g'(t) - 2e^{2t} g(t) = 3e^{4t}$. Dividing both sides by e^{2t}, we find that g must satisfy the ODE $g' - 2g = 3e^{2t}$. Hence $g(t) = 3t\,e^{2t} + c\,e^{2t}$.

19. $W(f,g) = fg' - f'g$. Also, $W(u,v) = W(2f - g, f + 2g)$. Upon evaluation, $W(u,v) = 5fg' - 5f'g = 5W(f,g)$.

20. $W(f,g) = fg' - f'g = t\cos t - \sin t$, and $W(u,v) = -4fg' + 4f'g$. Hence $W(u,v) = -4t\cos t + 4\sin t$.

21. We compute
$$W(a_1 y_1 + a_2 y_2, b_1 y_1 + b_2 y_2) = \begin{vmatrix} a_1 y_1 + a_2 y_2 & b_1 y_1 + b_2 y_2 \\ a_1 y_1' + a_2 y_2' & b_1 y_1' + b_2 y_2' \end{vmatrix} =$$
$$= (a_1 y_1 + a_2 y_2)(b_1 y_1' + b_2 y_2') - (b_1 y_1 + b_2 y_2)(a_1 y_1' + a_2 y_2') =$$
$$= a_1 b_2 (y_1 y_2' - y_1' y_2) - a_2 b_1 (y_1 y_2' - y_1' y_2) = (a_1 b_2 - a_2 b_1) W(y_1, y_2).$$
This now readily shows that y_3 and y_4 form a fundamental set of solutions if and only if $a_1 b_2 - a_2 b_1 \neq 0$.

23. The general solution is $y = c_1 e^{-3t} + c_2 e^{-t}$. $W(e^{-3t}, e^{-t}) = 2e^{-4t}$, and hence the exponentials form a fundamental set of solutions. On the other hand, the fundamental solutions must also satisfy the conditions $y_1(1) = 1$, $y_1'(1) = 0$; $y_2(1) = 0$, $y_2'(1) = 1$. For y_1, the initial conditions require $c_1 + c_2 = e$, $-3c_1 - c_2 = 0$. The

coefficients are $c_1 = -e^3/2$, $c_2 = 3e/2$. For the solution y_2, the initial conditions require $c_1 + c_2 = 0$, $-3c_1 - c_2 = e$. The coefficients are $c_1 = -e^3/2$, $c_2 = e/2$. Hence the fundamental solutions are

$$y_1 = -\frac{1}{2}e^{-3(t-1)} + \frac{3}{2}e^{-(t-1)} \quad \text{and} \quad y_2 = -\frac{1}{2}e^{-3(t-1)} + \frac{1}{2}e^{-(t-1)}.$$

24. Yes. $y_1'' = -4\cos 2t$; $y_2'' = -4\sin 2t$. $W(\cos 2t, \sin 2t) = 2$.

25. Clearly, $y_1 = e^t$ is a solution. $y_2' = (1+t)e^t$, $y_2'' = (2+t)e^t$. Substitution into the ODE results in $(2+t)e^t - 2(1+t)e^t + te^t = 0$. Furthermore, $W(e^t, te^t) = e^{2t}$. Hence the solutions form a fundamental set of solutions.

27. Clearly, $y_1 = x$ is a solution. $y_2' = \cos x$, $y_2'' = -\sin x$. Substitution into the ODE results in $(1 - x\cot x)(-\sin x) - x(\cos x) + \sin x = 0$. We can compute that $W(y_1, y_2) = x\cos x - \sin x$, which is nonzero for $0 < x < \pi$. Hence $\{x, \sin x\}$ is a fundamental set of solutions.

30. Writing the equation in standard form, we find that $P(t) = \sin t / \cos t$. Hence the Wronskian is $W(t) = ce^{-\int (\sin t/\cos t)\,dt} = ce^{\ln|\cos t|} = c\cos t$, in which c is some constant.

31. After writing the equation in standard form, we have $P(x) = 1/x$. The Wronskian is $W(x) = ce^{-\int (1/x)\,dx} = ce^{-\ln|x|} = c/x$, in which c is some constant.

32. Writing the equation in standard form, we find that $P(x) = -2x/(1-x^2)$. The Wronskian is $W(x) = ce^{-\int -2x/(1-x^2)\,dx} = ce^{-\ln|1-x^2|} = c/(1-x^2)$, in which c is some constant.

33. Rewrite the equation as $p(t)y'' + p'(t)y' + q(t)y = 0$. After writing the equation in standard form, we have $P(t) = p'(t)/p(t)$. Hence the Wronskian is

$$W(t) = ce^{-\int p'(t)/p(t)\,dt} = ce^{-\ln p(t)} = c/p(t).$$

35. The Wronskian associated with the solutions of the differential equation is given by $W(t) = ce^{-\int -2/t^2\,dt} = ce^{-2/t}$. Since $W(2) = 3$, it follows that for the hypothesized set of solutions, $c = 3e$. Hence $W(4) = 3\sqrt{e}$.

36. For the given differential equation, the Wronskian satisfies the first order differential equation $W' + p(t)W = 0$. Given that W is constant, it is necessary that $p(t) \equiv 0$.

37. Direct calculation shows that $W(fg, fh) = (fg)(fh)' - (fg)'(fh) = (fg)(f'h + fh') - (f'g + fg')(fh) = f^2 W(g, h)$.

39. Since y_1 and y_2 are solutions, they are differentiable. The hypothesis can thus be restated as $y_1'(t_0) = y_2'(t_0) = 0$ at some point t_0 in the interval of definition.

This implies that $W(y_1, y_2)(t_0) = 0$. But $W(y_1, y_2)(t_0) = ce^{-\int p(t)dt}$, which cannot be equal to zero, unless $c = 0$. Hence $W(y_1, y_2) \equiv 0$, which is ruled out for a fundamental set of solutions.

42. $P = 1$, $Q = x$, $R = 1$. We have $P'' - Q' + R = 0$. The equation is exact. Note that $(y')' + (xy)' = 0$. Hence $y' + xy = c_1$. This equation is linear, with integrating factor $\mu = e^{x^2/2}$. Therefore the general solution is

$$y(x) = c_1 e^{-x^2/2} \int_{x_0}^{x} e^{u^2/2} du + c_2 e^{-x^2/2}.$$

43. $P = 1$, $Q = 3x^2$, $R = x$. Note that $P'' - Q' + R = -5x$, and therefore the differential equation is not exact.

45. $P = x^2$, $Q = x$, $R = -1$. We have $P'' - Q' + R = 0$. The equation is exact. Write the equation as $(x^2 y')' - (xy)' = 0$. After integration, we conclude that $x^2 y' - xy = c$. Divide both sides of the differential equation by x^2. The resulting equation is linear, with integrating factor $\mu = 1/x$. Hence $(y/x)' = cx^{-3}$. The solution is $y(t) = c_1 x^{-1} + c_2 x$.

47. $P = x^2$, $Q = x$, $R = x^2 - \nu^2$. Hence the coefficients are $2P' - Q = 3x$ and $P'' - Q' + R = x^2 + 1 - \nu^2$. The adjoint of the original differential equation is given by $x^2 \mu'' + 3x \mu' + (x^2 + 1 - \nu^2)\mu = 0$.

49. $P = 1$, $Q = 0$, $R = -x$. Hence the coefficients are given by $2P' - Q = 0$ and $P'' - Q' + R = -x$. Therefore the adjoint of the original equation is $\mu'' - x\mu = 0$.

3.3

2. $e^{2-3i} = e^2 e^{-3i} = e^2(\cos 3 - i \sin 3)$.

3. $e^{i\pi} = \cos \pi + i \sin \pi = -1$.

4. $e^{2-(\pi/2)i} = e^2(\cos(\pi/2) - i \sin(\pi/2)) = -e^2 i$.

6. $\pi^{-1+2i} = e^{(-1+2i)\ln \pi} = e^{-\ln \pi} e^{2 \ln \pi \, i} = (\cos(2 \ln \pi) + i \sin(2 \ln \pi))/\pi$.

8. The characteristic equation is $r^2 - 2r + 6 = 0$, with roots $r = 1 \pm i\sqrt{5}$. Hence the general solution is $y = c_1 e^t \cos \sqrt{5} t + c_2 e^t \sin \sqrt{5} t$.

9. The characteristic equation is $r^2 + 2r - 8 = 0$, with roots $r = -4, 2$. The roots are real and different, hence the general solution is $y = c_1 e^{-4t} + c_2 e^{2t}$.

10. The characteristic equation is $r^2 + 2r + 2 = 0$, with roots $r = -1 \pm i$. Hence the general solution is $y = c_1 e^{-t} \cos t + c_2 e^{-t} \sin t$.

12. The characteristic equation is $4r^2 + 9 = 0$, with roots $r = \pm(3/2)i$. Hence the general solution is $y = c_1 \cos(3t/2) + c_2 \sin(3t/2)$.

13. The characteristic equation is $r^2 + 2r + 1.25 = 0$, with roots $r = -1 \pm i/2$. Hence the general solution is $y = c_1 e^{-t} \cos(t/2) + c_2 e^{-t} \sin(t/2)$.

15. The characteristic equation is $r^2 + r + 1.25 = 0$, with roots $r = -(1/2) \pm i$. Hence the general solution is $y = c_1 e^{-t/2} \cos t + c_2 e^{-t/2} \sin t$.

16. The characteristic equation is $r^2 + 4r + 6.25 = 0$, with roots $r = -2 \pm (3/2)i$. Hence the general solution is $y = c_1 e^{-2t} \cos(3t/2) + c_2 e^{-2t} \sin(3t/2)$.

17. The characteristic equation is $r^2 + 4 = 0$, with roots $r = \pm 2i$. Hence the general solution is $y = c_1 \cos 2t + c_2 \sin 2t$. Now $y' = -2c_1 \sin 2t + 2c_2 \cos 2t$. Based on the first condition, $y(0) = 0$, we require that $c_1 = 0$. In order to satisfy the condition $y'(0) = 1$, we find that $2c_2 = 1$. The constants are $c_1 = 0$ and $c_2 = 1/2$. Hence the specific solution is $y(t) = \sin 2t /2$. The solution is periodic.

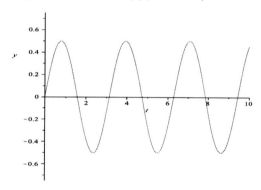

19. The characteristic equation is $r^2 - 2r + 5 = 0$, with roots $r = 1 \pm 2i$. Hence the general solution is $y = c_1 e^t \cos 2t + c_2 e^t \sin 2t$. Based on the initial condition $y(\pi/2) = 0$, we require that $c_1 = 0$. It follows that $y = c_2 e^t \sin 2t$, and so the first derivative is $y' = c_2 e^t \sin 2t + 2c_2 e^t \cos 2t$. In order to satisfy the condition $y'(\pi/2) = 2$, we find that $-2e^{\pi/2} c_2 = 2$. Hence we have $c_2 = -e^{-\pi/2}$. Therefore the specific solution is $y(t) = -e^{t-\pi/2} \sin 2t$. The solution oscillates with an exponentially growing amplitude.

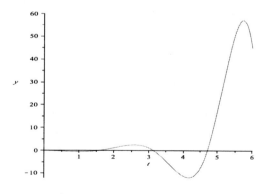

20. The characteristic equation is $r^2 + 1 = 0$, with roots $r = \pm i$. Hence the general solution is $y = c_1 \cos t + c_2 \sin t$. Its derivative is $y' = -c_1 \sin t + c_2 \cos t$. Based on the first condition, $y(\pi/3) = 2$, we require that $c_1 + \sqrt{3} c_2 = 4$. In order to satisfy the condition $y'(\pi/3) = -4$, we find that $-\sqrt{3} c_1 + c_2 = -8$. Solving these for the constants, $c_1 = 1 + 2\sqrt{3}$ and $c_2 = \sqrt{3} - 2$. Hence the specific solution is a steady oscillation, given by $y(t) = (1 + 2\sqrt{3}) \cos t + (\sqrt{3} - 2) \sin t$.

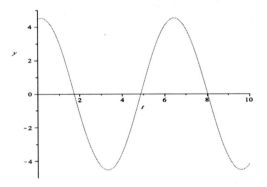

21. From Problem 15, the general solution is $y = c_1 e^{-t/2} \cos t + c_2 e^{-t/2} \sin t$. Invoking the first initial condition, $y(0) = 3$, which implies that $c_1 = 3$. Substituting, it follows that $y = 3e^{-t/2} \cos t + c_2 e^{-t/2} \sin t$, and so the first derivative is

$$y' = -\frac{3}{2} e^{-t/2} \cos t - 3 e^{-t/2} \sin t + c_2 e^{-t/2} \cos t - \frac{c_2}{2} e^{-t/2} \sin t.$$

Invoking the initial condition, $y'(0) = 1$, we find that $-3/2 + c_2 = 1$, and so $c_2 = 5/2$. Hence the specific solution is $y(t) = 3 e^{-t/2} \cos t + (5/2) e^{-t/2} \sin t$. It oscillates with an exponentially decreasing amplitude.

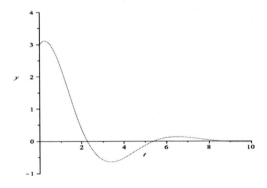

24.(a) The characteristic equation is $5r^2 + 2r + 7 = 0$, with roots $r = -(1 \pm i\sqrt{34})/5$. The solution is $u = c_1 e^{-t/5} \cos \sqrt{34} t/5 + c_2 e^{-t/5} \sin \sqrt{34} t/5$. Invoking the given initial conditions, we obtain the equations for the coefficients: $c_1 = 2$, $-2 + \sqrt{34} c_2 = 5$. That is, $c_1 = 2$, $c_2 = 7/\sqrt{34}$. Hence the specific solution is

$$u(t) = 2e^{-t/5} \cos \frac{\sqrt{34}}{5} t + \frac{7}{\sqrt{34}} e^{-t/5} \sin \frac{\sqrt{34}}{5} t.$$

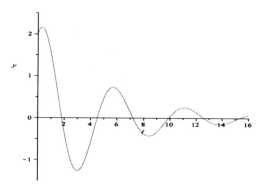

(b) Based on the graph of $u(t)$, T is in the interval $14 < t < 16$. A numerical solution on that interval yields $T \approx 14.5115$.

26.(a) The characteristic equation is $r^2 + 2ar + (a^2 + 1) = 0$, with roots $r = -a \pm i$. Hence the general solution is $y(t) = c_1 e^{-at} \cos t + c_2 e^{-at} \sin t$. Based on the initial conditions, we find that $c_1 = 1$ and $c_2 = a$. Therefore the specific solution is given by $y(t) = e^{-at} \cos t + a e^{-at} \sin t = \sqrt{1 + a^2}\, e^{-at} \cos(t - \phi)$, in which $\phi = \arctan(a)$.

(b) For estimation, note that $|y(t)| \leq \sqrt{1 + a^2}\, e^{-at}$. Now consider the inequality $\sqrt{1 + a^2}\, e^{-at} \leq 1/10$. The inequality holds for $t \geq (1/a) \ln(10\sqrt{1 + a^2})$. Therefore $T \leq (1/a) \ln(10\sqrt{1 + a^2})$. Setting $a = 1$, the numerical value is $T \approx 1.8763$.

(c) Similarly, $T_{1/4} \approx 7.4284$, $T_{1/2} \approx 4.3003$, $T_2 \approx 1.5116$.

(d)

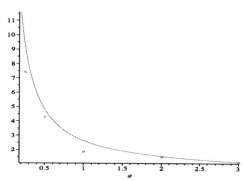

Note that the estimates T_a approach the graph of $(1/a) \ln(10\sqrt{1 + a^2})$ as a gets large.

27. Direct calculation gives the result. On the other hand, it was shown in Problem 3.2.37 that $W(fg, fh) = f^2 W(g, h)$. Hence $W(e^{\lambda t} \cos \mu t, e^{\lambda t} \sin \mu t) = e^{2\lambda t} W(\cos \mu t, \sin \mu t) = e^{2\lambda t} [\cos \mu t (\sin \mu t)' - (\cos \mu t)' \sin \mu t] = \mu e^{2\lambda t}$.

28.(a) Clearly, y_1 and y_2 are solutions. Also, $W(\cos t, \sin t) = \cos^2 t + \sin^2 t = 1$.

(b) $y' = ie^{it}$, $y'' = i^2 e^{it} = -e^{it}$. Evidently, y is a solution and so $y = c_1 y_1 + c_2 y_2$.

(c) Setting $t = 0$, $1 = c_1 \cos 0 + c_2 \sin 0$, and $c_1 = 1$.

(d) Differentiating, $ie^{it} = c_2 \cos t$. Setting $t = 0$, $i = c_2 \cos 0$ and hence $c_2 = i$. Therefore $e^{it} = \cos t + i \sin t$.

29. Euler's formula is $e^{it} = \cos t + i \sin t$. It follows that $e^{-it} = \cos t - i \sin t$. Adding these equation, $e^{it} + e^{-it} = 2 \cos t$. Subtracting the two equations results in $e^{it} - e^{-it} = 2i \sin t$.

30. Let $r_1 = \lambda_1 + i\mu_1$, and $r_2 = \lambda_2 + i\mu_2$. Then
$$e^{(r_1+r_2)t} = e^{(\lambda_1+\lambda_2)t+i(\mu_1+\mu_2)t} = e^{(\lambda_1+\lambda_2)t}[\cos(\mu_1+\mu_2)t + i\sin(\mu_1+\mu_2)t] =$$
$$= e^{(\lambda_1+\lambda_2)t}[(\cos \mu_1 t + i \sin \mu_1 t)(\cos \mu_2 t + i \sin \mu_2 t)] =$$
$$= e^{\lambda_1 t}(\cos \mu_1 t + i \sin \mu_1 t) \cdot e^{\lambda_2 t}(\cos \mu_1 t + i \sin \mu_1 t) = e^{r_1 t} e^{r_2 t}.$$

Hence $e^{(r_1+r_2)t} = e^{r_1 t} e^{r_2 t}$.

32. Clearly, $u' = \lambda e^{\lambda t} \cos \mu t - \mu e^{\lambda t} \sin \mu t = e^{\lambda t}(\lambda \cos \mu t - \mu \sin \mu t)$ and then $u'' = \lambda e^{\lambda t}(\lambda \cos \mu t - \mu \sin \mu t) + e^{\lambda t}(-\lambda \mu \sin \mu t - \mu^2 \cos \mu t)$. Plugging these into the differential equation, dividing by $e^{\lambda t} \neq 0$ and arranging the sine and cosine terms we obtain that the identity to prove is
$$(a(\lambda^2 - \mu^2) + b\lambda + c) \cos \mu t + (-2\lambda \mu a - b\mu) \sin \mu t = 0.$$

We know that $\lambda \pm i\mu$ solves the characteristic equation $ar^2 + br + c = 0$, so $a(\lambda - i\mu)^2 + b(\lambda - i\mu) + c = a(\lambda^2 - \mu^2) + b\lambda + c + i(-2\lambda\mu a - \mu b) = 0$. If this complex number is zero, then both the real and imaginary parts of it are zero, but those are the coefficients of $\cos \mu t$ and $\sin \mu t$ in the above identity, which proves that $au'' + bu' + cu = 0$. The solution for v is analogous.

35. The equation transforms into $y'' + y = 0$. The characteristic roots are $r = \pm i$. The solution is $y = c_1 \cos(x) + c_2 \sin(x) = c_1 \cos(\ln t) + c_2 \sin(\ln t)$.

37. The equation transforms into $y'' + 2y' + 1.25y = 0$. The characteristic roots are $r = -1 \pm i/2$. The solution is
$$y = c_1 e^{-x} \cos(x/2) + c_2 e^{-x} \sin(x/2) = c_1 \frac{\cos(\frac{1}{2} \ln t)}{t} + c_2 \frac{\sin(\frac{1}{2} \ln t)}{t}.$$

38. The equation transforms into $y'' - 5y' - 6y = 0$. The characteristic roots are $r = -1, 6$. The solution is $y = c_1 e^{-x} + c_2 e^{6x} = c_1 e^{-\ln t} + c_2 e^{6 \ln t} = c_1/t + c_2 t^6$.

39. The equation transforms into $y'' - 5y' + 6y = 0$. The characteristic roots are $r = 2, 3$. The solution is $y = c_1 e^{2x} + c_2 e^{3x} = c_1 e^{2 \ln t} + c_2 e^{3 \ln t} = c_1 t^2 + c_2 t^3$.

41. The equation transforms into $y'' + 2y' - 3y = 0$. The characteristic roots are $r = 1, -3$. The solution is $y = c_1 e^x + c_2 e^{-3x} = c_1 e^{\ln t} + c_2 e^{-3 \ln t} = c_1 t + c_2/t^3$.

42. The equation transforms into $y'' + 6y' + 10y = 0$. The characteristic roots are $r = -3 \pm i$. The solution is

$$y = c_1 e^{-3x}\cos(x) + c_2 e^{-3x}\sin(x) = c_1 \frac{1}{t^3}\cos(\ln t) + c_2 \frac{1}{t^3}\sin(\ln t).$$

43.(a) By the chain rule, $y'(x) = (dy/dx)x'$. In general, $dz/dt = (dz/dx)(dx/dt)$. Setting $z = (dy/dt)$, we have

$$\frac{d^2y}{dt^2} = \frac{dz}{dx}\frac{dx}{dt} = \frac{d}{dx}\left[\frac{dy}{dx}\frac{dx}{dt}\right]\frac{dx}{dt} = \left[\frac{d^2y}{dx^2}\frac{dx}{dt}\right]\frac{dx}{dt} + \frac{dy}{dx}\frac{d}{dx}\left[\frac{dx}{dt}\right]\frac{dx}{dt}.$$

However,

$$\frac{d}{dx}\left[\frac{dx}{dt}\right]\frac{dx}{dt} = \left[\frac{d^2x}{dt^2}\right]\frac{dt}{dx}\cdot\frac{dx}{dt} = \frac{d^2x}{dt^2}.$$

Hence

$$\frac{d^2y}{dt^2} = \frac{d^2y}{dx^2}\left[\frac{dx}{dt}\right]^2 + \frac{dy}{dx}\frac{d^2x}{dt^2}.$$

(b) Substituting the results in part (a) into the general differential equation, $y'' + p(t)y' + q(t)y = 0$, we find that

$$\frac{d^2y}{dx^2}\left[\frac{dx}{dt}\right]^2 + \frac{dy}{dx}\frac{d^2x}{dt^2} + p(t)\frac{dy}{dx}\frac{dx}{dt} + q(t)y = 0.$$

Collecting the terms,

$$\left[\frac{dx}{dt}\right]^2\frac{d^2y}{dx^2} + \left[\frac{d^2x}{dt^2} + p(t)\frac{dx}{dt}\right]\frac{dy}{dx} + q(t)y = 0.$$

(c) Assuming $(dx/dt)^2 = k\,q(t)$, and $q(t) > 0$, we find that $dx/dt = \sqrt{k\,q(t)}$, which can be integrated. That is, $x = u(t) = \int \sqrt{k\,q(t)}\,dt = \int \sqrt{q(t)}\,dt$, since $k = 1$.

(d) Let $k = 1$. It follows that $d^2x/dt^2 + p(t)dx/dt = du/dt + p(t)u(t) = q'/2\sqrt{q} + p\sqrt{q}$. Hence

$$\left[\frac{d^2x}{dt^2} + p(t)\frac{dx}{dt}\right] \bigg/ \left[\frac{dx}{dt}\right]^2 = \frac{q'(t) + 2p(t)q(t)}{2\,[q(t)]^{3/2}}.$$

As long as $dx/dt \neq 0$, the differential equation can be expressed as

$$\frac{d^2y}{dx^2} + \left[\frac{q'(t) + 2p(t)q(t)}{2\,[q(t)]^{3/2}}\right]\frac{dy}{dx} + y = 0.$$

For the case $q(t) < 0$, write $q(t) = -[-q(t)]$, and set $(dx/dt)^2 = -q(t)$.

45. $p(t) = 3t$ and $q(t) = t^2$. We have $x = \int t\,dt = t^2/2$. Furthermore,

$$\frac{q'(t) + 2p(t)q(t)}{2\,[q(t)]^{3/2}} = \frac{1 + 3t^2}{t^2}.$$

The ratio is not constant, and therefore the equation cannot be transformed.

46. $p(t) = t - 1/t$ and $q(t) = t^2$. We have $x = \int t\, dt = t^2/2$. Furthermore,
$$\frac{q'(t) + 2p(t)q(t)}{2\left[q(t)\right]^{3/2}} = 1.$$

The ratio is constant, and therefore the equation can be transformed. From Problem 43, the transformed equation is
$$\frac{d^2 y}{dx^2} + \frac{dy}{dx} + y = 0.$$

Based on the methods in this section, the characteristic equation is $r^2 + r + 1 = 0$, with roots $r = (-1 \pm i\sqrt{3})/2$. The general solution is $y(x) = c_1 e^{-x/2} \cos \sqrt{3}\, x/2 + c_2 e^{-x/2} \sin \sqrt{3}\, x/2$. Since $x = t^2/2$, the solution in the original variable t is
$$y(t) = e^{-t^2/4}\left[c_1 \cos\left(\sqrt{3}\, t^2/4\right) + c_2 \sin\left(\sqrt{3}\, t^2/4\right)\right].$$

3.4

2. The characteristic equation is $9r^2 + 6r + 1 = 0$, with the double root $r = -1/3$. The general solution is $y(t) = c_1 e^{-t/3} + c_2 t\, e^{-t/3}$.

3. The characteristic equation is $4r^2 - 4r - 3 = 0$, with roots $r = -1/2,\, 3/2$. The general solution is $y(t) = c_1 e^{-t/2} + c_2 e^{3t/2}$.

4. The characteristic equation is $4r^2 + 12r + 9 = 0$, with double root $r = -3/2$. The general solution is $y(t) = (c_1 + c_2\, t)e^{-3t/2}$.

6. The characteristic equation is $r^2 - 6r + 9 = 0$, with the double root $r = 3$. The general solution is $y(t) = c_1 e^{3t} + c_2 t\, e^{3t}$.

7. The characteristic equation is $4r^2 + 17r + 4 = 0$, with roots $r = -1/4,\, -4$. The general solution is $y(t) = c_1 e^{-t/4} + c_2 e^{-4t}$.

8. The characteristic equation is $16r^2 + 24r + 9 = 0$, with double root $r = -3/4$. The general solution is $y(t) = c_1 e^{-3t/4} + c_2 t\, e^{-3t/4}$.

10. The characteristic equation is $2r^2 + 2r + 1 = 0$. We obtain the complex roots $r = (-1 \pm i)/2$. The general solution is $y(t) = c_1 e^{-t/2} \cos(t/2) + c_2 e^{-t/2} \sin(t/2)$.

11. The characteristic equation is $9r^2 - 12r + 4 = 0$, with the double root $r = 2/3$. The general solution is $y(t) = c_1 e^{2t/3} + c_2 t\, e^{2t/3}$. Invoking the first initial condition, it follows that $c_1 = 2$. Now $y'(t) = (4/3 + c_2)e^{2t/3} + 2c_2 t\, e^{2t/3}/3$. Invoking the second initial condition, $4/3 + c_2 = -1$, or $c_2 = -7/3$. Hence we obtain the solution $y(t) = 2e^{2t/3} - (7/3)te^{2t/3}$. Since the second term dominates for large t, $y(t) \to -\infty$.

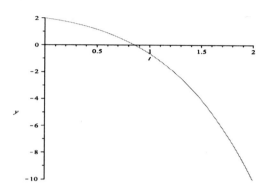

13. The characteristic equation is $9r^2 + 6r + 82 = 0$. We obtain the complex roots $r = -1/3 \pm 3i$. The general solution is $y(t) = c_1 e^{-t/3}\cos 3t + c_2 e^{-t/3}\sin 3t$. Based on the first initial condition, $c_1 = -1$. Invoking the second initial condition, we conclude that $1/3 + 3c_2 = 2$, or $c_2 = 5/9$. Hence $y(t) = -e^{-t/3}\cos 3t + (5/9)e^{-t/3}\sin 3t$. The solution oscillates with an exponentially decreasing amplitude.

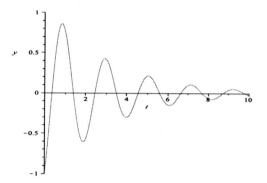

15.(a) The characteristic equation is $4r^2 + 12r + 9 = 0$, with double root $r = -3/2$. The general solution is $y(t) = c_1 e^{-3t/2} + c_2 t\, e^{-3t/2}$. Invoking the first initial condition, it follows that $c_1 = 1$. Now $y'(t) = (-3/2 + c_2)e^{-3t/2} - (3/2)c_2 t\, e^{-3t/2}$. The second initial condition requires that $-3/2 + c_2 = -4$, or $c_2 = -5/2$. Hence the specific solution is $y(t) = e^{-3t/2} - (5/2)t\, e^{-3t/2}$.

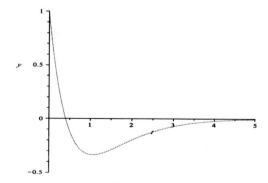

(b) The solution crosses the x-axis at $t = 2/5$.

(c) The solution has a minimum at the point $(16/15, -5e^{-8/5}/3)$.

(d) Given that $y'(0) = b$, we have $-3/2 + c_2 = b$, or $c_2 = b + 3/2$. Hence the solution is $y(t) = e^{-3t/2} + (b+3/2)t\, e^{-3t/2}$. Since the second term dominates, the long-term solution depends on the sign of the coefficient $b + 3/2$. The critical value is $b = -3/2$.

16. The characteristic roots are $r_1 = r_2 = 1/2$. Hence the general solution is given by $y(t) = c_1 e^{t/2} + c_2 t\, e^{t/2}$. Invoking the initial conditions, we require that $c_1 = 2$, and that $1 + c_2 = b$. The specific solution is $y(t) = 2e^{t/2} + (b-1)t\, e^{t/2}$. Since the second term dominates, the long-term solution depends on the sign of the coefficient $b - 1$. The critical value is $b = 1$.

18.(a) The characteristic roots are $r_1 = r_2 = -2/3$. Therefore the general solution is given by $y(t) = c_1 e^{-2t/3} + c_2 t\, e^{-2t/3}$. Invoking the initial conditions, we require that $c_1 = a$, and that $-2a/3 + c_2 = -1$. After solving for the coefficients, the specific solution is $y(t) = ae^{-2t/3} + (2a/3 - 1)t\, e^{-2t/3}$.

(b) Since the second term dominates, the long-term solution depends on the sign of the coefficient $2a/3 - 1$. The critical value is $a = 3/2$.

20.(a) The characteristic equation is $r^2 + 2a\,r + a^2 = (r+a)^2 = 0$.

(b) With $p(t) = 2a$, Abel's Formula becomes $W(y_1, y_2) = c\, e^{-\int 2a\, dt} = c\, e^{-2at}$.

(c) $y_1(t) = e^{-at}$ is a solution. From part (b), with $c = 1$, $e^{-at} y_2'(t) + a\, e^{-at} y_2(t) = e^{-2at}$, which can be written as $(e^{at} y_2(t))' = 1$, resulting in $e^{at} y_2(t) = t$.

22.(a) If the characteristic equation $ar^2 + br + c$ has equal roots r_1, then $ar_1^2 + br_1 + c = a(r - r_1)^2 = 0$. Then clearly $L[e^{rt}] = (ar^2 + br + c)e^{rt} = a(r - r_1)^2 e^{rt}$. This gives immediately that $L[e^{r_1 t}] = 0$.

(b) Differentiating the identity in part (a) with respect to r we get $(2ar + b)e^{rt} + (ar^2 + br + c)te^{rt} = 2a(r - r_1)e^{rt} + a(r - r_1)^2 te^{rt}$. Again, this gives $L[te^{r_1 t}] = 0$.

23. Set $y_2(t) = t^2 v(t)$. Substitution into the differential equation results in
$$t^2(t^2 v'' + 4tv' + 2v) - 4t(t^2 v' + 2tv) + 6t^2 v = 0.$$
After collecting terms, we end up with $t^4 v'' = 0$. Hence $v(t) = c_1 + c_2 t$, and thus $y_2(t) = c_1 t^2 + c_2 t^3$. Setting $c_1 = 0$ and $c_2 = 1$, we obtain $y_2(t) = t^3$.

24. Set $y_2(t) = t\, v(t)$. Substitution into the differential equation results in
$$t^2(tv'' + 2v') + 2t(tv' + v) - 2tv = 0.$$
After collecting terms, we end up with $t^3 v'' + 4t^2 v' = 0$. This equation is linear in the variable $w = v'$. It follows that $v'(t) = c\, t^{-4}$, and $v(t) = c_1 t^{-3} + c_2$. Thus

$y_2(t) = c_1 t^{-2} + c_2 t$. Setting $c_1 = 1$ and $c_2 = 0$, we obtain $y_2(t) = t^{-2}$.

26. Set $y_2(t) = tv(t)$. Substitution into the differential equation results in $v'' - v' = 0$. This equation is linear in the variable $w = v'$. It follows that $v'(t) = c_1 e^t$, and $v(t) = c_1 e^t + c_2$. Thus $y_2(t) = c_1 t e^t + c_2 t$. Setting $c_1 = 1$ and $c_2 = 0$, we obtain $y_2(t) = te^t$.

28. Set $y_2(x) = e^x v(x)$. Substitution into the differential equation results in $v'' + (x-2)/(x-1) v' = 0$. This equation is linear in the variable $w = v'$. An integrating factor is $\mu = e^{\int (x-2)/(x-1)\,dx} = e^x/(x-1)$. Rewrite the equation as $[e^x v'/(x-1)]' = 0$, from which it follows that $v'(x) = c(x-1)e^{-x}$. Hence $v(x) = c_1 x e^{-x} + c_2$ and $y_2(x) = c_1 x + c_2 e^x$. Setting $c_1 = 1$ and $c_2 = 0$, we obtain $y_2(x) = x$.

29. Set $y_2(x) = y_1(x) v(x)$, in which $y_1(x) = x^{1/4} e^{2\sqrt{x}}$. It can be verified that y_1 is a solution of the differential equation, that is, $x^2 y_1'' - (x - 0.1875) y_1 = 0$. Substitution of the given form of y_2 results in the differential equation $2x^{9/4} v'' + (4x^{7/4} + x^{5/4}) v' = 0$. This equation is linear in the variable $w = v'$. An integrating factor is $\mu = e^{\int [2x^{-1/2} + 1/(2x)]\,dx} = \sqrt{x}\, e^{4\sqrt{x}}$. Rewrite the equation as $[\sqrt{x}\, e^{4\sqrt{x}} v']' = 0$, from which it follows that $v'(x) = c e^{-4\sqrt{x}}/\sqrt{x}$. Integrating, $v(x) = c_1 e^{-4\sqrt{x}} + c_2$ and as a result, $y_2(x) = c_1 x^{1/4} e^{-2\sqrt{x}} + c_2 x^{1/4} e^{2\sqrt{x}}$. Setting $c_1 = 1$ and $c_2 = 0$, we obtain $y_2(x) = x^{1/4} e^{-2\sqrt{x}}$.

31. Direct substitution verifies that $y_1(t) = e^{-\delta x^2/2}$ is a solution of the differential equation. Now set $y_2(x) = y_1(x) v(x)$. Substitution of y_2 into the equation results in $v'' - \delta x v' = 0$. This equation is linear in the variable $w = v'$. An integrating factor is $\mu = e^{-\delta x^2/2}$. Rewrite the equation as $[e^{-\delta x^2/2} v']' = 0$, from which it follows that $v'(x) = c_1 e^{\delta x^2/2}$. Integrating, we obtain

$$v(x) = c_1 \int_0^x e^{\delta u^2/2} du + v(0).$$

Hence

$$y_2(x) = c_1 e^{-\delta x^2/2} \int_0^x e^{\delta u^2/2} du + c_2 e^{-\delta x^2/2}.$$

Setting $c_2 = 0$, we obtain a second independent solution.

33. After writing the differential equation in standard form, we have $p(t) = 3/t$. Based on Abel's identity, $W(y_1, y_2) = c_1 e^{-\int 3/t\,dt} = c_1 t^{-3}$. As shown in Problem 32, two solutions of a second order linear equation satisfy $(y_2/y_1)' = W(y_1, y_2)/y_1^2$. In the given problem, $y_1(t) = t^{-1}$. Hence $(t y_2)' = c_1 t^{-1}$. Integrating both sides of the equation, $y_2(t) = c_1 t^{-1} \ln t + c_2 t^{-1}$. Setting $c_1 = 1$ and $c_2 = 0$ we obtain $y_2(t) = t^{-1} \ln t$.

35. After writing the differential equation in standard form, we have $p(x) = -x/(x-1)$. Based on Abel's identity, $W(y_1, y_2) = c e^{\int x/(x-1)\,dx} = c e^x (x-1)$. Two solutions of a second order linear equation satisfy $(y_2/y_1)' = W(y_1, y_2)/y_1^2$. In the given problem, $y_1(x) = e^x$. Hence $(e^{-x} y_2)' = c e^{-x} (x-1)$. Integrating both

sides of the equation, $y_2(x) = c_1 x + c_2 e^x$. Setting $c_1 = 1$ and $c_2 = 0$, we obtain $y_2(x) = x$.

36. Write the differential equation in standard form to find $p(x) = 1/x$. Based on Abel's identity, $W(y_1, y_2) = c e^{-\int 1/x\, dx} = c x^{-1}$. Two solutions of a second order linear differential equation satisfy $(y_2/y_1)' = W(y_1, y_2)/y_1^2$. In the given problem, $y_1(x) = x^{-1/2} \sin x$. Hence

$$\left(\frac{\sqrt{x}}{\sin x} y_2\right)' = c \frac{1}{\sin^2 x}.$$

Integrating both sides of the equation, $y_2(x) = c_1 x^{-1/2} \cos x + c_2 x^{-1/2} \sin x$. Setting $c_1 = 1$ and $c_2 = 0$, we obtain $y_2(x) = x^{-1/2} \cos x$.

38.(a) The characteristic equation is $ar^2 + c = 0$. If $a, c > 0$, then the roots are $r = \pm i \sqrt{c/a}$. The general solution is

$$y(t) = c_1 \cos \sqrt{\frac{c}{a}}\, t + c_2 \sin \sqrt{\frac{c}{a}}\, t,$$

which is bounded.

(b) The characteristic equation is $ar^2 + br = 0$. The roots are $r = 0, -b/a$, and hence the general solution is $y(t) = c_1 + c_2 e^{-bt/a}$. Clearly, $y(t) \to c_1$. With the given initial conditions, $c_1 = y_0 + (a/b) y_0'$.

39. Note that $2 \cos t \sin t = \sin 2t$. Then $1 - k \cos t \sin t = 1 - (k/2) \sin 2t$. Now if $0 < k < 2$, then $(k/2) \sin 2t < |\sin 2t|$ and $-(k/2) \sin 2t > -|\sin 2t|$. Hence

$$1 - k \cos t \sin t = 1 - \frac{k}{2} \sin 2t > 1 - |\sin 2t| \geq 0.$$

40. The equation transforms into $y'' - 4y' + 4y = 0$. We obtain a double root $r = 2$. The solution is $y = c_1 e^{2x} + c_2 x e^{2x} = c_1 e^{2 \ln t} + c_2 \ln t e^{2 \ln t} = c_1 t^2 + c_2 t^2 \ln t$.

42. The equation transforms into $y'' - 7y'/2 + 5y/2 = 0$. The characteristic roots are $r = 1, 5/2$, so the solution is $y = c_1 e^x + c_2 e^{5x/2} = c_1 e^{\ln t} + c_2 e^{5 \ln t/2} = c_1 t + c_2 t^{5/2}$.

43. The equation transforms into $y'' + 2y' + y = 0$. We get a double root $r = -1$. The solution is $y = c_1 e^{-x} + c_2 x e^{-x} = c_1 e^{-\ln t} + c_2 \ln t e^{-\ln t} = c_1 t^{-1} + c_2 t^{-1} \ln t$.

44. The equation transforms into $y'' - 3y' + 9y/4 = 0$. We obtain the double root $r = 3/2$. The solution is $y = c_1 e^{3x/2} + c_2 x e^{3x/2} = c_1 e^{3 \ln t/2} + c_2 \ln t e^{3 \ln t/2} = c_1 t^{3/2} + c_2 t^{3/2} \ln t$.

3.5

2. The characteristic equation for the homogeneous problem is $r^2 + 2r + 5 = 0$, with complex roots $r = -1 \pm 2i$. Hence $y_c(t) = c_1 e^{-t} \cos 2t + c_2 e^{-t} \sin 2t$. Since the function $g(t) = 3\sin 2t$ is not proportional to the solutions of the homogeneous equation, set $Y = A\cos 2t + B\sin 2t$. Substitution into the given differential equation, and comparing the coefficients, results in the system of equations $B - 4A = 3$ and $A + 4B = 0$. Hence $Y = -(12/17)\cos 2t + (3/17)\sin 2t$. The general solution is $y(t) = y_c(t) + Y$.

3. The characteristic equation for the homogeneous problem is $r^2 - r - 2 = 0$, with roots $r = -1, 2$. Hence $y_c(t) = c_1 e^{-t} + c_2 e^{2t}$. Set $Y = At^2 + Bt + C$. Substitution into the given differential equation, and comparing the coefficients, results in the system of equations $-2A = 4$, $-2A - 2B = -2$ and $2A - B - 2C = 0$. Hence $Y = -2t^2 + 3t - 7/2$. The general solution is $y(t) = y_c(t) + Y$.

4. The characteristic equation for the homogeneous problem is $r^2 + r - 6 = 0$, with roots $r = -3, 2$. Hence $y_c(t) = c_1 e^{-3t} + c_2 e^{2t}$. Set $Y = Ae^{3t} + Be^{-2t}$. Substitution into the given differential equation, and comparing the coefficients, results in the system of equations $6A = 12$ and $-4B = 12$. Hence $Y = 2e^{3t} - 3e^{-2t}$. The general solution is $y(t) = y_c(t) + Y$.

5. The characteristic equation for the homogeneous problem is $r^2 - 2r - 3 = 0$, with roots $r = -1, 3$. Hence $y_c(t) = c_1 e^{-t} + c_2 e^{3t}$. Note that the assignment $Y = Ate^{-t}$ is not sufficient to match the coefficients. Try $Y = Ate^{-t} + Bt^2 e^{-t}$. Substitution into the differential equation, and comparing the coefficients, results in the system of equations $-4A + 2B = 0$ and $-8B = -3$. This implies that $Y = (3/16)te^{-t} + (3/8)t^2 e^{-t}$. The general solution is $y(t) = y_c(t) + Y$.

7. The characteristic equation for the homogeneous problem is $r^2 + 9 = 0$, with complex roots $r = \pm 3i$. Hence $y_c(t) = c_1 \cos 3t + c_2 \sin 3t$. To simplify the analysis, set $g_1(t) = 6$ and $g_2(t) = t^2 e^{3t}$. By inspection, we have $Y_1 = 2/3$. Based on the form of g_2, set $Y_2 = Ae^{3t} + Bte^{3t} + Ct^2 e^{3t}$. Substitution into the differential equation, and comparing the coefficients, results in the system of equations $18A + 6B + 2C = 0$, $18B + 12C = 0$, and $18C = 1$. Hence

$$Y_2 = \frac{1}{162}e^{3t} - \frac{1}{27}te^{3t} + \frac{1}{18}t^2 e^{3t}.$$

The general solution is $y(t) = y_c(t) + Y_1 + Y_2$.

9. The characteristic equation for the homogeneous problem is $2r^2 + 3r + 1 = 0$, with roots $r = -1, -1/2$. Hence $y_c(t) = c_1 e^{-t} + c_2 e^{-t/2}$. To simplify the analysis, set $g_1(t) = t^2$ and $g_2(t) = 3\sin t$. Based on the form of g_1, set $Y_1 = A + Bt + Ct^2$. Substitution into the differential equation, and comparing the coefficients, results in the system of equations $A + 3B + 4C = 0$, $B + 6C = 0$, and $C = 1$. Hence we obtain $Y_1 = 14 - 6t + t^2$. On the other hand, set $Y_2 = D\cos t + E\sin t$. After substitution into the ODE, we find that $D = -9/10$ and $E = -3/10$. The general solution is $y(t) = y_c(t) + Y_1 + Y_2$.

11. The characteristic equation for the homogeneous problem is $r^2 + \omega_0^2 = 0$, with complex roots $r = \pm \omega_0 i$. Hence $y_c(t) = c_1 \cos \omega_0 t + c_2 \sin \omega_0 t$. Since $\omega \neq \omega_0$, set $Y = A \cos \omega t + B \sin \omega t$. Substitution into the ODE and comparing the coefficients results in the system of equations $(\omega_0^2 - \omega^2)A = 1$ and $(\omega_0^2 - \omega^2)B = 0$. Hence

$$Y = \frac{1}{\omega_0^2 - \omega^2} \cos \omega t.$$

The general solution is $y(t) = y_c(t) + Y$.

12. From Problem 11, $y_c(t)$ is known. Since $\cos \omega_0 t$ is a solution of the homogeneous problem, set $Y = At \cos \omega_0 t + Bt \sin \omega_0 t$. Substitution into the given ODE and comparing the coefficients results in $A = 0$ and $B = 1/2\omega_0$. Hence the general solution is $y(t) = c_1 \cos \omega_0 t + c_2 \sin \omega_0 t + t \sin \omega_0 t/(2\omega_0)$.

14. The characteristic equation for the homogeneous problem is $r^2 - r - 2 = 0$, with roots $r = -1, 2$. Hence $y_c(t) = c_1 e^{-t} + c_2 e^{2t}$. Based on the form of the right hand side, that is, $\cosh(2t) = (e^{2t} + e^{-2t})/2$, set $Y = At e^{2t} + Be^{-2t}$. Substitution into the given ODE and comparing the coefficients results in $A = 1/6$ and $B = 1/8$. Hence the general solution is $y(t) = c_1 e^{-t} + c_2 e^{2t} + t e^{2t}/6 + e^{-2t}/8$.

16. The characteristic equation for the homogeneous problem is $r^2 + 4 = 0$, with roots $r = \pm 2i$. Hence $y_c(t) = c_1 \cos 2t + c_2 \sin 2t$. Set $Y_1 = A + Bt + Ct^2$. Comparing the coefficients of the respective terms, we find that $A = -1/8$, $B = 0$, $C = 1/4$. Now set $Y_2 = De^t$, and obtain $D = 3/5$. Hence the general solution is $y(t) = c_1 \cos 2t + c_2 \sin 2t - 1/8 + t^2/4 + 3e^t/5$. Invoking the initial conditions, we require that $19/40 + c_1 = 0$ and $3/5 + 2c_2 = 2$. Hence $c_1 = -19/40$ and $c_2 = 7/10$.

17. The characteristic equation for the homogeneous problem is $r^2 - 2r + 1 = 0$, with a double root $r = 1$. Hence $y_c(t) = c_1 e^t + c_2 t e^t$. Consider $g_1(t) = te^t$. Note that g_1 is a solution of the homogeneous problem. Set $Y_1 = At^2 e^t + Bt^3 e^t$ (the first term is not sufficient for a match). Upon substitution, we obtain $Y_1 = t^3 e^t/6$. By inspection, $Y_2 = 4$. Hence the general solution is $y(t) = c_1 e^t + c_2 t e^t + t^3 e^t/6 + 4$. Invoking the initial conditions, we require that $c_1 + 4 = 1$ and $c_1 + c_2 = 1$. Hence $c_1 = -3$ and $c_2 = 4$.

19. The characteristic equation for the homogeneous problem is $r^2 + 4 = 0$, with roots $r = \pm 2i$. Hence $y_c(t) = c_1 \cos 2t + c_2 \sin 2t$. Since the function $\sin 2t$ is a solution of the homogeneous problem, set $Y = At \cos 2t + Bt \sin 2t$. Upon substitution, we obtain $Y = -3t \cos 2t /4$. Hence the general solution is $y(t) = c_1 \cos 2t + c_2 \sin 2t - 3t \cos 2t/4$. Invoking the initial conditions, we require that $c_1 = 2$ and $2c_2 - (3/4) = -1$. Hence $c_1 = 2$ and $c_2 = -1/8$.

20. The characteristic equation for the homogeneous problem is $r^2 + 2r + 5 = 0$, with complex roots $r = -1 \pm 2i$. Hence $y_c(t) = c_1 e^{-t} \cos 2t + c_2 e^{-t} \sin 2t$. Based on the form of $g(t)$, set $Y = At e^{-t} \cos 2t + Bt e^{-t} \sin 2t$. After comparing coefficients, we obtain $Y = te^{-t} \sin 2t$. Hence the general solution is $y(t) = c_1 e^{-t} \cos 2t + c_2 e^{-t} \sin 2t + t e^{-t} \sin 2t$. Invoking the initial conditions, we require

that $c_1 = 1$ and $-c_1 + 2c_2 = 0$. Hence $c_1 = 1$ and $c_2 = 1/2$.

22.(a) The characteristic equation for the homogeneous problem is $r^2 + 1 = 0$, with complex roots $r = \pm i$. Hence $y_c(t) = c_1 \cos t + c_2 \sin t$. Let $g_1(t) = t \sin t$ and $g_2(t) = t$. By inspection, it is easy to see that $Y_2(t) = t$. Based on the form of $g_1(t)$, set $Y_1(t) = At \cos t + Bt \sin t + Ct^2 \cos t + Dt^2 \sin t$.

(b) Substitution into the equation and comparing the coefficients results in $A = 0$, $B = 1/4$, $C = -1/4$, and $D = 0$. Hence $Y(t) = t + t \sin t/4 - t^2 \cos t/4$.

23.(a) The characteristic equation for the homogeneous problem is $r^2 - 5r + 6 = 0$, with roots $r = 2, 3$. Hence $y_c(t) = c_1 e^{2t} + c_2 e^{3t}$. Consider $g_1(t) = e^{2t}(3t + 4) \sin t$, and $g_2(t) = e^t \cos 2t$. Based on the form of these functions on the right hand side of the ODE, set $Y_2(t) = e^t(A_1 \cos 2t + A_2 \sin 2t)$ and $Y_1(t) = (B_1 + B_2 t)e^{2t} \sin t + (C_1 + C_2 t)e^{2t} \cos t$.

(b) Substitution into the equation and comparing the coefficients results in

$$Y(t) = -\frac{1}{20}(e^t \cos 2t + 3e^t \sin 2t) + \frac{3}{2}te^{2t}(\cos t - \sin t) + e^{2t}(\frac{1}{2}\cos t - 5 \sin t).$$

25.(a) We obtain the double characteristic root $r = 2$. Hence $y_c(t) = c_1 e^{2t} + c_2 t e^{2t}$. Consider the functions $g_1(t) = 2t^2$, $g_2(t) = 4te^{2t}$, and $g_3(t) = t \sin 2t$. The corresponding forms of the respective parts of the particular solution are $Y_1(t) = A_0 + A_1 t + A_2 t^2$, $Y_2(t) = e^{2t}(B_2 t^2 + B_3 t^3)$, and $Y_3(t) = t(C_1 \cos 2t + C_2 \sin 2t) + (D_1 \cos 2t + D_2 \sin 2t)$.

(b) Substitution into the equation and comparing the coefficients results in

$$Y(t) = \frac{1}{4}(3 + 4t + 2t^2) + \frac{2}{3}t^3 e^{2t} + \frac{1}{8}t \cos 2t + \frac{1}{16}(\cos 2t - \sin 2t).$$

26.(a) The homogeneous solution is $y_c(t) = c_1 \cos 2t + c_2 \sin 2t$. Since $\cos 2t$ and $\sin 2t$ are both solutions of the homogeneous equation, set

$$Y(t) = t(A_0 + A_1 t + A_2 t^2) \cos 2t + t(B_0 + B_1 t + B_2 t^2) \sin 2t.$$

(b) Substitution into the equation and comparing the coefficients results in

$$Y(t) = (\frac{13}{32}t - \frac{1}{12}t^3) \cos 2t + \frac{1}{16}(28t + 13t^2) \sin 2t.$$

27.(a) The homogeneous solution is $y_c(t) = c_1 e^{-t} + c_2 t e^{-2t}$. None of the functions on the right hand side are solutions of the homogenous equation. In order to include all possible combinations of the derivatives, consider

$$Y(t) = e^t(A_0 + A_1 t + A_2 t^2) \cos 2t + e^t(B_0 + B_1 t + B_2 t^2) \sin 2t +$$
$$+ e^{-t}(C_1 \cos t + C_2 \sin t) + De^t.$$

(b) Substitution into the differential equation and comparing the coefficients results in

$$Y(t) = e^t(A_0 + A_1 t + A_2 t^2)\cos 2t + e^t(B_0 + B_1 t + B_2 t^2)\sin 2t$$
$$+ e^{-t}\left(-\frac{3}{2}\cos t + \frac{3}{2}\sin t\right) + 2e^t/3,$$

in which $A_0 = -4105/35152$, $A_1 = 73/676$, $A_2 = -5/52$, $B_0 = -1233/35152$, $B_1 = 10/169$, $B_2 = 1/52$.

28.(a) The homogeneous solution is $y_c(t) = c_1 e^{-t}\cos 2t + c_2 e^{-t}\sin 2t$. None of the terms on the right hand side are solutions of the homogenous equation. In order to include the appropriate combinations of derivatives, consider

$$Y(t) = e^{-t}(A_1 t + A_2 t^2)\cos 2t + e^{-t}(B_1 t + B_2 t^2)\sin 2t +$$
$$+ e^{-2t}(C_0 + C_1 t)\cos 2t + e^{-2t}(D_0 + D_1 t)\sin 2t.$$

(b) Substitution into the differential equation and comparing the coefficients results in

$$Y(t) = \frac{3}{16} t e^{-t}\cos 2t + \frac{3}{8} t^2 e^{-t}\sin 2t$$
$$- \frac{1}{25} e^{-2t}(7 + 10t)\cos 2t + \frac{1}{25} e^{-2t}(1 + 5t)\sin 2t.$$

30. The homogeneous solution is $y_c(t) = c_1 \cos \lambda t + c_2 \sin \lambda t$. Since the differential operator does not contain a first derivative (and $\lambda \neq m\pi$), we can set

$$Y(t) = \sum_{m=1}^{N} C_m \sin m\pi t.$$

Substitution into the differential equation yields

$$-\sum_{m=1}^{N} m^2 \pi^2 C_m \sin m\pi t + \lambda^2 \sum_{m=1}^{N} C_m \sin m\pi t = \sum_{m=1}^{N} a_m \sin m\pi t.$$

Equating coefficients of the individual terms, we obtain

$$C_m = \frac{a_m}{\lambda^2 - m^2 \pi^2}, \quad m = 1, 2 \ldots N.$$

32. The homogeneous solution is $y_c(t) = c_1 e^{-t}\cos 2t + c_2 e^{-t}\sin 2t$. The input function is independent of the homogeneous solutions, on any interval. Since the right hand side is piecewise constant, it follows by inspection that

$$Y(t) = \begin{cases} 1/5, & 0 \leq t \leq \pi/2 \\ 0, & t > \pi/2 \end{cases}.$$

For $0 \leq t \leq \pi/2$, the general solution is $y(t) = c_1 e^{-t} \cos 2t + c_2 e^{-t} \sin 2t + 1/5$. Invoking the initial conditions $y(0) = y'(0) = 0$, we require that $c_1 = -1/5$, and that $c_2 = -1/10$. Hence

$$y(t) = \frac{1}{5} - \frac{1}{10}(2e^{-t} \cos 2t + e^{-t} \sin 2t)$$

on the interval $0 \leq t \leq \pi/2$. We now have the values $y(\pi/2) = (1 + e^{-\pi/2})/5$, and $y'(\pi/2) = 0$. For $t > \pi/2$, the general solution is $y(t) = d_1 e^{-t} \cos 2t + d_2 e^{-t} \sin 2t$. It follows that $y(\pi/2) = -e^{-\pi/2} d_1$ and $y'(\pi/2) = e^{-\pi/2} d_1 - 2 e^{-\pi/2} d_2$. Since the solution is continuously differentiable, we require that $-e^{-\pi/2} d_1 = (1 + e^{-\pi/2})/5$ and $e^{-\pi/2} d_1 - 2 e^{-\pi/2} d_2 = 0$. Solving for the coefficients, $d_1 = 2 d_2 = -(e^{\pi/2} + 1)/5$.

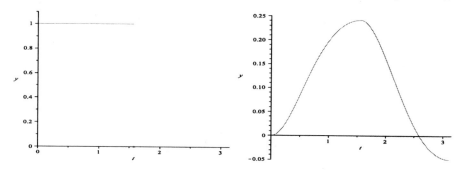

34. Since $a, b, c > 0$, the roots of the characteristic equation have negative real parts. That is, $r = \alpha \pm \beta i$, where $\alpha < 0$. Hence the homogeneous solution is

$$y_c(t) = c_1 e^{\alpha t} \cos \beta t + c_2 e^{\alpha t} \sin \beta t.$$

If $g(t) = d$, then the general solution is

$$y(t) = d/c + c_1 e^{\alpha t} \cos \beta t + c_2 e^{\alpha t} \sin \beta t.$$

Since $\alpha < 0$, $y(t) \to d/c$ as $t \to \infty$. If $c = 0$, then the characteristic roots are $r = 0$ and $r = -b/a$. The ODE becomes $ay'' + by' = d$. Integrating both sides, we find that $ay' + by = dt + c_1$. The general solution can be expressed as

$$y(t) = dt/b + c_1 + c_2 e^{-bt/a}.$$

In this case, the solution grows without bound. If $b = 0$, also, then the differential equation can be written as $y'' = d/a$, which has general solution $y(t) = dt^2/2a + c_1 + c_2$. Hence the assertion is true only if the coefficients are positive.

35.(a) Since D is a linear operator, $D^2 y + bDy + cy = D^2 y - (r_1 + r_2) Dy + r_1 r_2 y = D^2 y - r_2 Dy - r_1 Dy + r_1 r_2 y = D(Dy - r_2 y) - r_1 (Dy - r_2 y) = (D - r_1)(D - r_2) y$.

(b) Let $u = (D - r_2) y$. Then the ODE (i) can be written as $(D - r_1) u = g(t)$, that is, $u' - r_1 u = g(t)$. The latter is a linear first order equation in u. Its general solution is

$$u(t) = e^{r_1 t} \int_{t_0}^{t} e^{-r_1 \tau} g(\tau) d\tau + c_1 e^{r_1 t}.$$

From above, we have $y' - r_2 y = u(t)$. This equation is also a first order ODE. Hence the general solution of the original second order equation is

$$y(t) = e^{r_2 t} \int_{t_0}^{t} e^{-r_2 \tau} u(\tau) d\tau + c_2 e^{r_2 t}.$$

Note that the solution $y(t)$ contains two arbitrary constants.

37. Note that $(2D^2 + 3D + 1)y = (2D + 1)(D + 1)y$. Let $u = (D + 1)y$, and solve the ODE $2u' + u = t^2 + 3 \sin t$. This equation is a linear first order ODE, with solution

$$u(t) = e^{-t/2} \int_{t_0}^{t} e^{\tau/2} \left[\tau^2/2 + \frac{3}{2} \sin \tau \right] d\tau + c e^{-t/2} =$$

$$= t^2 - 4t + 8 - \frac{6}{5} \cos t + \frac{3}{5} \sin t + c e^{-t/2}.$$

Now consider the ODE $y' + y = u(t)$. The general solution of this first order ODE is

$$y(t) = e^{-t} \int_{t_0}^{t} e^{\tau} u(\tau) d\tau + c_2 e^{-t},$$

in which $u(t)$ is given above. Substituting for $u(t)$ and performing the integration,

$$y(t) = t^2 - 6t + 14 - \frac{9}{10} \cos t - \frac{3}{10} \sin t + c_1 e^{-t/2} + c_2 e^{-t}.$$

38. We have $(D^2 + 2D + 1)y = (D + 1)(D + 1)y$. Let $u = (D + 1)y$, and consider the ODE $u' + u = 2e^{-t}$. The general solution is $u(t) = 2t e^{-t} + c e^{-t}$. We therefore have the first order equation $u' + u = 2t e^{-t} + c_1 e^{-t}$. The general solution of the latter differential equation is

$$y(t) = e^{-t} \int_{t_0}^{t} [2\tau + c_1] d\tau + c_2 e^{-t} = e^{-t}(t^2 + c_1 t + c_2).$$

39. We have $(D^2 + 2D)y = D(D + 2)y$. Let $u = (D + 2)y$, and consider the equation $u' = 3 + 4 \sin 2t$. Direct integration results in $u(t) = 3t - 2 \cos 2t + c$. The problem is reduced to solving the ODE $y' + 2y = 3t - 2 \cos 2t + c$. The general solution of this first order differential equation is

$$y(t) = e^{-2t} \int_{t_0}^{t} e^{2\tau} [3\tau - 2 \cos 2\tau + c] d\tau + c_2 e^{-2t} =$$

$$= \frac{3}{2} t - \frac{1}{2} (\cos 2t + \sin 2t) + c_1 + c_2 e^{-2t}.$$

3.6

1. The solution of the homogeneous equation is $y_c(t) = c_1 e^{2t} + c_2 e^{3t}$. The functions $y_1(t) = e^{2t}$ and $y_2(t) = e^{3t}$ form a fundamental set of solutions. The Wronskian

of these functions is $W(y_1, y_2) = e^{5t}$. Using the method of variation of parameters, the particular solution is given by $Y(t) = u_1(t) y_1(t) + u_2(t) y_2(t)$, in which

$$u_1(t) = -\int \frac{e^{3t}(2e^t)}{W(t)} dt = 2e^{-t} \quad \text{and} \quad u_2(t) = \int \frac{e^{2t}(2e^t)}{W(t)} dt = -e^{-2t}.$$

Hence the particular solution is $Y(t) = 2e^t - e^t = e^t$.

3. The solution of the homogeneous equation is $y_c(t) = c_1 e^{-t} + c_2 t e^{-t}$. The functions $y_1(t) = e^{-t}$ and $y_2(t) = te^{-t}$ form a fundamental set of solutions. The Wronskian of these functions is $W(y_1, y_2) = e^{-2t}$. Using the method of variation of parameters, the particular solution is given by $Y(t) = u_1(t) y_1(t) + u_2(t) y_2(t)$, in which

$$u_1(t) = -\int \frac{te^{-t}(3e^{-t})}{W(t)} dt = -3t^2/2 \quad \text{and} \quad u_2(t) = \int \frac{e^{-t}(3e^{-t})}{W(t)} dt = 3t.$$

Hence the particular solution is $Y(t) = -3t^2 e^{-t}/2 + 3t^2 e^{-t} = 3t^2 e^{-t}/2$.

4. The functions $y_1(t) = e^{t/2}$ and $y_2(t) = te^{t/2}$ form a fundamental set of solutions. The Wronskian of these functions is $W(y_1, y_2) = e^t$. First write the equation in standard form, so that $g(t) = 4e^{t/2}$. Using the method of variation of parameters, the particular solution is given by $Y(t) = u_1(t) y_1(t) + u_2(t) y_2(t)$, in which

$$u_1(t) = -\int \frac{te^{t/2}(4e^{t/2})}{W(t)} dt = -2t^2 \quad \text{and} \quad u_2(t) = \int \frac{e^{t/2}(4e^{t/2})}{W(t)} dt = 4t.$$

Hence the particular solution is $Y(t) = -2t^2 e^{t/2} + 4t^2 e^{t/2} = 2t^2 e^{t/2}$.

6. The solution of the homogeneous equation is $y_c(t) = c_1 \cos 3t + c_2 \sin 3t$. The two functions $y_1(t) = \cos 3t$ and $y_2(t) = \sin 3t$ form a fundamental set of solutions, with $W(y_1, y_2) = 3$. The particular solution is given by $Y(t) = u_1(t) y_1(t) + u_2(t) y_2(t)$, in which

$$u_1(t) = -\int \frac{\sin 3t (9 \sec^2 3t)}{W(t)} dt = -\csc 3t$$

$$u_2(t) = \int \frac{\cos 3t (9 \sec^2 3t)}{W(t)} dt = \ln(\sec 3t + \tan 3t),$$

since $0 < t < \pi/6$. Hence $Y(t) = -1 + (\sin 3t) \ln(\sec 3t + \tan 3t)$. The general solution is given by

$$y(t) = c_1 \cos 3t + c_2 \sin 3t + (\sin 3t) \ln(\sec 3t + \tan 3t) - 1.$$

7. The functions $y_1(t) = e^{-2t}$ and $y_2(t) = te^{-2t}$ form a fundamental set of solutions. The Wronskian of these functions is $W(y_1, y_2) = e^{-4t}$. The particular solution is given by $Y(t) = u_1(t) y_1(t) + u_2(t) y_2(t)$, in which

$$u_1(t) = -\int \frac{te^{-2t}(t^{-2}e^{-2t})}{W(t)} dt = -\ln t \quad \text{and} \quad u_2(t) = \int \frac{e^{-2t}(t^{-2}e^{-2t})}{W(t)} dt = -1/t.$$

Hence the particular solution is $Y(t) = -e^{-2t}\ln t - e^{-2t}$. Since the second term is a solution of the homogeneous equation, the general solution is given by

$$y(t) = c_1 e^{-2t} + c_2 t e^{-2t} - e^{-2t}\ln t.$$

8. The solution of the homogeneous equation is $y_c(t) = c_1 \cos 2t + c_2 \sin 2t$. The two functions $y_1(t) = \cos 2t$ and $y_2(t) = \sin 2t$ form a fundamental set of solutions, with $W(y_1, y_2) = 2$. The particular solution is given by $Y(t) = u_1(t)\, y_1(t) + u_2(t)\, y_2(t)$, in which

$$u_1(t) = -\int \frac{\sin 2t(3 \csc 2t)}{W(t)} dt = -3t/2$$

$$u_2(t) = \int \frac{\cos 2t(3 \csc 2t)}{W(t)} dt = \frac{3}{4}\ln(\sin 2t),$$

since $0 < t < \pi/2$. Hence $Y(t) = -(3/2)t \cos 2t + (3/4)(\sin 2t)\ln(\sin 2t)$. The general solution is given by

$$y(t) = c_1 \cos 2t + c_2 \sin 2t - \frac{3}{2} t \cos 2t + \frac{3}{4}(\sin 2t)\ln(\sin 2t).$$

9. The functions $y_1(t) = \cos(t/2)$ and $y_2(t) = \sin(t/2)$ form a fundamental set of solutions. The Wronskian of these functions is $W(y_1, y_2) = 1/2$. First write the ODE in standard form, so that $g(t) = \sec(t/2)/2$. The particular solution is given by $Y(t) = u_1(t)\, y_1(t) + u_2(t)\, y_2(t)$, in which

$$u_1(t) = -\int \frac{\cos(t/2)\,[\sec(t/2)]}{2W(t)} dt = 2\ln(\cos(t/2))$$

$$u_2(t) = \int \frac{\sin(t/2)\,[\sec(t/2)]}{2W(t)} dt = t.$$

The particular solution is $Y(t) = 2\cos(t/2)\ln(\cos(t/2)) + t \sin(t/2)$. The general solution is given by

$$y(t) = c_1 \cos(t/2) + c_2 \sin(t/2) + 2\cos(t/2)\ln(\cos(t/2)) + t \sin(t/2).$$

10. The solution of the homogeneous equation is $y_c(t) = c_1 e^t + c_2 t e^t$. The functions $y_1(t) = e^t$ and $y_2(t) = t e^t$ form a fundamental set of solutions, with $W(y_1, y_2) = e^{2t}$. The particular solution is given by $Y(t) = u_1(t)\, y_1(t) + u_2(t)\, y_2(t)$, in which

$$u_1(t) = -\int \frac{t e^t (e^t)}{W(t)(1+t^2)} dt = -\frac{1}{2}\ln(1+t^2)$$

$$u_2(t) = \int \frac{e^t (e^t)}{W(t)(1+t^2)} dt = \arctan t.$$

The particular solution is $Y(t) = -(1/2)e^t \ln(1+t^2) + t e^t \arctan(t)$. Hence the general solution is given by

$$y(t) = c_1 e^t + c_2 t e^t - \frac{1}{2} e^t \ln(1+t^2) + t e^t \arctan(t).$$

12. The functions $y_1(t) = \cos 2t$ and $y_2(t) = \sin 2t$ form a fundamental set of solutions, with $W(y_1, y_2) = 2$. The particular solution is given by $Y(t) = u_1(t) y_1(t) + u_2(t) y_2(t)$, in which

$$u_1(t) = -\frac{1}{2} \int_{t_0}^{t} g(s) \sin 2s \, ds \quad \text{and} \quad u_2(t) = \frac{1}{2} \int_{t_0}^{t} g(s) \cos 2s \, ds.$$

Hence the particular solution is

$$Y(t) = -\frac{1}{2} \cos 2t \int_{t_0}^{t} g(s) \sin 2s \, ds + \frac{1}{2} \sin 2t \int_{t_0}^{t} g(s) \cos 2s \, ds.$$

Note that $\sin 2t \cos 2s - \cos 2t \sin 2s = \sin(2t - 2s)$. It follows that

$$Y(t) = \frac{1}{2} \int_{t_0}^{t} g(s) \sin(2t - 2s) ds.$$

The general solution of the differential equation is given by

$$y(t) = c_1 \cos 2t + c_2 \sin 2t + \frac{1}{2} \int_{t_0}^{t} g(s) \sin(2t - 2s) ds.$$

13. Note first that $p(t) = 0$, $q(t) = -2/t^2$ and $g(t) = (3t^2 - 1)/t^2$. The functions $y_1(t)$ and $y_2(t)$ are solutions of the homogeneous equation, verified by substitution. The Wronskian of these two functions is $W(y_1, y_2) = -3$. Using the method of variation of parameters, the particular solution is $Y(t) = u_1(t) y_1(t) + u_2(t) y_2(t)$, in which

$$u_1(t) = -\int \frac{t^{-1}(3t^2 - 1)}{t^2 W(t)} dt = t^{-2}/6 + \ln t$$

$$u_2(t) = \int \frac{t^2(3t^2 - 1)}{t^2 W(t)} dt = -t^3/3 + t/3.$$

Therefore $Y(t) = 1/6 + t^2 \ln t - t^2/3 + 1/3$.

15. Observe that $g(t) = t e^{2t}$. The functions $y_1(t)$ and $y_2(t)$ are a fundamental set of solutions. The Wronskian of these two functions is $W(y_1, y_2) = t e^t$. Using the method of variation of parameters, the particular solution is $Y(t) = u_1(t) y_1(t) + u_2(t) y_2(t)$, in which

$$u_1(t) = -\int \frac{e^t(t e^{2t})}{W(t)} dt = -e^{2t}/2 \quad \text{and} \quad u_2(t) = \int \frac{(1+t)(t e^{2t})}{W(t)} dt = t e^t.$$

Therefore $Y(t) = -(1+t)e^{2t}/2 + t e^{2t} = -e^{2t}/2 + t e^{2t}/2$.

16. Observe that $g(t) = 2(1-t) e^{-t}$. Direct substitution of $y_1(t) = e^t$ and $y_2(t) = t$ verifies that they are solutions of the homogeneous equation. The Wronskian of the two solutions is $W(y_1, y_2) = (1-t) e^t$. Using the method of variation of parameters, the particular solution is $Y(t) = u_1(t) y_1(t) + u_2(t) y_2(t)$, in which

$$u_1(t) = -\int \frac{2t(1-t)e^{-t}}{W(t)} dt = te^{-2t} + e^{-2t}/2$$

$$u_2(t) = \int \frac{2(1-t)}{W(t)} dt = -2e^{-t}.$$

Therefore $Y(t) = te^{-t} + e^{-t}/2 - 2te^{-t} = -te^{-t} + e^{-t}/2$.

17. Note that $g(x) = \ln x$. The functions $y_1(x) = x^2$ and $y_2(x) = x^2 \ln x$ are solutions of the homogeneous equation, as verified by substitution. The Wronskian of the solutions is $W(y_1, y_2) = x^3$. Using the method of variation of parameters, the particular solution is $Y(x) = u_1(x) y_1(x) + u_2(x) y_2(x)$, in which

$$u_1(x) = -\int \frac{x^2 \ln x (\ln x)}{W(x)} dx = -(\ln x)^3/3$$

$$u_2(x) = \int \frac{x^2 (\ln x)}{W(x)} dx = (\ln x)^2/2.$$

Therefore $Y(x) = -x^2 (\ln x)^3/3 + x^2 (\ln x)^3/2 = x^2 (\ln x)^3/6$.

19. First write the equation in standard form. Note that the forcing function becomes $g(x)/(1-x)$. The functions $y_1(x) = e^x$ and $y_2(x) = x$ are a fundamental set of solutions, as verified by substitution. The Wronskian of the solutions is $W(y_1, y_2) = (1-x)e^x$. Using the method of variation of parameters, the particular solution is $Y(x) = u_1(x) y_1(x) + u_2(x) y_2(x)$, in which

$$u_1(x) = -\int_{x_0}^{x} \frac{\tau(g(\tau))}{(1-\tau)W(\tau)} d\tau \quad \text{and} \quad u_2(x) = \int_{x_0}^{x} \frac{e^\tau (g(\tau))}{(1-\tau)W(\tau)} d\tau.$$

Therefore

$$Y(x) = -e^x \int_{x_0}^{x} \frac{\tau(g(\tau))}{(1-\tau)W(\tau)} d\tau + x \int_{x_0}^{x} \frac{e^\tau (g(\tau))}{(1-\tau)W(\tau)} d\tau =$$

$$= \int_{x_0}^{x} \frac{(xe^\tau - e^x \tau) g(\tau)}{(1-\tau)^2 e^\tau} d\tau.$$

20. First write the equation in standard form. The forcing function becomes $g(x)/x^2$. The functions $y_1(x) = x^{-1/2} \sin x$ and $y_2(x) = x^{-1/2} \cos x$ are a fundamental set of solutions. The Wronskian of the solutions is $W(y_1, y_2) = -1/x$. Using the method of variation of parameters, the particular solution is $Y(x) = u_1(x) y_1(x) + u_2(x) y_2(x)$, in which

$$u_1(x) = \int_{x_0}^{x} \frac{\cos \tau (g(\tau))}{\tau \sqrt{\tau}} d\tau \quad \text{and} \quad u_2(x) = -\int_{x_0}^{x} \frac{\sin \tau (g(\tau))}{\tau \sqrt{\tau}} d\tau.$$

Therefore

$$Y(x) = \frac{\sin x}{\sqrt{x}} \int_{x_0}^{x} \frac{\cos \tau (g(\tau))}{\tau \sqrt{\tau}} dt - \frac{\cos x}{\sqrt{x}} \int_{x_0}^{x} \frac{\sin \tau (g(\tau))}{\tau \sqrt{\tau}} d\tau =$$

$$= \frac{1}{\sqrt{x}} \int_{x_0}^{x} \frac{\sin(x-\tau) g(\tau)}{\tau \sqrt{\tau}} d\tau.$$

21. Let $y_1(t)$ and $y_2(t)$ be a fundamental set of solutions, and $W(t) = W(y_1, y_2)$ be the corresponding Wronskian. Any solution, $u(t)$, of the homogeneous equation is

a linear combination $u(t) = \alpha_1 y_1(t) + \alpha_2 y_2(t)$. Invoking the initial conditions, we require that

$$y_0 = \alpha_1 y_1(t_0) + \alpha_2 y_2(t_0)$$
$$y_0' = \alpha_1 y_1'(t_0) + \alpha_2 y_2'(t_0)$$

Note that this system of equations has a unique solution, since $W(t_0) \neq 0$. Now consider the nonhomogeneous problem, $L[v] = g(t)$, with homogeneous initial conditions. Using the method of variation of parameters, the particular solution is given by

$$Y(t) = -y_1(t) \int_{t_0}^{t} \frac{y_2(s) \, g(s)}{W(s)} ds + y_2(t) \int_{t_0}^{t} \frac{y_1(s) \, g(s)}{W(s)} ds.$$

The general solution of the IVP (iii) is

$$v(t) = \beta_1 y_1(t) + \beta_2 y_2(t) + Y(t) = \beta_1 y_1(t) + \beta_2 y_2(t) + y_1(t) u_1(t) + y_2(t) u_2(t)$$

in which u_1 and u_2 are defined above. Invoking the initial conditions, we require that

$$0 = \beta_1 y_1(t_0) + \beta_2 y_2(t_0) + Y(t_0)$$
$$0 = \beta_1 y_1'(t_0) + \beta_2 y_2'(t_0) + Y'(t_0)$$

Based on the definition of u_1 and u_2, $Y(t_0) = 0$. Furthermore, since $y_1 u_1' + y_2 u_2' = 0$, it follows that $Y'(t_0) = 0$. Hence the only solution of the above system of equations is the trivial solution. Therefore $v(t) = Y(t)$. Now consider the function $y = u + v$. Then $L[y] = L[u + v] = L[u] + L[v] = g(t)$. That is, $y(t)$ is a solution of the nonhomogeneous problem. Further, $y(t_0) = u(t_0) + v(t_0) = y_0$, and similarly, $y'(t_0) = y_0'$. By the uniqueness theorems, $y(t)$ is the unique solution of the initial value problem.

23.(a) A fundamental set of solutions is $y_1(t) = \cos t$ and $y_2(t) = \sin t$. The Wronskian $W(t) = y_1 y_2' - y_1' y_2 = 1$. By the result in Problem 22,

$$Y(t) = \int_{t_0}^{t} \frac{\cos(s)\sin(t) - \cos(t)\sin(s)}{W(s)} g(s) ds$$
$$= \int_{t_0}^{t} [\cos(s)\sin(t) - \cos(t)\sin(s)] g(s) ds.$$

Finally, we have $\cos(s)\sin(t) - \cos(t)\sin(s) = \sin(t - s)$.

(b) Using Problem 21 and part (a), the solution is

$$y(t) = y_0 \cos t + y_0' \sin t + \int_0^t \sin(t-s) g(s) ds.$$

24. A fundamental set of solutions is $y_1(t) = e^{at}$ and $y_2(t) = e^{bt}$. The Wronskian $W(t) = y_1 y_2' - y_1' y_2 = (b-a) e^{(a+b)t}$. By the result in Problem 22,

$$Y(t) = \int_{t_0}^{t} \frac{e^{as} e^{bt} - e^{at} e^{bs}}{W(s)} g(s) ds = \frac{1}{b-a} \int_{t_0}^{t} \frac{e^{as} e^{bt} - e^{at} e^{bs}}{e^{(a+b)s}} g(s) ds.$$

Hence the particular solution is

$$Y(t) = \frac{1}{b-a} \int_{t_0}^{t} \left[e^{b(t-s)} - e^{a(t-s)}\right] g(s) ds.$$

26. A fundamental set of solutions is $y_1(t) = e^{at}$ and $y_2(t) = te^{at}$. The Wronskian $W(t) = y_1 y_2' - y_1' y_2 = e^{2at}$. By the result in Problem 22,

$$Y(t) = \int_{t_0}^{t} \frac{te^{as+at} - se^{at+as}}{W(s)} g(s) ds = \int_{t_0}^{t} \frac{(t-s)e^{as+at}}{e^{2as}} g(s) ds.$$

Hence the particular solution is

$$Y(t) = \int_{t_0}^{t} (t-s)e^{a(t-s)} g(s) ds.$$

27. The form of the kernel depends on the characteristic roots. If the roots are real and distinct,

$$K(t-s) = \frac{e^{b(t-s)} - e^{a(t-s)}}{b-a}.$$

If the roots are real and identical,

$$K(t-s) = (t-s)e^{a(t-s)}.$$

If the roots are complex conjugates,

$$K(t-s) = \frac{e^{\lambda(t-s)} \sin \mu(t-s)}{\mu}.$$

28. Let $y(t) = v(t)y_1(t)$, in which $y_1(t)$ is a solution of the homogeneous equation. Substitution into the given ODE results in

$$v''y_1 + 2v'y_1' + vy_1'' + p(t)[v'y_1 + vy_1'] + q(t)vy_1 = g(t).$$

By assumption, $y_1'' + p(t)y_1 + q(t)y_1 = 0$, hence $v(t)$ must be a solution of the ODE

$$v''y_1 + [2y_1' + p(t)y_1] v' = g(t).$$

Setting $w = v'$, we also have $w'y_1 + [2y_1' + p(t)y_1] w = g(t)$.

30. First write the equation as $y'' + 7t^{-1}y' + 5t^{-2} y = t^{-1}$. As shown in Problem 28, the function $y(t) = t^{-1}v(t)$ is a solution of the given ODE as long as v is a solution of

$$t^{-1}v'' + \left[-2t^{-2} + 7t^{-2}\right] v' = t^{-1},$$

that is, $v'' + 5t^{-1} v' = 1$. This ODE is linear and first order in v'. The integrating factor is $\mu = t^5$. The solution is $v' = t/6 + ct^{-5}$. Direct integration now results in $v(t) = t^2/12 + c_1 t^{-4} + c_2$. Hence $y(t) = t/12 + c_1 t^{-5} + c_2 t^{-1}$.

31. Write the equation as $y'' - t^{-1}(1+t)y' + t^{-1} y = te^{2t}$. As shown in Problem 28, the function $y(t) = (1+t)v(t)$ is a solution of the given ODE as long as v is a solution of

$$(1+t) v'' + \left[2 - t^{-1}(1+t)^2\right] v' = te^{2t},$$

that is,
$$v'' - \frac{1+t^2}{t(t+1)} v' = \frac{t}{t+1} e^{2t}.$$

This equation is first order linear in v', with integrating factor $\mu = t^{-1}(1+t)^2 e^{-t}$. The solution is $v' = (t^2 e^{2t} + c_1 t e^t)/(1+t)^2$. Integrating, we obtain $v(t) = e^{2t}/2 - e^{2t}/(t+1) + c_1 e^t/(t+1) + c_2$. Hence the solution of the original ODE is $y(t) = (t-1)e^{2t}/2 + c_1 e^t + c_2(t+1)$.

32. Write the equation as $y'' + t(1-t)^{-1} y - (1-t)^{-1} y = 2(1-t) e^{-t}$. The function $y(t) = e^t v(t)$ is a solution to the given ODE as long as v is a solution of
$$e^t v'' + \left[2e^t + t(1-t)^{-1} e^t\right] v' = 2(1-t) e^{-t},$$
that is, $v'' + [(2-t)/(1-t)] v' = 2(1-t) e^{-2t}$. This equation is first order linear in v', with integrating factor $\mu = e^t/(t-1)$. The solution is
$$v' = (t-1)(2e^{-2t} + c_1 e^{-t}).$$
Integrating, we obtain $v(t) = (1/2 - t)e^{-2t} - c_1 t e^{-t} + c_2$. Hence the solution of the original ODE is $y(t) = (1/2 - t)e^{-t} - c_1 t + c_2 e^t$.

3.7

1. $R \cos \delta = 3$ and $R \sin \delta = 4$, so $R = \sqrt{25} = 5$ and $\delta = \arctan(4/3)$. We obtain that $u = 5 \cos(2t - \arctan(4/3))$.

3. $R \cos \delta = 4$ and $R \sin \delta = -2$, so $R = \sqrt{20} = 2\sqrt{5}$ and $\delta = -\arctan(1/2)$. We obtain that $u = 2\sqrt{5} \cos(3t + \arctan(1/2))$.

4. $R \cos \delta = -2$ and $R \sin \delta = -3$, so $R = \sqrt{13}$ and $\delta = \pi + \arctan(3/2)$. We obtain that $u = \sqrt{13} \cos(\pi t - \pi - \arctan(3/2))$.

5. The spring constant is $k = 2/(1/2) = 4$ lb/ft. Mass $m = 2/32 = 1/16$ lb-s^2/ft. Since there is no damping, the equation of motion is $u''/16 + 4u = 0$, that is, $u'' + 64u = 0$. The initial conditions are $u(0) = 1/4$ ft, $u'(0) = 0$ ft/s. The general solution is $u(t) = A \cos 8t + B \sin 8t$. Invoking the initial conditions, we have $u(t) = \cos 8t /4$. $R = 1/4$ ft, $\delta = 0$ rad, $\omega_0 = 8$ rad/s, and $T = \pi/4$ s.

7. The spring constant is $k = 3/(1/4) = 12$ lb/ft. Mass $m = 3/32$ lb-s^2/ft. Since there is no damping, the equation of motion is $3u''/32 + 12u = 0$, that is, $u'' + 128u = 0$. The initial conditions are $u(0) = -1/12$ ft, $u'(0) = 2$ ft/s. The general solution is $u(t) = A \cos 8\sqrt{2} t + B \sin 8\sqrt{2} t$. Invoking the initial conditions, we have
$$u(t) = -\frac{1}{12} \cos 8\sqrt{2} t + \frac{1}{4\sqrt{2}} \sin 8\sqrt{2} t.$$
$R = \sqrt{11/288}$ ft, $\delta = \pi - \arctan(3/\sqrt{2})$ rad, $\omega_0 = 8\sqrt{2}$ rad/s, $T = \pi/(4\sqrt{2})$ s.

10. The spring constant is $k = 16/(1/4) = 64$ lb/ft. Mass $m = 1/2$ lb-s^2/ft. The damping coefficient is $\gamma = 2$ lb-s/ft. Hence the equation of motion is $u''/2 + 2u' + 64u = 0$, that is, $u'' + 4u' + 128u = 0$. The initial conditions are $u(0) = 0$ ft, $u'(0) = 1/4$ ft/s. The general solution is $u(t) = A\cos 2\sqrt{31}\,t + B\sin 2\sqrt{31}\,t$. Invoking the initial conditions, we have

$$u(t) = \frac{1}{8\sqrt{31}} e^{-2t} \sin 2\sqrt{31}\,t.$$

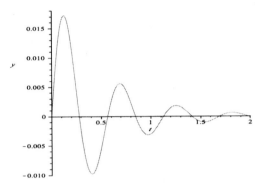

Solving $u(t) = 0$, on the interval $[0.2, 0.4]$, we obtain $t = \pi/2\sqrt{31} = 0.2821$ s. Based on the graph, and the solution of $u(t) = -0.01/12$ ft, we have $|u(t)| \leq 0.01$ for $t \geq \tau = 1.5927$.

11. The spring constant is $k = 3/(.1) = 30$ N/m. The damping coefficient is given as $\gamma = 3/5$ N-s/m. Hence the equation of motion is $2u'' + 3u'/5 + 30u = 0$, that is, $u'' + 0.3u' + 15u = 0$. The initial conditions are $u(0) = 0.05$ m and $u'(0) = 0.01$ m/s. The general solution is $u(t) = A\cos\mu t + B\sin\mu t$, in which $\mu = 3.87008$ rad/s. Invoking the initial conditions, we have $u(t) = e^{-0.15t}(0.05\cos\mu t + 0.00452\sin\mu t)$. Also, $\mu/\omega_0 = 3.87008/\sqrt{15} \approx 0.99925$.

13. The frequency of the undamped motion is $\omega_0 = 1$. The quasi frequency of the damped motion is $\mu = \sqrt{4-\gamma^2}/2$. Setting $\mu = 2\omega_0/3$, we obtain $\gamma = 2\sqrt{5}/3$.

14. The spring constant is $k = mg/L$. The equation of motion for an undamped system is $mu'' + mgu/L = 0$. Hence the natural frequency of the system is $\omega_0 = \sqrt{g/L}$. The period is $T = 2\pi/\omega_0$.

15. The general solution of the system is $u(t) = A\cos\gamma(t-t_0) + B\sin\gamma(t-t_0)$. Invoking the initial conditions, we have $u(t) = u_0\cos\gamma(t-t_0) + (u_0'/\gamma)\sin\gamma(t-t_0)$. Clearly, the functions $v = u_0\cos\gamma(t-t_0)$ and $w = (u_0'/\gamma)\sin\gamma(t-t_0)$ satisfy the given criteria.

16. Note that $r\sin(\omega_0 t - \theta) = r\sin\omega_0 t\cos\theta - r\cos\omega_0 t\sin\theta$. Comparing the given expressions, we have $A = -r\sin\theta$ and $B = r\cos\theta$. That is, $r = R = \sqrt{A^2 + B^2}$, and $\tan\theta = -A/B = -1/\tan\delta$. The latter relation is also $\tan\theta + \cot\delta = 1$.

18. The system is critically damped, when $R = 2\sqrt{L/C}$. Here $R = 1000$ ohms.

21.(a) Let $u = Re^{-\gamma t/2m}\cos(\mu t - \delta)$. Then attains a maximum when $\mu t_k - \delta = 2k\pi$. Hence $T_d = t_{k+1} - t_k = 2\pi/\mu$.

(b) $u(t_k)/u(t_{k+1}) = e^{-\gamma t_k/2m}/e^{-\gamma t_{k+1}/2m} = e^{(\gamma t_{k+1} - \gamma t_k)/2m}$. Hence $u(t_k)/u(t_{k+1}) = e^{\gamma(2\pi/\mu)/2m} = e^{\gamma T_d/2m}$.

(c) $\Delta = \ln[u(t_k)/u(t_{k+1})] = \gamma(2\pi/\mu)/2m = \pi\gamma/\mu m$.

22. The spring constant is $k = 16/(1/4) = 64$ lb/ft. Mass $m = 1/2$ lb-s^2/ft. The damping coefficient is $\gamma = 2$ lb-s/ft. The quasi frequency is $\mu = 2\sqrt{31}$ rad/s. Hence $\Delta = 2\pi/\sqrt{31} \approx 1.1285$.

25.(a) The solution of the IVP is $u(t) = e^{-t/8}(2\cos 3\sqrt{7}\,t/8 + (2\sqrt{7}/21)\sin 3\sqrt{7}\,t/8)$.

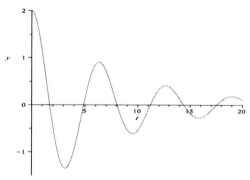

Using the plot, and numerical analysis, $\tau \approx 41.715$.

(b) For $\gamma = 0.5$, $\tau \approx 20.402$; for $\gamma = 1.0$, $\tau \approx 9.168$; for $\gamma = 1.5$, $\tau \approx 7.184$.

(c)

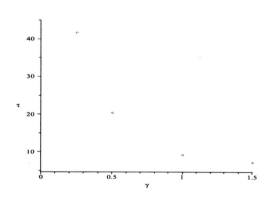

(d) For $\gamma = 1.6$, $\tau \approx 7.218$; for $\gamma = 1.7$, $\tau \approx 6.767$; for $\gamma = 1.8$, $\tau \approx 5.473$; for $\gamma = 1.9$, $\tau \approx 6.460$. τ steadily decreases to about $\tau_{min} \approx 4.873$, corresponding to the critical value $\gamma_0 \approx 1.73$.

(e) We can rewrite the solution as $u(t) = Re^{-\gamma t/2}\cos(\mu t - \delta)$, where $R = 4/\sqrt{4-\gamma^2}$.

Neglecting the cosine factor, we can approximate τ by solving $Re^{-\gamma\tau/2} = 1/100$, thus finding
$$\tau \approx \frac{2}{\gamma}\ln(100R) = \frac{2}{\gamma}\ln\left(\frac{400}{\sqrt{4-\gamma^2}}\right).$$

For $\gamma = 0.25$, $\tau \approx 42.4495$; for $\gamma = 0.5$, $\tau \approx 21.3223$; for $\gamma = 1.0$, $\tau \approx 10.8843$; for $\gamma = 1.5$, $\tau \approx 7.61554$; for $\gamma = 1.6$, $\tau \approx 7.26143$; for $\gamma = 1.7$, $\tau \approx 6.98739$; for $\gamma = 1.8$, $\tau \approx 6.80965$; for $\gamma = 1.9$, $\tau \approx 6.80239$.

26.(a) The characteristic equation is $mr^2 + \gamma r + k = 0$. Since $\gamma^2 < 4km$, the roots are $r_{1,2} = (-\gamma \pm i\sqrt{4mk - \gamma^2})/2m$. The general solution is
$$u(t) = e^{-\gamma t/2m}\left[A\cos\frac{\sqrt{4mk-\gamma^2}}{2m}t + B\sin\frac{\sqrt{4mk-\gamma^2}}{2m}t\right].$$

Invoking the initial conditions, $A = u_0$ and $B = (2mv_0 - \gamma u_0)/\sqrt{4mk - \gamma^2}$.

(b) We can write $u(t) = Re^{-\gamma t/2m}\cos(\mu t - \delta)$, in which
$$R = \sqrt{u_0^2 + \frac{(2mv_0 - \gamma u_0)^2}{4mk - \gamma^2}} \quad \text{and} \quad \delta = \arctan\left[\frac{(2mv_0 - \gamma u_0)}{u_0\sqrt{4mk - \gamma^2}}\right].$$

(c)
$$R = \sqrt{u_0^2 + \frac{(2mv_0 - \gamma u_0)^2}{4mk - \gamma^2}} = 2\sqrt{\frac{m(ku_0^2 + \gamma u_0 v_0 + mv_0^2)}{4mk - \gamma^2}} = \sqrt{\frac{a + b\gamma}{4mk - \gamma^2}}.$$

It is evident that R increases (monotonically) without bound as $\gamma \to (2\sqrt{mk})^-$.

28.(a) The general solution is $u(t) = A\cos\sqrt{2}\,t + B\sin\sqrt{2}\,t$. Invoking the initial conditions, we have $u(t) = \sqrt{2}\sin\sqrt{2}\,t$.

(b)

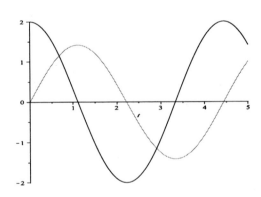

(c)

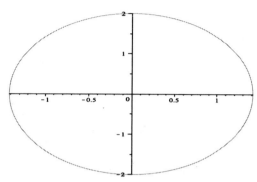

The condition $u'(0) = 2$ implies that $u(t)$ initially increases. Hence the phase point travels clockwise.

31. Based on Newton's second law, with the positive direction to the right, $\sum F = mu''$, where $\sum F = -ku - \gamma u'$. Hence the equation of motion is $mu'' + \gamma u' + ku = 0$. The only difference in this problem is that the equilibrium position is located at the unstretched configuration of the spring.

32.(a) The restoring force exerted by the spring is $F_s = -(ku + \epsilon u^3)$. The opposing viscous force is $F_d = -\gamma u'$. Based on Newton's second law, with the positive direction to the right, $F_s + F_d = mu''$. Hence the equation of motion is $mu'' + \gamma u' + ku + \epsilon u^3 = 0$.

(b) With the specified parameter values, the equation of motion is $u'' + u = 0$. The general solution of this ODE is $u(t) = A\cos t + B\sin t$. Invoking the initial conditions, the specific solution is $u(t) = \sin t$. Clearly, the amplitude is $R = 1$, and the period of the motion is $T = 2\pi$.

(c) Given $\epsilon = 0.1$, the equation of motion is $u'' + u + 0.1\,u^3 = 0$. A solution of the IVP can be generated numerically. We estimate $A = 0.98$ and $T = 6.07$.

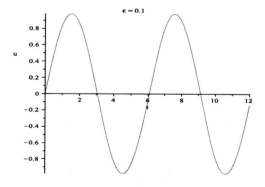

(d) For $\epsilon = 0.2$, $A = 0.96$ and $T = 5.90$. For $\epsilon = 0.3$, $A = 0.94$ and $T = 5.74$.

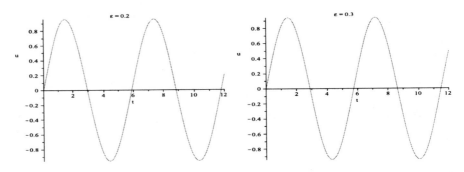

(e) The amplitude and period both seem to decrease.

(f) For $\epsilon = -0.1$, $A = 1.03$ and $T = 6.55$. For $\epsilon = -0.2$, $A = 1.06$ and $T = 6.90$. For $\epsilon = -0.3$, $A = 1.11$ and $T = 7.41$. The amplitude and period both seem to increase.

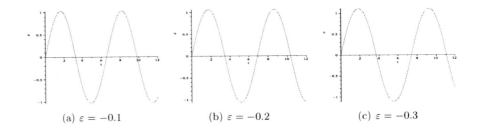

(a) $\varepsilon = -0.1$ (b) $\varepsilon = -0.2$ (c) $\varepsilon = -0.3$

3.8

2. We have $\sin(\alpha \pm \beta) = \sin \alpha \cos \beta \pm \cos \alpha \sin \beta$. Subtracting the two identities, we obtain $\sin(\alpha + \beta) - \sin(\alpha - \beta) = 2 \cos \alpha \sin \beta$. Setting $\alpha + \beta = 7t$ and $\alpha - \beta = 6t$, we get that $\alpha = 6.5t$ and $\beta = 0.5t$. This implies that $\sin 7t - \sin 6t = 2 \sin(t/2) \cos(13t/2)$.

3. Consider the trigonometric identities $\cos(\alpha \pm \beta) = \cos \alpha \cos \beta \mp \sin \alpha \sin \beta$. Adding the two identities, we get $\cos(\alpha - \beta) + \cos(\alpha + \beta) = 2 \cos \alpha \cos \beta$. Comparing the expressions, set $\alpha + \beta = 2\pi t$ and $\alpha - \beta = \pi t$. This means $\alpha = 3\pi t/2$ and $\beta = \pi t/2$. Upon substitution, we have $\cos(\pi t) + \cos(2\pi t) = 2 \cos(3\pi t/2) \cos(\pi t/2)$.

4. Adding the two identities $\sin(\alpha \pm \beta) = \sin \alpha \cos \beta \pm \cos \alpha \sin \beta$, it follows that $\sin(\alpha - \beta) + \sin(\alpha + \beta) = 2 \sin \alpha \cos \beta$. Setting $\alpha + \beta = 4t$ and $\alpha - \beta = 3t$, we have $\alpha = 7t/2$ and $\beta = t/2$. Hence $\sin 3t + \sin 4t = 2 \sin(7t/2) \cos(t/2)$.

6. Using MKS units, the spring constant is $k = 5(9.8)/0.1 = 490$ N/m, and the damping coefficient is $\gamma = 2/0.04 = 50$ N-s/m. The equation of motion is

$$5u'' + 50u' + 490u = 10 \sin(t/2).$$

The initial conditions are $u(0) = 0$ m and $u'(0) = 0.03$ m/s.

8.(a) The homogeneous solution is $u_c(t) = Ae^{-5t} \cos\sqrt{73}\,t + Be^{-5t} \sin\sqrt{73}\,t$. Based on the method of undetermined coefficients, the particular solution is

$$U(t) = \frac{1}{153281}\left[-160 \cos(t/2) + 3128 \sin(t/2)\right].$$

Hence the general solution of the ODE is $u(t) = u_c(t) + U(t)$. Invoking the initial conditions, we find that

$$A = 160/153281 \text{ and } B = 383443\sqrt{73}/1118951300.$$

Hence the response is

$$u(t) = \frac{1}{153281}\left[160\, e^{-5t} \cos\sqrt{73}\,t + \frac{383443\sqrt{73}}{7300} e^{-5t} \sin\sqrt{73}\,t\right] + U(t).$$

(b) $u_c(t)$ is the transient part and $U(t)$ is the steady state part of the response.

(c)

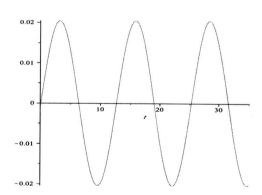

(d) The amplitude of the forced response is given by $R = 2/\Delta$, in which

$$\Delta = \sqrt{25(98 - \omega^2)^2 + 2500\,\omega^2}.$$

The maximum amplitude is attained when Δ is a minimum. Hence the amplitude is maximum at $\omega = 4\sqrt{3}$ rad/s.

9. The spring constant is $k = 12$ lb/ft and hence the equation of motion is

$$\frac{6}{32}u'' + 12u = 4 \cos 7t,$$

that is, $u'' + 64u = (64/3)\cos 7t$. The initial conditions are $u(0) = 0$ ft, $u'(0) = 0$ ft/s. The general solution is $u(t) = A\cos 8t + B\sin 8t + (64/45)\cos 7t$. Invoking the initial conditions, we have $u(t) = -(64/45)\cos 8t + (64/45)\cos 7t = (128/45)\sin(t/2)\sin(15t/2)$.

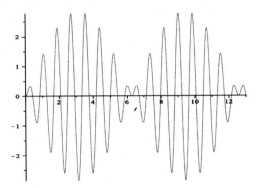

12. The equation of motion is $2u'' + u' + 3u = 3\cos 3t - 2\sin 3t$. Since the system is damped, the steady state response is equal to the particular solution. Using the method of undetermined coefficients, we obtain $u_{ss}(t) = (\sin 3t - \cos 3t)/6$. Further, we find that $R = \sqrt{2}/6$ and $\delta = \arctan(-1) = 3\pi/4$. Hence we can write $u_{ss}(t) = (\sqrt{2}/6)\cos(3t - 3\pi/4)$.

13.(a,b) Plug in $u(t) = R\cos(\omega t - \delta)$ into the equation $mu'' + \gamma u' + ku = F_0 \cos\omega t$, then use trigonometric identities and compare the coefficients of $\cos\omega t$ and $\sin\omega t$. The result follows.

(c) The amplitude of the steady-state response is given by

$$R = \frac{F_0}{\sqrt{m^2(\omega_0^2 - \omega^2)^2 + \gamma^2 \omega^2}}.$$

Since F_0 is constant, the amplitude is maximum when the denominator of R is minimum. Let $z = \omega^2$, and consider the function $f(z) = m^2(\omega_0^2 - z)^2 + \gamma^2 z$. Note that $f(z)$ is a quadratic, with minimum at $z = \omega_0^2 - \gamma^2/2m^2$. Hence the amplitude R attains a maximum at $\omega_{max}^2 = \omega_0^2 - \gamma^2/2m^2$. Furthermore, since $\omega_0^2 = k/m$,

$$\omega_{max}^2 = \omega_0^2 \left[1 - \frac{\gamma^2}{2km}\right].$$

Substituting $\omega^2 = \omega_{max}^2$ into the expression for the amplitude,

$$R = \frac{F_0}{\sqrt{\gamma^4/4m^2 + \gamma^2(\omega_0^2 - \gamma^2/2m^2)}} = \frac{F_0}{\sqrt{\omega_0^2\gamma^2 - \gamma^4/4m^2}} = \frac{F_0}{\gamma\omega_0\sqrt{1 - \gamma^2/4mk}}.$$

17.(a) The steady state part of the solution $U(t) = A\cos\omega t + B\sin\omega t$ may be found by substituting this expression into the differential equation and solving for A and B. We find that

$$A = \frac{32(2-\omega^2)}{64 - 63\omega^2 + 16\omega^4}, \qquad B = \frac{8\omega}{64 - 63\omega^2 + 16\omega^4}.$$

(b) The amplitude is
$$A = \frac{8}{\sqrt{64 - 63\omega^2 + 16\omega^4}}.$$

(c)

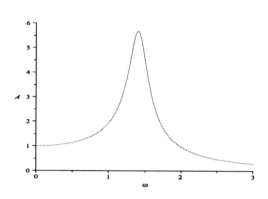

(d) See Problem 13. The amplitude is maximum when the denominator of A is minimum. That is, when $\omega = \omega_{max} = 3\sqrt{14}/8 \approx 1.4031$. Hence $A = 64/\sqrt{127}$.

18.(a) The homogeneous solution is $u_c(t) = A\cos t + B\sin t$. Based on the method of undetermined coefficients, the particular solution is
$$U(t) = \frac{3}{1 - \omega^2} \cos \omega t.$$

Hence the general solution of the ODE is $u(t) = u_c(t) + U(t)$. Invoking the initial conditions, we find that $A = 3/(\omega^2 - 1)$ and $B = 0$. Hence the response is
$$u(t) = \frac{3}{1 - \omega^2}[\cos \omega t - \cos t].$$

(b)

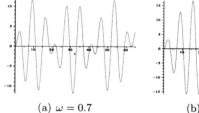 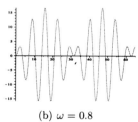 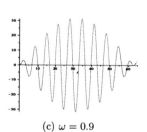

(a) $\omega = 0.7$ (b) $\omega = 0.8$ (c) $\omega = 0.9$

Note that
$$u(t) = \frac{6}{1 - \omega^2} \sin\left[\frac{(1-\omega)t}{2}\right] \sin\left[\frac{(\omega+1)t}{2}\right].$$

19.(a) The homogeneous solution is $u_c(t) = A\cos t + B\sin t$. Based on the method of undetermined coefficients, the particular solution is

$$U(t) = \frac{3}{1-\omega^2}\cos\omega t.$$

Hence the general solution is $u(t) = u_c(t) + U(t)$. Invoking the initial conditions, we find that $A = (\omega^2+2)/(\omega^2-1)$ and $B = 1$. Hence the response is

$$u(t) = \frac{1}{1-\omega^2}\left[3\cos\omega t - (\omega^2+2)\cos t\right] + \sin t.$$

(b)

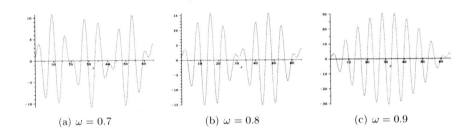

(a) $\omega = 0.7$ (b) $\omega = 0.8$ (c) $\omega = 0.9$

Note that

$$u(t) = \frac{6}{1-\omega^2}\sin\left[\frac{(1-\omega)t}{2}\right]\sin\left[\frac{(\omega+1)t}{2}\right] + \cos t + \sin t.$$

20.

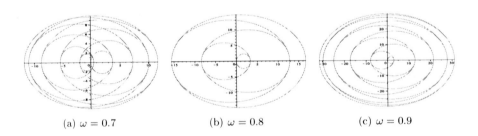

(a) $\omega = 0.7$ (b) $\omega = 0.8$ (c) $\omega = 0.9$

21.(a)

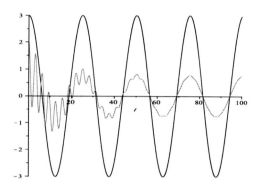

(b) Phase plot - u' vs u :

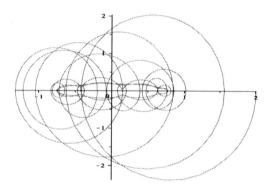

23.(a)

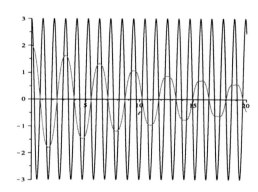

(b) Phase plot - u' vs u :

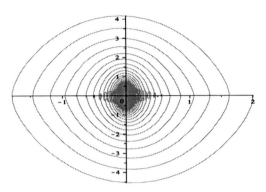

25.(a)

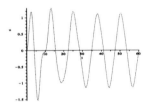
(a) $\omega = 0.5$

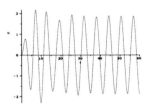
(b) $\omega = 0.75$

(c) $\omega = 1$

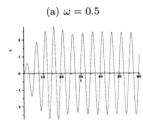
(d) $\omega = 1.25$

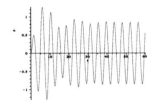
(e) $\omega = 1.5$

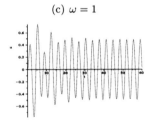
(f) $\omega = 1.75$

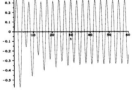
(g) $\omega = 2$

(b) R vs ω

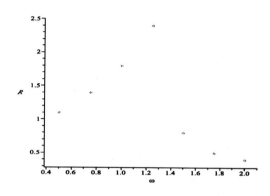

(c) The amplitude for a similar system with a linear spring is given by
$$R = \frac{5}{\sqrt{25 - 49\omega^2 + 25\omega^4}}.$$

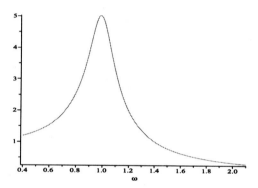

CHAPTER 4

Higher Order Linear Equations

4.1

1. The differential equation is in standard form. Its coefficients, as well as the function $g(t) = t$, are continuous everywhere. Hence solutions are valid on the entire real line.

3. Writing the equation in standard form, the coefficients are rational functions with singularities at $t = 0$ and $t = 1$. Hence the solutions are valid on the intervals $(-\infty, 0)$, $(0, 1)$, and $(1, \infty)$.

5. Writing the equation in standard form, the coefficients are rational functions with a singularity at $x_0 = 1$. Furthermore, $p_4(x) = \tan x/(x-1)$ is undefined, and hence not continuous, at $x_k = \pm(2k+1)\pi/2$, $k = 0, 1, 2, \ldots$. Hence solutions are defined on any interval that does not contain x_0 or x_k.

6. Writing the equation in standard form, the coefficients are rational functions with singularities at $x = \pm 2$. Hence the solutions are valid on the intervals $(-\infty, -2)$, $(-2, 2)$, and $(2, \infty)$.

7. Evaluating the Wronskian of the three functions, $W(f_1, f_2, f_3) = -14$. Hence the functions are linearly independent.

9. Evaluating the Wronskian of the four functions, $W(f_1, f_2, f_3, f_4) = 0$. Hence the functions are linearly dependent. To find a linear relation among the functions,

109

we need to find constants c_1, c_2, c_3, c_4, not all zero, such that

$$c_1 f_1(t) + c_2 f_2(t) + c_3 f_3(t) + c_4 f_4(t) = 0.$$

Collecting the common terms, we obtain

$$(c_2 + 2c_3 + c_4)t^2 + (2c_1 - c_3 + c_4)t + (-3c_1 + c_2 + c_4) = 0,$$

which results in three equations in four unknowns. Arbitrarily setting $c_4 = -1$, we can solve the equations $c_2 + 2c_3 = 1$, $2c_1 - c_3 = 1$, $-3c_1 + c_2 = 1$, to find that $c_1 = 2/7$, $c_2 = 13/7$, $c_3 = -3/7$. Hence

$$2f_1(t) + 13f_2(t) - 3f_3(t) - 7f_4(t) = 0.$$

10. Evaluating the Wronskian of the three functions, $W(f_1, f_2, f_3) = 156$. Hence the functions are linearly independent.

11. Substitution verifies that the functions are solutions of the differential equation. Furthermore, we have $W(1, \cos t, \sin t) = 1$.

12. Substitution verifies that the functions are solutions of the differential equation. Furthermore, we have $W(1, t, \cos t, \sin t) = 1$.

14. Substitution verifies that the functions are solutions of the differential equation. Furthermore, we have $W(1, t, e^{-t}, t e^{-t}) = e^{-2t}$.

15. Substitution verifies that the functions are solutions of the differential equation. Furthermore, we have $W(1, x, x^3) = 6x$.

16. Substitution verifies that the functions are solutions of the differential equation. Furthermore, we have $W(x, x^2, 1/x) = 6/x$.

18. The operation of taking a derivative is linear, and hence $(c_1 y_1 + c_2 y_2)^{(k)} = c_1 y_1^{(k)} + c_2 y_2^{(k)}$. It follows that

$$L[c_1 y_1 + c_2 y_2] = c_1 y_1^{(n)} + c_2 y_2^{(n)} + p_1(c_1 y_1^{(n-1)} + c_2 y_2^{(n-1)}) + \ldots + p_n(c_1 y_1 + c_2 y_2).$$

Rearranging the terms, we obtain $L[c_1 y_1 + c_2 y_2] = c_1 L[y_1] + c_2 L[y_2]$. Since y_1 and y_2 are solutions, $L[c_1 y_1 + c_2 y_2] = 0$. The rest follows by induction.

20.(a) Let $f(t)$ and $g(t)$ be arbitrary functions. Then $W(f, g) = fg' - f'g$. Hence $W'(f, g) = f'g' + fg'' - f''g - f'g' = fg'' - f''g$. That is,

$$W'(f, g) = \begin{vmatrix} f & g \\ f'' & g'' \end{vmatrix}.$$

Now expand the 3-by-3 determinant as

$$W(y_1, y_2, y_3) = y_1 \begin{vmatrix} y_2' & y_3' \\ y_2'' & y_3'' \end{vmatrix} - y_2 \begin{vmatrix} y_1' & y_3' \\ y_1'' & y_3'' \end{vmatrix} + y_3 \begin{vmatrix} y_1' & y_2' \\ y_1'' & y_2'' \end{vmatrix}.$$

Differentiating, we obtain

$$W'(y_1, y_2, y_3) = y_1' \begin{vmatrix} y_2' & y_3' \\ y_2'' & y_3'' \end{vmatrix} - y_2' \begin{vmatrix} y_1' & y_3' \\ y_1'' & y_3'' \end{vmatrix} + y_3' \begin{vmatrix} y_1' & y_2' \\ y_1'' & y_2'' \end{vmatrix} +$$
$$+ y_1 \begin{vmatrix} y_2' & y_3' \\ y_2''' & y_3''' \end{vmatrix} - y_2 \begin{vmatrix} y_1' & y_3' \\ y_1''' & y_3''' \end{vmatrix} + y_3 \begin{vmatrix} y_1' & y_2' \\ y_1''' & y_2''' \end{vmatrix}.$$

The second line follows from the observation above. Now we find that

$$W'(y_1, y_2, y_3) = \begin{vmatrix} y_1' & y_2' & y_3' \\ y_1' & y_2' & y_3' \\ y_1'' & y_2'' & y_3'' \end{vmatrix} + \begin{vmatrix} y_1 & y_2 & y_3 \\ y_1' & y_2' & y_3' \\ y_1''' & y_2''' & y_3''' \end{vmatrix}.$$

Hence the assertion is true, since the first determinant is equal to zero.

(b) Based on the properties of determinants,

$$p_2(t)p_3(t)W' = \begin{vmatrix} p_3 y_1 & p_3 y_2 & p_3 y_3 \\ p_2 y_1' & p_2 y_2' & p_2 y_3' \\ y_1''' & y_2''' & y_3''' \end{vmatrix}.$$

Adding the first two rows to the third row does not change the value of the determinant. Since the functions are assumed to be solutions of the given ODE, addition of the rows results in

$$p_2(t)p_3(t)W' = \begin{vmatrix} p_3 y_1 & p_3 y_2 & p_3 y_3 \\ p_2 y_1' & p_2 y_2' & p_2 y_3' \\ -p_1 y_1'' & -p_1 y_2'' & -p_1 y_3'' \end{vmatrix}.$$

It follows that $p_2(t)p_3(t)W' = -p_1(t)p_2(t)p_3(t)W$. As long as the coefficients are not zero, we obtain $W' = -p_1(t)W$.

(c) The first order equation $W' = -p_1(t)W$ is linear, with integrating factor $\mu(t) = e^{\int p_1(t)dt}$. Hence $W(t) = ce^{-\int p_1(t)dt}$. Furthermore, $W(t)$ is zero only if $c = 0$.

(d) It can be shown, by mathematical induction, that

$$W'(y_1, y_2, \ldots, y_n) = \begin{vmatrix} y_1 & y_2 & \cdots & y_{n-1} & y_n \\ y_1' & y_2' & \cdots & y_{n-1}' & y_n' \\ \vdots & & & & \vdots \\ y_1^{(n-2)} & y_2^{(n-2)} & \cdots & y_{n-1}^{(n-2)} & y_n^{(n-2)} \\ y_1^{(n)} & y_2^{(n)} & \cdots & y_{n-1}^{(n)} & y_n^{(n)} \end{vmatrix}.$$

Based on the reasoning in part (b), it follows that

$$p_2(t)p_3(t)\ldots p_n(t)W' = -p_1(t)p_2(t)p_3(t)\ldots p_n(t)W,$$

and hence $W' = -p_1(t)W$.

21. Inspection of the coefficients reveals that $p_1(t) = 2$. Based on Problem 20, we find that $W' = -2W$, and hence $W = ce^{-2t}$.

22. Inspection of the coefficients reveals that $p_1(t) = 0$. Based on Problem 20, we find that $W' = 0$, and hence $W = c$.

24. Writing the equation in standard form, we find that $p_1(t) = 1/t$. Using Abel's formula, the Wronskian has the form $W(t) = c\,e^{-\int 1/t\,dt} = c e^{-\ln t} = c/t$.

26. Let $y(t) = y_1(t)v(t)$. Then $y' = y_1'v + y_1v'$, $y'' = y_1''v + 2y_1'v' + y_1v''$, and $y''' = y_1'''v + 3y_1''v' + 3y_1'v'' + y_1v'''$. Substitution into the ODE results in

$$y_1'''v + 3y_1''v' + 3y_1'v'' + y_1v''' + p_1\left[y_1''v + 2y_1'v' + y_1v''\right] +$$
$$+ p_2\left[y_1'v + y_1v'\right] + p_3 y_1 v = 0.$$

Since y_1 is assumed to be a solution, all terms containing the factor $v(t)$ vanish. Hence

$$y_1 v''' + \left[p_1 y_1 + 3y_1'\right]v'' + \left[3y_1'' + 2p_1 y_1' + p_2 y_1\right]v' = 0,$$

which is a second order ODE in the variable $u = v'$.

28. First write the equation in standard form:

$$y''' - 3\frac{t+2}{t(t+3)}y'' + 6\frac{t+1}{t^2(t+3)}y' - \frac{6}{t^2(t+3)}y = 0.$$

Let $y(t) = t^2 v(t)$. Substitution into the given ODE results in

$$t^2 v''' + 3\frac{t(t+4)}{t+3}v'' = 0.$$

Set $w = v''$. Then w is a solution of the first order differential equation

$$w' + 3\frac{t+4}{t(t+3)}w = 0.$$

This equation is linear, with integrating factor $\mu(t) = t^4/(t+3)$. The general solution is $w = c(t+3)/t^4$. Integrating twice, $v(t) = c_1 t^{-1} + c_1 t^{-2} + c_2 t + c_3$. Hence $y(t) = c_1 t + c_1 + c_2 t^3 + c_3 t^2$. Finally, since $y_1(t) = t^2$ and $y_2(t) = t^3$ are given solutions, the third independent solution is $y_3(t) = c_1 t + c_1$.

4.2

1. The magnitude of $1 + i$ is $R = \sqrt{2}$ and the polar angle is $\pi/4$. Hence the polar form is given by $1 + i = \sqrt{2}\,e^{i\pi/4}$.

3. The magnitude of -3 is $R = 3$ and the polar angle is π. Hence $-3 = 3\,e^{i\pi}$.

4. The magnitude of $-i$ is $R = 1$ and the polar angle is $3\pi/2$. Hence $-i = e^{3\pi i/2}$.

5. The magnitude of $\sqrt{3} - i$ is $R = 2$ and the polar angle is $-\pi/6 = 11\pi/6$. Hence the polar form is given by $\sqrt{3} - i = 2\,e^{11\pi i/6}$.

6. The magnitude of $-1-i$ is $R = \sqrt{2}$ and the polar angle is $5\pi/4$. Hence the polar form is given by $-1-i = \sqrt{2}\, e^{5\pi i/4}$.

7. Writing the complex number in polar form, $1 = e^{2m\pi i}$, where m may be any integer. Thus $1^{1/3} = e^{2m\pi i/3}$. Setting $m = 0, 1, 2$ successively, we obtain the three roots as $1^{1/3} = 1$, $1^{1/3} = e^{2\pi i/3}$, $1^{1/3} = e^{4\pi i/3}$. Equivalently, the roots can also be written as 1, $\cos(2\pi/3) + i\sin(2\pi/3) = (-1+i\sqrt{3})/2$, $\cos(4\pi/3) + i\sin(4\pi/3) = (-1-i\sqrt{3})/2$.

9. Writing the complex number in polar form, $1 = e^{2m\pi i}$, where m may be any integer. Thus $1^{1/4} = e^{2m\pi i/4}$. Setting $m = 0, 1, 2, 3$ successively, we obtain the three roots as $1^{1/4} = 1$, $1^{1/4} = e^{\pi i/2}$, $1^{1/4} = e^{\pi i}$, $1^{1/4} = e^{3\pi i/2}$. Equivalently, the roots can also be written as 1, $\cos(\pi/2) + i\sin(\pi/2) = i$, $\cos(\pi) + i\sin(\pi) = -1$, $\cos(3\pi/2) + i\sin(3\pi/2) = -i$.

10. In polar form, $2(\cos \pi/3 + i \sin \pi/3) = 2\, e^{i(\pi/3 + 2m\pi)}$, in which m is any integer. Thus $[2(\cos \pi/3 + i \sin \pi/3)]^{1/2} = 2^{1/2}\, e^{i(\pi/6 + m\pi)}$. With $m = 0$, one square root is given by $2^{1/2}\, e^{i\pi/6} = (\sqrt{3} + i)/\sqrt{2}$. With $m = 1$, the other root is given by $2^{1/2}\, e^{i7\pi/6} = (-\sqrt{3} - i)/\sqrt{2}$.

11. The characteristic equation is $r^3 - r^2 - r + 1 = 0$. The roots are $r = -1, 1, 1$. One root is repeated, hence the general solution is $y = c_1 e^{-t} + c_2 e^t + c_3 t e^t$.

13. The characteristic equation is $r^3 - 2r^2 - r + 2 = 0$, with roots $r = -1, 1, 2$. The roots are real and distinct, so the general solution is $y = c_1 e^{-t} + c_2 e^t + c_3 e^{2t}$.

14. The characteristic equation can be written as $r^2(r^2 - 4r + 4) = 0$. The roots are $r = 0, 0, 2, 2$. There are two repeated roots, and hence the general solution is given by $y = c_1 + c_2 t + c_3 e^{2t} + c_4 t e^{2t}$.

16. The characteristic equation can be written as $(r^2 - 1)(r^2 - 4) = 0$. The roots are given by $r = \pm 1, \pm 2$. The roots are real and distinct, hence the general solution is $y = c_1 e^{-t} + c_2 e^t + c_3 e^{-2t} + c_4 e^{2t}$.

17. The characteristic equation can be written as $(r^2 - 1)^3 = 0$. The roots are given by $r = \pm 1$, each with multiplicity three. Hence the general solution is $y = c_1 e^{-t} + c_2 t e^{-t} + c_3 t^2 e^{-t} + c_4 e^t + c_5 t e^t + c_6 t^2 e^t$.

18. The characteristic equation can be written as $r^2(r^4 - 1) = 0$. The roots are given by $r = 0, 0, \pm 1, \pm i$. The general solution is $y = c_1 + c_2 t + c_3 e^{-t} + c_4 e^t + c_5 \cos t + c_6 \sin t$.

19. The characteristic equation can be written as $r(r^4 - 3r^3 + 3r^2 - 3r + 2) = 0$. Examining the coefficients, it follows that $r^4 - 3r^3 + 3r^2 - 3r + 2 = (r-1)(r-2)(r^2+1)$. Hence the roots are $r = 0, 1, 2, \pm i$. The general solution of the ODE is given by $y = c_1 + c_2 e^t + c_3 e^{2t} + c_4 \cos t + c_5 \sin t$.

20. The characteristic equation can be written as $r(r^3 - 8) = 0$, with roots $r = 0$,

$2\,e^{2m\pi i/3}$, $m = 0, 1, 2$. That is, $r = 0, 2, -1 \pm i\sqrt{3}$. Hence the general solution is $y = c_1 + c_2 e^{2t} + e^{-t}\left[c_3 \cos\sqrt{3}\,t + c_4 \sin\sqrt{3}\,t\right]$.

21. The characteristic equation can be written as $(r^4 + 4)^2 = 0$. The roots of the equation $r^4 + 4 = 0$ are $r = 1 \pm i$, $-1 \pm i$. Each of these roots has multiplicity two. The general solution is $y = e^t\left[c_1 \cos t + c_2 \sin t\right] + te^t\left[c_3 \cos t + c_4 \sin t\right] + e^{-t}\left[c_5 \cos t + c_6 \sin t\right] + te^{-t}\left[c_7 \cos t + c_8 \sin t\right]$.

22. The characteristic equation can be written as $(r^2 + 1)^2 = 0$. The roots are given by $r = \pm i$, each with multiplicity two. The general solution is $y = c_1 \cos t + c_2 \sin t + t\left[c_3 \cos t + c_4 \sin t\right]$.

24. The characteristic equation is $r^3 + 5r^2 + 6r + 2 = 0$. Examining the coefficients, we find that $r^3 + 5r^2 + 6r + 2 = (r+1)(r^2 + 4r + 2)$. Hence the roots are deduced as $r = -1, -2 \pm \sqrt{2}$. The general solution is $y = c_1 e^{-t} + c_2 e^{(-2+\sqrt{2})t} + c_3 e^{(-2-\sqrt{2})t}$.

25. The characteristic equation is $18r^3 + 21r^2 + 14r + 4 = 0$. By examining the first and last coefficients, we find that $18r^3 + 21r^2 + 14r + 4 = (2r+1)(9r^2 + 6r + 4)$. Hence the roots are $r = -1/2, (-1 \pm \sqrt{3}i)/3$. The general solution of the ODE is given by $y = c_1 e^{-t/2} + e^{-t/3}\left[c_2 \cos(t/\sqrt{3}) + c_3 \sin(t/\sqrt{3})\right]$.

26. The characteristic equation is $r^4 - 7r^3 + 6r^2 + 30r - 36 = 0$. By examining the first and last coefficients, we find that $r^4 - 7r^3 + 6r^2 + 30r - 36 = (r-3)(r+2)(r^2 - 6r + 6)$. The roots are $r = -2, 3, 3 \pm \sqrt{3}$. The general solution is $y = c_1 e^{-2t} + c_2 e^{3t} + c_3 e^{(3-\sqrt{3})t} + c_4 e^{(3+\sqrt{3})t}$.

28. The characteristic equation is $r^4 + 6r^3 + 17r^2 + 22r + 14 = 0$. It can be shown that $r^4 + 6r^3 + 17r^2 + 22r + 14 = (r^2 + 2r + 2)(r^2 + 4r + 7)$. Hence the roots are $r = -1 \pm i$, $-2 \pm i\sqrt{3}$. The general solution of the euqation is $y = e^{-t}(c_1 \cos t + c_2 \sin t) + e^{-2t}(c_3 \cos\sqrt{3}\,t + c_4 \sin\sqrt{3}\,t)$.

32. The characteristic equation is $r^3 - r^2 + r - 1 = 0$, with roots $r = 1$, $\pm i$. Hence the general solution is $y(t) = c_1 e^t + c_2 \cos t + c_3 \sin t$. Invoking the initial conditions, we obtain the system of equations $c_1 + c_2 = 2$, $c_1 + c_3 = -1$, $c_1 - c_2 = -2$, with solution $c_1 = 0$, $c_2 = 2$, $c_3 = -1$. Therefore the solution of the initial value problem is $y(t) = 2\cos t - \sin t$, which oscillates as $t \to \infty$.

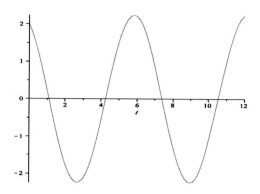

33. The characteristic equation is $2r^4 - r^3 - 9r^2 + 4r + 4 = 0$, with roots $r = -1/2, 1, \pm 2$. Hence the general solution is $y(t) = c_1 e^{-t/2} + c_2 e^t + c_3 e^{-2t} + c_4 e^{2t}$. Applying the initial conditions, we obtain the system of equations $c_1 + c_2 + c_3 + c_4 = -2$, $-c_1/2 + c_2 - 2c_3 + 2c_4 = 0$, $c_1/4 + c_2 + 4c_3 + 4c_4 = -2$, $-c_1/8 + c_2 - 8c_3 + 8c_4 = 0$, with solution $c_1 = -16/15$, $c_2 = -2/3$, $c_3 = -1/6$, $c_4 = -1/10$. Therefore the solution of the initial value problem is $y(t) = -(16/15)e^{-t/2} - (2/3)e^t - e^{-2t}/6 - e^{2t}/10$. The solution decreases without bound.

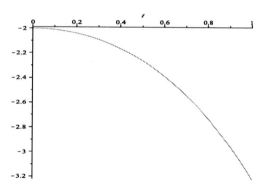

35. The characteristic equation is $6r^3 + 5r^2 + r = 0$, with roots $r = 0, -1/3, -1/2$. The general solution is $y(t) = c_1 + c_2 e^{-t/3} + c_3 e^{-t/2}$. Invoking the initial conditions, we require that $c_1 + c_2 + c_3 = -2$, $-c_2/3 - c_3/2 = 2$, $c_2/9 + c_3/4 = 0$. The solution is $c_1 = 8$, $c_2 = -18$, $c_3 = 8$. Therefore the solution of the initial value problem is $y(t) = 8 - 18e^{-t/3} + 8e^{-t/2}$. It approaches 8 as $t \to \infty$.

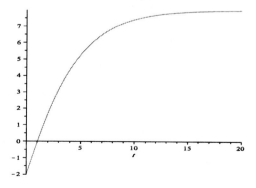

36. The general solution is derived in Problem 28 as

$$y(t) = e^{-t}\left[c_1 \cos t + c_2 \sin t\right] + e^{-2t}\left[c_3 \cos \sqrt{3}\,t + c_4 \sin \sqrt{3}\,t\right].$$

Invoking the initial conditions, we obtain the system of equations

$$c_1 + c_3 = 1$$
$$-c_1 + c_2 - 2c_3 + \sqrt{3}\,c_4 = -2$$
$$-2c_2 + c_3 - 4\sqrt{3}\,c_4 = 0$$
$$2c_1 + 2c_2 + 10c_3 + 9\sqrt{3}\,c_4 = 3$$

with solution $c_1 = 21/13$, $c_2 = -38/13$, $c_3 = -8/13$, $c_4 = 17\sqrt{3}/39$.

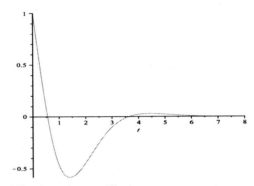

The solution is a rapidly decaying oscillation.

40.(a) Suppose that $c_1 e^{r_1 t} + c_2 e^{r_2 t} + \ldots + c_n e^{r_n t} = 0$, and each of the r_k are real and different. Multiplying this equation by $e^{-r_1 t}$, we obtain that $c_1 + c_2 e^{(r_2-r_1)t} + \ldots + c_n e^{(r_n-r_1)t} = 0$. Differentiation results in

$$c_2(r_2 - r_1)e^{(r_2-r_1)t} + \ldots + c_n(r_n - r_1)e^{(r_n-r_1)t} = 0.$$

(b) Now multiplying the latter equation by $e^{-(r_2-r_1)t}$, and differentiating, we obtain

$$c_3(r_3 - r_2)(r_3 - r_1)e^{(r_3-r_2)t} + \ldots + c_n(r_n - r_2)(r_n - r_1)e^{(r_n-r_2)t} = 0.$$

(c) Following the above steps in a similar manner, it follows that

$$c_n(r_n - r_{n-1})\ldots(r_n - r_1)e^{(r_n-r_{n-1})t} = 0.$$

Since these equations hold for all t, and all the r_k are different, we have $c_n = 0$. Hence $c_1 e^{r_1 t} + c_2 e^{r_2 t} + \ldots + c_{n-1} e^{r_{n-1} t} = 0$, $-\infty < t < \infty$.

(d) The same procedure can now be repeated, successively, to show that $c_1 = c_2 = \ldots = c_n = 0$.

41.(a) Recall the derivative formula

$$\frac{d^n}{dx^n}(uv) = \binom{n}{0} v \frac{d^n u}{dx^n} + \binom{n}{1} \frac{dv}{dx} \frac{d^{n-1} u}{dx^{n-1}} + \ldots + \binom{n}{n} \frac{d^n v}{dx^n} u.$$

Let $u = (r - r_1)^s$ and $v = q(r)$. Note that

$$\frac{d^n}{dr^n}[(r-r_1)^s] = s \cdot (s-1) \ldots (s-n+1)(r-r_1)^{s-n}$$

and

$$\frac{d^s}{dr^s}[(r-r_1)^s] = s\,!\,.$$

Therefore

$$\frac{d^n}{dr^n}[(r-r_1)^s q(r)]\Big|_{r=r_1} = 0$$

only if $n < s$, since it is assumed that $q(r_1) \neq 0$.

(b) Differential operators commute, so that

$$\frac{\partial}{\partial r}\left(\frac{d^k}{dt^k} e^{rt}\right) = \frac{d^k}{dt^k}\left(\frac{\partial e^{rt}}{\partial r}\right) = \frac{d^k}{dt^k}(t\,e^{rt}).$$

Likewise,

$$\frac{\partial^j}{\partial r^j}\left(\frac{d^k}{dt^k} e^{rt}\right) = \frac{d^k}{dt^k}\left(\frac{\partial^j e^{rt}}{\partial r^j}\right) = \frac{d^k}{dt^k}(t^j\,e^{rt}).$$

It follows that

$$\frac{\partial^j}{\partial r^j} L\left[e^{rt}\right] = L\left[t^j\,e^{rt}\right].$$

(c) From Eq. (i), we have

$$\frac{\partial^j}{\partial r^j}\left[e^{rt}\,Z(r)\right] = L\left[t^j\,e^{rt}\right].$$

Based on the product formula in part (a),

$$\frac{\partial^j}{\partial r^j}\left[e^{rt}\,Z(r)\right]\Big|_{r=r_1} = 0$$

if $j < s$. Therefore $L\left[t^j\,e^{r_1 t}\right] = 0$ if $j < s$.

4.3

2. The general solution of the homogeneous equation is $y_c = c_1 e^t + c_2 e^{-t} + c_3 \cos t + c_4 \sin t$. Let $g_1(t) = 3t$ and $g_2(t) = \cos t$. By inspection, we find that $Y_1(t) = -3t$. Since $g_2(t)$ is a solution of the homogeneous equation, set $Y_2(t) = t(A \cos t + B \sin t)$. Substitution into the given ODE and comparing the coefficients of similar term results in $A = 0$ and $B = -1/4$. Hence the general solution of the nonhomogeneous problem is $y(t) = y_c(t) - 3t - t \sin t /4$.

3. The characteristic equation corresponding to the homogeneous problem can be written as $(r+1)(r^2+1) = 0$. The solution of the homogeneous equation is $y_c = c_1 e^{-t} + c_2 \cos t + c_3 \sin t$. Let $g_1(t) = e^{-t}$ and $g_2(t) = 4t$. Since $g_1(t)$ is a solution of the homogeneous equation, set $Y_1(t) = Ate^{-t}$. Substitution into the ODE results in $A = 1/2$. Now let $Y_2(t) = Bt + C$. We find that $B = -C = 4$. Hence the general solution of the nonhomogeneous problem is $y(t) = y_c(t) + te^{-t}/2 + 4(t-1)$.

4. The characteristic equation corresponding to the homogeneous problem can be written as $r(r+1)(r-1) = 0$. The solution of the homogeneous equation is $y_c = c_1 + c_2 e^t + c_3 e^{-t}$. Since $g(t) = 2 \sin t$ is not a solution of the homogeneous problem, we can set $Y(t) = A \cos t + B \sin t$. Substitution into the ODE results in $A = 1$ and $B = 0$. Thus the general solution is $y(t) = c_1 + c_2 e^t + c_3 e^{-t} + \cos t$.

6. The characteristic equation corresponding to the homogeneous problem can be written as $(r^2+1)^2 = 0$. It follows that $y_c = c_1 \cos t + c_2 \sin t + t(c_3 \cos t + c_4 \sin t)$. Since $g(t)$ is not a solution of the homogeneous problem, set $Y(t) = A + B \cos 2t + C \sin 2t$. Substitution into the ODE results in $A = 3$, $B = 1/9$, $C = 0$. Thus the general solution is $y(t) = y_c(t) + 3 + \cos 2t /9$.

7. The characteristic equation corresponding to the homogeneous problem can be written as $r^3(r^3+1) = 0$. Thus the homogeneous solution is

$$y_c = c_1 + c_2 t + c_3 t^2 + c_4 e^{-t} + e^{t/2}\left[c_5 \cos(\sqrt{3}\,t/2) + c_5 \sin(\sqrt{3}\,t/2)\right].$$

Note the $g(t) = t$ is a solution of the homogenous problem. Consider a particular solution of the form $Y(t) = t^3(At + B)$. Substitution into the ODE gives us that $A = 1/24$ and $B = 0$. Thus the general solution is $y(t) = y_c(t) + t^4/24$.

8. The characteristic equation corresponding to the homogeneous problem can be written as $r^3(r+1) = 0$. Hence the homogeneous solution is $y_c = c_1 + c_2 t + c_3 t^2 + c_4 e^{-t}$. Since $g(t)$ is not a solution of the homogeneous problem, set $Y(t) = A \cos 2t + B \sin 2t$. Substitution into the ODE results in $A = 1/40$ and $B = 1/20$. Thus the general solution is $y(t) = y_c(t) + (\cos 2t + 2 \sin 2t)/40$.

10. From Problem 22 in Section 4.2, the homogeneous solution is $y_c = c_1 \cos t + c_2 \sin t + t[c_3 \cos t + c_4 \sin t]$. Since $g(t)$ is not a solution of the homogeneous problem, substitute $Y(t) = At + B$ into the ODE to obtain $A = 3$ and $B = 4$. Thus the general solution is $y(t) = y_c(t) + 3t + 4$. Invoking the initial conditions, we find that $c_1 = -4$, $c_2 = -4$, $c_3 = 1$, $c_4 = -3/2$. Therefore the solution of the initial value problem is $y(t) = (t-4) \cos t - (3t/2+4) \sin t + 3t + 4$.

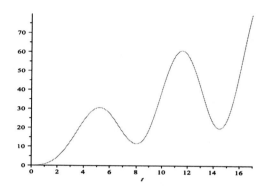

11. The characteristic equation can be written as $r(r^2 - 3r + 2) = 0$. Hence the homogeneous solution is $y_c = c_1 + c_2 e^t + c_3 e^{2t}$. Let $g_1(t) = e^t$ and $g_2(t) = t$. Note that g_1 is a solution of the homogeneous problem. Set $Y_1(t) = Ate^t$. Substitution into the ODE results in $A = -1$. Now let $Y_2(t) = Bt^2 + Ct$. Substitution into the ODE results in $B = 1/4$ and $C = 3/4$. Therefore the general solution is $y(t) = c_1 + c_2 e^t + c_3 e^{2t} - te^t + (t^2 + 3t)/4$. Invoking the initial conditions, we find that $c_1 = 1$, $c_2 = c_3 = 0$. The solution of the initial value problem is $y(t) = 1 - te^t + (t^2 + 3t)/4$.

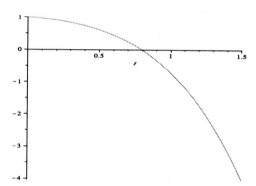

12. The characteristic equation can be written as $(r - 1)(r + 3)(r^2 + 4) = 0$. Hence the homogeneous solution is $y_c = c_1 e^t + c_2 e^{-3t} + c_3 \cos 2t + c_4 \sin 2t$. None of the terms in $g(t)$ is a solution of the homogeneous problem. Therefore we can assume a form $Y(t) = Ae^{-t} + B \cos t + C \sin t$. Substitution into the ODE results in the values $A = 1/20$, $B = -2/5$, $C = -4/5$. Hence the general solution is $y(t) = c_1 e^t + c_2 e^{-3t} + c_3 \cos 2t + c_4 \sin 2t + e^{-t}/20 - (2 \cos t + 4 \sin t)/5$. Invoking the initial conditions, we find that $c_1 = 81/40$, $c_2 = 73/520$, $c_3 = 77/65$, $c_4 = -49/130$.

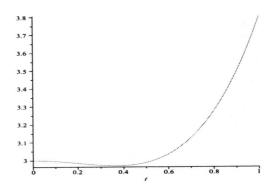

14. From Problem 4, the homogeneous solution is $y_c = c_1 + c_2 e^t + c_3 e^{-t}$. Consider the terms $g_1(t) = te^{-t}$ and $g_2(t) = 2\cos t$. Note that since $r = -1$ is a simple root of the characteristic equation, we set $Y_1(t) = t(At+B)e^{-t}$. The function $2\cos t$ is not a solution of the homogeneous equation. We set $Y_2(t) = C\cos t + D\sin t$. Hence the particular solution has the form $Y(t) = t(At+B)e^{-t} + C\cos t + D\sin t$.

15. The characteristic equation can be written as $(r^2 - 1)^2 = 0$. The roots are given as $r = \pm 1$, each with multiplicity two. Hence the solution of the homogeneous problem is $y_c = c_1 e^t + c_2 t e^t + c_3 e^{-t} + c_4 t e^{-t}$. Let $g_1(t) = e^t$ and $g_2(t) = \sin t$. The function e^t is a solution of the homogeneous problem. Since $r = 1$ has multiplicity two, we set $Y_1(t) = At^2 e^t$. The function $\sin t$ is not a solution of the homogeneous equation. We can set $Y_2(t) = B\cos t + C\sin t$. Hence the particular solution has the form $Y(t) = At^2 e^t + B\cos t + C\sin t$.

16. The characteristic equation can be written as $r^2(r^2 + 4) = 0$, and the roots are $r = 0, \pm 2i$. The root $r = 0$ has multiplicity *two*, hence the homogeneous solution is $y_c = c_1 + c_2 t + c_3 \cos 2t + c_4 \sin 2t$. The functions $g_1(t) = \sin 2t$ and $g_2(t) = 4$ are solutions of the homogenous equation. The complex roots have multiplicity one, therefore we need to set $Y_1(t) = At\cos 2t + Bt\sin 2t$. Now $g_2(t) = 4$ is associated with the double root $r = 0$, so we set $Y_2(t) = Ct^2$. Finally, $g_3(t) = te^t$ (and its derivatives) is independent of the homogeneous solution. Therefore set $Y_3(t) = (Dt + E)e^t$. Conclude that the particular solution has the form $Y(t) = At\cos 2t + Bt\sin 2t + Ct^2 + (Dt + E)e^t$.

18. The characteristic equation can be written as $r^2(r^2 + 2r + 2) = 0$, with roots $r = 0$, with multiplicity two, and $r = -1 \pm i$. This means that the homogeneous solution is $y_c = c_1 + c_2 t + c_3 e^{-t}\cos t + c_4 e^{-t}\sin t$. The function $g_1(t) = 3e^t + 2te^{-t}$, and all of its derivatives, is independent of the homogeneous solution. Therefore set $Y_1(t) = Ae^t + (Bt + C)e^{-t}$. Now $g_2(t) = e^{-t}\sin t$ is a solution of the homogeneous equation, associated with the complex roots. We need to set $Y_2(t) = t(De^{-t}\cos t + Ee^{-t}\sin t)$. It follows that the particular solution has the form $Y(t) = Ae^t + (Bt + C)e^{-t} + t(De^{-t}\cos t + Ee^{-t}\sin t)$.

19. Differentiating $y = u(t)v(t)$, successively, we have
$$y' = u'v + uv'$$
$$y'' = u''v + 2u'v' + uv''$$
$$\vdots$$
$$y^{(n)} = \sum_{j=0}^{n} \binom{n}{j} u^{(n-j)} v^{(j)}$$

Setting $v(t) = e^{\alpha t}$, $v^{(j)} = \alpha^j e^{\alpha t}$. So for any $p = 1, 2, \ldots, n$,
$$y^{(p)} = e^{\alpha t} \sum_{j=0}^{p} \binom{p}{j} \alpha^j u^{(p-j)}.$$

It follows that
$$L\left[e^{\alpha t} u\right] = e^{\alpha t} \sum_{p=0}^{n} \left[a_{n-p} \sum_{j=0}^{p} \binom{p}{j} \alpha^j u^{(p-j)} \right] \qquad (*).$$

It is evident that the right hand side of Eq. (*) is of the form
$$e^{\alpha t} \left[k_0 u^{(n)} + k_1 u^{(n-1)} + \ldots + k_{n-1} u' + k_n u \right].$$

Hence the operator equation $L\left[e^{\alpha t} u\right] = e^{\alpha t}(b_0 t^m + b_1 t^{m-1} + \ldots + b_{m-1} t + b_m)$ can be written as
$$k_0 u^{(n)} + k_1 u^{(n-1)} + \ldots + k_{n-1} u' + k_n u = b_0 t^m + b_1 t^{m-1} + \ldots + b_{m-1} t + b_m.$$

The coefficients k_i, $i = 0, 1, \ldots, n$ can be determined by collecting the like terms in the double summation in Eq. (*). For example, k_0 is the coefficient of $u^{(n)}$. The only term that contains $u^{(n)}$ is when $p = n$ and $j = 0$. Hence $k_0 = a_0$. On the other hand, k_n is the coefficient of $u(t)$. The inner summation in (*) contains terms with u, given by $\alpha^p u$ (when $j = p$), for each $p = 0, 1, \ldots, n$. Hence
$$k_n = \sum_{p=0}^{n} a_{n-p} \alpha^p.$$

21.(a) Clearly, e^{2t} is a solution of $y' - 2y = 0$, and te^{-t} is a solution of the differential equation $y'' + 2y' + y = 0$. The latter ODE has characteristic equation $(r+1)^2 = 0$. Hence $(D-2)\left[3e^{2t}\right] = 3(D-2)\left[e^{2t}\right] = 0$ and $(D+1)^2\left[te^{-t}\right] = 0$. Furthermore, we have $(D-2)(D+1)^2\left[te^{-t}\right] = (D-2)[0] = 0$, and $(D-2)(D+1)^2\left[3e^{2t}\right] = (D+1)^2(D-2)\left[3e^{2t}\right] = (D+1)^2[0] = 0$.

(b) Based on part (a),
$$(D-2)(D+1)^2 \left[(D-2)^3(D+1)Y\right] = (D-2)(D+1)^2 \left[3e^{2t} - te^{-t}\right] = 0,$$

since the operators are linear. The implied operations are associative and commutative. Hence $(D-2)^4(D+1)^3 Y = 0$. The operator equation corresponds to the solution of a linear homogeneous ODE with characteristic equation $(r-2)^4(r+1)^3 = 0$. The roots are $r = 2$, with multiplicity 4 and $r = -1$, with multiplicity 3. It follows that the given homogeneous solution is $Y(t) = c_1 e^{2t} + c_2 t e^{2t} + c_3 t^2 e^{2t} + c_4 t^3 e^{2t} + c_5 e^{-t} + c_6 t e^{-t} + c_7 t^2 e^{-t}$, which is a linear combination of seven independent solutions.

22. (15) Observe that $(D-1)[e^t] = 0$ and $(D^2+1)[\sin t] = 0$. Hence the operator $H(D) = (D-1)(D^2+1)$ is an annihilator of $e^t + \sin t$. The operator corresponding to the left hand side of the given ODE is $(D^2-1)^2$. It follows that
$$(D+1)^2(D-1)^3(D^2+1)Y = 0.$$
The resulting ODE is homogeneous, with solution $Y(t) = c_1 e^{-t} + c_2 t e^{-t} + c_3 e^t + c_4 t e^t + c_5 t^2 e^t + c_6 \cos t + c_7 \sin t$. After examining the homogeneous solution of Problem 15, and eliminating duplicate terms, we have $Y(t) = c_5 t^2 e^t + c_6 \cos t + c_7 \sin t$.

22. (16) We find that $D[4] = 0$, $(D-1)^2[te^t] = 0$, and $(D^2+4)[\sin 2t] = 0$. The operator $H(D) = D(D-1)^2(D^2+4)$ is an annihilator of $4 + te^t + \sin 2t$. The operator corresponding to the left hand side of the ODE is $D^2(D^2+4)$. It follows that
$$D^3(D-1)^2(D^2+4)^2 Y = 0.$$
The resulting ODE is homogeneous, with solution $Y(t) = c_1 + c_2 t + c_3 t^2 + c_4 e^t + c_5 t e^t + c_6 \cos 2t + c_7 \sin 2t + c_8 t \cos 2t + c_9 t \sin 2t$. After examining the homogeneous solution of Problem 16, and eliminating duplicate terms, we have $Y(t) = c_3 t^2 + c_4 e^t + c_5 t e^t + c_8 t \cos 2t + c_9 t \sin 2t$.

22. (18) Observe that $(D-1)[e^t] = 0$, $(D+1)^2[te^{-t}] = 0$. The function $e^{-t} \sin t$ is a solution of a second order ODE with characteristic roots $r = -1 \pm i$. It follows that $(D^2 + 2D + 2)[e^{-t} \sin t] = 0$. Therefore the operator
$$H(D) = (D-1)(D+1)^2(D^2+2D+2)$$
is an annihilator of $3e^t + 2te^{-t} + e^{-t} \sin t$. The operator corresponding to the left hand side of the given ODE is $D^2(D^2+2D+2)$. It follows that
$$D^2(D-1)(D+1)^2(D^2+2D+2)^2 Y = 0.$$
The resulting ODE is homogeneous, with solution $Y(t) = c_1 + c_2 t + c_3 e^t + c_4 e^{-t} + c_5 t e^{-t} + e^{-t}(c_6 \cos t + c_7 \sin t) + te^{-t}(c_8 \cos t + c_9 \sin t)$. After examining the homogeneous solution of Problem 18, and eliminating duplicate terms, we have $Y(t) = c_3 e^t + c_4 e^{-t} + c_5 t e^{-t} + te^{-t}(c_8 \cos t + c_9 \sin t)$.

4.4

2. The characteristic equation is $r(r^2 - 1) = 0$. Hence the homogeneous solution is $y_c(t) = c_1 + c_2 e^t + c_3 e^{-t}$. The Wronskian is evaluated as $W(1, e^t, e^{-t}) = 2$. Now compute the three determinants

$$W_1(t) = \begin{vmatrix} 0 & e^t & e^{-t} \\ 0 & e^t & -e^{-t} \\ 1 & e^t & e^{-t} \end{vmatrix} = -2, \quad W_2(t) = \begin{vmatrix} 1 & 0 & e^{-t} \\ 0 & 0 & -e^{-t} \\ 0 & 1 & e^{-t} \end{vmatrix} = e^{-t},$$

$$W_3(t) = \begin{vmatrix} 1 & e^t & 0 \\ 0 & e^t & 0 \\ 0 & e^t & 1 \end{vmatrix} = e^t.$$

The solution of the system of equations (10) is

$$u_1'(t) = \frac{t\,W_1(t)}{W(t)} = -t, \quad u_2'(t) = \frac{t\,W_2(t)}{W(t)} = te^{-t}/2,$$

$$u_3'(t) = \frac{t\,W_3(t)}{W(t)} = te^t/2.$$

Hence $u_1(t) = -t^2/2$, $u_2(t) = -e^{-t}(t+1)/2$, $u_3(t) = e^t(t-1)/2$. The particular solution becomes $Y(t) = -t^2/2 - (t+1)/2 + (t-1)/2 = -t^2/2 - 1$. The constant is a solution of the homogeneous equation, therefore the general solution is

$$y(t) = c_1 + c_2 e^t + c_3 e^{-t} - t^2/2.$$

3. From Problem 13 in Section 4.2, $y_c(t) = c_1 e^{-t} + c_2 e^t + c_3 e^{2t}$. The Wronskian is evaluated as $W(e^{-t}, e^t, e^{2t}) = 6 e^{2t}$. Now compute the three determinants

$$W_1(t) = \begin{vmatrix} 0 & e^t & e^{2t} \\ 0 & e^t & 2e^{2t} \\ 1 & e^t & 4e^{2t} \end{vmatrix} = e^{3t}, \quad W_2(t) = \begin{vmatrix} e^{-t} & 0 & e^{2t} \\ -e^{-t} & 0 & 2e^{2t} \\ e^{-t} & 1 & 4e^{2t} \end{vmatrix} = -3e^t,$$

$$W_3(t) = \begin{vmatrix} e^{-t} & e^t & 0 \\ -e^{-t} & e^t & 0 \\ e^{-t} & e^t & 1 \end{vmatrix} = 2.$$

Hence $u_1'(t) = e^{5t}/6$, $u_2'(t) = -e^{3t}/2$, $u_3'(t) = e^{2t}/3$. Therefore the particular solution can be expressed as $Y(t) = e^{-t}\left[e^{5t}/30\right] - e^t\left[e^{3t}/6\right] + e^{2t}\left[e^{2t}/6\right] = e^{4t}/30$.

6. From Problem 22 in Section 4.2, $y_c(t) = c_1 \cos t + c_2 \sin t + t[c_3 \cos t + c_4 \sin t]$. The Wronskian is evaluated as $W(\cos t, \sin t, t \cos t, t \sin t) = 4$. Now compute the four auxiliary determinants

$$W_1(t) = \begin{vmatrix} 0 & \sin t & t \cos t & t \sin t \\ 0 & \cos t & \cos t - t \sin t & \sin t + t \cos t \\ 0 & -\sin t & -2\sin t - t \cos t & 2\cos t - t \sin t \\ 1 & -\cos t & -3\cos t + t \sin t & -3\sin t - t \cos t \end{vmatrix} =$$

$$= -2\sin t + 2t \cos t,$$

$$W_2(t) = \begin{vmatrix} \cos t & 0 & t \cos t & t \sin t \\ -\sin t & 0 & \cos t - t \sin t & \sin t + t \cos t \\ -\cos t & 0 & -2\sin t - t \cos t & 2\cos t - t \sin t \\ \sin t & 1 & -3\cos t + t \sin t & -3\sin t - t \cos t \end{vmatrix} =$$

$$= 2t \sin t + 2\cos t,$$

$$W_3(t) = \begin{vmatrix} \cos t & \sin t & 0 & t \sin t \\ -\sin t & \cos t & 0 & \sin t + t \cos t \\ -\cos t & -\sin t & 0 & 2\cos t - t \sin t \\ \sin t & -\cos t & 1 & -3\sin t - t \cos t \end{vmatrix} = -2\cos t,$$

$$W_4(t) = \begin{vmatrix} \cos t & \sin t & t\cos t & 0 \\ -\sin t & \cos t & \cos t - t\sin t & 0 \\ -\cos t & -\sin t & -2\sin t - t\cos t & 0 \\ \sin t & -\cos t & -3\cos t + t\sin t & 1 \end{vmatrix} = -2\sin t.$$

It follows that

$$u_1'(t) = \left[-\sin^2 t + t\sin t \cos t\right]/2, \quad u_2'(t) = \left[t\sin^2 t + \sin t, \cos t\right]/2,$$

$$u_3'(t) = -\sin t \cos t/2, \quad \text{and} \quad u_4'(t) = -\sin^2 t/2.$$

Hence

$$u_1(t) = (3\sin t \cos t - 2t\cos^2 t - t)/8, \quad u_2(t) = (\sin^2 t - 2\cos^2 t - 2t\sin t \cos t + t^2)/8,$$

$$u_3(t) = -\sin^2 t/4, \quad \text{and} \quad u_4(t) = [\cos t \sin t - t]/4.$$

Therefore the particular solution can be expressed as $Y(t) = u_1(t)\cos t + u_2(t)\sin t + u_3(t)t\cos t + u_4(t)t\sin t = (\sin t - 3t\cos t - t^2 \sin t)/8$. Note that only the last term is not a solution of the homogeneous equation. Hence the general solution is $y(t) = c_1 \cos t + c_2 \sin t + t\left[c_3 \cos t + c_4 \sin t\right] - t^2 \sin t/8$.

8. Based on the results in Problem 2, $y_c(t) = c_1 + c_2 e^t + c_3 e^{-t}$. It was also shown that $W(1, e^t, e^{-t}) = 2$, with $W_1(t) = -2$, $W_2(t) = e^{-t}$, $W_3(t) = e^t$. Therefore we have $u_1'(t) = -\csc t$, $u_2'(t) = e^{-t}\csc t/2$, $u_3'(t) = e^t \csc t/2$. The particular solution can be expressed as $Y(t) = [u_1(t)] + e^{-t}[u_2(t)] + e^t[u_3(t)]$. More specifically,

$$Y(t) = \ln|\csc(t) + \cot(t)| + \frac{e^t}{2}\int_{t_0}^{t} e^{-s}\csc(s)ds + \frac{e^{-t}}{2}\int_{t_0}^{t} e^{s}\csc(s)ds$$

$$= \ln|\csc(t) + \cot(t)| + \int_{t_0}^{t} \cosh(t-s)\csc(s)ds.$$

9. Based on Problem 4, $u_1'(t) = \sec t$, $u_2'(t) = -1$, $u_3'(t) = -\tan t$. The particular solution can be expressed as $Y(t) = [u_1(t)] + \cos t \, [u_2(t)] + \sin t \, [u_3(t)]$. That is, $Y(t) = \ln|\sec(t) + \tan(t)| - t\cos t + \sin t \ln|\cos(t)|$. Hence the general solution of the initial value problem is $y(t) = c_1 + c_2 \cos t + c_3 \sin t + \ln|\sec(t) + \tan(t)| - t\cos t + \sin t \ln|\cos(t)|$. Invoking the initial conditions, we require that $c_1 + c_2 = 2$, $c_3 = 1$, $-c_2 = -2$. Therefore $y(t) = 2\cos t + \sin t + \ln|\sec(t) + \tan(t)| - t\cos t + \sin t \ln|\cos(t)|$. Since $-\pi/2 < t < \pi/2$, the absolute value signs may be removed.

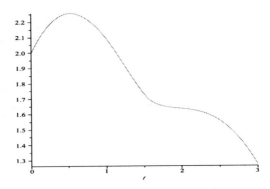

10. From Problem 6, $y(t) = c_1 \cos t + c_2 \sin t + c_3 t \cos t + c_4 t \sin t - t^2 \sin t/8$. In order to satisfy the initial conditions, we require that $c_1 = 2$, $c_2 + c_3 = 0$, $-c_1 + 2c_4 = -1$, $-3/4 - c_2 - 3c_3 = 1$. Therefore $y(t) = (7\sin t - 7t \cos t + 4t \sin t - t^2 \sin t)/8 + 2\cos t$.

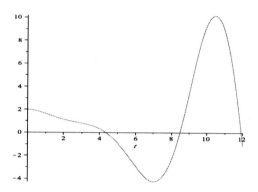

12. From Problem 8, the general solution of the initial value problem is

$$y(t) = c_1 + c_2 e^t + c_3 e^{-t} + \ln|\csc(t) + \cot(t)| +$$

$$+ \frac{e^t}{2} \int_{t_0}^{t} e^{-s} \csc(s)ds + \frac{e^{-t}}{2} \int_{t_0}^{t} e^{s} \csc(s)ds.$$

In this case, $t_0 = \pi/2$. Observe that $y(\pi/2) = y_c(\pi/2)$, $y'(\pi/2) = y_c'(\pi/2)$, and $y''(\pi/2) = y_c''(\pi/2)$. Therefore we obtain the system of equations

$$c_1 + c_2 e^{\pi/2} + c_3 e^{-\pi/2} = 2,$$
$$c_2 e^{\pi/2} - c_3 e^{-\pi/2} = 1,$$
$$c_2 e^{\pi/2} + c_3 e^{-\pi/2} = -1.$$

Hence the solution of the initial value problem is

$$y(t) = 3 - e^{-t+\pi/2} + \ln|\csc(t) + \cot(t)| + \int_{\pi/2}^{t} \cosh(t-s) \csc(s)ds.$$

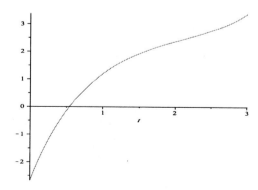

13. First write the equation as $y''' + x^{-1}y'' - 2x^{-2}y' + 2x^{-3}y = 2x$. The Wronskian is evaluated as $W(x, x^2, 1/x) = 6/x$. Now compute the three determinants

$$W_1(x) = \begin{vmatrix} 0 & x^2 & 1/x \\ 0 & 2x & -1/x^2 \\ 1 & 2 & 2/x^3 \end{vmatrix} = -3, \quad W_2(x) = \begin{vmatrix} x & 0 & 1/x \\ 1 & 0 & -1/x^2 \\ 0 & 1 & 2/x^3 \end{vmatrix} = 2/x,$$

$$W_3(x) = \begin{vmatrix} x & x^2 & 0 \\ 1 & 2x & 0 \\ 0 & 2 & 1 \end{vmatrix} = x^2.$$

Hence $u_1'(x) = -x^2$, $u_2'(x) = 2x/3$, $u_3'(x) = x^4/3$. Therefore the particular solution can be expressed as

$$Y(x) = x\left[-x^3/3\right] + x^2\left[x^2/3\right] + \frac{1}{x}\left[x^5/15\right] = x^4/15.$$

15. The homogeneous solution is $y_c(t) = c_1 \cos t + c_2 \sin t + c_3 \cosh t + c_4 \sinh t$. The Wronskian is evaluated as $W(\cos t, \sin t, \cosh t, \sinh t) = 4$. Now the four additional determinants are given by $W_1(t) = 2\sin t$, $W_2(t) = -2\cos t$, $W_3(t) = -2\sinh t$, $W_4(t) = 2\cosh t$. If follows that

$$u_1'(t) = g(t)\sin(t)/2, \quad u_2'(t) = -g(t)\cos(t)/2,$$
$$u_3'(t) = -g(t)\sinh(t)/2, \quad u_4'(t) = g(t)\cosh(t)/2.$$

Therefore the particular solution can be expressed as

$$Y(t) = \frac{\cos(t)}{2}\int_{t_0}^{t} g(s)\sin(s)\,ds - \frac{\sin(t)}{2}\int_{t_0}^{t} g(s)\cos(s)\,ds -$$
$$- \frac{\cosh(t)}{2}\int_{t_0}^{t} g(s)\sinh(s)\,ds + \frac{\sinh(t)}{2}\int_{t_0}^{t} g(s)\cosh(s)\,ds.$$

Using the appropriate identities, the integrals can be combined to obtain

$$Y(t) = \frac{1}{2}\int_{t_0}^{t} g(s)\sinh(t-s)\,ds - \frac{1}{2}\int_{t_0}^{t} g(s)\sin(t-s)\,ds.$$

17. First write the equation as $y''' - 3x^{-1}y'' + 6x^{-2}y' - 6x^{-3}y = g(x)/x^3$. It can be shown that $y_c(x) = c_1 x + c_2 x^2 + c_3 x^3$ is a solution of the homogeneous equation. The Wronskian of this fundamental set of solutions is $W(x, x^2, x^3) = 2x^3$. The three additional determinants are given by $W_1(x) = x^4$, $W_2(x) = -2x^3$, $W_3(x) = x^2$. Hence $u_1'(x) = g(x)/2x^2$, $u_2'(x) = -g(x)/x^3$, $u_3'(x) = g(x)/2x^4$. Now the particular solution can be expressed as

$$Y(x) = x\int_{x_0}^{x} \frac{g(t)}{2t^2}\,dt - x^2\int_{x_0}^{x} \frac{g(t)}{t^3}\,dt + x^3\int_{x_0}^{x} \frac{g(t)}{2t^4}\,dt =$$
$$= \frac{1}{2}\int_{x_0}^{x}\left[\frac{x}{t^2} - \frac{2x^2}{t^3} + \frac{x^3}{t^4}\right] g(t)\,dt.$$

CHAPTER 5

Series Solutions of Second Order Linear Equations

5.1

1. Apply the ratio test:

$$\lim_{n \to \infty} \frac{|(x-3)^{n+1}|}{|(x-3)^n|} = \lim_{n \to \infty} |x-3| = |x-3|.$$

Hence the series converges absolutely for $|x-3| < 1$. The radius of convergence is $\rho = 1$. The series diverges for $x = 2$ and $x = 4$, since the n-th term does not approach zero.

3. Applying the ratio test,

$$\lim_{n \to \infty} \frac{|n!\, x^{2n+2}|}{|(n+1)!\, x^{2n}|} = \lim_{n \to \infty} \frac{x^2}{n+1} = 0.$$

The series converges absolutely for all values of x. Thus the radius of convergence is $\rho = \infty$.

4. Apply the ratio test:

$$\lim_{n \to \infty} \frac{|2^{n+1} x^{n+1}|}{|2^n x^n|} = \lim_{n \to \infty} 2\,|x| = 2\,|x|.$$

Hence the series converges absolutely for $2\,|x| < 1$, or $|x| < 1/2$. The radius of convergence is $\rho = 1/2$. The series diverges for $x = \pm 1/2$, since the n-th term does not approach zero.

127

6. Applying the ratio test,

$$\lim_{n\to\infty} \frac{|n(x-x_0)^{n+1}|}{|(n+1)(x-x_0)^n|} = \lim_{n\to\infty} \frac{n}{n+1} |(x-x_0)| = |(x-x_0)|.$$

Hence the series converges absolutely for $|(x-x_0)| < 1$. The radius of convergence is $\rho = 1$. At $x = x_0 + 1$, we obtain the harmonic series, which is divergent. At the other endpoint, $x = x_0 - 1$, we obtain

$$\sum_{n=1}^{\infty} \frac{(-1)^n}{n},$$

which is conditionally convergent.

7. Apply the ratio test :

$$\lim_{n\to\infty} \frac{|3^n(n+1)^2(x+2)^{n+1}|}{|3^{n+1}n^2(x+2)^n|} = \lim_{n\to\infty} \frac{(n+1)^2}{3n^2} |(x+2)| = \frac{1}{3}|(x+2)|.$$

Hence the series converges absolutely for $\frac{1}{3}|x+2| < 1$, or $|x+2| < 3$. The radius of convergence is $\rho = 3$. At $x = -5$ and $x = +1$, the series diverges, since the n-th term does not approach zero.

8. Applying the ratio test,

$$\lim_{n\to\infty} \frac{|n^n(n+1)!\, x^{n+1}|}{|(n+1)^{n+1}n!\, x^n|} = \lim_{n\to\infty} \frac{n^n}{(n+1)^n} |x| = \frac{1}{e}|x|,$$

since

$$\lim_{n\to\infty} \frac{n^n}{(n+1)^n} = \lim_{n\to\infty} (1+\frac{1}{n})^{-n} = e^{-1}.$$

Hence the series converges absolutely for $|x| < e$. The radius of convergence is $\rho = e$. At $x = \pm e$, the series diverges, since the n-th term does not approach zero. This follows from the fact that

$$\lim_{n\to\infty} \frac{n!\, e^n}{n^n\sqrt{2\pi n}} = 1.$$

10. We have $f(x) = e^x$, with $f^{(n)}(x) = e^x$, for $n = 1, 2, \ldots$. Therefore $f^{(n)}(0) = 1$. Hence the Taylor expansion about $x_0 = 0$ is

$$e^x = \sum_{n=0}^{\infty} \frac{x^n}{n!}.$$

Applying the ratio test,

$$\lim_{n\to\infty} \frac{|n!\,x^{n+1}|}{|(n+1)!\,x^n|} = \lim_{n\to\infty} \frac{1}{n+1} |x| = 0.$$

The radius of convergence is $\rho = \infty$.

11. We have $f(x) = x$, with $f'(x) = 1$ and $f^{(n)}(x) = 0$, for $n = 2, \ldots$. Clearly, $f(1) = 1$ and $f'(1) = 1$, with all other derivatives equal to zero. Hence the Taylor expansion about $x_0 = 1$ is
$$x = 1 + (x - 1).$$
Since the series has only a finite number of terms, it converges absolutely for all x.

14. We have $f(x) = 1/(1 + x)$, $f'(x) = -1/(1 + x)^2$, $f''(x) = 2/(1 + x)^3, \ldots$ with $f^{(n)}(x) = (-1)^n n!/(1 + x)^{n+1}$, for $n \geq 1$. It follows that $f^{(n)}(0) = (-1)^n n!$ for $n \geq 0$. Hence the Taylor expansion about $x_0 = 0$ is
$$\frac{1}{1+x} = \sum_{n=0}^{\infty} (-1)^n x^n.$$
Applying the ratio test,
$$\lim_{n \to \infty} \frac{|x^{n+1}|}{|x^n|} = \lim_{n \to \infty} |x| = |x|.$$
The series converges absolutely for $|x| < 1$, but diverges at $x = \pm 1$.

15. We have $f(x) = 1/(1-x)$, $f'(x) = 1/(1-x)^2$, $f''(x) = 2/(1-x)^3, \ldots$ with $f^{(n)}(x) = n!/(1-x)^{n+1}$, for $n \geq 1$. It follows that $f^{(n)}(0) = n!$, for $n \geq 0$. Hence the Taylor expansion about $x_0 = 0$ is
$$\frac{1}{1-x} = \sum_{n=0}^{\infty} x^n.$$
Applying the ratio test,
$$\lim_{n \to \infty} \frac{|x^{n+1}|}{|x^n|} = \lim_{n \to \infty} |x| = |x|.$$
The series converges absolutely for $|x| < 1$, but diverges at $x = \pm 1$.

16. We have $f(x) = 1/(1-x)$, $f'(x) = 1/(1-x)^2$, $f''(x) = 2/(1-x)^3, \ldots$ with $f^{(n)}(x) = n!/(1-x)^{n+1}$, for $n \geq 1$. It follows that $f^{(n)}(2) = (-1)^{n+1} n!$ for $n \geq 0$. Hence the Taylor expansion about $x_0 = 2$ is
$$\frac{1}{1-x} = -\sum_{n=0}^{\infty} (-1)^n (x-2)^n.$$
Applying the ratio test,
$$\lim_{n \to \infty} \frac{|(x-2)^{n+1}|}{|(x-2)^n|} = \lim_{n \to \infty} |x-2| = |x-2|.$$
The series converges absolutely for $|x - 2| < 1$, but diverges at $x = 1$ and $x = 3$.

17. Applying the ratio test,
$$\lim_{n \to \infty} \frac{|(n+1)x^{n+1}|}{|n x^n|} = \lim_{n \to \infty} \frac{n+1}{n} |x| = |x|.$$

The series converges absolutely for $|x| < 1$. Term-by-term differentiation results in

$$y' = \sum_{n=1}^{\infty} n^2 x^{n-1} = 1 + 4x + 9x^2 + 16x^3 + \ldots$$

$$y'' = \sum_{n=2}^{\infty} n^2(n-1) x^{n-2} = 4 + 18x + 48x^2 + 100x^3 + \ldots$$

Shifting the indices, we can also write

$$y' = \sum_{n=0}^{\infty} (n+1)^2 x^n \quad \text{and} \quad y'' = \sum_{n=0}^{\infty} (n+2)^2(n+1) x^n.$$

20. Shifting the index in the second series, that is, setting $n = k+1$,

$$\sum_{k=0}^{\infty} a_k x^{k+1} = \sum_{n=1}^{\infty} a_{n-1} x^n.$$

Hence

$$\sum_{k=0}^{\infty} a_{k+1} x^k + \sum_{k=0}^{\infty} a_k x^{k+1} = \sum_{k=0}^{\infty} a_{k+1} x^k + \sum_{k=1}^{\infty} a_{k-1} x^k$$

$$= a_1 + \sum_{k=1}^{\infty} (a_{k+1} + a_{k-1}) x^{k+1}.$$

21. Shifting the index by 2, that is, setting $m = n - 2$,

$$\sum_{n=2}^{\infty} n(n-1) a_n x^{n-2} = \sum_{m=0}^{\infty} (m+2)(m+1) a_{m+2} x^m$$

$$= \sum_{n=0}^{\infty} (n+2)(n+1) a_{n+2} x^n.$$

22. Shift the index down by 2, that is, set $m = n + 2$. It follows that

$$\sum_{n=0}^{\infty} a_n x^{n+2} = \sum_{m=2}^{\infty} a_{m-2} x^m = \sum_{n=2}^{\infty} a_{n-2} x^n.$$

24. Clearly,

$$(1-x^2) \sum_{n=2}^{\infty} n(n-1) a_n x^{n-2} = \sum_{n=2}^{\infty} n(n-1) a_n x^{n-2} - \sum_{n=2}^{\infty} n(n-1) a_n x^n.$$

Shifting the index in the first series, that is, setting $k = n - 2$,

$$\sum_{n=2}^{\infty} n(n-1) a_n x^{n-2} = \sum_{k=0}^{\infty} (k+2)(k+1) a_{k+2} x^k$$

$$= \sum_{n=0}^{\infty} (n+2)(n+1) a_{n+2} x^n.$$

Hence
$$(1-x^2)\sum_{n=2}^{\infty} n(n-1)a_n x^{n-2} = \sum_{n=0}^{\infty}(n+2)(n+1)a_{n+2}\, x^n - \sum_{n=2}^{\infty} n(n-1)a_n\, x^n.$$

Note that when $n=0$ and $n=1$, the coefficients in the second series are zero. So
$$(1-x^2)\sum_{n=2}^{\infty} n(n-1)a_n x^{n-2} = \sum_{n=0}^{\infty}[(n+2)(n+1)a_{n+2} - n(n-1)a_n]\, x^n.$$

26. Clearly,
$$\sum_{n=1}^{\infty} na_n\, x^{n-1} + x\sum_{n=0}^{\infty} a_n\, x^n = \sum_{n=1}^{\infty} na_n\, x^{n-1} + \sum_{n=0}^{\infty} a_n\, x^{n+1}.$$

Shifting the index in the first series, that is, setting $k = n-1$,
$$\sum_{n=1}^{\infty} na_n\, x^{n-1} = \sum_{k=0}^{\infty}(k+1)a_{k+1}x^k.$$

Shifting the index in the second series, that is, setting $k = n+1$,
$$\sum_{n=0}^{\infty} a_n\, x^{n+1} = \sum_{k=1}^{\infty} a_{k-1}x^k.$$

Combining the series, and starting the summation at $n=1$,
$$\sum_{n=1}^{\infty} na_n\, x^{n-1} + x\sum_{n=0}^{\infty} a_n\, x^n = a_1 + \sum_{n=1}^{\infty}[(n+1)a_{n+1} + a_{n-1}]\, x^n.$$

27. We note that
$$x\sum_{n=2}^{\infty} n(n-1)a_n\, x^{n-2} + \sum_{n=0}^{\infty} a_n\, x^n = \sum_{n=2}^{\infty} n(n-1)a_n\, x^{n-1} + \sum_{n=0}^{\infty} a_n\, x^n.$$

Shifting the index in the first series, that is, setting $k = n-1$,
$$\sum_{n=2}^{\infty} n(n-1)a_n\, x^{n-1} = \sum_{k=1}^{\infty} k(k+1)a_{k+1}x^k = \sum_{k=0}^{\infty} k(k+1)a_{k+1}x^k,$$

since the coefficient of the term associated with $k=0$ is zero. Combining the series,
$$x\sum_{n=2}^{\infty} n(n-1)a_n\, x^{n-2} + \sum_{n=0}^{\infty} a_n\, x^n = \sum_{n=0}^{\infty}[n(n+1)a_{n+1} + a_n]\, x^n.$$

5.2

1.(a,b,d) Let $y = a_0 + a_1 x + a_2 x^2 + \ldots + a_n x^n + \ldots$. Then

$$y'' = \sum_{n=2}^{\infty} n(n-1) a_n x^{n-2} = \sum_{n=0}^{\infty} (n+2)(n+1) a_{n+2} x^n.$$

Substitution into the ODE results in

$$\sum_{n=0}^{\infty} (n+2)(n+1) a_{n+2} x^n - \sum_{n=0}^{\infty} a_n x^n = 0$$

or

$$\sum_{n=0}^{\infty} \left[(n+2)(n+1) a_{n+2} - a_n \right] x^n = 0.$$

Equating all the coefficients to zero,

$$(n+2)(n+1) a_{n+2} - a_n = 0, \qquad n = 0, 1, 2, \ldots.$$

We obtain the recurrence relation

$$a_{n+2} = \frac{a_n}{(n+1)(n+2)}, \qquad n = 0, 1, 2, \ldots.$$

The subscripts differ by two, so for $k = 1, 2, \ldots$

$$a_{2k} = \frac{a_{2k-2}}{(2k-1)2k} = \frac{a_{2k-4}}{(2k-3)(2k-2)(2k-1)2k} = \ldots = \frac{a_0}{(2k)!}$$

and

$$a_{2k+1} = \frac{a_{2k-1}}{2k(2k+1)} = \frac{a_{2k-3}}{(2k-2)(2k-1)2k(2k+1)} = \ldots = \frac{a_1}{(2k+1)!}.$$

Hence

$$y = a_0 \sum_{k=0}^{\infty} \frac{x^{2k}}{(2k)!} + a_1 \sum_{k=0}^{\infty} \frac{x^{2k+1}}{(2k+1)!}.$$

The linearly independent solutions are

$$y_1 = 1 + \frac{x^2}{2!} + \frac{x^4}{4!} + \frac{x^6}{6!} + \ldots = \cosh x$$

$$y_2 = x + \frac{x^3}{3!} + \frac{x^5}{5!} + \frac{x^7}{7!} + \ldots = \sinh x.$$

(c) The Wronskian at 0 is 1.

4.(a,b,d) Let $y = a_0 + a_1 x + a_2 x^2 + \ldots + a_n x^n + \ldots$. Then

$$y'' = \sum_{n=2}^{\infty} n(n-1) a_n x^{n-2} = \sum_{n=0}^{\infty} (n+2)(n+1) a_{n+2} x^n.$$

Substitution into the ODE results in

$$\sum_{n=0}^{\infty}(n+2)(n+1)a_{n+2}\,x^n + k^2 x^2 \sum_{n=0}^{\infty} a_n x^n = 0.$$

Rewriting the second summation,

$$\sum_{n=0}^{\infty}(n+2)(n+1)a_{n+2}\,x^n + \sum_{n=2}^{\infty} k^2 a_{n-2}\,x^n = 0,$$

that is,

$$2a_2 + 3\cdot 2\,a_3 x + \sum_{n=2}^{\infty}\left[(n+2)(n+1)a_{n+2} + k^2 a_{n-2}\right] x^n = 0.$$

Setting the coefficients equal to zero, we have $a_2 = 0$, $a_3 = 0$, and

$$(n+2)(n+1)a_{n+2} + k^2 a_{n-2} = 0, \quad \text{for } n = 2, 3, 4, \ldots.$$

The recurrence relation can be written as

$$a_{n+2} = -\frac{k^2 a_{n-2}}{(n+2)(n+1)}, \quad n = 2, 3, 4, \ldots.$$

The indices differ by four, so $a_4, a_8, a_{12}, \ldots$ are defined by

$$a_4 = -\frac{k^2 a_0}{4\cdot 3}, \quad a_8 = -\frac{k^2 a_4}{8\cdot 7}, \quad a_{12} = -\frac{k^2 a_8}{12\cdot 11}, \ldots.$$

Similarly, $a_5, a_9, a_{13}, \ldots$ are defined by

$$a_5 = -\frac{k^2 a_1}{5\cdot 4}, \quad a_9 = -\frac{k^2 a_5}{9\cdot 8}, \quad a_{13} = -\frac{k^2 a_9}{13\cdot 12}, \ldots.$$

The remaining coefficients are zero. Therefore the general solution is

$$y = a_0\left[1 - \frac{k^2}{4\cdot 3}x^4 + \frac{k^4}{8\cdot 7\cdot 4\cdot 3}x^8 - \frac{k^6}{12\cdot 11\cdot 8\cdot 7\cdot 4\cdot 3}x^{12} + \ldots\right] +$$

$$+ a_1\left[x - \frac{k^2}{5\cdot 4}x^5 + \frac{k^4}{9\cdot 8\cdot 5\cdot 4}x^9 - \frac{k^6}{13\cdot 12\cdot 9\cdot 8\cdot 4\cdot 4}x^{13} + \ldots\right].$$

Note that for the even coefficients,

$$a_{4m} = -\frac{k^2 a_{4m-4}}{(4m-1)4m}, \quad m = 1, 2, 3, \ldots$$

and for the odd coefficients,

$$a_{4m+1} = -\frac{k^2 a_{4m-3}}{4m(4m+1)}, \quad m = 1, 2, 3, \ldots.$$

Hence the linearly independent solutions are

$$y_1(x) = 1 + \sum_{m=0}^{\infty} \frac{(-1)^{m+1}(k^2 x^4)^{m+1}}{3\cdot 4\cdot 7\cdot 8 \ldots (4m+3)(4m+4)}$$

$$y_2(x) = x\left[1 + \sum_{m=0}^{\infty} \frac{(-1)^{m+1}(k^2 x^4)^{m+1}}{4\cdot 5\cdot 8\cdot 9 \ldots (4m+4)(4m+5)}\right].$$

(c) The Wronskian at 0 is 1.

6.(a,b) Let $y = a_0 + a_1 x + a_2 x^2 + \ldots + a_n x^n + \ldots$. Then

$$y' = \sum_{n=1}^{\infty} n a_n x^{n-1} = \sum_{n=0}^{\infty} (n+1) a_{n+1} x^n$$

and

$$y'' = \sum_{n=2}^{\infty} n(n-1) a_n x^{n-2} = \sum_{n=0}^{\infty} (n+2)(n+1) a_{n+2} x^n.$$

Substitution into the ODE results in

$$(2 + x^2) \sum_{n=0}^{\infty} (n+2)(n+1) a_{n+2} x^n - x \sum_{n=0}^{\infty} (n+1) a_{n+1} x^n + 4 \sum_{n=0}^{\infty} a_n x^n = 0.$$

Before proceeding, write

$$x^2 \sum_{n=0}^{\infty} (n+2)(n+1) a_{n+2} x^n = \sum_{n=2}^{\infty} n(n-1) a_n x^n$$

and

$$x \sum_{n=0}^{\infty} (n+1) a_{n+1} x^n = \sum_{n=1}^{\infty} n a_n x^n.$$

It follows that

$$4a_0 + 4a_2 + (3a_1 + 12 a_3)x +$$

$$+ \sum_{n=2}^{\infty} [2(n+2)(n+1) a_{n+2} + n(n-1) a_n - n a_n + 4 a_n] x^n = 0.$$

Equating the coefficients to zero, we find that $a_2 = -a_0$, $a_3 = -a_1/4$, and

$$a_{n+2} = -\frac{n^2 - 2n + 4}{2(n+2)(n+1)} a_n, \quad n = 0, 1, 2, \ldots.$$

The indices differ by two, so for $k = 0, 1, 2, \ldots$

$$a_{2k+2} = -\frac{(2k)^2 - 4k + 4}{2(2k+2)(2k+1)} a_{2k}$$

and

$$a_{2k+3} = -\frac{(2k+1)^2 - 4k + 2}{2(2k+3)(2k+2)} a_{2k+1}.$$

Hence the linearly independent solutions are

$$y_1(x) = 1 - x^2 + \frac{x^4}{6} - \frac{x^6}{30} + \ldots$$

$$y_2(x) = x - \frac{x^3}{4} + \frac{7x^5}{160} - \frac{19 x^7}{1920} + \ldots.$$

(c) The Wronskian at 0 is 1.

7.(a,b,d) Let $y = a_0 + a_1 x + a_2 x^2 + \ldots + a_n x^n + \ldots$. Then

$$y' = \sum_{n=1}^{\infty} n a_n x^{n-1} = \sum_{n=0}^{\infty} (n+1) a_{n+1} x^n$$

and

$$y'' = \sum_{n=2}^{\infty} n(n-1) a_n x^{n-2} = \sum_{n=0}^{\infty} (n+2)(n+1) a_{n+2} x^n.$$

Substitution into the ODE results in

$$\sum_{n=0}^{\infty} (n+2)(n+1) a_{n+2} x^n + x \sum_{n=0}^{\infty} (n+1) a_{n+1} x^n + 2 \sum_{n=0}^{\infty} a_n x^n = 0.$$

First write

$$x \sum_{n=0}^{\infty} (n+1) a_{n+1} x^n = \sum_{n=1}^{\infty} n a_n x^n.$$

We then obtain

$$2a_2 + 2a_0 + \sum_{n=1}^{\infty} [(n+2)(n+1) a_{n+2} + n a_n + 2 a_n] x^n = 0.$$

It follows that $a_2 = -a_0$ and $a_{n+2} = -a_n/(n+1)$, $n = 0, 1, 2, \ldots$. Note that the indices differ by two, so for $k = 1, 2, \ldots$

$$a_{2k} = -\frac{a_{2k-2}}{2k-1} = \frac{a_{2k-4}}{(2k-3)(2k-1)} = \ldots = \frac{(-1)^k a_0}{1 \cdot 3 \cdot 5 \ldots (2k-1)}$$

and

$$a_{2k+1} = -\frac{a_{2k-1}}{2k} = \frac{a_{2k-3}}{(2k-2)2k} = \ldots = \frac{(-1)^k a_1}{2 \cdot 4 \cdot 6 \ldots (2k)}.$$

Hence the linearly independent solutions are

$$y_1(x) = 1 - \frac{x^2}{1} + \frac{x^4}{1 \cdot 3} - \frac{x^6}{1 \cdot 3 \cdot 5} + \ldots = 1 + \sum_{n=1}^{\infty} \frac{(-1)^n x^{2n}}{1 \cdot 3 \cdot 5 \ldots (2n-1)}$$

$$y_2(x) = x - \frac{x^3}{2} + \frac{x^5}{2 \cdot 4} - \frac{x^7}{2 \cdot 4 \cdot 6} + \ldots = x + \sum_{n=1}^{\infty} \frac{(-1)^n x^{2n+1}}{2 \cdot 4 \cdot 6 \ldots (2n)}.$$

(c) The Wronskian at 0 is 1.

9.(a,b,d) Let $y = a_0 + a_1 x + a_2 x^2 + \ldots + a_n x^n + \ldots$. Then

$$y' = \sum_{n=1}^{\infty} n a_n x^{n-1} = \sum_{n=0}^{\infty} (n+1) a_{n+1} x^n$$

and

$$y'' = \sum_{n=2}^{\infty} n(n-1) a_n x^{n-2} = \sum_{n=0}^{\infty} (n+2)(n+1) a_{n+2} x^n.$$

Substitution into the ODE results in

$$(1+x^2)\sum_{n=0}^{\infty}(n+2)(n+1)a_{n+2}x^n - 4x\sum_{n=0}^{\infty}(n+1)a_{n+1}x^n + 6\sum_{n=0}^{\infty}a_n x^n = 0.$$

Before proceeding, write

$$x^2\sum_{n=0}^{\infty}(n+2)(n+1)a_{n+2}x^n = \sum_{n=2}^{\infty}n(n-1)a_n x^n$$

and

$$x\sum_{n=0}^{\infty}(n+1)a_{n+1}x^n = \sum_{n=1}^{\infty}n\,a_n x^n.$$

It follows that

$$6a_0 + 2a_2 + (2a_1 + 6a_3)x +$$

$$+ \sum_{n=2}^{\infty}[(n+2)(n+1)a_{n+2} + n(n-1)a_n - 4n\,a_n + 6a_n]\,x^n = 0.$$

Setting the coefficients equal to zero, we obtain $a_2 = -3a_0$, $a_3 = -a_1/3$, and

$$a_{n+2} = -\frac{(n-2)(n-3)}{(n+1)(n+2)}a_n, \quad n = 0, 1, 2, \ldots.$$

Observe that for $n = 2$ and $n = 3$, we obtain $a_4 = a_5 = 0$. Since the indices differ by two, we also have $a_n = 0$ for $n \geq 4$. Therefore the general solution is a polynomial

$$y = a_0 + a_1 x - 3a_0 x^2 - a_1 x^3/3.$$

Hence the linearly independent solutions are

$$y_1(x) = 1 - 3x^2 \quad \text{and} \quad y_2(x) = x - x^3/3.$$

(c) The Wronskian is $(x^2+1)^2$. At $x = 0$ it is 1.

10.(a,b,d) Let $y = a_0 + a_1 x + a_2 x^2 + \ldots + a_n x^n + \ldots$. Then

$$y'' = \sum_{n=2}^{\infty}n(n-1)a_n x^{n-2} = \sum_{n=0}^{\infty}(n+2)(n+1)a_{n+2}x^n.$$

Substitution into the ODE results in

$$(4-x^2)\sum_{n=0}^{\infty}(n+2)(n+1)a_{n+2}x^n + 2\sum_{n=0}^{\infty}a_n x^n = 0.$$

First write

$$x^2\sum_{n=0}^{\infty}(n+2)(n+1)a_{n+2}x^n = \sum_{n=2}^{\infty}n(n-1)a_n x^n.$$

It follows that

$$2a_0 + 8a_2 + (2a_1 + 24a_3)x +$$

$$+ \sum_{n=2}^{\infty} [4(n+2)(n+1)a_{n+2} - n(n-1)a_n + 2a_n]x^n = 0.$$

We obtain $a_2 = -a_0/4$, $a_3 = -a_1/12$ and

$$4(n+2)a_{n+2} = (n-2)a_n, \quad n = 0, 1, 2, \ldots.$$

Note that for $n = 2$, $a_4 = 0$. Since the indices differ by two, we also have $a_{2k} = 0$ for $k = 2, 3, \ldots$. On the other hand, for $k = 1, 2, \ldots$,

$$a_{2k+1} = \frac{(2k-3)a_{2k-1}}{4(2k+1)} = \frac{(2k-5)(2k-3)a_{2k-3}}{4^2(2k-1)(2k+1)} = \cdots = \frac{-a_1}{4^k(2k-1)(2k+1)}.$$

Therefore the general solution is

$$y = a_0 + a_1 x - a_0 \frac{x^2}{4} - a_1 \sum_{n=1}^{\infty} \frac{x^{2n+1}}{4^n(2n-1)(2n+1)}.$$

Hence the linearly independent solutions are $y_1(x) = 1 - x^2/4$ and

$$y_2(x) = x - \frac{x^3}{12} - \frac{x^5}{240} - \frac{x^7}{2240} - \cdots = x - \sum_{n=1}^{\infty} \frac{x^{2n+1}}{4^n(2n-1)(2n+1)}.$$

(c) The Wronskian at 0 is 1.

11.(a,b,d) Let $y = a_0 + a_1 x + a_2 x^2 + \ldots + a_n x^n + \ldots$. Then

$$y' = \sum_{n=1}^{\infty} n a_n x^{n-1} = \sum_{n=0}^{\infty} (n+1)a_{n+1} x^n$$

and

$$y'' = \sum_{n=2}^{\infty} n(n-1)a_n x^{n-2} = \sum_{n=0}^{\infty} (n+2)(n+1)a_{n+2} x^n.$$

Substitution into the ODE results in

$$(3-x^2)\sum_{n=0}^{\infty}(n+2)(n+1)a_{n+2}x^n - 3x\sum_{n=0}^{\infty}(n+1)a_{n+1}x^n - \sum_{n=0}^{\infty} a_n x^n = 0.$$

Before proceeding, write

$$x^2 \sum_{n=0}^{\infty}(n+2)(n+1)a_{n+2}x^n = \sum_{n=2}^{\infty} n(n-1)a_n x^n$$

and

$$x \sum_{n=0}^{\infty}(n+1)a_{n+1}x^n = \sum_{n=1}^{\infty} n a_n x^n.$$

It follows that

$$6a_2 - a_0 + (-4a_1 + 18a_3)x +$$

$$+ \sum_{n=2}^{\infty} [3(n+2)(n+1)a_{n+2} - n(n-1)a_n - 3n a_n - a_n]x^n = 0.$$

We obtain $a_2 = a_0/6$, $2a_3 = a_1/9$, and
$$3(n+2)a_{n+2} = (n+1)a_n, \quad n = 0, 1, 2, \ldots.$$
The indices differ by two, so for $k = 1, 2, \ldots$
$$a_{2k} = \frac{(2k-1)a_{2k-2}}{3(2k)} = \frac{(2k-3)(2k-1)a_{2k-4}}{3^2(2k-2)(2k)} = \cdots = \frac{3 \cdot 5 \ldots (2k-1)\, a_0}{3^k \cdot 2 \cdot 4 \ldots (2k)}$$
and
$$a_{2k+1} = \frac{(2k)a_{2k-1}}{3(2k+1)} = \frac{(2k-2)(2k)a_{2k-3}}{3^2(2k-1)(2k+1)} = \cdots = \frac{2 \cdot 4 \cdot 6 \ldots (2k)\, a_1}{3^k \cdot 3 \cdot 5 \ldots (2k+1)}.$$
Hence the linearly independent solutions are
$$y_1(x) = 1 + \frac{x^2}{6} + \frac{x^4}{24} + \frac{5x^6}{432} + \ldots = 1 + \sum_{n=1}^{\infty} \frac{3 \cdot 5 \ldots (2n-1)\, x^{2n}}{3^n \cdot 2 \cdot 4 \ldots (2n)}$$
$$y_2(x) = x + \frac{2x^3}{9} + \frac{8x^5}{135} + \frac{16x^7}{945} + \ldots = x + \sum_{n=1}^{\infty} \frac{2 \cdot 4 \cdot 6 \ldots (2n)\, x^{2n+1}}{3^n \cdot 3 \cdot 5 \ldots (2n+1)}.$$

(c) The Wronskian at 0 is 1.

12.(a,b,d) Let $y = a_0 + a_1 x + a_2 x^2 + \ldots + a_n x^n + \ldots$. Then
$$y' = \sum_{n=1}^{\infty} n a_n x^{n-1} = \sum_{n=0}^{\infty} (n+1)a_{n+1} x^n$$
and
$$y'' = \sum_{n=2}^{\infty} n(n-1) a_n x^{n-2} = \sum_{n=0}^{\infty} (n+2)(n+1) a_{n+2}\, x^n.$$
Substitution into the ODE results in
$$(1-x) \sum_{n=0}^{\infty} (n+2)(n+1) a_{n+2}\, x^n + x \sum_{n=0}^{\infty} (n+1)a_{n+1} x^n - \sum_{n=0}^{\infty} a_n x^n = 0.$$
Before proceeding, write
$$x \sum_{n=0}^{\infty} (n+2)(n+1) a_{n+2}\, x^n = \sum_{n=1}^{\infty} (n+1) n\, a_{n+1} x^n$$
and
$$x \sum_{n=0}^{\infty} (n+1) a_{n+1} x^n = \sum_{n=1}^{\infty} n\, a_n x^n.$$
It follows that
$$2a_2 - a_0 + \sum_{n=1}^{\infty} [(n+2)(n+1) a_{n+2} - (n+1) n\, a_{n+1} + n\, a_n - a_n] x^n = 0.$$
We obtain $a_2 = a_0/2$ and
$$(n+2)(n+1) a_{n+2} - (n+1) n\, a_{n+1} + (n-1) a_n = 0$$

for $n = 0, 1, 2, \ldots$. Writing out the individual equations,

$$3 \cdot 2\, a_3 - 2 \cdot 1\, a_2 = 0$$
$$4 \cdot 3\, a_4 - 3 \cdot 2\, a_3 + a_2 = 0$$
$$5 \cdot 4\, a_5 - 4 \cdot 3\, a_4 + 2\, a_3 = 0$$
$$6 \cdot 5\, a_6 - 5 \cdot 4\, a_5 + 3\, a_4 = 0$$
$$\vdots$$

The coefficients are calculated successively as $a_3 = a_0/(2 \cdot 3)$, $a_4 = a_3/2 - a_2/12 = a_0/24$, $a_5 = 3a_4/5 - a_3/10 = a_0/120$, …. We can now see that for $n \geq 2$, a_n is proportional to a_0. In fact, for $n \geq 2$, $a_n = a_0/(n!)$. Therefore the general solution is

$$y = a_0 + a_1 x + \frac{a_0 x^2}{2!} + \frac{a_0 x^3}{3!} + \frac{a_0 x^4}{4!} + \ldots.$$

Hence the linearly independent solutions are $y_2(x) = x$ and

$$y_1(x) = 1 + \sum_{n=2}^{\infty} \frac{x^n}{n!} = e^x - x.$$

(c) The Wronskian is $e^x(1 - x)$. At $x = 0$ it is 1.

13.(a,b,d) Let $y = a_0 + a_1 x + a_2 x^2 + \ldots + a_n x^n + \ldots$. Then

$$y' = \sum_{n=1}^{\infty} n a_n x^{n-1} = \sum_{n=0}^{\infty} (n+1) a_{n+1} x^n$$

and

$$y'' = \sum_{n=2}^{\infty} n(n-1) a_n x^{n-2} = \sum_{n=0}^{\infty} (n+2)(n+1) a_{n+2}\, x^n.$$

Substitution into the ODE results in

$$2 \sum_{n=0}^{\infty} (n+2)(n+1) a_{n+2}\, x^n + x \sum_{n=0}^{\infty} (n+1) a_{n+1} x^n + 3 \sum_{n=0}^{\infty} a_n x^n = 0.$$

First write

$$x \sum_{n=0}^{\infty} (n+1) a_{n+1} x^n = \sum_{n=1}^{\infty} n\, a_n x^n.$$

We then obtain

$$4a_2 + 3a_0 + \sum_{n=1}^{\infty} \left[2(n+2)(n+1) a_{n+2} + n\, a_n + 3 a_n\right] x^n = 0.$$

It follows that $a_2 = -3a_0/4$ and

$$2(n+2)(n+1) a_{n+2} + (n+3) a_n = 0$$

for $n = 0, 1, 2, \ldots$. The indices differ by two, so for $k = 1, 2, \ldots$

$$a_{2k} = -\frac{(2k+1)a_{2k-2}}{2(2k-1)(2k)} = \frac{(2k-1)(2k+1)a_{2k-4}}{2^2(2k-3)(2k-2)(2k-1)(2k)} = \cdots$$

$$= \frac{(-1)^k 3 \cdot 5 \ldots (2k+1)}{2^k (2k)!} a_0.$$

and

$$a_{2k+1} = -\frac{(2k+2)a_{2k-1}}{2(2k)(2k+1)} = \frac{(2k)(2k+2)a_{2k-3}}{2^2(2k-2)(2k-1)(2k)(2k+1)} = \cdots$$

$$= \frac{(-1)^k 4 \cdot 6 \ldots (2k)(2k+2)}{2^k (2k+1)!} a_1.$$

Hence the linearly independent solutions are

$$y_1(x) = 1 - \frac{3}{4}x^2 + \frac{5}{32}x^4 - \frac{7}{384}x^6 + \ldots = \sum_{n=0}^{\infty} \frac{(-1)^n 3 \cdot 5 \ldots (2n+1)}{2^n (2n)!} x^{2n}$$

$$y_2(x) = x - \frac{1}{3}x^3 + \frac{1}{20}x^5 - \frac{1}{210}x^7 + \ldots = x + \sum_{n=1}^{\infty} \frac{(-1)^n 4 \cdot 6 \ldots (2n+2)}{2^n (2n+1)!} x^{2n+1}.$$

(c) The Wronskian at 0 is 1.

15.(a) From Problem 2, we have

$$y_1(x) = \sum_{n=0}^{\infty} \frac{x^{2n}}{2^n n!} \quad \text{and} \quad y_2(x) = \sum_{n=0}^{\infty} \frac{2^n n! \, x^{2n+1}}{(2n+1)!}.$$

Since $a_0 = y(0)$ and $a_1 = y'(0)$, we have $y(x) = 2y_1(x) + y_2(x)$. That is,

$$y(x) = 2 + x + x^2 + \frac{1}{3}x^3 + \frac{1}{4}x^4 + \frac{1}{15}x^5 + \frac{1}{24}x^6 + \ldots.$$

The four- and five-term polynomial approximations are

$$p_4 = 2 + x + x^2 + x^3/3, \text{ and } p_5 = 2 + x + x^2 + x^3/3 + x^4/4.$$

(b)

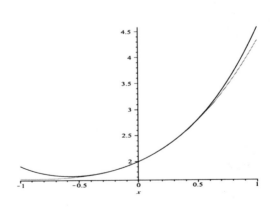

(c) The four-term approximation p_4 appears to be reasonably accurate (within 10%) on the interval $|x| < 0.7$.

17.(a) From Problem 7, the linearly independent solutions are
$$y_1(x) = 1 + \sum_{n=1}^{\infty} \frac{(-1)^n x^{2n}}{1 \cdot 3 \cdot 5 \ldots (2n-1)} \quad \text{and} \quad y_2(x) = x + \sum_{n=1}^{\infty} \frac{(-1)^n x^{2n+1}}{2 \cdot 4 \cdot 6 \ldots (2n)}.$$

Since $a_0 = y(0)$ and $a_1 = y'(0)$, we have $y(x) = 4\,y_1(x) - y_2(x)$. That is,
$$y(x) = 4 - x - 4x^2 + \frac{1}{2}x^3 + \frac{4}{3}x^4 - \frac{1}{8}x^5 - \frac{4}{15}x^6 + \ldots.$$

The four- and five-term polynomial approximations are
$$p_4 = 4 - x - 4x^2 + \frac{1}{2}x^3, \text{ and } p_5 = 4 - x - 4x^2 + \frac{1}{2}x^3 + \frac{4}{3}x^4.$$

(b)

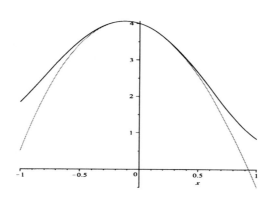

(c) The four-term approximation p_4 appears to be reasonably accurate (within 10%) on the interval $|x| < 0.5$.

18.(a) From Problem 12, we have
$$y_1(x) = 1 + \sum_{n=2}^{\infty} \frac{x^n}{n!} \quad \text{and} \quad y_2(x) = x.$$

Since $a_0 = y(0)$ and $a_1 = y'(0)$, we have $y(x) = -3\,y_1(x) + 2\,y_2(x)$. That is,
$$y(x) = -3 + 2x - \frac{3}{2}x^2 - \frac{1}{2}x^3 - \frac{1}{8}x^4 - \frac{1}{40}x^5 - \frac{1}{240}x^6 + \ldots.$$

The four- and five-term polynomial approximations are
$$p_4 = -3 + 2x - \frac{3}{2}x^2 - \frac{1}{2}x^3, \text{ and } p_5 = -3 + 2x - \frac{3}{2}x^2 - \frac{1}{2}x^3 - \frac{1}{8}x^4.$$

(b)

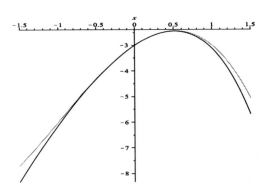

(c) The four-term approximation p_4 appears to be reasonably accurate (within 10%) on the interval $|x| < 0.9$.

20. Two linearly independent solutions of Airy's equation (about $x_0 = 0$) are

$$y_1(x) = 1 + \sum_{n=1}^{\infty} \frac{x^{3n}}{2 \cdot 3 \ldots (3n-1)(3n)}$$

$$y_2(x) = x + \sum_{n=1}^{\infty} \frac{x^{3n+1}}{3 \cdot 4 \ldots (3n)(3n+1)}.$$

Applying the ratio test to the terms of $y_1(x)$,

$$\lim_{n \to \infty} \frac{|2 \cdot 3 \ldots (3n-1)(3n)\, x^{3n+3}|}{|2 \cdot 3 \ldots (3n+2)(3n+3)\, x^{3n}|} = \lim_{n \to \infty} \frac{1}{(3n+1)(3n+2)(3n+3)} |x|^3 = 0.$$

Similarly, applying the ratio test to the terms of $y_2(x)$,

$$\lim_{n \to \infty} \frac{|3 \cdot 4 \ldots (3n)(3n+1)\, x^{3n+4}|}{|3 \cdot 4 \ldots (3n+3)(3n+4)\, x^{3n+1}|} = \lim_{n \to \infty} \frac{1}{(3n+2)(3n+3)(3n+4)} |x|^3 = 0.$$

Hence both series converge absolutely for all x.

21. Let $y = a_0 + a_1 x + a_2 x^2 + \ldots + a_n x^n + \ldots$. Then

$$y' = \sum_{n=1}^{\infty} n a_n x^{n-1} = \sum_{n=0}^{\infty} (n+1) a_{n+1} x^n$$

and

$$y'' = \sum_{n=2}^{\infty} n(n-1) a_n x^{n-2} = \sum_{n=0}^{\infty} (n+2)(n+1) a_{n+2} x^n.$$

Substitution into the ODE results in

$$\sum_{n=0}^{\infty} (n+2)(n+1) a_{n+2} x^n - 2x \sum_{n=0}^{\infty} (n+1) a_{n+1} x^n + \lambda \sum_{n=0}^{\infty} a_n x^n = 0.$$

First write
$$x\sum_{n=0}^{\infty}(n+1)a_{n+1}x^n = \sum_{n=1}^{\infty}n\,a_n x^n.$$

We then obtain
$$2a_2 + \lambda a_0 + \sum_{n=1}^{\infty}[(n+2)(n+1)a_{n+2} - 2n\,a_n + \lambda a_n]x^n = 0.$$

Setting the coefficients equal to zero, it follows that
$$a_{n+2} = \frac{(2n-\lambda)}{(n+1)(n+2)}a_n$$

for $n = 0, 1, 2, \ldots$. Note that the indices differ by two, so for $k = 1, 2, \ldots$

$$a_{2k} = \frac{(4k-4-\lambda)a_{2k-2}}{(2k-1)2k} = \frac{(4k-8-\lambda)(4k-4-\lambda)a_{2k-4}}{(2k-3)(2k-2)(2k-1)2k}$$
$$= (-1)^k \frac{\lambda\ldots(\lambda-4k+8)(\lambda-4k+4)}{(2k)!}a_0.$$

and
$$a_{2k+1} = \frac{(4k-2-\lambda)a_{2k-1}}{2k(2k+1)} = \frac{(4k-6-\lambda)(4k-2-\lambda)a_{2k-3}}{(2k-2)(2k-1)2k(2k+1)}$$
$$= (-1)^k \frac{(\lambda-2)\ldots(\lambda-4k+6)(\lambda-4k+2)}{(2k+1)!}a_1.$$

Hence the linearly independent solutions of the Hermite equation (about $x_0 = 0$) are

$$y_1(x) = 1 - \frac{\lambda}{2!}x^2 + \frac{\lambda(\lambda-4)}{4!}x^4 - \frac{\lambda(\lambda-4)(\lambda-8)}{6!}x^6 + \ldots$$

$$y_2(x) = x - \frac{\lambda-2}{3!}x^3 + \frac{(\lambda-2)(\lambda-6)}{5!}x^5 - \frac{(\lambda-2)(\lambda-6)(\lambda-10)}{7!}x^7 + \ldots.$$

(b) Based on the recurrence relation
$$a_{n+2} = \frac{(2n-\lambda)}{(n+1)(n+2)}a_n,$$

the series solution will terminate as long as λ is a nonnegative even integer. If $\lambda = 2m$, then one or the other of the solutions in part (b) will contain at most $m/2 + 1$ terms. In particular, we obtain the polynomial solutions corresponding to $\lambda = 0, 2, 4, 6, 8, 10$:

$\lambda = 0$	$y_1(x) = 1$
$\lambda = 2$	$y_2(x) = x$
$\lambda = 4$	$y_1(x) = 1 - 2x^2$
$\lambda = 6$	$y_2(x) = x - 2x^3/3$
$\lambda = 8$	$y_1(x) = 1 - 4x^2 + 4x^4/3$
$\lambda = 10$	$y_2(x) = x - 4x^3/3 + 4x^5/15$

(c) Observe that if $\lambda = 2n$, and $a_0 = a_1 = 1$, then

$$a_{2k} = (-1)^k \frac{2n \ldots (2n - 4k + 8)(2n - 4k + 4)}{(2k)!}$$

and

$$a_{2k+1} = (-1)^k \frac{(2n - 2) \ldots (2n - 4k + 6)(2n - 4k + 2)}{(2k + 1)!}.$$

for $k = 1, 2, \ldots [n/2]$. It follows that the coefficient of x^n, in y_1 and y_2, is

$$a_n = \begin{cases} (-1)^k \frac{4^k k!}{(2k)!} & \text{for } n = 2k \\ (-1)^k \frac{4^k k!}{(2k+1)!} & \text{for } n = 2k + 1 \end{cases}$$

Then by definition,

$$H_n(x) = \begin{cases} (-1)^k 2^n \frac{(2k)!}{4^k k!} y_1(x) = (-1)^k \frac{(2k)!}{k!} y_1(x) & \text{for } n = 2k \\ (-1)^k 2^n \frac{(2k+1)!}{4^k k!} y_2(x) = (-1)^k \frac{2(2k+1)!}{k!} y_2(x) & \text{for } n = 2k + 1 \end{cases}$$

Therefore the first six Hermite polynomials are

$$H_0(x) = 1$$
$$H_1(x) = 2x$$
$$H_2(x) = 4x^2 - 2$$
$$H_3(x) = 8x^8 - 12x$$
$$H_4(x) = 16x^4 - 48x^2 + 12$$
$$H_5(x) = 32x^5 - 160x^3 + 120x$$

24. The series solution is given by

$$y(x) = 1 - x^2 + \frac{x^4}{6} - \frac{x^6}{30} + \frac{x^8}{120} + \ldots.$$

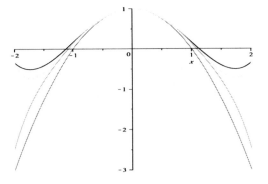

25. The series solution is given by

$$y(x) = x - \frac{x^3}{2} + \frac{x^5}{2 \cdot 4} - \frac{x^7}{2 \cdot 4 \cdot 6} + \frac{x^9}{2 \cdot 4 \cdot 6 \cdot 8} - \ldots.$$

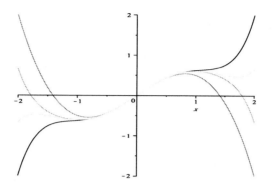

27. The series solution is given by

$$y(x) = 1 - \frac{x^4}{12} + \frac{x^8}{672} - \frac{x^{12}}{88704} + \ldots .$$

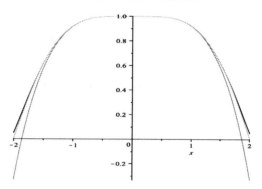

28. Let $y = a_0 + a_1 x + a_2 x^2 + \ldots + a_n x^n + \ldots$. Then

$$y' = \sum_{n=1}^{\infty} n a_n x^{n-1} = \sum_{n=0}^{\infty} (n+1) a_{n+1} x^n$$

and

$$y'' = \sum_{n=2}^{\infty} n(n-1) a_n x^{n-2} = \sum_{n=0}^{\infty} (n+2)(n+1) a_{n+2} x^n.$$

Substitution into the ODE results in

$$(1-x) \sum_{n=0}^{\infty} (n+2)(n+1) a_{n+2} x^n + x \sum_{n=0}^{\infty} (n+1) a_{n+1} x^n - 2 \sum_{n=0}^{\infty} a_n x^n = 0.$$

After appropriately shifting the indices, it follows that

$$2a_2 - 2a_0 + \sum_{n=1}^{\infty} \left[(n+2)(n+1) a_{n+2} - (n+1)n \, a_{n+1} + n \, a_n - 2 \, a_n \right] x^n = 0.$$

We find that $a_2 = a_0$ and

$$(n+2)(n+1) a_{n+2} - (n+1)n \, a_{n+1} + (n-2) a_n = 0$$

for $n = 1, 2, \ldots$. Writing out the individual equations,
$$3 \cdot 2\, a_3 - 2 \cdot 1\, a_2 - a_1 = 0$$
$$4 \cdot 3\, a_4 - 3 \cdot 2\, a_3 = 0$$
$$5 \cdot 4\, a_5 - 4 \cdot 3\, a_4 + a_3 = 0$$
$$6 \cdot 5\, a_6 - 5 \cdot 4\, a_5 + 2\, a_4 = 0$$
$$\vdots$$

Since $a_0 = 0$ and $a_1 = 1$, the remaining coefficients satisfy the equations
$$3 \cdot 2\, a_3 - 1 = 0$$
$$4 \cdot 3\, a_4 - 3 \cdot 2\, a_3 = 0$$
$$5 \cdot 4\, a_5 - 4 \cdot 3\, a_4 + a_3 = 0$$
$$6 \cdot 5\, a_6 - 5 \cdot 4\, a_5 + 2\, a_4 = 0$$
$$\vdots$$

That is, $a_3 = 1/6$, $a_4 = 1/12$, $a_5 = 1/24$, $a_6 = 1/45, \ldots$. Hence the series solution of the initial value problem is
$$y(x) = x + \frac{1}{6}x^3 + \frac{1}{12}x^4 + \frac{1}{24}x^5 + \frac{1}{45}x^6 + \frac{13}{1008}x^7 + \cdots.$$

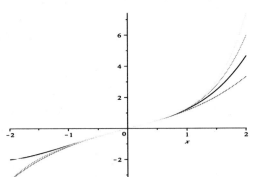

5.3

2. Let $y = \phi(x)$ be a solution of the initial value problem. First note that
$$y'' = -(\sin x)y' - (\cos x)y.$$
Differentiating twice,
$$y''' = -(\sin x)y'' - 2(\cos x)y' + (\sin x)y$$
$$y^{(4)} = -(\sin x)y''' - 3(\cos x)y'' + 3(\sin x)y' + (\cos x)y.$$

Given that $\phi(0) = 0$ and $\phi'(0) = 1$, the first equation gives $\phi''(0) = 0$ and the last two equations give $\phi'''(0) = -2$ and $\phi^{(4)}(0) = 0$.

4. Let $y = \phi(x)$ be a solution of the initial value problem. First note that
$$y'' = -x^2 y' - (\sin x)y.$$
Differentiating twice,
$$y''' = -x^2 y'' - (2x + \sin x)y' - (\cos x)y$$
$$y^{(4)} = -x^2 y''' - (4x + \sin x)y'' - (2 + 2\cos x)y' + (\sin x)y.$$
Given that $\phi(0) = a_0$ and $\phi'(0) = a_1$, the first equation gives $\phi''(0) = 0$ and the last two equations give $\phi'''(0) = -a_0$ and $\phi^{(4)}(0) = -4a_1$.

5. Clearly, $p(x) = 4$ and $q(x) = 6x$ are analytic for all x. Hence the series solutions converge everywhere.

8. The only root of $P(x) = x$ is zero. Hence $\rho_{min} = 1$.

12. The Taylor series expansion of e^x, about $x_0 = 0$, is
$$e^x = \sum_{n=0}^{\infty} \frac{x^n}{n!}.$$
Let $y = a_0 + a_1 x + a_2 x^2 + \ldots + a_n x^n + \ldots$. Substituting into the ODE,
$$\left[\sum_{n=0}^{\infty} \frac{x^n}{n!}\right] \left[\sum_{n=0}^{\infty} (n+2)(n+1)a_{n+2} x^n\right] + x \sum_{n=0}^{\infty} a_n x^n = 0.$$
First note that
$$x \sum_{n=0}^{\infty} a_n x^n = \sum_{n=1}^{\infty} a_{n-1} x^n = a_0 x + a_1 x^2 + a_2 x^3 + \ldots + a_{n-1} x^n + \ldots.$$
The coefficient of x^n in the product of the two series is
$$c_n = 2a_2 \frac{1}{n!} + 6a_3 \frac{1}{(n-1)!} + 12a_4 \frac{1}{(n-2)!} + \ldots$$
$$\ldots + (n+1)n\, a_{n+1} + (n+2)(n+1)a_{n+2}.$$
Expanding the individual series, it follows that
$$2a_2 + (2a_2 + 6a_3)x + (a_2 + 6a_3 + 12a_4)x^2 + (a_2 + 6a_3 + 12a_4 + 20a_5)x^3 + \ldots$$
$$\ldots + a_0 x + a_1 x^2 + a_2 x^3 + \ldots = 0.$$
Setting the coefficients equal to zero, we obtain the system $2a_2 = 0$, $2a_2 + 6a_3 + a_0 = 0$, $a_2 + 6a_3 + 12a_4 + a_1 = 0$, $a_2 + 6a_3 + 12a_4 + 20a_5 + a_2 = 0, \ldots$. Hence the general solution is
$$y(x) = a_0 + a_1 x - a_0 \frac{x^3}{6} + (a_0 - a_1)\frac{x^4}{12} + (2a_1 - a_0)\frac{x^5}{40} + (\frac{4}{3}a_0 - 2a_1)\frac{x^6}{120} + \ldots.$$
We find that two linearly independent solutions ($W(y_1, y_2)(0) = 1$) are
$$y_1(x) = 1 - \frac{x^3}{6} + \frac{x^4}{12} - \frac{x^5}{40} + \ldots$$

$$y_2(x) = x - \frac{x^4}{12} + \frac{x^5}{20} - \frac{x^6}{60} + \ldots$$

Since $p(x) = 0$ and $q(x) = xe^{-x}$ converge everywhere, $\rho = \infty$.

13. The Taylor series expansion of $\cos x$, about $x_0 = 0$, is

$$\cos x = \sum_{n=0}^{\infty} \frac{(-1)^n x^{2n}}{(2n)!}.$$

Let $y = a_0 + a_1 x + a_2 x^2 + \ldots + a_n x^n + \ldots$. Substituting into the ODE,

$$\left[\sum_{n=0}^{\infty} \frac{(-1)^n x^{2n}}{(2n)!}\right]\left[\sum_{n=0}^{\infty}(n+2)(n+1)a_{n+2}\,x^n\right] + \sum_{n=1}^{\infty} na_n x^n - 2\sum_{n=0}^{\infty} a_n x^n = 0.$$

The coefficient of x^n in the product of the two series is

$$c_n = 2a_2 b_n + 6a_3 b_{n-1} + 12a_4 b_{n-2} + \ldots + (n+1)na_{n+1}b_1 + (n+2)(n+1)a_{n+2}b_0,$$

in which $\cos x = b_0 + b_1 x + b_2 x^2 + \ldots + b_n x^n + \ldots$. It follows that

$$2a_2 - 2a_0 + \sum_{n=1}^{\infty} c_n x^n + \sum_{n=1}^{\infty}(n-2)a_n x^n = 0.$$

Expanding the product of the series, it follows that

$$2a_2 - 2a_0 + 6a_3 x + (-a_2 + 12a_4)x^2 + (-3a_3 + 20a_5)x^3 + \ldots$$
$$\ldots - a_1 x + a_3 x^3 + 2a_4 x^4 + \ldots = 0.$$

Setting the coefficients equal to zero, $a_2 - a_0 = 0$, $6a_3 - a_1 = 0$, $-a_2 + 12a_4 = 0$, $-3a_3 + 20a_5 + a_3 = 0$, ... Hence the general solution is

$$y(x) = a_0 + a_1 x + a_0 x^2 + a_1 \frac{x^3}{6} + a_0 \frac{x^4}{12} + a_1 \frac{x^5}{60} + a_0 \frac{x^6}{120} + a_1 \frac{x^7}{560} + \ldots.$$

We find that two linearly independent solutions $(W(y_1, y_2)(0) = 1)$ are

$$y_1(x) = 1 + x^2 + \frac{x^4}{12} + \frac{x^6}{120} + \ldots$$

$$y_2(x) = x + \frac{x^3}{6} + \frac{x^5}{60} + \frac{x^7}{560} + \ldots$$

The nearest zero of $P(x) = \cos x$ is at $x = \pm\pi/2$. Hence $\rho_{min} = \pi/2$.

14. The Taylor series expansion of $\ln(1 + x)$, about $x_0 = 0$, is

$$\ln(1 + x) = \sum_{n=1}^{\infty} \frac{(-1)^{n+1} x^n}{n}.$$

Let $y = a_0 + a_1 x + a_2 x^2 + \ldots + a_n x^n + \ldots$. Substituting into the ODE,

$$\left[\sum_{n=0}^{\infty} \frac{(-1)^n x^n}{n!}\right]\sum_{n=0}^{\infty}(n+2)(n+1)a_{n+2}\,x^n$$

$$+ \left[\sum_{n=1}^{\infty} \frac{(-1)^{n+1}x^n}{n}\right] \sum_{n=0}^{\infty} (n+1)a_{n+1}x^n - x\sum_{n=0}^{\infty} a_n x^n = 0.$$

The first product is the series
$$2a_2 + (-2a_2 + 6a_3)x + (a_2 - 6a_3 + 12a_4)x^2 + (-a_2 + 6a_3 - 12a_4 + 20a_5)x^3 + \ldots$$

The second product is the series
$$a_1 x + (2a_2 - a_1/2)x^2 + (3a_3 - a_2 + a_1/3)x^3 + (4a_4 - 3a_3/2 + 2a_2/3 - a_1/4)x^3 + \ldots$$

Combining the series and equating the coefficients to zero, we obtain
$$2a_2 = 0$$
$$-2a_2 + 6a_3 + a_1 - a_0 = 0$$
$$12a_4 - 6a_3 + 3a_2 - 3a_1/2 = 0$$
$$20a_5 - 12a_4 + 9a_3 - 3a_2 + a_1/3 = 0$$
$$\vdots$$

Hence the general solution is
$$y(x) = a_0 + a_1 x + (a_0 - a_1)\frac{x^3}{6} + (2a_0 + a_1)\frac{x^4}{24} + a_1\frac{7x^5}{120} + (\frac{5}{3}a_1 - a_0)\frac{x^6}{120} + \ldots$$

We find that two linearly independent solutions ($W(y_1, y_2)(0) = 1$) are
$$y_1(x) = 1 + \frac{x^3}{6} + \frac{x^4}{12} - \frac{x^6}{120} + \ldots$$
$$y_2(x) = x - \frac{x^3}{6} + \frac{x^4}{24} + \frac{7x^5}{120} + \ldots$$

The coefficient $p(x) = e^x \ln(1+x)$ is analytic at $x_0 = 0$, but its power series has a radius of convergence $\rho = 1$.

15. If $y_1 = x$ and $y_2 = x^2$ are solutions, then substituting y_2 into the ODE results in
$$2P(x) + 2x\,Q(x) + x^2 R(x) = 0.$$

Setting $x = 0$, we find that $P(0) = 0$. Similarly, substituting y_1 into the ODE results in $Q(0) = 0$. Therefore $P(x)/Q(x)$ and $R(x)/P(x)$ may not be analytic. If they were, Theorem 3.2.1 would guarantee that y_1 and y_2 were the only two solutions. But note that an arbitrary value of $y(0)$ cannot be a linear combination of $y_1(0)$ and $y_2(0)$. Hence $x_0 = 0$ must be a singular point.

16. Let $y = a_0 + a_1 x + a_2 x^2 + \ldots + a_n x^n + \ldots$. Substituting into the ODE,
$$\sum_{n=0}^{\infty} (n+1)a_{n+1} x^n - \sum_{n=0}^{\infty} a_n x^n = 0.$$

That is,
$$\sum_{n=0}^{\infty} [(n+1)a_{n+1} - a_n] x^n = 0.$$

Setting the coefficients equal to zero, we obtain

$$a_{n+1} = \frac{a_n}{n+1}$$

for $n = 0, 1, 2, \ldots$. It is easy to see that $a_n = a_0/(n!)$. Therefore the general solution is

$$y(x) = a_0\left[1 + x + \frac{x^2}{2!} + \frac{x^3}{3!} + \ldots\right] = a_0 e^x.$$

The coefficient $a_0 = y(0)$, which can be arbitrary.

17. Let $y = a_0 + a_1 x + a_2 x^2 + \ldots + a_n x^n + \ldots$. Substituting into the ODE,

$$\sum_{n=0}^{\infty}(n+1)a_{n+1}x^n - x\sum_{n=0}^{\infty}a_n x^n = 0.$$

That is,

$$\sum_{n=0}^{\infty}(n+1)a_{n+1}x^n - \sum_{n=1}^{\infty}a_{n-1}x^n = 0.$$

Combining the series, we have

$$a_1 + \sum_{n=1}^{\infty}[(n+1)a_{n+1} - a_{n-1}]\,x^n = 0.$$

Setting the coefficient equal to zero, $a_1 = 0$ and $a_{n+1} = a_{n-1}/(n+1)$ for $n = 1, 2, \ldots$. Note that the indices differ by two, so for $k = 1, 2, \ldots$

$$a_{2k} = \frac{a_{2k-2}}{(2k)} = \frac{a_{2k-4}}{(2k-2)(2k)} = \ldots = \frac{a_0}{2\cdot 4\ldots(2k)}$$

and

$$a_{2k+1} = 0.$$

Hence the general solution is

$$y(x) = a_0\left[1 + \frac{x^2}{2} + \frac{x^4}{2^2\,2!} + \frac{x^6}{2^3\,3!} + \ldots + \frac{x^{2n}}{2^n\,n!} + \ldots\right] = a_0 e^{x^2/2}.$$

The coefficient $a_0 = y(0)$, which can be arbitrary.

19. Let $y = a_0 + a_1 x + a_2 x^2 + \ldots + a_n x^n + \ldots$. Substituting into the ODE,

$$(1-x)\sum_{n=0}^{\infty}(n+1)a_{n+1}x^n - \sum_{n=0}^{\infty}a_n x^n = 0.$$

That is,

$$\sum_{n=0}^{\infty}(n+1)a_{n+1}x^n - \sum_{n=1}^{\infty}n\,a_n x^n - \sum_{n=0}^{\infty}a_n x^n = 0.$$

Combining the series, we have

$$a_1 - a_0 + \sum_{n=1}^{\infty}[(n+1)a_{n+1} - n\,a_n - a_n]\,x^n = 0.$$

Setting the coefficients equal to zero, $a_1 = a_0$ and $a_{n+1} = a_n$ for $n = 0, 1, 2, \ldots$. Hence the general solution is

$$y(x) = a_0 \left[1 + x + x^2 + x^3 + \ldots + x^n + \ldots\right] = a_0 \frac{1}{1-x}.$$

The coefficient $a_0 = y(0)$, which can be arbitrary.

21. Let $y = a_0 + a_1 x + a_2 x^2 + \ldots + a_n x^n + \ldots$. Substituting into the ODE,

$$\sum_{n=0}^{\infty} (n+1)a_{n+1} x^n + x \sum_{n=0}^{\infty} a_n x^n = 1 + x.$$

That is,

$$\sum_{n=0}^{\infty} (n+1)a_{n+1} x^n + \sum_{n=1}^{\infty} a_{n-1} x^n = 1 + x.$$

Combining the series, and the nonhomogeneous terms, we have

$$(a_1 - 1) + (2a_2 + a_0 - 1)x + \sum_{n=2}^{\infty} \left[(n+1)a_{n+1} + a_{n-1}\right] x^n = 0.$$

Setting the coefficients equal to zero, we obtain $a_1 = 1$, $2a_2 + a_0 - 1 = 0$, and

$$a_n = -\frac{a_{n-2}}{n}, \quad n = 3, 4, \ldots.$$

The indices differ by two, so for $k = 2, 3, \ldots$

$$a_{2k} = -\frac{a_{2k-2}}{(2k)} = \frac{a_{2k-4}}{(2k-2)(2k)} = \ldots = \frac{(-1)^{k-1} a_2}{4 \cdot 6 \ldots (2k)} = \frac{(-1)^k (a_0 - 1)}{2 \cdot 4 \cdot 6 \ldots (2k)},$$

and for $k = 1, 2, \ldots$

$$a_{2k+1} = -\frac{a_{2k-1}}{(2k+1)} = \frac{a_{2k-3}}{(2k-1)(2k+1)} = \ldots = \frac{(-1)^k}{3 \cdot 5 \ldots (2k+1)}.$$

Hence the general solution is

$$y(x) = a_0 + x + \frac{1-a_0}{2} x^2 - \frac{x^3}{3} + a_0 \frac{x^4}{2^2 \, 2!} + \frac{x^5}{3 \cdot 5} - a_0 \frac{x^6}{2^3 \, 3!} - \ldots$$

Collecting the terms containing a_0,

$$y(x) = a_0 \left[1 - \frac{x^2}{2} + \frac{x^4}{2^2 \, 2!} - \frac{x^6}{2^3 \, 3!} + \ldots\right]$$

$$+ \left[x + \frac{x^2}{2} - \frac{x^3}{3} - \frac{x^4}{2^2 \, 2!} + \frac{x^5}{3 \cdot 5} + \frac{x^6}{2^3 \, 3!} - \frac{x^7}{3 \cdot 5 \cdot 7} + \ldots\right].$$

Upon inspection, we find that

$$y(x) = a_0 e^{-x^2/2} + \left[x + \frac{x^2}{2} - \frac{x^3}{3} - \frac{x^4}{2^2 \, 2!} + \frac{x^5}{3 \cdot 5} + \frac{x^6}{2^3 \, 3!} - \frac{x^7}{3 \cdot 5 \cdot 7} + \ldots\right].$$

Note that the given ODE is first order linear, with integrating factor $\mu(x) = e^{x^2/2}$. The general solution is given by

$$y(x) = e^{-x^2/2} \int_0^x e^{u^2/2} du + (y(0) - 1)e^{-x^2/2} + 1.$$

23. If $\alpha = 0$, then $y_1(x) = 1$. If $\alpha = 2n$, then $a_{2m} = 0$ for $m \geq n+1$. As a result,

$$y_1(x) = 1 + \sum_{m=1}^{n} (-1)^m \frac{2^m n(n-1)\ldots(n-m+1)(2n+1)(2n+3)\ldots(2n+2m-1)}{(2m)!} x^{2m}.$$

$\alpha = 0$	1
$\alpha = 2$	$1 - 3x^2$
$\alpha = 4$	$1 - 10x^2 + \frac{35}{3}x^4$

If $\alpha = 2n+1$, then $a_{2m+1} = 0$ for $m \geq n+1$. As a result,

$$y_2(x) = x + \sum_{m=1}^{n} (-1)^m \frac{2^m n(n-1)\ldots(n-m+1)(2n+3)(2n+5)\ldots(2n+2m+1)}{(2m+1)!} x^{2m+1}.$$

$\alpha = 1$	x
$\alpha = 3$	$x - \frac{5}{3}x^3$
$\alpha = 5$	$x - \frac{14}{3}x^3 + \frac{21}{5}x^5$

24.(a) Based on Problem 23,

$\alpha = 0$	1	$y_1(1) = 1$
$\alpha = 2$	$1 - 3x^2$	$y_1(1) = -2$
$\alpha = 4$	$1 - 10x^2 + \frac{35}{3}x^4$	$y_1(1) = \frac{8}{3}$

Normalizing the polynomials, we obtain

$$P_0(x) = 1$$

$$P_2(x) = -\frac{1}{2} + \frac{3}{2}x^2$$

$$P_4(x) = \frac{3}{8} - \frac{15}{4}x^2 + \frac{35}{8}x^4$$

$\alpha = 1$	x	$y_2(1) = 1$
$\alpha = 3$	$x - \frac{5}{3}x^3$	$y_2(1) = -\frac{2}{3}$
$\alpha = 5$	$x - \frac{14}{3}x^3 + \frac{21}{5}x^5$	$y_2(1) = \frac{8}{15}$

Similarly,

$$P_1(x) = x$$

$$P_3(x) = -\frac{3}{2}x + \frac{5}{2}x^3$$

$$P_5(x) = \frac{15}{8}x - \frac{35}{4}x^3 + \frac{63}{8}x^5$$

(b)

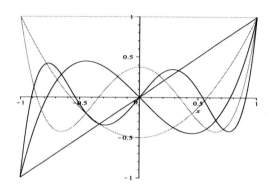

(c) $P_0(x)$ has no roots. $P_1(x)$ has one root at $x = 0$. The zeros of $P_2(x)$ are at $x = \pm 1/\sqrt{3}$. The zeros of $P_3(x)$ are $x = 0, \pm\sqrt{3/5}$. The roots of $P_4(x)$ are given by $x^2 = (15 + 2\sqrt{30})/35$, $(15 - 2\sqrt{30})/35$. The roots of $P_5(x)$ are given by $x = 0$ and $x^2 = (35 + 2\sqrt{70})/63$, $(35 - 2\sqrt{70})/63$.

25. Observe that
$$P_n(-1) = \frac{(-1)^n}{2^n} \sum_{k=0}^{[n/2]} \frac{(-1)^k (2n - 2k)!}{k!(n-k)!(n-2k)!} = (-1)^n P_n(1).$$

But $P_n(1) = 1$ for all nonnegative integers n.

27. We have
$$(x^2 - 1)^n = \sum_{k=0}^{n} \frac{(-1)^{n-k} n!}{k!(n-k)!} x^{2k},$$

which is a polynomial of degree $2n$. Differentiating n times,
$$\frac{d^n}{dx^n}(x^2 - 1)^n = \sum_{k=\mu}^{n} \frac{(-1)^{n-k} n!}{k!(n-k)!}(2k)(2k-1)\ldots(2k-n+1)x^{2k-n},$$

in which the lower index is $\mu = [n/2] + 1$. Note that if $n = 2m + 1$, then $\mu = m + 1$. Now shift the index, by setting $k = n - j$. Hence
$$\frac{d^n}{dx^n}(x^2 - 1)^n = \sum_{j=0}^{[n/2]} \frac{(-1)^j n!}{(n-j)!j!}(2n - 2j)(2n - 2j - 1)\ldots(n - 2j + 1)x^{n-2j}$$
$$= n! \sum_{j=0}^{[n/2]} \frac{(-1)^j (2n - 2j)!}{(n-j)!j!(n-2j)!} x^{n-2j}.$$

Based on Problem 25,
$$\frac{d^n}{dx^n}(x^2 - 1)^n = n! \, 2^n P_n(x).$$

29. Since the $n+1$ polynomials $P_0, P_1, \ldots, P_n$ are linearly independent, and the degree of P_k is k, any polynomial f of degree n can be expressed as a linear combination

$$f(x) = \sum_{k=0}^{n} a_k P_k(x).$$

Multiplying both sides by P_m and integrating,

$$\int_{-1}^{1} f(x) P_m(x) dx = \sum_{k=0}^{n} a_k \int_{-1}^{1} P_k(x) P_m(x) dx.$$

Based on Problem 28,

$$\int_{-1}^{1} P_k(x) P_m(x) dx = \frac{2}{2m+1} \delta_{km}.$$

Hence

$$\int_{-1}^{1} f(x) P_m(x) dx = \frac{2}{2m+1} a_m.$$

5.4

1. Substitution of $y = x^r$ results in the quadratic equation $F(r) = 0$, where $F(r) = r(r-1) + 4r + 2 = r^2 + 3r + 2$. The roots are $r = -2, -1$. Hence the general solution, for $x \neq 0$, is $y = c_1 x^{-2} + c_2 x^{-1}$.

3. Substitution of $y = x^r$ results in the quadratic equation $F(r) = 0$, where $F(r) = r(r-1) - 3r + 4 = r^2 - 4r + 4$. The root is $r = 2$, with multiplicity two. Hence the general solution, for $x \neq 0$, is $y = (c_1 + c_2 \ln |x|) x^2$.

5. Substitution of $y = x^r$ results in the quadratic equation $F(r) = 0$, where $F(r) = r(r-1) - r + 1 = r^2 - 2r + 1$. The root is $r = 1$, with multiplicity two. Hence the general solution, for $x \neq 0$, is $y = (c_1 + c_2 \ln |x|) x$.

6. Substitution of $y = (x-1)^r$ results in the quadratic equation $F(r) = 0$, where $F(r) = r^2 + 7r + 12$. The roots are $r = -3, -4$. Hence the general solution, for $x \neq 1$, is $y = c_1 (x-1)^{-3} + c_2 (x-1)^{-4}$.

7. Substitution of $y = x^r$ results in the quadratic equation $F(r) = 0$, where $F(r) = r^2 + 5r - 1$. The roots are $r = -(5 \pm \sqrt{29})/2$. Hence the general solution, for $x \neq 0$, is $y = c_1 |x|^{-(5+\sqrt{29})/2} + c_2 |x|^{-(5-\sqrt{29})/2}$.

8. Substitution of $y = x^r$ results in the quadratic equation $F(r) = 0$, where $F(r) = r^2 - 3r + 3$. The roots are complex, with $r = (3 \pm i\sqrt{3})/2$. Hence the general solution, for $x \neq 0$, is

$$y = c_1 |x|^{3/2} \cos(\frac{\sqrt{3}}{2} \ln |x|) + c_2 |x|^{3/2} \sin(\frac{\sqrt{3}}{2} \ln |x|).$$

10. Substitution of $y = (x-2)^r$ results in the quadratic equation $F(r) = 0$, where $F(r) = r^2 + 4r + 8$. The roots are complex, with $r = -2 \pm 2i$. Hence the general solution, for $x \neq 2$, is $y = c_1 (x-2)^{-2} \cos(2 \ln|x-2|) + c_2 (x-2)^{-2} \sin(2 \ln|x-2|)$.

11. Substitution of $y = x^r$ results in the quadratic equation $F(r) = 0$, where $F(r) = r^2 + r + 4$. The roots are complex, with $r = -(1 \pm i\sqrt{15})/2$. Hence the general solution, for $x \neq 0$, is

$$y = c_1 |x|^{-1/2} \cos(\frac{\sqrt{15}}{2} \ln|x|) + c_2 |x|^{-1/2} \sin(\frac{\sqrt{15}}{2} \ln|x|).$$

12. Substitution of $y = x^r$ results in the quadratic equation $F(r) = 0$, where $F(r) = r^2 - 5r + 4$. The roots are $r = 1, 4$. Hence the general solution is $y = c_1 x + c_2 x^4$.

14. Substitution of $y = x^r$ results in the quadratic equation $F(r) = 0$, where $F(r) = 4r^2 + 4r + 17$. The roots are complex, with $r = -1/2 \pm 2i$. Hence the general solution, for $x > 0$, is $y = c_1 x^{-1/2} \cos(2 \ln x) + c_2 x^{-1/2} \sin(2 \ln x)$. Invoking the initial conditions, we obtain the system of equations

$$c_1 = 2, \qquad -\frac{1}{2} c_1 + 2c_2 = -3.$$

Hence the solution of the initial value problem is

$$y(x) = 2 x^{-1/2} \cos(2 \ln x) - x^{-1/2} \sin(2 \ln x).$$

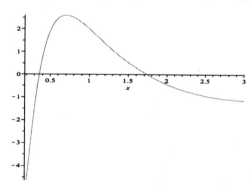

As $x \to 0^+$, the solution decreases without bound.

15. Substitution of $y = x^r$ results in the quadratic equation $F(r) = 0$, where $F(r) = r^2 - 4r + 4$. The root is $r = 2$, with multiplicity two. Hence the general solution, for $x < 0$, is $y = (c_1 + c_2 \ln|x|) x^2$. Invoking the initial conditions, we obtain the system of equations

$$c_1 = 2, \qquad -2c_1 - c_2 = 3.$$

Hence the solution of the initial value problem is

$$y(x) = (2 - 7 \ln|x|) x^2.$$

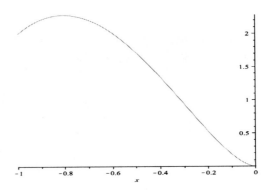

We find that $y(x) \to 0$ as $x \to 0^-$.

18. We see that $P(x) = 0$ when $x = 0$ and 1. Since the three coefficients have no factors in common, both of these points are singular points. Near $x = 0$,

$$\lim_{x \to 0} x\,p(x) = \lim_{x \to 0} x \frac{2x}{x^2(1-x)^2} = 2.$$

$$\lim_{x \to 0} x^2 q(x) = \lim_{x \to 0} x^2 \frac{4}{x^2(1-x)^2} = 4.$$

The singular point $x = 0$ is regular. Considering $x = 1$,

$$\lim_{x \to 1} (x-1)p(x) = \lim_{x \to 1} (x-1)\frac{2x}{x^2(1-x)^2}.$$

The latter limit does not exist. Hence $x = 1$ is an irregular singular point.

19. $P(x) = 0$ when $x = 0$ and 1. Since the three coefficients have no common factors, both of these points are singular points. Near $x = 0$,

$$\lim_{x \to 0} x\,p(x) = \lim_{x \to 0} x \frac{x-2}{x^2(1-x)}.$$

The limit does not exist, and so $x = 0$ is an irregular singular point. Considering $x = 1$,

$$\lim_{x \to 1} (x-1)p(x) = \lim_{x \to 1} (x-1)\frac{x-2}{x^2(1-x)} = 1.$$

$$\lim_{x \to 1} (x-1)^2 q(x) = \lim_{x \to 1} (x-1)^2 \frac{-3x}{x^2(1-x)} = 0.$$

Hence $x = 1$ is a regular singular point.

20. $P(x) = 0$ when $x = 0$ and ± 1. Since the three coefficients have no common factors, both of these points are singular points. Near $x = 0$,

$$\lim_{x \to 0} x\,p(x) = \lim_{x \to 0} x \frac{2}{x^3(1-x^2)}.$$

The limit does not exist, and so $x = 0$ is an irregular singular point. Near $x = -1$,

$$\lim_{x \to -1} (x+1)p(x) = \lim_{x \to -1} (x+1)\frac{2}{x^3(1-x^2)} = -1.$$

$$\lim_{x \to -1}(x+1)^2 q(x) = \lim_{x \to -1}(x+1)^2 \frac{2}{x^3(1-x^2)} = 0.$$

Hence $x = -1$ is a regular singular point. At $x = 1$,

$$\lim_{x \to 1}(x-1)p(x) = \lim_{x \to 1}(x-1)\frac{2}{x^3(1-x^2)} = -1.$$

$$\lim_{x \to 1}(x-1)^2 q(x) = \lim_{x \to 1}(x-1)^2 \frac{2}{x^3(1-x^2)} = 0.$$

Hence $x = 1$ is a regular singular point.

22. The only singular point is at $x = 0$. We find that

$$\lim_{x \to 0} x\, p(x) = \lim_{x \to 0} x\frac{x}{x^2} = 1.$$

$$\lim_{x \to 0} x^2 q(x) = \lim_{x \to 0} x^2 \frac{x^2 - \nu^2}{x^2} = -\nu^2.$$

Hence $x = 0$ is a regular singular point.

23. The only singular point is at $x = -3$. We find that

$$\lim_{x \to -3}(x+3)p(x) = \lim_{x \to -3}(x+3)\frac{-2x}{x+3} = 6.$$

$$\lim_{x \to -3}(x+3)^2 q(x) = \lim_{x \to -3}(x+3)^2 \frac{1-x^2}{x+3} = 0.$$

Hence $x = -3$ is a regular singular point.

24. Dividing the ODE by $x(1-x^2)^3$, we find that

$$p(x) = \frac{1}{x(1-x^2)} \quad \text{and} \quad q(x) = \frac{2}{x(1+x)^2(1-x)^3}.$$

The singular points are at $x = 0$ and ± 1. For $x = 0$,

$$\lim_{x \to 0} x\, p(x) = \lim_{x \to 0} x\frac{1}{x(1-x^2)} = 1.$$

$$\lim_{x \to 0} x^2 q(x) = \lim_{x \to 0} x^2 \frac{2}{x(1+x)^2(1-x)^3} = 0.$$

Hence $x = 0$ is a regular singular point. For $x = -1$,

$$\lim_{x \to -1}(x+1)p(x) = \lim_{x \to -1}(x+1)\frac{1}{x(1-x^2)} = -\frac{1}{2}.$$

$$\lim_{x \to -1}(x+1)^2 q(x) = \lim_{x \to -1}(x+1)^2 \frac{2}{x(1+x)^2(1-x)^3} = -\frac{1}{4}.$$

Hence $x = -1$ is a regular singular point. For $x = 1$,

$$\lim_{x \to 1}(x-1)p(x) = \lim_{x \to 1}(x-1)\frac{1}{x(1-x^2)} = -\frac{1}{2}.$$

$$\lim_{x \to 1}(x-1)^2 q(x) = \lim_{x \to 1}(x-1)^2 \frac{2}{x(1+x)^2(1-x)^3} \,.$$

The latter limit does not exist. Hence $x = 1$ is an irregular singular point.

25. Dividing the ODE by $(x+2)^2(x-1)$, we find that

$$p(x) = \frac{3}{(x+2)^2} \quad \text{and} \quad q(x) = \frac{-2}{(x+2)(x-1)} \,.$$

The singular points are at $x = -2$ and 1. For $x = -2$,

$$\lim_{x \to -2}(x+2)p(x) = \lim_{x \to -2}(x+2)\frac{3}{(x+2)^2} \,.$$

The limit does not exist. Hence $x = -2$ is an irregular singular point. For $x = 1$,

$$\lim_{x \to 1}(x-1)p(x) = \lim_{x \to 1}(x-1)\frac{3}{(x+2)^2} = 0 \,.$$

$$\lim_{x \to 1}(x-1)^2 q(x) = \lim_{x \to 1}(x-1)^2 \frac{-2}{(x+2)(x-1)} = 0 \,.$$

Hence $x = 1$ is a regular singular point.

26. $P(x) = 0$ when $x = 0$ and 3. Since the three coefficients have no common factors, both of these points are singular points. Near $x = 0$,

$$\lim_{x \to 0} x p(x) = \lim_{x \to 0} x \frac{x+1}{x(3-x)} = \frac{1}{3} \,.$$

$$\lim_{x \to 0} x^2 q(x) = \lim_{x \to 0} x^2 \frac{-2}{x(3-x)} = 0 \,.$$

Hence $x = 0$ is a regular singular point. For $x = 3$,

$$\lim_{x \to 3}(x-3)p(x) = \lim_{x \to 3}(x-3)\frac{x+1}{x(3-x)} = -\frac{4}{3} \,.$$

$$\lim_{x \to 3}(x-3)^2 q(x) = \lim_{x \to 3}(x-3)^2 \frac{-2}{x(3-x)} = 0 \,.$$

Hence $x = 3$ is a regular singular point.

27. Dividing the ODE by $(x^2 + x - 2)$, we find that

$$p(x) = \frac{x+1}{(x+2)(x-1)} \quad \text{and} \quad q(x) = \frac{2}{(x+2)(x-1)} \,.$$

The singular points are at $x = -2$ and 1. For $x = -2$,

$$\lim_{x \to -2}(x+2)p(x) = \lim_{x \to -2}\frac{x+1}{x-1} = \frac{1}{3} \,.$$

$$\lim_{x \to -2}(x+2)^2 q(x) = \lim_{x \to -2}\frac{2(x+2)}{x-1} = 0 \,.$$

Hence $x = -2$ is a regular singular point. For $x = 1$,

$$\lim_{x \to 1}(x-1)p(x) = \lim_{x \to 1}\frac{x+1}{x+2} = \frac{2}{3}.$$

$$\lim_{x \to 1}(x-1)^2 q(x) = \lim_{x \to 1}\frac{2(x-1)}{(x+2)} = 0.$$

Hence $x = 1$ is a regular singular point.

29. Note that $p(x) = \ln|x|$ and $q(x) = 3x$. Evidently, $p(x)$ is not analytic at $x_0 = 0$. Furthermore, the function $x p(x) = x \ln|x|$ does not have a Taylor series about $x_0 = 0$. Hence $x = 0$ is an irregular singular point.

30. $P(x) = 0$ when $x = 0$. Since the three coefficients have no common factors, $x = 0$ is a singular point. The Taylor series of $e^x - 1$, about $x = 0$, is

$$e^x - 1 = x + x^2/2 + x^3/6 + \ldots.$$

Hence the function $x p(x) = 2(e^x - 1)/x$ is analytic at $x = 0$. Similarly, the Taylor series of $e^{-x} \cos x$, about $x = 0$, is

$$e^{-x} \cos x = 1 - x + x^3/3 - x^4/6 + \ldots.$$

The function $x^2 q(x) = e^{-x} \cos x$ is also analytic at $x = 0$. Hence $x = 0$ is a regular singular point.

31. $P(x) = 0$ when $x = 0$. Since the three coefficients have no common factors, $x = 0$ is a singular point. The Taylor series of $\sin x$, about $x = 0$, is

$$\sin x = x - x^3/3! + x^5/5! - \ldots.$$

Hence the function $x p(x) = -3 \sin x / x$ is analytic at $x = 0$. On the other hand, $q(x)$ is a rational function, with

$$\lim_{x \to 0} x^2 q(x) = \lim_{x \to 0} x^2 \frac{1+x^2}{x^2} = 1.$$

Hence $x = 0$ is a regular singular point.

32. $P(x) = 0$ when $x = 0$. Since the three coefficients have no common factors, $x = 0$ is a singular point. We find that

$$\lim_{x \to 0} x p(x) = \lim_{x \to 0} x \frac{1}{x} = 1.$$

Although the function $R(x) = \cot x$ does not have a Taylor series about $x = 0$, note that $x^2 q(x) = x \cot x = 1 - x^2/3 - x^4/45 - 2x^6/945 - \ldots$. Hence $x = 0$ is a regular singular point. Furthermore, $q(x) = \cot x / x^2$ is undefined at $x = \pm n\pi$. Therefore the points $x = \pm n\pi$ are also singular points. First note that

$$\lim_{x \to \pm n\pi}(x \mp n\pi) p(x) = \lim_{x \to \pm n\pi}(x \mp n\pi)\frac{1}{x} = 0.$$

Furthermore, since $\cot x$ has period π,

$$q(x) = \cot x/x = \cot(x \mp n\pi)/x = \cot(x \mp n\pi)\frac{1}{(x \mp n\pi) \pm n\pi}.$$

Therefore

$$(x \mp n\pi)^2 q(x) = (x \mp n\pi)\cot(x \mp n\pi)\left[\frac{(x \mp n\pi)}{(x \mp n\pi) \pm n\pi}\right].$$

From above,

$$(x \mp n\pi)\cot(x \mp n\pi) = 1 - (x \mp n\pi)^2/3 - (x \mp n\pi)^4/45 - \ldots.$$

Note that the function in brackets is analytic near $x = \pm n\pi$. It follows that the function $(x \mp n\pi)^2 q(x)$ is also analytic near $x = \pm n\pi$. Hence all the singular points are regular.

34. The singular points are located at $x = \pm n\pi$, $n = 0, 1, \ldots$. Dividing the ODE by $x \sin x$, we find that $x\,p(x) = 3\csc x$ and $x^2 q(x) = x^2 \csc x$. Evidently, $x\,p(x)$ is not even defined at $x = 0$. Hence $x = 0$ is an irregular singular point. On the other hand, the Taylor series of $x \csc x$, about $x = 0$, is

$$x \csc x = 1 + x^2/6 + 7x^4 360 + \ldots.$$

Noting that $\csc(x \mp n\pi) = (-1)^n \csc x$,

$$(x \mp n\pi)p(x) = 3(-1)^n(x \mp n\pi)\csc(x \mp n\pi)/x$$
$$= 3(-1)^n(x \mp n\pi)\csc(x \mp n\pi)\left[\frac{1}{(x \mp n\pi) \pm n\pi}\right].$$

It is apparent that $(x \mp n\pi)p(x)$ is analytic at $x = \pm n\pi$. Similarly,

$$(x \mp n\pi)^2 q(x) = (x \mp n\pi)^2 \csc x = (-1)^n(x \mp n\pi)^2 \csc(x \mp n\pi),$$

which is also analytic at $x = \pm n\pi$. Hence all other singular points are regular.

36. Substitution of $y = x^r$ results in the quadratic equation $r^2 - r + \beta = 0$. The roots are

$$r = \frac{1 \pm \sqrt{1 - 4\beta}}{2}.$$

If $\beta > 1/4$, the roots are complex, with $r_{1,2} = (1 \pm i\sqrt{4\beta - 1})/2$. Hence the general solution, for $x \neq 0$, is

$$y = c_1 |x|^{1/2} \cos(\frac{1}{2}\sqrt{4\beta - 1}\,\ln|x|) + c_2 |x|^{1/2} \sin(\frac{1}{2}\sqrt{4\beta - 1}\,\ln|x|).$$

Since the trigonometric factors are bounded, $y(x) \to 0$ as $x \to 0$. If $\beta = 1/4$, the roots are equal, and

$$y = c_1 |x|^{1/2} + c_2 |x|^{1/2} \ln|x|.$$

Since $\lim_{x \to 0} \sqrt{|x|}\,\ln|x| = 0$, $y(x) \to 0$ as $x \to 0$. If $\beta < 1/4$, the roots are real, with $r_{1,2} = (1 \pm \sqrt{1 - 4\beta})/2$. Hence the general solution, for $x \neq 0$, is

$$y = c_1 |x|^{1/2 + \sqrt{1-4\beta}/2} + c_2 |x|^{1/2 - \sqrt{1-4\beta}/2}.$$

Evidently, solutions approach zero as long as $1/2 - \sqrt{1-4\beta}/2 > 0$. That is,
$$0 < \beta < 1/4.$$
Hence all solutions approach zero for $\beta > 0$.

37. Substitution of $y = x^r$ results in the quadratic equation $r^2 - r - 2 = 0$. The roots are $r = -1, 2$. Hence the general solution, for $x \neq 0$, is $y = c_1 x^{-1} + c_2 x^2$. Invoking the initial conditions, we obtain the system of equations
$$c_1 + c_2 = 1, \quad -c_1 + 2c_2 = \gamma$$
Hence the solution of the initial value problem is
$$y(x) = \frac{2-\gamma}{3}x^{-1} + \frac{1+\gamma}{3}x^2.$$
The solution is bounded, as $x \to 0$, if $\gamma = 2$.

38. Substitution of $y = x^r$ results in the quadratic equation $r^2 + (\alpha - 1)r + 5/2 = 0$. Formally, the roots are given by
$$r = \frac{1 - \alpha \pm \sqrt{\alpha^2 - 2\alpha - 9}}{2} = \frac{1 - \alpha \pm \sqrt{(\alpha - 1 - \sqrt{10})(\alpha - 1 + \sqrt{10})}}{2}.$$

(i) The roots will be complex if $|1 - \alpha| < \sqrt{10}$. For solutions to approach zero, as $x \to \infty$, we need $-\sqrt{10} < 1 - \alpha < 0$.

(ii) The roots will be equal if $|1 - \alpha| = \sqrt{10}$. In this case, all solutions approach zero as long as $1 - \alpha = -\sqrt{10}$.

(iii) The roots will be real and distinct if $|1 - \alpha| > \sqrt{10}$. It follows that
$$r_{max} = \frac{1 - \alpha + \sqrt{\alpha^2 - 2\alpha - 9}}{2}.$$
For solutions to approach zero, we need $1 - \alpha + \sqrt{\alpha^2 - 2\alpha - 9} < 0$. That is, $1 - \alpha < -\sqrt{10}$. Hence all solutions approach zero, as $x \to \infty$, as long as $\alpha > 1$.

42. $x = 0$ is the only singular point. Dividing the ODE by $2x^2$, we have $p(x) = 3/(2x)$ and $q(x) = -x^{-2}(1+x)/2$. It follows that
$$\lim_{x \to 0} x\, p(x) = \lim_{x \to 0} x \frac{3}{2x} = \frac{3}{2},$$
$$\lim_{x \to 0} x^2 q(x) = \lim_{x \to 0} x^2 \frac{-(1+x)}{2x^2} = -\frac{1}{2},$$
so $x = 0$ is a regular singular point. Let $y = a_0 + a_1 x + a_2 x^2 + \ldots + a_n x^n + \ldots$. Substitution into the ODE results in
$$2x^2 \sum_{n=0}^{\infty} (n+2)(n+1)a_{n+2}x^n + 3x \sum_{n=0}^{\infty} (n+1)a_{n+1}x^n - (1+x)\sum_{n=0}^{\infty} a_n x^n = 0.$$

That is,

$$2\sum_{n=2}^{\infty} n(n-1)a_n x^n + 3\sum_{n=1}^{\infty} n a_n x^n - \sum_{n=0}^{\infty} a_n x^n - \sum_{n=1}^{\infty} a_{n-1} x^n = 0.$$

It follows that

$$-a_0 + (2a_1 - a_0)x + \sum_{n=2}^{\infty} [2n(n-1)a_n + 3n a_n - a_n - a_{n-1}] x^n = 0.$$

Equating the coefficients to zero, we find that $a_0 = 0$, $2a_1 - a_0 = 0$, and

$$(2n-1)(n+1)a_n = a_{n-1}, \quad n = 2, 3, \ldots.$$

We conclude that all the a_n are equal to zero. Hence $y(x) = 0$ is the only solution that can be obtained.

44. Based on Problem 43, the change of variable, $x = 1/\xi$, transforms the ODE into the form

$$\xi^4 \frac{d^2 y}{d\xi^2} + 2\xi^3 \frac{dy}{d\xi} + y = 0.$$

Evidently, $\xi = 0$ is a singular point. Now $p(\xi) = 2/\xi$ and $q(\xi) = 1/\xi^4$. Since the value of $\lim_{\xi \to 0} \xi^2 q(\xi)$ does not exist, $\xi = 0$ ($x = \infty$) is an irregular singular point.

46. Under the transformation $x = 1/\xi$, the ODE becomes

$$\xi^4 (1 - \frac{1}{\xi^2}) \frac{d^2 y}{d\xi^2} + \left[2\xi^3 (1 - \frac{1}{\xi^2}) + 2\xi^2 \frac{1}{\xi} \right] \frac{dy}{d\xi} + \alpha(\alpha+1)y = 0,$$

that is,

$$(\xi^4 - \xi^2) \frac{d^2 y}{d\xi^2} + 2\xi^3 \frac{dy}{d\xi} + \alpha(\alpha+1)y = 0.$$

Therefore $\xi = 0$ is a singular point. Note that

$$p(\xi) = \frac{2\xi}{\xi^2 - 1} \quad \text{and} \quad q(\xi) = \frac{\alpha(\alpha+1)}{\xi^2(\xi^2 - 1)}.$$

It follows that

$$\lim_{\xi \to 0} \xi p(\xi) = \lim_{\xi \to 0} \xi \frac{2\xi}{\xi^2 - 1} = 0,$$

$$\lim_{\xi \to 0} \xi^2 q(\xi) = \lim_{\xi \to 0} \xi^2 \frac{\alpha(\alpha+1)}{\xi^2(\xi^2 - 1)} = -\alpha(\alpha+1).$$

Hence $\xi = 0$ ($x = \infty$) is a regular singular point.

48. Under the transformation $x = 1/\xi$, the ODE becomes

$$\xi^4 \frac{d^2 y}{d\xi^2} + \left[2\xi^3 + 2\xi^2 \frac{1}{\xi} \right] \frac{dy}{d\xi} + \lambda y = 0,$$

that is,

$$\xi^4 \frac{d^2 y}{d\xi^2} + 2(\xi^3 + \xi) \frac{dy}{d\xi} + \lambda y = 0.$$

Therefore $\xi = 0$ is a singular point. Note that
$$p(\xi) = \frac{2(\xi^2 + 1)}{\xi^3} \text{ and } q(\xi) = \frac{\lambda}{\xi^4}.$$
It immediately follows that the limit $\lim_{\xi \to 0} \xi p(\xi)$ does not exist. Hence $\xi = 0$ ($x = \infty$) is an irregular singular point.

49. Under the transformation $x = 1/\xi$, the ODE becomes
$$\xi^4 \frac{d^2 y}{d\xi^2} + 2\xi^3 \frac{dy}{d\xi} - \frac{1}{\xi} y = 0.$$
Therefore $\xi = 0$ is a singular point. Note that
$$p(\xi) = \frac{2}{\xi} \text{ and } q(\xi) = \frac{-1}{\xi^5}.$$
We find that
$$\lim_{\xi \to 0} \xi p(\xi) = \lim_{\xi \to 0} \xi \frac{2}{\xi} = 2,$$
but
$$\lim_{\xi \to 0} \xi^2 q(\xi) = \lim_{\xi \to 0} \xi^2 \frac{(-1)}{\xi^5}.$$
The latter limit does not exist. Hence $\xi = 0$ ($x = \infty$) is an irregular singular point.

5.5

1.(a) $P(x) = 0$ when $x = 0$. Since the three coefficients have no common factors, $x = 0$ is a singular point. Near $x = 0$,
$$\lim_{x \to 0} x p(x) = \lim_{x \to 0} x \frac{1}{2x} = \frac{1}{2}.$$
$$\lim_{x \to 0} x^2 q(x) = \lim_{x \to 0} x^2 \frac{1}{2} = 0.$$
Hence $x = 0$ is a regular singular point.

(b) Let
$$y = x^r (a_0 + a_1 x + a_2 x^2 + \ldots + a_n x^n + \ldots) = \sum_{n=0}^{\infty} a_n x^{r+n}.$$
Then
$$y' = \sum_{n=0}^{\infty} (r+n) a_n x^{r+n-1}$$
and
$$y'' = \sum_{n=0}^{\infty} (r+n)(r+n-1) a_n x^{r+n-2}.$$

Substitution into the ODE results in

$$2\sum_{n=0}^{\infty}(r+n)(r+n-1)a_n x^{r+n-1} + \sum_{n=0}^{\infty}(r+n)a_n x^{r+n-1} + \sum_{n=0}^{\infty}a_n x^{r+n+1} = 0.$$

That is,

$$2\sum_{n=0}^{\infty}(r+n)(r+n-1)a_n x^{r+n} + \sum_{n=0}^{\infty}(r+n)a_n x^{r+n} + \sum_{n=2}^{\infty}a_{n-2} x^{r+n} = 0.$$

It follows that

$$a_0\left[2r(r-1)+r\right]x^r + a_1\left[2(r+1)r + r + 1\right]x^{r+1}$$

$$+ \sum_{n=2}^{\infty}\left[2(r+n)(r+n-1)a_n + (r+n)a_n + a_{n-2}\right]x^{r+n} = 0.$$

Assuming that $a_0 \neq 0$, we obtain the indicial equation $2r^2 - r = 0$, with roots $r_1 = 1/2$ and $r_2 = 0$. It immediately follows that $a_1 = 0$. Setting the remaining coefficients equal to zero, we have

$$a_n = \frac{-a_{n-2}}{(r+n)\left[2(r+n) - 1\right]}, \quad n = 2, 3, \ldots.$$

(c) For $r = 1/2$, the recurrence relation becomes

$$a_n = \frac{-a_{n-2}}{n(1+2n)}, \quad n = 2, 3, \ldots.$$

Since $a_1 = 0$, the odd coefficients are zero. Furthermore, for $k = 1, 2, \ldots$,

$$a_{2k} = \frac{-a_{2k-2}}{2k(1+4k)} = \frac{a_{2k-4}}{(2k-2)(2k)(4k-3)(4k+1)} = \frac{(-1)^k a_0}{2^k k!\, 5 \cdot 9 \cdot 13 \ldots (4k+1)}.$$

(d) For $r = 0$, the recurrence relation becomes

$$a_n = \frac{-a_{n-2}}{n(2n-1)}, \quad n = 2, 3, \ldots.$$

Since $a_1 = 0$, the odd coefficients are zero, and for $k = 1, 2, \ldots$,

$$a_{2k} = \frac{-a_{2k-2}}{2k(4k-1)} = \frac{a_{2k-4}}{(2k-2)(2k)(4k-5)(4k-1)} = \frac{(-1)^k a_0}{2^k k!\, 3 \cdot 7 \cdot 11 \ldots (4k-1)}.$$

The two linearly independent solutions are

$$y_1(x) = \sqrt{x}\left[1 + \sum_{k=1}^{\infty}\frac{(-1)^k x^{2k}}{2^k k!\, 5 \cdot 9 \cdot 13 \ldots (4k+1)}\right]$$

$$y_2(x) = 1 + \sum_{k=1}^{\infty}\frac{(-1)^k x^{2k}}{2^k k!\, 3 \cdot 7 \cdot 11 \ldots (4k-1)}.$$

3.(a) Note that $x\,p(x) = 0$ and $x^2 q(x) = x$, which are both analytic at $x = 0$.

(b) Set $y = x^r(a_0 + a_1 x + a_2 x^2 + \ldots + a_n x^n + \ldots)$. Substitution into the ODE results in

$$\sum_{n=0}^{\infty} (r+n)(r+n-1)a_n x^{r+n-1} + \sum_{n=0}^{\infty} a_n x^{r+n} = 0,$$

and after multiplying both sides of the equation by x,

$$\sum_{n=0}^{\infty} (r+n)(r+n-1)a_n x^{r+n} + \sum_{n=1}^{\infty} a_{n-1} x^{r+n} = 0.$$

It follows that

$$a_0 \left[r(r-1)\right] x^r + \sum_{n=1}^{\infty} \left[(r+n)(r+n-1)a_n + a_{n-1}\right] x^{r+n} = 0.$$

Setting the coefficients equal to zero, the indicial equation is $r(r-1) = 0$. The roots are $r_1 = 1$ and $r_2 = 0$. Here $r_1 - r_2 = 1$. The recurrence relation is

$$a_n = \frac{-a_{n-1}}{(r+n)(r+n-1)}, \quad n = 1, 2, \ldots.$$

(c) For $r = 1$,

$$a_n = \frac{-a_{n-1}}{n(n+1)}, \quad n = 1, 2, \ldots.$$

Hence for $n \geq 1$,

$$a_n = \frac{-a_{n-1}}{n(n+1)} = \frac{a_{n-2}}{(n-1)n^2(n+1)} = \ldots = \frac{(-1)^n a_0}{n!(n+1)!}.$$

Therefore one solution is

$$y_1(x) = x \sum_{n=0}^{\infty} \frac{(-1)^n x^n}{n!(n+1)!}.$$

5.(a) Here $xp(x) = 2/3$ and $x^2 q(x) = x^2/3$, which are both analytic at $x = 0$.

(b) Set $y = x^r(a_0 + a_1 x + a_2 x^2 + \ldots + a_n x^n + \ldots)$. Substitution into the ODE results in

$$3 \sum_{n=0}^{\infty} (r+n)(r+n-1)a_n x^{r+n} + 2 \sum_{n=0}^{\infty} (r+n)a_n x^{r+n} + \sum_{n=0}^{\infty} a_n x^{r+n+2} = 0.$$

It follows that

$$a_0 \left[3r(r-1) + 2r\right] x^r + a_1 \left[3(r+1)r + 2(r+1)\right] x^{r+1}$$

$$+ \sum_{n=2}^{\infty} \left[3(r+n)(r+n-1)a_n + 2(r+n)a_n + a_{n-2}\right] x^{r+n} = 0.$$

Assuming $a_0 \neq 0$, the indicial equation is $3r^2 - r = 0$, with roots $r_1 = 1/3$, $r_2 = 0$. Setting the remaining coefficients equal to zero, we have $a_1 = 0$, and

$$a_n = \frac{-a_{n-2}}{(r+n)\left[3(r+n) - 1\right]}, \quad n = 2, 3, \ldots.$$

It immediately follows that the odd coefficients are equal to zero.

(c) For $r = 1/3$,
$$a_n = \frac{-a_{n-2}}{n(1+3n)}, \quad n = 2, 3, \ldots.$$

So for $k = 1, 2, \ldots$,
$$a_{2k} = \frac{-a_{2k-2}}{2k(6k+1)} = \frac{a_{2k-4}}{(2k-2)(2k)(6k-5)(6k+1)} = \frac{(-1)^k a_0}{2^k \, k! \, 7 \cdot 13 \ldots (6k+1)}.$$

(d) For $r = 0$,
$$a_n = \frac{-a_{n-2}}{n(3n-1)}, \quad n = 2, 3, \ldots.$$

So for $k = 1, 2, \ldots$,
$$a_{2k} = \frac{-a_{2k-2}}{2k(6k-1)} = \frac{a_{2k-4}}{(2k-2)(2k)(6k-7)(6k-1)} = \frac{(-1)^k a_0}{2^k \, k! \, 5 \cdot 11 \ldots (6k-1)}.$$

The two linearly independent solutions are
$$y_1(x) = x^{1/3} \left[1 + \sum_{k=1}^{\infty} \frac{(-1)^k}{k! \, 7 \cdot 13 \ldots (6k+1)} \left(\frac{x^2}{2}\right)^k \right]$$

$$y_2(x) = 1 + \sum_{k=1}^{\infty} \frac{(-1)^k}{k! \, 5 \cdot 11 \ldots (6k-1)} \left(\frac{x^2}{2}\right)^k.$$

6.(a) Note that $x \, p(x) = 1$ and $x^2 q(x) = x - 2$, which are both analytic at $x = 0$.

(b) Set $y = x^r (a_0 + a_1 x + a_2 x^2 + \ldots + a_n x^n + \ldots)$. Substitution into the ODE results in
$$\sum_{n=0}^{\infty} (r+n)(r+n-1) a_n \, x^{r+n} + \sum_{n=0}^{\infty} (r+n) a_n \, x^{r+n}$$
$$+ \sum_{n=0}^{\infty} a_n x^{r+n+1} - 2 \sum_{n=0}^{\infty} a_n x^{r+n} = 0.$$

After adjusting the indices in the second-to-last series, we obtain
$$a_0 \left[r(r-1) + r - 2 \right] x^r$$
$$+ \sum_{n=1}^{\infty} \left[(r+n)(r+n-1) a_n + (r+n) a_n - 2 a_n + a_{n-1} \right] x^{r+n} = 0.$$

Assuming $a_0 \neq 0$, the indicial equation is $r^2 - 2 = 0$, with roots $r = \pm \sqrt{2}$. Setting the remaining coefficients equal to zero, the recurrence relation is
$$a_n = \frac{-a_{n-1}}{(r+n)^2 - 2}, \quad n = 1, 2, \ldots.$$

Note that $(r+n)^2 - 2 = (r+n+\sqrt{2})(r+n-\sqrt{2})$.

(c) For $r = \sqrt{2}$,
$$a_n = \frac{-a_{n-1}}{n(n+2\sqrt{2})}, \quad n = 1, 2, \ldots.$$

It follows that
$$a_n = \frac{(-1)^n a_0}{n!(1+2\sqrt{2})(2+2\sqrt{2})\ldots(n+2\sqrt{2})}, \quad n = 1, 2, \ldots.$$

(d) For $r = -\sqrt{2}$,
$$a_n = \frac{-a_{n-1}}{n(n-2\sqrt{2})}, \quad n = 1, 2, \ldots,$$

and therefore
$$a_n = \frac{(-1)^n a_0}{n!(1-2\sqrt{2})(2-2\sqrt{2})\ldots(n-2\sqrt{2})}, \quad n = 1, 2, \ldots.$$

The two linearly independent solutions are
$$y_1(x) = x^{\sqrt{2}}\left[1 + \sum_{n=1}^{\infty} \frac{(-1)^n x^n}{n!(1+2\sqrt{2})(2+2\sqrt{2})\ldots(n+2\sqrt{2})}\right]$$

$$y_2(x) = x^{-\sqrt{2}}\left[1 + \sum_{n=1}^{\infty} \frac{(-1)^n x^n}{n!(1-2\sqrt{2})(2-2\sqrt{2})\ldots(n-2\sqrt{2})}\right].$$

7.(a) Here $xp(x) = 1 - x$ and $x^2 q(x) = -x$, which are both analytic at $x = 0$.

(b) Set $y = x^r(a_0 + a_1 x + a_2 x^2 + \ldots + a_n x^n + \ldots)$. Substitution into the ODE results in
$$\sum_{n=0}^{\infty}(r+n)(r+n-1)a_n x^{r+n-1} + \sum_{n=0}^{\infty}(r+n)a_n x^{r+n-1}$$
$$\sum_{n=0}^{\infty}(r+n)a_n x^{r+n} - \sum_{n=0}^{\infty} a_n x^{r+n} = 0.$$

After multiplying both sides by x,
$$\sum_{n=0}^{\infty}(r+n)(r+n-1)a_n x^{r+n} + \sum_{n=0}^{\infty}(r+n)a_n x^{r+n}$$
$$- \sum_{n=0}^{\infty}(r+n)a_n x^{r+n+1} - \sum_{n=0}^{\infty} a_n x^{r+n+1} = 0.$$

After adjusting the indices in the last two series, we obtain
$$a_0[r(r-1)+r]x^r$$
$$+ \sum_{n=1}^{\infty}[(r+n)(r+n-1)a_n + (r+n)a_n - (r+n)a_{n-1}]x^{r+n} = 0.$$

Assuming $a_0 \neq 0$, the indicial equation is $r^2 = 0$, with roots $r_1 = r_2 = 0$. Setting the remaining coefficients equal to zero, the recurrence relation is

$$a_n = \frac{a_{n-1}}{r+n}, \quad n = 1, 2, \ldots.$$

(c) With $r = 0$,

$$a_n = \frac{a_{n-1}}{n}, \quad n = 1, 2, \ldots.$$

Hence one solution is

$$y_1(x) = 1 + \frac{x}{1!} + \frac{x^2}{2!} + \ldots + \frac{x^n}{n!} + \ldots = e^x.$$

8.(a) Note that $xp(x) = 3/2$ and $x^2 q(x) = x^2 - 1/2$, which are both analytic at $x = 0$.

(b) Set $y = x^r(a_0 + a_1 x + a_2 x^2 + \ldots + a_n x^n + \ldots)$. Substitution into the ODE results in

$$2 \sum_{n=0}^{\infty}(r+n)(r+n-1)a_n x^{r+n} + 3 \sum_{n=0}^{\infty}(r+n)a_n x^{r+n}$$

$$+ 2 \sum_{n=0}^{\infty} a_n x^{r+n+2} - \sum_{n=0}^{\infty} a_n x^{r+n} = 0.$$

After adjusting the indices in the second-to-last series, we obtain

$$a_0 \left[2r(r-1) + 3r - 1 \right] x^r + a_1 \left[2(r+1)r + 3(r+1) - 1 \right]$$

$$+ \sum_{n=2}^{\infty} \left[2(r+n)(r+n-1)a_n + 3(r+n)a_n - a_n + 2 a_{n-2} \right] x^{r+n} = 0.$$

Assuming $a_0 \neq 0$, the indicial equation is $2r^2 + r - 1 = 0$, with roots $r_1 = 1/2$ and $r_2 = -1$. Setting the remaining coefficients equal to zero, the recurrence relation is

$$a_n = \frac{-2a_{n-2}}{(r+n+1)\left[2(r+n) - 1\right]}, \quad n = 2, 3, \ldots.$$

Setting the remaining coefficients equal to zero, we have $a_1 = 0$, which implies that all of the odd coefficients are zero.

(c) With $r = 1/2$,

$$a_n = \frac{-2a_{n-2}}{n(2n+3)}, \quad n = 2, 3, \ldots.$$

So for $k = 1, 2, \ldots$,

$$a_{2k} = \frac{-a_{2k-2}}{k(4k+3)} = \frac{a_{2k-4}}{(k-1)k(4k-5)(4k+3)} = \frac{(-1)^k a_0}{k! \, 7 \cdot 11 \, \ldots \, (4k+3)}.$$

(d) With $r = -1$,

$$a_n = \frac{-2a_{n-2}}{n(2n-3)}, \quad n = 2, 3, \ldots.$$

So for $k = 1, 2, \ldots$,

$$a_{2k} = \frac{-a_{2k-2}}{k(4k-3)} = \frac{a_{2k-4}}{(k-1)k(4k-11)(4k-3)} = \frac{(-1)^k a_0}{k! \, 5 \cdot 9 \ldots (4k-3)}.$$

The two linearly independent solutions are

$$y_1(x) = x^{1/2} \left[1 + \sum_{n=1}^{\infty} \frac{(-1)^n x^{2n}}{n! \, 7 \cdot 11 \ldots (4n+3)} \right]$$

$$y_2(x) = x^{-1} \left[1 + \sum_{n=1}^{\infty} \frac{(-1)^n x^{2n}}{n! \, 5 \cdot 9 \ldots (4n-3)} \right].$$

9.(a) Note that $x p(x) = -x - 3$ and $x^2 q(x) = x + 3$, which are both analytic at $x = 0$.

(b) Set $y = x^r(a_0 + a_1 x + a_2 x^2 + \ldots + a_n x^n + \ldots)$. Substitution into the ODE results in

$$\sum_{n=0}^{\infty} (r+n)(r+n-1) a_n x^{r+n} - \sum_{n=0}^{\infty} (r+n) a_n x^{r+n+1} - 3 \sum_{n=0}^{\infty} (r+n) a_n x^{r+n}$$

$$+ \sum_{n=0}^{\infty} a_n x^{r+n+1} + 3 \sum_{n=0}^{\infty} a_n x^{r+n} = 0.$$

After adjusting the indices in the second-to-last series, we obtain

$$a_0 \left[r(r-1) - 3r + 3 \right] x^r$$

$$+ \sum_{n=1}^{\infty} \left[(r+n)(r+n-1) a_n - (r+n-2) a_{n-1} - 3(r+n-1) a_n \right] x^{r+n} = 0.$$

Assuming $a_0 \neq 0$, the indicial equation is $r^2 - 4r + 3 = 0$, with roots $r_1 = 3$ and $r_2 = 1$. Setting the remaining coefficients equal to zero, the recurrence relation is

$$a_n = \frac{(r+n-2) a_{n-1}}{(r+n-1)(r+n-3)}, \quad n = 1, 2, \ldots.$$

(c) With $r = 3$,

$$a_n = \frac{(n+1) a_{n-1}}{n(n+2)}, \quad n = 1, 2, \ldots.$$

It follows that for $n \geq 1$,

$$a_n = \frac{(n+1) a_{n-1}}{n(n+2)} = \frac{a_{n-2}}{(n-1)(n+2)} = \ldots = \frac{2 a_0}{n! \, (n+2)}.$$

Therefore one solution is

$$y_1(x) = x^3 \left[1 + \sum_{n=1}^{\infty} \frac{2 x^n}{n! \, (n+2)} \right].$$

10.(a) Here $x p(x) = 0$ and $x^2 q(x) = x^2 + 1/4$, which are both analytic at $x = 0$.

(b) Set $y = x^r(a_0 + a_1 x + a_2 x^2 + \ldots + a_n x^n + \ldots)$. Substitution into the ODE results in

$$\sum_{n=0}^{\infty}(r+n)(r+n-1)a_n x^{r+n} + \sum_{n=0}^{\infty} a_n x^{r+n+2} + \frac{1}{4}\sum_{n=0}^{\infty} a_n x^{r+n} = 0.$$

After adjusting the indices in the second series, we obtain

$$a_0\left[r(r-1) + \frac{1}{4}\right]x^r + a_1\left[(r+1)r + \frac{1}{4}\right]x^{r+1}$$
$$+ \sum_{n=2}^{\infty}\left[(r+n)(r+n-1)a_n + \frac{1}{4}a_n + a_{n-2}\right]x^{r+n} = 0.$$

Assuming $a_0 \neq 0$, the indicial equation is $r^2 - r + \frac{1}{4} = 0$, with roots $r_1 = r_2 = 1/2$. Setting the remaining coefficients equal to zero, we find that $a_1 = 0$. The recurrence relation is

$$a_n = \frac{-4a_{n-2}}{(2r+2n-1)^2}, \quad n = 2, 3, \ldots.$$

(c) With $r = 1/2$,

$$a_n = \frac{-a_{n-2}}{n^2}, \quad n = 2, 3, \ldots.$$

Since $a_1 = 0$, the odd coefficients are zero. So for $k \geq 1$,

$$a_{2k} = \frac{-a_{2k-2}}{4k^2} = \frac{a_{2k-4}}{4^2(k-1)^2 k^2} = \ldots = \frac{(-1)^k a_0}{4^k (k!)^2}.$$

Therefore one solution is

$$y_1(x) = \sqrt{x}\left[1 + \sum_{n=1}^{\infty}\frac{(-1)^n x^{2n}}{2^{2n}(n!)^2}\right].$$

12.(a) Dividing through by the leading coefficient, the ODE can be written as

$$y'' - \frac{x}{1-x^2}y' + \frac{\alpha^2}{1-x^2}y = 0.$$

For $x = 1$,

$$p_0 = \lim_{x\to 1}(x-1)p(x) = \lim_{x\to 1}\frac{x}{x+1} = \frac{1}{2}.$$

$$q_0 = \lim_{x\to 1}(x-1)^2 q(x) = \lim_{x\to 1}\frac{\alpha^2(1-x)}{x+1} = 0.$$

For $x = -1$,

$$p_0 = \lim_{x\to -1}(x+1)p(x) = \lim_{x\to -1}\frac{x}{x-1} = \frac{1}{2}.$$

$$q_0 = \lim_{x\to -1}(x+1)^2 q(x) = \lim_{x\to -1}\frac{\alpha^2(x+1)}{(1-x)} = 0.$$

Hence $x = -1$ and $x = 1$ are regular singular points. As shown in Example 1, the indicial equation is given by

$$r(r-1) + p_0 r + q_0 = 0.$$

In this case, both sets of roots are $r_1 = 1/2$ and $r_2 = 0$.

(b) Let $t = x - 1$, and $u(t) = y(t+1)$. Under this change of variable, the differential equation becomes

$$(t^2 + 2t)u'' + (t+1)u' - \alpha^2 u = 0.$$

Based on part (a), $t = 0$ is a regular singular point. Set $u = \sum_{n=0}^{\infty} a_n t^{r+n}$. Substitution into the ODE results in

$$\sum_{n=0}^{\infty}(r+n)(r+n-1)a_n t^{r+n} + 2\sum_{n=0}^{\infty}(r+n)(r+n-1)a_n t^{r+n-1}$$

$$+ \sum_{n=0}^{\infty}(r+n)a_n t^{r+n} + \sum_{n=0}^{\infty}(r+n)a_n t^{r+n-1} - \alpha^2 \sum_{n=0}^{\infty} a_n t^{r+n} = 0.$$

Upon inspection, we can also write

$$\sum_{n=0}^{\infty}(r+n)^2 a_n t^{r+n} + 2\sum_{n=0}^{\infty}(r+n)(r+n-\tfrac{1}{2})a_n t^{r+n-1} - \alpha^2 \sum_{n=0}^{\infty} a_n t^{r+n} = 0.$$

After adjusting the indices in the second series, it follows that

$$a_0\left[2r(r-\tfrac{1}{2})\right]t^{r-1}$$

$$+ \sum_{n=0}^{\infty}\left[(r+n)^2 a_n + 2(r+n+1)(r+n+\tfrac{1}{2})a_{n+1} - \alpha^2 a_n\right]t^{r+n} = 0.$$

Assuming that $a_0 \neq 0$, the indicial equation is $2r^2 - r = 0$, with roots $r = 0, 1/2$. The recurrence relation is

$$(r+n)^2 a_n + 2(r+n+1)(r+n+\tfrac{1}{2})a_{n+1} - \alpha^2 a_n = 0, \quad n = 0, 1, 2, \ldots.$$

With $r_1 = 1/2$, we find that for $n \geq 1$,

$$a_n = \frac{4\alpha^2 - (2n-1)^2}{4n(2n+1)}a_{n-1} = (-1)^n \frac{\left[1-4\alpha^2\right]\left[9-4\alpha^2\right]\ldots\left[(2n-1)^2 - 4\alpha^2\right]}{2^n(2n+1)!} a_0.$$

With $r_2 = 0$, we find that for $n \geq 1$,

$$a_n = \frac{\alpha^2 - (n-1)^2}{n(2n-1)}a_{n-1} = (-1)^n \frac{\alpha(-\alpha)\left[1-\alpha^2\right]\left[4-\alpha^2\right]\ldots\left[(n-1)^2 - \alpha^2\right]}{n! \cdot 3 \cdot 5 \ldots (2n-1)} a_0.$$

The two linearly independent solutions of the Chebyshev equation are

$$y_1(x) = |x-1|^{1/2}\left(1 + \sum_{n=1}^{\infty}(-1)^n \frac{(1-4\alpha^2)(9-4\alpha^2)\ldots((2n-1)^2-4\alpha^2)}{2^n(2n+1)!}(x-1)^n\right)$$

$$y_2(x) = 1 + \sum_{n=1}^{\infty}(-1)^n \frac{\alpha(-\alpha)(1-\alpha^2)(4-\alpha^2)\ldots((n-1)^2-\alpha^2)}{n! \cdot 3 \cdot 5 \ldots (2n-1)}(x-1)^n.$$

13.(a) Here $xp(x) = 1 - x$ and $x^2 q(x) = \lambda x$, which are both analytic at $x = 0$. In fact,
$$p_0 = \lim_{x \to 0} x\, p(x) = 1 \quad \text{and} \quad q_0 = \lim_{x \to 0} x^2 q(x) = 0.$$

(b) The indicial equation is $r(r-1) + r = 0$, with roots $r_{1,2} = 0$.

(c) Set
$$y = a_0 + a_1 x + a_2 x^2 + \ldots + a_n x^n + \ldots.$$
Substitution into the ODE results in
$$\sum_{n=2}^{\infty} n(n-1) a_n x^{n-1} + \sum_{n=1}^{\infty} n a_n x^{n-1} - \sum_{n=0}^{\infty} n a_n x^n + \lambda \sum_{n=0}^{\infty} a_n x^n = 0.$$
That is,
$$\sum_{n=1}^{\infty} n(n+1) a_{n+1} x^n + \sum_{n=0}^{\infty} (n+1) a_{n+1} x^n - \sum_{n=1}^{\infty} n a_n x^n + \lambda \sum_{n=0}^{\infty} a_n x^n = 0.$$
It follows that
$$a_1 + \lambda a_0 + \sum_{n=1}^{\infty} \left[(n+1)^2 a_{n+1} - (n - \lambda) a_n \right] x^n = 0.$$
Setting the coefficients equal to zero, we find that $a_1 = -\lambda a_0$, and
$$a_n = \frac{(n - 1 - \lambda)}{n^2} a_{n-1}, \quad n = 2, 3, \ldots.$$
That is, for $n \geq 2$,
$$a_n = \frac{(n - 1 - \lambda)}{n^2} a_{n-1} = \ldots = \frac{(-\lambda)(1 - \lambda) \ldots (n - 1 - \lambda)}{(n!)^2} a_0.$$
Therefore one solution of the Laguerre equation is
$$y_1(x) = 1 + \sum_{n=1}^{\infty} \frac{(-\lambda)(1 - \lambda) \ldots (n - 1 - \lambda)}{(n!)^2} x^n.$$
Note that if $\lambda = m$, a positive integer, then $a_n = 0$ for $n \geq m + 1$. In that case, the solution is a polynomial
$$y_1(x) = 1 + \sum_{n=1}^{m} \frac{(-\lambda)(1 - \lambda) \ldots (n - 1 - \lambda)}{(n!)^2} x^n.$$

5.6

2.(a) $P(x) = 0$ only for $x = 0$. Furthermore, $xp(x) = -2 - x$ and $x^2 q(x) = 2 + x^2$. It follows that
$$p_0 = \lim_{x \to 0}(-2 - x) = -2$$
$$q_0 = \lim_{x \to 0}(2 + x^2) = 2$$
and therefore $x = 0$ is a regular singular point.

(b) The indicial equation is given by $r(r-1) - 2r + 2 = 0$, that is, $r^2 - 3r + 2 = 0$, with roots $r_1 = 2$ and $r_2 = 1$.

4. The coefficients $P(x)$, $Q(x)$, and $R(x)$ are analytic for all $x \in \mathbb{R}$. Hence there are no singular points.

5.(a) $P(x) = 0$ only for $x = 0$. Furthermore, $xp(x) = 3\sin x / x$ and $x^2 q(x) = -2$. It follows that
$$p_0 = \lim_{x \to 0} 3 \frac{\sin x}{x} = 3$$
$$q_0 = \lim_{x \to 0} -2 = -2$$
and therefore $x = 0$ is a regular singular point.

(b) The indicial equation is given by $r(r-1) + 3r - 2 = 0$, that is, $r^2 + 2r - 2 = 0$, with roots $r_1 = -1 + \sqrt{3}$ and $r_2 = -1 - \sqrt{3}$.

6.(a) $P(x) = 0$ for $x = 0$ and $x = -2$. We note that $p(x) = x^{-1}(x+2)^{-1}/2$, and $q(x) = -(x+2)^{-1}/2$. For the singularity at $x = 0$,
$$p_0 = \lim_{x \to 0} \frac{1}{2(x+2)} = \frac{1}{4}$$
$$q_0 = \lim_{x \to 0} \frac{-x^2}{2(x+2)} = 0$$
and therefore $x = 0$ is a regular singular point.
For the singularity at $x = -2$,
$$p_0 = \lim_{x \to -2}(x+2)p(x) = \lim_{x \to -2} \frac{1}{2x} = -\frac{1}{4}$$
$$q_0 = \lim_{x \to -2}(x+2)^2 q(x) = \lim_{x \to -2} \frac{-(x+2)}{2} = 0$$
and therefore $x = -2$ is a regular singular point.

(b) For $x = 0$: the indicial equation is given by $r(r-1) + r/4 = 0$, that is, $r^2 - 3r/4 = 0$, with roots $r_1 = 3/4$ and $r_2 = 0$.
For $x = -2$: the indicial equation is given by $r(r-1) - r/4 = 0$, that is, $r^2 - 5r/4 = 0$, with roots $r_1 = 5/4$ and $r_2 = 0$.

7.(a) $P(x) = 0$ only for $x = 0$. Furthermore, $xp(x) = 1/2 + \sin x / 2x$ and $x^2 q(x) = 1$. It follows that

$$p_0 = \lim_{x \to 0} xp(x) = 1$$
$$q_0 = \lim_{x \to 0} x^2 q(x) = 1$$

and therefore $x = 0$ is a regular singular point.

(b) The indicial equation is given by

$$r(r-1) + r + 1 = 0,$$

that is, $r^2 + 1 = 0$, with complex conjugate roots $r = \pm i$.

8.(a) Note that $P(x) = 0$ only for $x = -1$. We find that $p(x) = 3(x-1)/(x+1)$, and $q(x) = 3/(x+1)^2$. It follows that

$$p_0 = \lim_{x \to -1} (x+1)p(x) = \lim_{x \to -1} 3(x-1) = -6$$
$$q_0 = \lim_{x \to -1} (x+1)^2 q(x) = \lim_{x \to -1} 3 = 3$$

and therefore $x = -1$ is a regular singular point.

(b) The indicial equation is given by

$$r(r-1) - 6r + 3 = 0,$$

that is, $r^2 - 7r + 3 = 0$, with roots $r_1 = (7 + \sqrt{37})/2$ and $r_2 = (7 - \sqrt{37})/2$.

10.(a) $P(x) = 0$ for $x = 2$ and $x = -2$. We note that $p(x) = 2x(x-2)^{-2}(x+2)^{-1}$, and $q(x) = 3(x-2)^{-1}(x+2)^{-1}$. For the singularity at $x = 2$,

$$\lim_{x \to 2} (x-2)p(x) = \lim_{x \to 2} \frac{2x}{x^2 - 4},$$

which is undefined. Therefore $x = 2$ is an irregular singular point. For the singularity at $x = -2$,

$$p_0 = \lim_{x \to -2} (x+2)p(x) = \lim_{x \to -2} \frac{2x}{(x-2)^2} = -\frac{1}{4}$$
$$q_0 = \lim_{x \to -2} (x+2)^2 q(x) = \lim_{x \to -2} \frac{3(x+2)}{x-2} = 0$$

and therefore $x = -2$ is a regular singular point.

(b) The indicial equation is given by $r(r-1) - r/4 = 0$, that is, $r^2 - 5r/4 = 0$, with roots $r_1 = 5/4$ and $r_2 = 0$.

11.(a) $P(x) = 0$ for $x = 2$ and $x = -2$. We note that $p(x) = 2x/(4-x^2)$, and $q(x) = 3/(4-x^2)$. For the singularity at $x = 2$,

$$p_0 = \lim_{x \to 2} (x-2)p(x) = \lim_{x \to 2} \frac{-2x}{x+2} = -1$$

$$q_0 = \lim_{x \to 2} (x-2)^2 q(x) = \lim_{x \to 2} \frac{3(2-x)}{x+2} = 0$$

and therefore $x = 2$ is a regular singular point.

For the singularity at $x = -2$,

$$p_0 = \lim_{x \to -2} (x+2)p(x) = \lim_{x \to -2} \frac{2x}{2-x} = -1$$

$$q_0 = \lim_{x \to -2} (x+2)^2 q(x) = \lim_{x \to -2} \frac{3(x+2)}{2-x} = 0$$

and therefore $x = -2$ is a regular singular point.

(b) For $x = 2$: the indicial equation is given by $r(r-1) - r = 0$, that is, $r^2 - 2r = 0$, with roots $r_1 = 2$ and $r_2 = 0$.

For $x = -2$: the indicial equation is given by $r(r-1) - r = 0$, that is, $r^2 - 2r = 0$, with roots $r_1 = 2$ and $r_2 = 0$.

12.(a) $P(x) = 0$ for $x = 0$ and $x = -3$. We note that $p(x) = -2x^{-1}(x+3)^{-1}$, and $q(x) = -1/(x+3)^2$. For the singularity at $x = 0$,

$$p_0 = \lim_{x \to 0} x p(x) = \lim_{x \to 0} \frac{-2}{x+3} = -\frac{2}{3}$$

$$q_0 = \lim_{x \to 0} x^2 q(x) = \lim_{x \to 0} \frac{-x^2}{(x+3)^2} = 0$$

and therefore $x = 0$ is a regular singular point.

For the singularity at $x = -3$,

$$p_0 = \lim_{x \to -3} (x+3)p(x) = \lim_{x \to -3} \frac{-2}{x} = \frac{2}{3}$$

$$q_0 = \lim_{x \to -3} (x+3)^2 q(x) = \lim_{x \to -3} (-1) = -1$$

and therefore $x = -3$ is a regular singular point.

(b) For $x = 0$: the indicial equation is given by $r(r-1) - 2r/3 = 0$, that is, $r^2 - 5r/3 = 0$, with roots $r_1 = 5/3$ and $r_2 = 0$.

For $x = -3$: the indicial equation is given by $r(r-1) + 2r/3 - 1 = 0$, that is, $r^2 - r/3 - 1 = 0$, with roots $r_1 = (1+\sqrt{37})/6$ and $r_2 = (1-\sqrt{37})/6$.

14.(a) Here $xp(x) = 2x$ and $x^2 q(x) = 6xe^x$. Both of these functions are analytic at $x = 0$, therefore $x = 0$ is a regular singular point. Note that $p_0 = q_0 = 0$.

(b) The indicial equation is given by $r(r-1) = 0$, that is, $r^2 - r = 0$, with roots $r_1 = 1$ and $r_2 = 0$.

(c) In order to find the solution corresponding to $r_1 = 1$, set $y = x \sum_{n=0}^{\infty} a_n x^n$. Upon substitution into the ODE, we have

$$\sum_{n=0}^{\infty} (n+2)(n+1)a_{n+1} x^{n+1} + 2 \sum_{n=0}^{\infty} (n+1)a_n x^{n+1} + 6 e^x \sum_{n=0}^{\infty} a_n x^{n+1} = 0.$$

After adjusting the indices in the first two series, and expanding the exponential function,

$$\sum_{n=1}^{\infty} n(n+1)a_n x^n + 2 \sum_{n=1}^{\infty} n a_{n-1} x^n + 6 a_0 x + (6a_0 + 6a_1)x^2$$

$$+ (6a_2 + 6a_1 + 3a_0)x^3 + (6a_3 + 6a_2 + 3a_1 + a_0)x^4 + \ldots = 0.$$

Equating the coefficients, we obtain the system of equations

$$2a_1 + 2a_0 + 6a_0 = 0$$
$$6a_2 + 4a_1 + 6a_0 + 6a_1 = 0$$
$$12a_3 + 6a_2 + 6a_2 + 6a_1 + 3a_0 = 0$$
$$20a_4 + 8a_3 + 6a_3 + 6a_2 + 3a_1 + a_0 = 0$$
$$\vdots$$

Setting $a_0 = 1$, solution of the system results in $a_1 = -4$, $a_2 = 17/3$, $a_3 = -47/12$, $a_4 = 191/120, \ldots$. Therefore one solution is

$$y_1(x) = x - 4x^2 + \frac{17}{3}x^3 - \frac{47}{12}x^4 + \ldots .$$

The exponents differ by an integer. So for a second solution, set

$$y_2(x) = a\, y_1(x) \ln x + 1 + c_1 x + c_2 x^2 + \ldots + c_n x^n + \ldots .$$

Substituting into the ODE, we obtain

$$a\, L[y_1(x)] \cdot \ln x + 2a\, y_1'(x) + 2a\, y_1(x) - a\frac{y_1(x)}{x} + L\left[1 + \sum_{n=1}^{\infty} c_n x^n\right] = 0.$$

Since $L[y_1(x)] = 0$, it follows that

$$L\left[1 + \sum_{n=1}^{\infty} c_n x^n\right] = -2a\, y_1'(x) - 2a\, y_1(x) + a\frac{y_1(x)}{x}.$$

More specifically,

$$\sum_{n=1}^{\infty} n(n+1)c_{n+1} x^n + 2 \sum_{n=1}^{\infty} n c_n x^n + 6 + (6 + 6c_1)x$$

$$+ (6c_2 + 6c_1 + 3)x^2 + \ldots = -a + 10ax - \frac{61}{3}ax^2 + \frac{193}{12}ax^3 + \ldots .$$

Equating the coefficients, we obtain the system of equations

$$6 = -a$$
$$2c_2 + 8c_1 + 6 = 10a$$
$$6c_3 + 10c_2 + 6c_1 + 3 = -\frac{61}{3}a$$
$$12c_4 + 12c_3 + 6c_2 + 3c_1 + 1 = \frac{193}{12}a$$
$$\vdots$$

Solving these equations for the coefficients, $a = -6$. In order to solve the remaining equations, set $c_1 = 0$. Then $c_2 = -33$, $c_3 = 449/6$, $c_4 = -1595/24, \ldots$. Therefore a second solution is

$$y_2(x) = -6\, y_1(x) \ln x + \left[1 - 33x^2 + \frac{449}{6}x^3 - \frac{1595}{24}x^4 + \ldots\right].$$

15.(a) Note the $p(x) = 6x/(x-1)$ and $q(x) = 3x^{-1}(x-1)^{-1}$. Furthermore, $x\,p(x) = 6x^2/(x-1)$ and $x^2 q(x) = 3x/(x-1)$. It follows that

$$p_0 = \lim_{x \to 0} \frac{6x^2}{x-1} = 0$$
$$q_0 = \lim_{x \to 0} \frac{3x}{x-1} = 0$$

and therefore $x = 0$ is a regular singular point.

(b) The indicial equation is given by $r(r-1) = 0$, that is, $r^2 - r = 0$, with roots $r_1 = 1$ and $r_2 = 0$.

(c) In order to find the solution corresponding to $r_1 = 1$, set $y = x\sum_{n=0}^{\infty} a_n x^n$. Upon substitution into the ODE, we have

$$\sum_{n=1}^{\infty} n(n+1)a_n x^{n+1} - \sum_{n=1}^{\infty} n(n+1)a_n x^n + 6\sum_{n=0}^{\infty}(n+1)a_n x^{n+2} + 3\sum_{n=0}^{\infty} a_n x^{n+1} = 0.$$

After adjusting the indices, it follows that

$$\sum_{n=2}^{\infty} n(n-1)a_{n-1} x^n - \sum_{n=1}^{\infty} n(n+1)a_n x^n + 6\sum_{n=2}^{\infty}(n-1)a_{n-2}x^n + 3\sum_{n=1}^{\infty} a_{n-1}x^n = 0.$$

That is,

$$-2a_1 + 3a_0 + \sum_{n=2}^{\infty}\left[-n(n+1)a_n + (n^2 - n + 3)a_{n-1} + 6(n-1)a_{n-2}\right]x^n = 0.$$

Setting the coefficients equal to zero, we have $a_1 = 3a_0/2$, and for $n \geq 2$,

$$n(n+1)a_n = (n^2 - n + 3)a_{n-1} + 6(n-1)a_{n-2}.$$

If we assign $a_0 = 1$, then we obtain $a_1 = 3/2$, $a_2 = 9/4$, $a_3 = 51/16$, ... Hence one solution is

$$y_1(x) = x + \frac{3}{2}x^2 + \frac{9}{4}x^3 + \frac{51}{16}x^4 + \frac{111}{40}x^5 + \ldots.$$

The exponents differ by an integer. So for a second solution, set

$$y_2(x) = a\, y_1(x) \ln x + 1 + c_1 x + c_2 x^2 + \ldots + c_n x^n + \ldots.$$

Substituting into the ODE, we obtain

$$2ax\, y_1'(x) - 2a\, y_1'(x) + 6ax\, y_1(x) - a\, y_1(x) + a\frac{y_1(x)}{x} + L\left[1 + \sum_{n=1}^{\infty} c_n x^n\right] = 0,$$

since $L\,[y_1(x)] = 0$. It follows that

$$L\left[1 + \sum_{n=1}^{\infty} c_n x^n\right] = 2a\, y_1'(x) - 2ax\, y_1'(x) + a\, y_1(x) - 6ax\, y_1(x) - a\frac{y_1(x)}{x}.$$

Now

$$L\left[1 + \sum_{n=1}^{\infty} c_n x^n\right] = 3 + (-2c_2 + 3c_1)x + (-6c_3 + 5c_2 + 6c_1)x^2 +$$
$$+ (-12c_4 + 9c_3 + 12c_2)x^3 + (-20c_5 + 15c_4 + 18c_3)x^4 + \ldots.$$

Substituting for $y_1(x)$, the right hand side of the ODE is

$$a + \frac{7}{2}ax + \frac{3}{4}ax^2 + \frac{33}{16}ax^3 - \frac{867}{80}ax^4 - \frac{441}{10}ax^5 + \ldots.$$

Equating the coefficients, we obtain the system of equations

$$3 = a$$
$$-2c_2 + 3c_1 = \frac{7}{2}a$$
$$-6c_3 + 5c_2 + 6c_1 = \frac{3}{4}a$$
$$-12c_4 + 9c_3 + 12c_2 = \frac{33}{16}a$$
$$\vdots$$

We find that $a = 3$. In order to solve the second equation, set $c_1 = 0$. Solution of the remaining equations results in $c_2 = -21/4$, $c_3 = -19/4$, $c_4 = -597/64$,.... Hence a second solution is

$$y_2(x) = 3\, y_1(x) \ln x + \left[1 - \frac{21}{4}x^2 - \frac{19}{4}x^3 - \frac{597}{64}x^4 + \ldots\right].$$

16.(a) After multiplying both sides of the ODE by x, we find that $x\,p(x) = 0$ and $x^2 q(x) = x$. Both of these functions are analytic at $x = 0$, hence $x = 0$ is a regular singular point.

(b) Furthermore, $p_0 = q_0 = 0$. So the indicial equation is $r(r-1) = 0$, with roots $r_1 = 1$ and $r_2 = 0$.

(c) In order to find the solution corresponding to $r_1 = 1$, set $y = x \sum_{n=0}^{\infty} a_n x^n$. Upon substitution into the ODE, we have

$$\sum_{n=1}^{\infty} n(n+1)a_n x^n + \sum_{n=0}^{\infty} a_n x^{n+1} = 0.$$

That is,

$$\sum_{n=1}^{\infty} [n(n+1)a_n + a_{n-1}] x^n = 0.$$

Setting the coefficients equal to zero, we find that for $n \geq 1$,

$$a_n = \frac{-a_{n-1}}{n(n+1)}.$$

It follows that

$$a_n = \frac{-a_{n-1}}{n(n+1)} = \frac{a_{n-2}}{(n-1)n^2(n+1)} = \ldots = \frac{(-1)^n a_0}{(n!)^2 (n+1)}.$$

Hence one solution is

$$y_1(x) = x - \frac{1}{2}x^2 + \frac{1}{12}x^3 - \frac{1}{144}x^4 + \frac{1}{2880}x^5 + \ldots.$$

The exponents differ by an integer. So for a second solution, set

$$y_2(x) = a\, y_1(x) \ln x + 1 + c_1 x + c_2 x^2 + \ldots + c_n x^n + \ldots.$$

Substituting into the ODE, we obtain

$$a L [y_1(x)] \cdot \ln x + 2a\, y_1'(x) - a\frac{y_1(x)}{x} + L\left[1 + \sum_{n=1}^{\infty} c_n x^n\right] = 0.$$

Since $L[y_1(x)] = 0$, it follows that

$$L\left[1 + \sum_{n=1}^{\infty} c_n x^n\right] = -2a\, y_1'(x) + a\frac{y_1(x)}{x}.$$

Now

$$L\left[1 + \sum_{n=1}^{\infty} c_n x^n\right] = 1 + (2c_2 + c_1)x + (6c_3 + c_2)x^2 + (12c_4 + c_3)x^3$$

$$+ (20c_5 + c_4)x^4 + (30c_6 + c_5)x^5 + \ldots.$$

Substituting for $y_1(x)$, the right hand side of the ODE is

$$-a + \frac{3}{2}ax - \frac{5}{12}ax^2 + \frac{7}{144}ax^3 - \frac{1}{320}ax^4 + \ldots.$$

Equating the coefficients, we obtain the system of equations

$$1 = -a$$
$$2c_2 + c_1 = \frac{3}{2}a$$
$$6c_3 + c_2 = -\frac{5}{12}a$$
$$12c_4 + c_3 = \frac{7}{144}a$$
$$\vdots$$

Evidently, $a = -1$. In order to solve the second equation, set $c_1 = 0$. We then find that $c_2 = -3/4$, $c_3 = 7/36$, $c_4 = -35/1728$, ... Therefore a second solution is

$$y_2(x) = -y_1(x)\ln x + \left[1 - \frac{3}{4}x^2 + \frac{7}{36}x^3 - \frac{35}{1728}x^4 + \dots\right].$$

19.(a) After dividing by the leading coefficient, we find that

$$p_0 = \lim_{x \to 0} x\,p(x) = \lim_{x \to 0} \frac{\gamma - (1 + \alpha + \beta)x}{1 - x} = \gamma.$$

$$q_0 = \lim_{x \to 0} x^2 q(x) = \lim_{x \to 0} \frac{-\alpha\beta\, x}{1 - x} = 0.$$

Hence $x = 0$ is a regular singular point. The indicial equation is $r(r-1) + \gamma r = 0$, with roots $r_1 = 1 - \gamma$ and $r_2 = 0$.

(b) For $x = 1$,

$$p_0 = \lim_{x \to 1}(x - 1)p(x) = \lim_{x \to 1} \frac{-\gamma + (1 + \alpha + \beta)x}{x} = 1 - \gamma + \alpha + \beta.$$

$$q_0 = \lim_{x \to 1}(x - 1)^2 q(x) = \lim_{x \to 1} \frac{\alpha\beta(x-1)}{x} = 0.$$

Hence $x = 1$ is a regular singular point. The indicial equation is

$$r^2 - (\gamma - \alpha - \beta)r = 0,$$

with roots $r_1 = \gamma - \alpha - \beta$ and $r_2 = 0$.

(c) Given that $r_1 - r_2$ is not a positive integer, we can set $y = \sum_{n=0}^{\infty} a_n x^n$. Substitution into the ODE results in

$$x(1-x)\sum_{n=2}^{\infty} n(n-1)a_n x^{n-2} + [\gamma - (1 + \alpha + \beta)x]\sum_{n=1}^{\infty} n a_n x^{n-1} - \alpha\beta\sum_{n=0}^{\infty} a_n x^n = 0.$$

That is,

$$\sum_{n=1}^{\infty} n(n+1)a_{n+1}x^n - \sum_{n=2}^{\infty} n(n-1)a_n x^n + \gamma\sum_{n=0}^{\infty}(n+1)a_{n+1}x^n$$

$$-(1+\alpha+\beta)\sum_{n=1}^{\infty} n a_n x^n - \alpha\beta \sum_{n=0}^{\infty} a_n x^n = 0.$$

Combining the series, we obtain

$$\gamma a_1 - \alpha\beta a_0 + [(2+2\gamma)a_2 - (1+\alpha+\beta+\alpha\beta)a_1]x + \sum_{n=2}^{\infty} A_n x^n = 0,$$

in which

$$A_n = (n+1)(n+\gamma)a_{n+1} - [n(n-1) + (1+\alpha+\beta)n + \alpha\beta]a_n.$$

Note that $n(n-1) + (1+\alpha+\beta)n + \alpha\beta = (n+\alpha)(n+\beta)$. Setting the coefficients equal to zero, we have $\gamma a_1 - \alpha\beta a_0 = 0$, and

$$a_{n+1} = \frac{(n+\alpha)(n+\beta)}{(n+1)(n+\gamma)} a_n$$

for $n \geq 1$. Hence one solution is

$$y_1(x) = 1 + \frac{\alpha\beta}{\gamma \cdot 1!}x + \frac{\alpha(\alpha+1)\beta(\beta+1)}{\gamma(\gamma+1) \cdot 2!}x^2 + \frac{\alpha(\alpha+1)(\alpha+2)\beta(\beta+1)(\beta+2)}{\gamma(\gamma+1)(\gamma+2) \cdot 3!}x^3 + \cdots.$$

Since the nearest other singularity is at $x = 1$, the radius of convergence of $y_1(x)$ will be at least $\rho = 1$.

(d) Given that $r_1 - r_2$ is not a positive integer, we can set $y = x^{1-\gamma}\sum_{n=0}^{\infty} b_n x^n$. Then substitution into the ODE results in

$$x(1-x)\sum_{n=0}^{\infty}(n+1-\gamma)(n-\gamma)a_n x^{n-\gamma-1}$$

$$+ [\gamma - (1+\alpha+\beta)x]\sum_{n=0}^{\infty}(n+1-\gamma)a_n x^{n-\gamma} - \alpha\beta \sum_{n=0}^{\infty} a_n x^{n+1-\gamma} = 0.$$

That is,

$$\sum_{n=0}^{\infty}(n+1-\gamma)(n-\gamma)a_n x^{n-\gamma} - \sum_{n=0}^{\infty}(n+1-\gamma)(n-\gamma)a_n x^{n+1-\gamma}$$

$$+ \gamma \sum_{n=0}^{\infty}(n+1-\gamma)a_n x^{n-\gamma} - (1+\alpha+\beta)\sum_{n=0}^{\infty}(n+1-\gamma)a_n x^{n+1-\gamma}$$

$$- \alpha\beta \sum_{n=0}^{\infty} a_n x^{n+1-\gamma} = 0.$$

After adjusting the indices,

$$\sum_{n=0}^{\infty}(n+1-\gamma)(n-\gamma)a_n x^{n-\gamma} - \sum_{n=1}^{\infty}(n-\gamma)(n-1-\gamma)a_{n-1} x^{n-\gamma}$$

$$+ \gamma \sum_{n=0}^{\infty}(n+1-\gamma)a_n x^{n-\gamma} - (1+\alpha+\beta)\sum_{n=1}^{\infty}(n-\gamma)a_{n-1} x^{n-\gamma} - \alpha\beta \sum_{n=1}^{\infty} a_{n-1} x^{n-\gamma} = 0.$$

Combining the series, we obtain

$$\sum_{n=1}^{\infty} B_n x^{n-\gamma} = 0,$$

in which

$$B_n = n(n+1-\gamma)b_n - [(n-\gamma)(n-\gamma+\alpha+\beta) + \alpha\beta]b_{n-1}.$$

Note that $(n-\gamma)(n-\gamma+\alpha+\beta) + \alpha\beta = (n+\alpha-\gamma)(n+\beta-\gamma)$. Setting $B_n = 0$, it follows that for $n \geq 1$,

$$b_n = \frac{(n+\alpha-\gamma)(n+\beta-\gamma)}{n(n+1-\gamma)} b_{n-1}.$$

Therefore a second solution is

$$y_2(x) = x^{1-\gamma}\left[1 + \frac{(1+\alpha-\gamma)(1+\beta-\gamma)}{(2-\gamma)1!}x \right.$$
$$\left. + \frac{(1+\alpha-\gamma)(2+\alpha-\gamma)(1+\beta-\gamma)(2+\beta-\gamma)}{(2-\gamma)(3-\gamma)2!}x^2 + \ldots\right].$$

(e) Under the transformation $x = 1/\xi$, the ODE becomes

$$\xi^4 \frac{1}{\xi}(1-\frac{1}{\xi})\frac{d^2y}{d\xi^2} + \left\{2\xi^3\frac{1}{\xi}(1-\frac{1}{\xi}) - \xi^2\left[\gamma - (1+\alpha+\beta)\frac{1}{\xi}\right]\right\}\frac{dy}{d\xi} - \alpha\beta\, y = 0.$$

That is,

$$(\xi^3 - \xi^2)\frac{d^2y}{d\xi^2} + [2\xi^2 - \gamma\xi^2 + (-1+\alpha+\beta)\xi]\frac{dy}{d\xi} - \alpha\beta\, y = 0.$$

Therefore $\xi = 0$ is a singular point. Note that

$$p(\xi) = \frac{(2-\gamma)\xi + (-1+\alpha+\beta)}{\xi^2 - \xi} \quad \text{and} \quad q(\xi) = \frac{-\alpha\beta}{\xi^3 - \xi^2}.$$

It follows that

$$p_0 = \lim_{\xi \to 0} \xi p(\xi) = \lim_{\xi \to 0} \frac{(2-\gamma)\xi + (-1+\alpha+\beta)}{\xi - 1} = 1 - \alpha - \beta,$$

$$q_0 = \lim_{\xi \to 0} \xi^2 q(\xi) = \lim_{\xi \to 0} \frac{-\alpha\beta}{\xi - 1} = \alpha\beta.$$

Hence $\xi = 0$ ($x = \infty$) is a regular singular point. The indicial equation is

$$r(r-1) + (1 - \alpha - \beta)r + \alpha\beta = 0,$$

or $r^2 - (\alpha + \beta)r + \alpha\beta = 0$. Evidently, the roots are $r = \alpha$ and $r = \beta$.

3. Here $xp(x) = 1$ and $x^2q(x) = 2x$, which are both analytic everywhere. We set $y = x^r(a_0 + a_1x + a_2x^2 + \ldots + a_nx^n + \ldots)$. Substitution into the ODE results in

$$\sum_{n=0}^{\infty}(r+n)(r+n-1)a_n x^{r+n} + \sum_{n=0}^{\infty}(r+n)a_n x^{r+n} + 2\sum_{n=0}^{\infty}a_n x^{r+n+1} = 0.$$

After adjusting the indices in the last series, we obtain

$$a_0[r(r-1) + r]x^r + \sum_{n=1}^{\infty}[(r+n)(r+n-1)a_n + (r+n)a_n + 2a_{n-1}]x^{r+n} = 0.$$

Assuming $a_0 \neq 0$, the indicial equation is $r^2 = 0$, with double root $r = 0$. Setting the remaining coefficients equal to zero, we have for $n \geq 1$,

$$a_n(r) = -\frac{2}{(n+r)^2}a_{n-1}(r).$$

It follows that

$$a_n(r) = \frac{(-1)^n 2^n}{[(n+r)(n+r-1)\ldots(1+r)]^2}a_0, \quad n \geq 1.$$

Since $r = 0$, one solution is given by

$$y_1(x) = \sum_{n=0}^{\infty}\frac{(-1)^n 2^n}{(n!)^2}x^n.$$

For a second linearly independent solution, we follow the discussion in Section 5.6. First note that

$$\frac{a_n'(r)}{a_n(r)} = -2\left[\frac{1}{n+r} + \frac{1}{n+r-1} + \ldots + \frac{1}{1+r}\right].$$

Setting $r = 0$,

$$a_n'(0) = -2H_n\, a_n(0) = -2H_n\frac{(-1)^n 2^n}{(n!)^2}.$$

Therefore,

$$y_2(x) = y_1(x)\ln x - 2\sum_{n=0}^{\infty}\frac{(-1)^n 2^n H_n}{(n!)^2}x^n.$$

4. Here $xp(x) = 4$ and $x^2q(x) = 2 + x$, which are both analytic everywhere. We set $y = x^r(a_0 + a_1x + a_2x^2 + \ldots + a_nx^n + \ldots)$. Substitution into the ODE results in

$$\sum_{n=0}^{\infty}(r+n)(r+n-1)a_n x^{r+n} + 4\sum_{n=0}^{\infty}(r+n)a_n x^{r+n}$$

$$+ \sum_{n=0}^{\infty}a_n x^{r+n+1} + 2\sum_{n=0}^{\infty}a_n x^{r+n} = 0.$$

After adjusting the indices in the second-to-last series, we obtain

$$a_0 \left[r(r-1) + 4r + 2\right] x^r$$

$$+ \sum_{n=1}^{\infty} \left[(r+n)(r+n-1)a_n + 4(r+n)a_n + 2\,a_n + a_{n-1}\right] x^{r+n} = 0.$$

Assuming $a_0 \neq 0$, the indicial equation is $r^2 + 3r + 2 = 0$, with roots $r_1 = -1$ and $r_2 = -2$. Setting the remaining coefficients equal to zero, we have for $n \geq 1$,

$$a_n(r) = -\frac{1}{(n+r+1)(n+r+2)}\, a_{n-1}(r).$$

It follows that

$$a_n(r) = \frac{(-1)^n}{\left[(n+r+1)(n+r)\ldots(2+r)\right]\left[(n+r+2)(n+r)\ldots(3+r)\right]}\, a_0, \quad n \geq 1.$$

Since $r_1 = -1$, one solution is given by

$$y_1(x) = x^{-1} \sum_{n=0}^{\infty} \frac{(-1)^n}{(n)!(n+1)!}\, x^n.$$

For a second linearly independent solution, we follow the discussion in Section 5.6. Since $r_1 - r_2 = N = 1$, we find that

$$a_1(r) = -\frac{1}{(r+2)(r+3)},$$

with $a_0 = 1$. Hence the leading coefficient in the solution is

$$a = \lim_{r \to -2} (r+2)\, a_1(r) = -1.$$

Further,

$$(r+2)\, a_n(r) = \frac{(-1)^n}{(n+r+2)\left[(n+r+1)(n+r)\ldots(3+r)\right]^2}.$$

Let $A_n(r) = (r+2)\, a_n(r)$. It follows that

$$\frac{A_n'(r)}{A_n(r)} = -\frac{1}{n+r+2} - 2\left[\frac{1}{n+r+1} + \frac{1}{n+r} + \ldots + \frac{1}{3+r}\right].$$

Setting $r = r_2 = -2$,

$$\frac{A_n'(-2)}{A_n(-2)} = -\frac{1}{n} - 2\left[\frac{1}{n-1} + \frac{1}{n-2} + \ldots + 1\right] = -H_n - H_{n-1}.$$

Hence

$$c_n(-2) = -(H_n + H_{n-1})\, A_n(-2) = -(H_n + H_{n-1})\frac{(-1)^n}{n!(n-1)!}.$$

Therefore,

$$y_2(x) = -y_1(x) \ln x + x^{-2}\left[1 - \sum_{n=1}^{\infty} \frac{(-1)^n (H_n + H_{n-1})}{n!(n-1)!}\, x^n\right].$$

6. Let $y(x) = v(x)/\sqrt{x}$. Then $y' = x^{-1/2} v' - x^{-3/2} v/2$ and $y'' = x^{-1/2} v'' - x^{-3/2} v' + 3 x^{-5/2} v/4$. Substitution into the ODE results in

$$\left[x^{3/2} v'' - x^{1/2} v' + 3 x^{-1/2} v/4 \right] + \left[x^{1/2} v' - x^{-1/2} v/2 \right] + (x^2 - \frac{1}{4}) x^{-1/2} v = 0.$$

Simplifying, we find that

$$v'' + v = 0,$$

with general solution $v(x) = c_1 \cos x + c_2 \sin x$. Hence

$$y(x) = c_1 x^{-1/2} \cos x + c_2 x^{-1/2} \sin x.$$

8. The absolute value of the ratio of consecutive terms is

$$\left| \frac{a_{2m+2} \, x^{2m+2}}{a_{2m} \, x^{2m}} \right| = \frac{|x|^{2m+2} \, 2^{2m} (m+1)! \, m!}{|x|^{2m} \, 2^{2m+2} (m+2)! (m+1)!} = \frac{|x|^2}{4(m+2)(m+1)}.$$

Applying the ratio test,

$$\lim_{m \to \infty} \left| \frac{a_{2m+2} \, x^{2m+2}}{a_{2m} \, x^{2m}} \right| = \lim_{m \to \infty} \frac{|x|^2}{4(m+2)(m+1)} = 0.$$

Hence the series for $J_1(x)$ converges absolutely for all values of x. Furthermore, since the series for $J_0(x)$ also converges absolutely for all x, term-by-term differentiation results in

$$J_0'(x) = \sum_{m=1}^{\infty} \frac{(-1)^m x^{2m-1}}{2^{2m-1} m!(m-1)!} = \sum_{m=0}^{\infty} \frac{(-1)^{m+1} x^{2m+1}}{2^{2m+1} (m+1)! \, m!} =$$

$$= -\frac{x}{2} \sum_{m=0}^{\infty} \frac{(-1)^m x^{2m}}{2^{2m} (m+1)! \, m!}.$$

Therefore, $J_0'(x) = -J_1(x)$.

9.(a) Note that $x \, p(x) = 1$ and $x^2 q(x) = x^2 - \nu^2$, which are both analytic at $x = 0$. Thus $x = 0$ is a regular singular point. Furthermore, $p_0 = 1$ and $q_0 = -\nu^2$. Hence the indicial equation is $r^2 - \nu^2 = 0$, with roots $r_1 = \nu$ and $r_2 = -\nu$.

(b) Set $y = x^r (a_0 + a_1 x + a_2 x^2 + \ldots + a_n x^n + \ldots)$. Substitution into the ODE results in

$$\sum_{n=0}^{\infty} (r+n)(r+n-1) a_n \, x^{r+n} + \sum_{n=0}^{\infty} (r+n) a_n \, x^{r+n}$$

$$+ \sum_{n=0}^{\infty} a_n x^{r+n+2} - \nu^2 \sum_{n=0}^{\infty} a_n x^{r+n} = 0.$$

After adjusting the indices in the second-to-last series, we obtain

$$a_0 \left[r(r-1) + r - \nu^2 \right] x^r + a_1 \left[(r+1)r + (r+1) - \nu^2 \right]$$

$$+ \sum_{n=2}^{\infty} \left[(r+n)(r+n-1) a_n + (r+n) a_n - \nu^2 a_n + a_{n-2} \right] x^{r+n} = 0.$$

Setting the coefficients equal to zero, we find that $a_1 = 0$, and

$$a_n = \frac{-1}{(r+n)^2 - \nu^2} a_{n-2},$$

for $n \geq 2$. It follows that $a_3 = a_5 = \ldots = a_{2m+1} = \ldots = 0$. Furthermore, with $r = \nu$,

$$a_n = \frac{-1}{n(n+2\nu)} a_{n-2}.$$

So for $m = 1, 2, \ldots$,

$$a_{2m} = \frac{-1}{2m(2m+2\nu)} a_{2m-2} = \frac{(-1)^m}{2^{2m} \, m!(1+\nu)(2+\nu)\ldots(m-1+\nu)(m+\nu)} a_0.$$

Hence one solution is

$$y_1(x) = x^\nu \left[1 + \sum_{m=1}^{\infty} \frac{(-1)^m}{m!(1+\nu)(2+\nu)\ldots(m-1+\nu)(m+\nu)} \left(\frac{x}{2}\right)^{2m} \right].$$

(c) Assuming that $r_1 - r_2 = 2\nu$ is not an integer, simply setting $r = -\nu$ in the above results in a second linearly independent solution

$$y_2(x) = x^{-\nu} \left[1 + \sum_{m=1}^{\infty} \frac{(-1)^m}{m!(1-\nu)(2-\nu)\ldots(m-1-\nu)(m-\nu)} \left(\frac{x}{2}\right)^{2m} \right].$$

(d) The absolute value of the ratio of consecutive terms in $y_1(x)$ is

$$\left| \frac{a_{2m+2} \, x^{2m+2}}{a_{2m} \, x^{2m}} \right| = \frac{|x|^{2m+2} \, 2^{2m} \, m!(1+\nu)\ldots(m+\nu)}{|x|^{2m} \, 2^{2m+2}(m+1)!(1+\nu)\ldots(m+1+\nu)}$$

$$= \frac{|x|^2}{4(m+1)(m+1+\nu)}.$$

Applying the ratio test,

$$\lim_{m \to \infty} \left| \frac{a_{2m+2} \, x^{2m+2}}{a_{2m} \, x^{2m}} \right| = \lim_{m \to \infty} \frac{|x|^2}{4(m+1)(m+1+\nu)} = 0.$$

Hence the series for $y_1(x)$ converges absolutely for all values of x. The same can be shown for $y_2(x)$. Note also, that if ν is a positive integer, then the coefficients in the series for $y_2(x)$ are undefined.

10.(a) It suffices to calculate $L[J_0(x) \ln x]$. Indeed,

$$[J_0(x) \ln x]' = J_0'(x) \ln x + \frac{J_0(x)}{x}$$

and

$$[J_0(x) \ln x]'' = J_0''(x) \ln x + 2\frac{J_0'(x)}{x} - \frac{J_0(x)}{x^2}.$$

Hence

$$L[J_0(x) \ln x] = x^2 J_0''(x) \ln x + 2x J_0'(x) - J_0(x)$$
$$+ x J_0'(x) \ln x + J_0(x) + x^2 J_0(x) \ln x.$$

Since $x^2 J_0''(x) + x J_0'(x) + x^2 J_0(x) = 0$,
$$L[J_0(x) \ln x] = 2x J_0'(x).$$

(b) Given that $L[y_2(x)] = 0$, after adjusting the indices in part (a), we have
$$b_1 x + 2^2 b_2 x^2 + \sum_{n=3}^{\infty} (n^2 b_n + b_{n-2}) x^n = -2x J_0'(x).$$

Using the series representation of $J_0'(x)$ in Problem 8,
$$b_1 x + 2^2 b_2 x^2 + \sum_{n=3}^{\infty} (n^2 b_n + b_{n-2}) x^n = -2 \sum_{n=1}^{\infty} \frac{(-1)^n (2n) x^{2n}}{2^{2n} (n!)^2}.$$

(c) Equating the coefficients on both sides of the equation, we find that
$$b_1 = b_3 = \ldots = b_{2m+1} = \ldots = 0.$$
Also, with $n = 1$, $2^2 b_2 = 1/(1!)^2$, that is, $b_2 = 1/\left[2^2 (1!)^2\right]$. Furthermore, for $m \geq 2$,
$$(2m)^2 b_{2m} + b_{2m-2} = -2 \frac{(-1)^m (2m)}{2^{2m} (m!)^2}.$$

More explicitly,
$$b_4 = -\frac{1}{2^2\, 4^2}\left(1 + \frac{1}{2}\right)$$
$$b_6 = \frac{1}{2^2\, 4^2\, 6^2}\left(1 + \frac{1}{2} + \frac{1}{3}\right)$$
$$\vdots$$

It can be shown, in general, that
$$b_{2m} = (-1)^{m+1} \frac{H_m}{2^{2m} (m!)^2}.$$

11. Bessel's equation of order one is
$$x^2 y'' + x y' + (x^2 - 1) y = 0.$$
Based on Problem 9, the roots of the indicial equation are $r_1 = 1$ and $r_2 = -1$. Set $y = x^r (a_0 + a_1 x + a_2 x^2 + \ldots + a_n x^n + \ldots)$. Substitution into the ODE results in
$$\sum_{n=0}^{\infty} (r+n)(r+n-1) a_n x^{r+n} + \sum_{n=0}^{\infty} (r+n) a_n x^{r+n}$$
$$+ \sum_{n=0}^{\infty} a_n x^{r+n+2} - \sum_{n=0}^{\infty} a_n x^{r+n} = 0.$$

After adjusting the indices in the second-to-last series, we obtain
$$a_0 \left[r(r-1) + r - 1\right] x^r + a_1 \left[(r+1)r + (r+1) - 1\right]$$

$$+ \sum_{n=2}^{\infty} [(r+n)(r+n-1)a_n + (r+n)a_n - a_n + a_{n-2}] x^{r+n} = 0.$$

Setting the coefficients equal to zero, we find that $a_1 = 0$, and

$$a_n(r) = \frac{-1}{(r+n)^2 - 1} a_{n-2}(r) = \frac{-1}{(n+r+1)(n+r-1)} a_{n-2}(r),$$

for $n \geq 2$. It follows that $a_3 = a_5 = \ldots = a_{2m+1} = \ldots = 0$. Solving the recurrence relation,

$$a_{2m}(r) = \frac{(-1)^m}{(2m+r+1)(2m+r-1)^2 \ldots (r+3)^2(r+1)} a_0.$$

With $r = r_1 = 1$,

$$a_{2m}(1) = \frac{(-1)^m}{2^{2m}(m+1)! \, m!} a_0.$$

For a second linearly independent solution, we follow the discussion in Section 5.6. Since $r_1 - r_2 = N = 2$, we find that

$$a_2(r) = -\frac{1}{(r+3)(r+1)},$$

with $a_0 = 1$. Hence the leading coefficient in the solution is

$$a = \lim_{r \to -1} (r+1) \, a_2(r) = -\frac{1}{2}.$$

Further,

$$(r+1) \, a_{2m}(r) = \frac{(-1)^m}{(2m+r+1) \left[(2m+r-1) \ldots (3+r)\right]^2}.$$

Let $A_n(r) = (r+1) \, a_n(r)$. It follows that

$$\frac{A'_{2m}(r)}{A_{2m}(r)} = -\frac{1}{2m+r+1} - 2\left[\frac{1}{2m+r-1} + \ldots + \frac{1}{3+r}\right].$$

Setting $r = r_2 = -1$, we calculate

$$c_{2m}(-1) = -\frac{1}{2}(H_m + H_{m-1}) A_{2m}(-1)$$

$$= -\frac{1}{2}(H_m + H_{m-1}) \frac{(-1)^m}{2m \left[(2m-2) \ldots 2\right]^2} = -\frac{1}{2}(H_m + H_{m-1}) \frac{(-1)^m}{2^{2m-1} \, m!(m-1)!}.$$

Note that $a_{2m+1}(r) = 0$ implies that $A_{2m+1}(r) = 0$, so

$$c_{2m+1}(-1) = \left[\frac{d}{dr} A_{2m+1}(r)\right]_{r=r_2} = 0.$$

Therefore,

$$y_2(x) = -\frac{1}{2} \left[x \sum_{m=0}^{\infty} \frac{(-1)^m}{(m+1)! \, m!} \left(\frac{x}{2}\right)^{2m} \right] \ln x$$

$$+\frac{1}{x}\left[1 - \sum_{m=1}^{\infty} \frac{(-1)^m(H_m + H_{m-1})}{m!(m-1)!}\left(\frac{x}{2}\right)^{2m}\right].$$

Based on the definition of $J_1(x)$,

$$y_2(x) = -J_1(x)\ln x + \frac{1}{x}\left[1 - \sum_{m=1}^{\infty} \frac{(-1)^m(H_m + H_{m-1})}{m!(m-1)!}\left(\frac{x}{2}\right)^{2m}\right].$$

CHAPTER 6

The Laplace Transform

6.1

3.

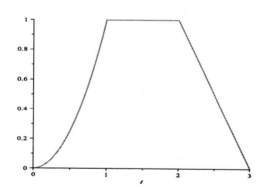

The function $f(t)$ is continuous.

4.

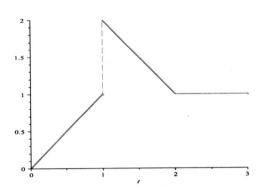

The function $f(t)$ has a jump discontinuity at $t = 1$, and is thus piecewise continuous.

7. Integration is a linear operation. It follows that

$$\int_0^A \cosh bt \cdot e^{-st} dt = \frac{1}{2}\int_0^A e^{bt} \cdot e^{-st} dt + \frac{1}{2}\int_0^A e^{-bt} \cdot e^{-st} dt =$$

$$= \frac{1}{2}\int_0^A e^{(b-s)t} dt + \frac{1}{2}\int_0^A e^{-(b+s)t} dt.$$

Hence

$$\int_0^A \cosh bt \cdot e^{-st} dt = \frac{1}{2}\left[\frac{1 - e^{(b-s)A}}{s - b}\right] + \frac{1}{2}\left[\frac{1 - e^{-(b+s)A}}{s + b}\right].$$

Taking a limit, as $A \to \infty$,

$$\int_0^\infty \cosh bt \cdot e^{-st} dt = \frac{1}{2}\left[\frac{1}{s - b}\right] + \frac{1}{2}\left[\frac{1}{s + b}\right] = \frac{s}{s^2 - b^2}.$$

Note that the above is valid for $s > |b|$.

8. Proceeding as in Problem 7,

$$\int_0^A \sinh bt \cdot e^{-st} dt = \frac{1}{2}\left[\frac{1 - e^{(b-s)A}}{s - b}\right] - \frac{1}{2}\left[\frac{1 - e^{-(b+s)A}}{s + b}\right].$$

Taking a limit, as $A \to \infty$,

$$\int_0^\infty \sinh bt \cdot e^{-st} dt = \frac{1}{2}\left[\frac{1}{s - b}\right] - \frac{1}{2}\left[\frac{1}{s + b}\right] = \frac{b}{s^2 - b^2}.$$

The limit exists as long as $s > |b|$.

10. Observe that $e^{at} \sinh bt = (e^{(a+b)t} - e^{(a-b)t})/2$. It follows that

$$\int_0^A e^{at} \sinh bt \cdot e^{-st} dt = \frac{1}{2}\left[\frac{1 - e^{(a+b-s)A}}{s - a + b}\right] - \frac{1}{2}\left[\frac{1 - e^{-(b-a+s)A}}{s + b - a}\right].$$

Taking a limit, as $A \to \infty$,

$$\int_0^\infty e^{at} \sinh bt \cdot e^{-st} dt = \frac{1}{2}\left[\frac{1}{s-a+b}\right] - \frac{1}{2}\left[\frac{1}{s+b-a}\right] = \frac{b}{(s-a)^2 - b^2}.$$

The limit exists as long as $s - a > |b|$.

11. Using the linearity of the Laplace transform,

$$\mathcal{L}[\sin bt] = \frac{1}{2i}\mathcal{L}[e^{ibt}] - \frac{1}{2i}\mathcal{L}[e^{-ibt}].$$

Since

$$\int_0^\infty e^{(a+ib)t} e^{-st} dt = \frac{1}{s-a-ib},$$

we have

$$\int_0^\infty e^{\pm ibt} e^{-st} dt = \frac{1}{s \mp ib}.$$

Therefore

$$\mathcal{L}[\sin bt] = \frac{1}{2i}\left[\frac{1}{s-ib} - \frac{1}{s+ib}\right] = \frac{b}{s^2 + b^2}.$$

The formula holds for $s > 0$.

12. Using the linearity of the Laplace transform,

$$\mathcal{L}[\cos bt] = \frac{1}{2}\mathcal{L}[e^{ibt}] + \frac{1}{2}\mathcal{L}[e^{-ibt}].$$

From Problem 11, we have

$$\int_0^\infty e^{\pm ibt} e^{-st} dt = \frac{1}{s \mp ib}.$$

Therefore

$$\mathcal{L}[\cos bt] = \frac{1}{2}\left[\frac{1}{s-ib} + \frac{1}{s+ib}\right] = \frac{s}{s^2 + b^2}.$$

The formula holds for $s > 0$.

14. Using the linearity of the Laplace transform,

$$\mathcal{L}[e^{at} \cos bt] = \frac{1}{2}\mathcal{L}[e^{(a+ib)t}] + \frac{1}{2}\mathcal{L}[e^{(a-ib)t}].$$

Based on the integration in Problem 11,

$$\int_0^\infty e^{(a \pm ib)t} e^{-st} dt = \frac{1}{s - a \mp ib}.$$

Therefore

$$\mathcal{L}[e^{at} \cos bt] = \frac{1}{2}\left[\frac{1}{s-a-ib} + \frac{1}{s-a+ib}\right] = \frac{s-a}{(s-a)^2 + b^2}.$$

The above is valid for $s > a$.

15. Integrating by parts,

$$\int_0^A te^{at} \cdot e^{-st}dt = -\frac{te^{(a-s)t}}{s-a}\Big|_0^A + \int_0^A \frac{1}{s-a}e^{(a-s)t}dt =$$

$$= \frac{1 - e^{A(a-s)} + A(a-s)e^{A(a-s)}}{(s-a)^2}.$$

Taking a limit, as $A \to \infty$,

$$\int_0^\infty te^{at} \cdot e^{-st}dt = \frac{1}{(s-a)^2}.$$

Note that the limit exists as long as $s > a$.

17. Observe that $t\cosh at = (te^{at} + te^{-at})/2$. For any value of c,

$$\int_0^A te^{ct} \cdot e^{-st}dt = -\frac{te^{(c-s)t}}{s-c}\Big|_0^A + \int_0^A \frac{1}{s-c}e^{(c-s)t}dt =$$

$$= \frac{1 - e^{A(c-s)} + A(c-s)e^{A(c-s)}}{(s-c)^2}.$$

Taking a limit, as $A \to \infty$,

$$\int_0^\infty te^{ct} \cdot e^{-st}dt = \frac{1}{(s-c)^2}.$$

Note that the limit exists as long as $s > |c|$. Therefore,

$$\int_0^\infty t\cosh at \cdot e^{-st}dt = \frac{1}{2}\left[\frac{1}{(s-a)^2} + \frac{1}{(s+a)^2}\right] = \frac{s^2 + a^2}{(s-a)^2(s+a)^2}.$$

18. Integrating by parts,

$$\int_0^A t^n e^{at} \cdot e^{-st}dt = -\frac{t^n e^{(a-s)t}}{s-a}\Big|_0^A + \int_0^A \frac{n}{s-a}t^{n-1}e^{(a-s)t}dt =$$

$$= -\frac{A^n e^{-(s-a)A}}{s-a} + \int_0^A \frac{n}{s-a}t^{n-1}e^{(a-s)t}dt.$$

Continuing to integrate by parts, it follows that

$$\int_0^A t^n e^{at} \cdot e^{-st}dt = -\frac{A^n e^{(a-s)A}}{s-a} - \frac{nA^{n-1}e^{(a-s)A}}{(s-a)^2} -$$

$$-\frac{n!Ae^{(a-s)A}}{(n-2)!(s-a)^3} - \cdots - \frac{n!(e^{(a-s)A} - 1)}{(s-a)^{n+1}}.$$

That is,

$$\int_0^A t^n e^{at} \cdot e^{-st}dt = p_n(A) \cdot e^{(a-s)A} + \frac{n!}{(s-a)^{n+1}},$$

in which $p_n(\xi)$ is a polynomial of degree n. For any given polynomial,

$$\lim_{A\to\infty} p_n(A) \cdot e^{-(s-a)A} = 0,$$

as long as $s > a$. Therefore,
$$\int_0^\infty t^n e^{at} \cdot e^{-st} dt = \frac{n!}{(s-a)^{n+1}}.$$

19. First observe that
$$\int \sin at \cdot e^{-st} dt = -\frac{\sin at \cdot e^{-st}}{s} + \frac{1}{s} \int a \cos at \cdot e^{-st} dt =$$
$$= -\frac{\sin at \cdot e^{-st}}{s} + \frac{a}{s}\left[-\frac{\cos at \cdot e^{-st}}{s} - \frac{a}{s}\int \sin at \cdot e^{-st} dt\right].$$

This implies that
$$\int \sin at \cdot e^{-st} dt = -\frac{(s \sin at + a \cos at)e^{-st}}{s^2 + a^2}.$$

Integrating by parts we obtain that
$$\int_0^A t^2 \sin at \cdot e^{-st} dt = -t^2 \frac{(s \sin at + a \cos at)e^{-st}}{s^2 + a^2}\bigg|_0^A +$$
$$+ \int_0^A 2t \frac{(s \sin at + a \cos at)e^{-st}}{s^2 + a^2} dt.$$

Taking the limit $A \to \infty$ and using the results of Problem 16 (from the Student Solutions Manual), we obtain that
$$\int_0^\infty t^2 \sin at \cdot e^{-st} dt = \frac{2s}{s^2+a^2} \frac{2as}{(s^2+a^2)^2} + \frac{2a}{s^2+a^2} \frac{s^2-a^2}{(s^2+a^2)^2} = \frac{2a(3s^2 - a^2)}{(s^2+a^2)^3}.$$

This is valid for $s > 0$.

20. Observe that $t^2 \sinh at = (t^2 e^{at} - t^2 e^{-at})/2$. Using the result in Problem 18,
$$\int_0^\infty t^2 \sinh at \cdot e^{-st} dt = \frac{1}{2}\left[\frac{2!}{(s-a)^3} - \frac{2!}{(s+a)^3}\right] = \frac{2a(3s^2+a^2)}{(s^2-a^2)^3}.$$

The above is valid for $s > |a|$.

22. Using the fact that $f(t) = 0$ when $t \geq 1$ and integration by parts, we obtain that
$$\mathcal{L}[f(t)] = \int_0^\infty e^{-st} f(t)\, dt = \int_0^1 e^{-st} t\, dt = \left[-\frac{e^{-st}}{s} t\right]_0^1 + \int_0^1 \frac{e^{-st}}{s} dt$$
$$= -\frac{e^{-s}}{s} + \left[-\frac{e^{-st}}{s^2}\right]_0^1 = -\frac{e^{-s}}{s} - \frac{e^{-s}}{s^2} + \frac{1}{s^2}.$$

23. Using the definition of the Laplace transform and Problem 22, we get that
$$\mathcal{L}[f(t)] = \int_0^\infty e^{-st} f(t)\, dt = \int_0^1 e^{-st} t\, dt + \int_1^\infty e^{-st}\, dt =$$

$$= -\frac{e^{-s}}{s} - \frac{e^{-s}}{s^2} + \frac{1}{s^2} + \frac{e^{-s}}{s} = -\frac{e^{-s}}{s^2} + \frac{1}{s^2}.$$

26. Integrating by parts,

$$\int_0^A t e^{-t} dt = -t e^{-t} \Big|_0^A + \int_0^A e^{-t} dt = 1 - e^{-A} - A e^{-A}.$$

Taking a limit, as $A \to \infty$,

$$\int_0^\infty t e^{-t} dt = 1.$$

Hence the integral converges.

27. Based on a series expansion, note that for $t > 0$, $e^t > 1 + t + t^2/2 > t^2/2$. It follows that for $t > 0$, $t^{-2} e^t > 1/2$. Hence for any finite $A > 1$,

$$\int_1^A t^{-2} e^t dt > \frac{A-1}{2}.$$

It is evident that the limit as $A \to \infty$ does not exist.

28. Using the fact that $|\cos t| \le 1$, and the fact that

$$\int_0^\infty e^{-t} dt = 1,$$

it follows that the given integral converges.

30.(a) Let $p > 0$. Integrating by parts,

$$\int_0^A e^{-x} x^p dx = -e^{-x} x^p \Big|_0^A + p \int_0^A e^{-x} x^{p-1} dx = -A^p e^{-A} + p \int_0^A e^{-x} x^{p-1} dx.$$

Taking a limit, as $A \to \infty$,

$$\int_0^\infty e^{-x} x^p dx = p \int_0^\infty e^{-x} x^{p-1} dx.$$

That is, $\Gamma(p+1) = p \Gamma(p)$.

(b) Setting $p = 0$,

$$\Gamma(1) = \int_0^\infty e^{-x} dx = 1.$$

(c) Let $p = n$. Using the result in part (a),

$$\Gamma(n+1) = n \Gamma(n)$$
$$= n(n-1) \Gamma(n-1)$$
$$\vdots$$
$$= n(n-1)(n-2) \cdots 2 \cdot 1 \cdot \Gamma(1).$$

Since $\Gamma(1) = 1$, $\Gamma(n+1) = n!$.

(d) Using the result in part (a),
$$\Gamma(p+n) = (p+n-1)\Gamma(p+n-1)$$
$$= (p+n-1)(p+n-2)\Gamma(p+n-2)$$
$$\vdots$$
$$= (p+n-1)(p+n-2)\cdots(p+1)p\,\Gamma(p).$$

Hence
$$\frac{\Gamma(p+n)}{\Gamma(p)} = p(p+1)(p+2)\cdots(p+n-1).$$

Given that $\Gamma(1/2) = \sqrt{\pi}$, it follows that
$$\Gamma(\tfrac{3}{2}) = \tfrac{1}{2}\Gamma(\tfrac{1}{2}) = \frac{\sqrt{\pi}}{2}$$

and
$$\Gamma(\tfrac{11}{2}) = \frac{9}{2}\cdot\frac{7}{2}\cdot\frac{5}{2}\cdot\frac{3}{2}\Gamma(\tfrac{3}{2}) = \frac{945\sqrt{\pi}}{32}.$$

6.2

1. Write the function as
$$\frac{3}{s^2+4} = \frac{3}{2}\frac{2}{s^2+4}.$$
Hence $\mathcal{L}^{-1}[Y(s)] = 3\sin 2t\,/2$.

3. Using partial fractions,
$$\frac{2}{s^2+3s-4} = \frac{2}{5}\left[\frac{1}{s-1} - \frac{1}{s+4}\right].$$
Hence $\mathcal{L}^{-1}[Y(s)] = 2(e^t - e^{-4t})/5$.

5. Note that the denominator $s^2 + 2s + 5$ is irreducible over the reals. Completing the square, $s^2 + 2s + 5 = (s+1)^2 + 4$. Now convert the function to a rational function of the variable $\xi = s + 1$. That is,
$$\frac{2s+2}{s^2+2s+5} = \frac{2(s+1)}{(s+1)^2+4}.$$
We know that
$$\mathcal{L}^{-1}\left[\frac{2\xi}{\xi^2+4}\right] = 2\cos 2t.$$
Using the fact that $\mathcal{L}[e^{at}f(t)] = \mathcal{L}[f(t)]_{s\to s-a}$,
$$\mathcal{L}^{-1}\left[\frac{2s+2}{s^2+2s+5}\right] = 2e^{-t}\cos 2t.$$

6. Using partial fractions,
$$\frac{2s-3}{s^2-4} = \frac{1}{4}\left[\frac{1}{s-2} + \frac{7}{s+2}\right].$$
Hence $\mathcal{L}^{-1}[Y(s)] = (e^{2t} + 7e^{-2t})/4$. Note that we can also write
$$\frac{2s-3}{s^2-4} = 2\frac{s}{s^2-4} - \frac{3}{2}\frac{2}{s^2-4}.$$

8. Using partial fractions,
$$\frac{8s^2 - 4s + 12}{s(s^2+4)} = 3\frac{1}{s} + 5\frac{s}{s^2+4} - 2\frac{2}{s^2+4}.$$
Hence $\mathcal{L}^{-1}[Y(s)] = 3 + 5\cos 2t - 2\sin 2t$.

9. The denominator $s^2 + 4s + 5$ is irreducible over the reals. Completing the square, $s^2 + 4s + 5 = (s+2)^2 + 1$. Now convert the function to a rational function of the variable $\xi = s + 2$. That is,
$$\frac{1-2s}{s^2+4s+5} = \frac{5-2(s+2)}{(s+2)^2+1}.$$
We find that
$$\mathcal{L}^{-1}\left[\frac{5}{\xi^2+1} - \frac{2\xi}{\xi^2+1}\right] = 5\sin t - 2\cos t.$$
Using the fact that $\mathcal{L}[e^{at}f(t)] = \mathcal{L}[f(t)]_{s\to s-a}$,
$$\mathcal{L}^{-1}\left[\frac{1-2s}{s^2+4s+5}\right] = e^{-2t}(5\sin t - 2\cos t).$$

10. Note that the denominator $s^2 + 2s + 10$ is irreducible over the reals. Completing the square, $s^2 + 2s + 10 = (s+1)^2 + 9$. Now convert the function to a rational function of the variable $\xi = s + 1$. That is,
$$\frac{2s-3}{s^2+2s+10} = \frac{2(s+1)-5}{(s+1)^2+9}.$$
We find that
$$\mathcal{L}^{-1}\left[\frac{2\xi}{\xi^2+9} - \frac{5}{\xi^2+9}\right] = 2\cos 3t - \frac{5}{3}\sin 3t.$$
Using the fact that $\mathcal{L}[e^{at}f(t)] = \mathcal{L}[f(t)]_{s\to s-a}$,
$$\mathcal{L}^{-1}\left[\frac{2s-3}{s^2+2s+10}\right] = e^{-t}\left(2\cos 3t - \frac{5}{3}\sin 3t\right).$$

12. Taking the Laplace transform of the ODE, we obtain
$$s^2 Y(s) - s\,y(0) - y'(0) + 3[sY(s) - y(0)] + 2Y(s) = 0.$$

Applying the initial conditions,
$$s^2 Y(s) + 3s Y(s) + 2 Y(s) - s - 3 = 0.$$

Solving for $Y(s)$, the transform of the solution is
$$Y(s) = \frac{s+3}{s^2 + 3s + 2}.$$

Using partial fractions,
$$\frac{s+3}{s^2+3s+2} = \frac{2}{s+1} - \frac{1}{s+2}.$$

Hence $y(t) = \mathcal{L}^{-1}[Y(s)] = 2e^{-t} - e^{-2t}$.

13. Taking the Laplace transform of the ODE, we obtain
$$s^2 Y(s) - s y(0) - y'(0) - 2[s Y(s) - y(0)] + 2 Y(s) = 0.$$

Applying the initial conditions,
$$s^2 Y(s) - 2s Y(s) + 2 Y(s) - 1 = 0.$$

Solving for $Y(s)$, the transform of the solution is
$$Y(s) = \frac{1}{s^2 - 2s + 2}.$$

Since the denominator is irreducible, write the transform as a function of $\xi = s - 1$. That is,
$$\frac{1}{s^2 - 2s + 2} = \frac{1}{(s-1)^2 + 1}.$$

First note that
$$\mathcal{L}^{-1}\left[\frac{1}{\xi^2 + 1}\right] = \sin t.$$

Using the fact that $\mathcal{L}[e^{at} f(t)] = \mathcal{L}[f(t)]_{s \to s-a}$,
$$\mathcal{L}^{-1}\left[\frac{1}{s^2 - 2s + 2}\right] = e^t \sin t.$$

Hence $y(t) = e^t \sin t$.

16. Taking the Laplace transform of the ODE, we obtain
$$s^2 Y(s) - s y(0) - y'(0) + 2[s Y(s) - y(0)] + 5 Y(s) = 0.$$

Applying the initial conditions,
$$s^2 Y(s) + 2s Y(s) + 5 Y(s) - 2s - 3 = 0.$$

Solving for $Y(s)$, the transform of the solution is
$$Y(s) = \frac{2s+3}{s^2 + 2s + 5}.$$

Since the denominator is irreducible, write the transform as a function of $\xi = s+1$. That is,
$$\frac{2s+3}{s^2+2s+5} = \frac{2(s+1)+1}{(s+1)^2+4}.$$

We know that
$$\mathcal{L}^{-1}\left[\frac{2\xi}{\xi^2+4} + \frac{1}{\xi^2+4}\right] = 2\cos 2t + \frac{1}{2}\sin 2t.$$

Using the fact that $\mathcal{L}[e^{at}f(t)] = \mathcal{L}[f(t)]_{s \to s-a}$, the solution of the IVP is
$$y(t) = \mathcal{L}^{-1}\left[\frac{2s+3}{s^2+2s+5}\right] = e^{-t}\left(2\cos 2t + \frac{1}{2}\sin 2t\right).$$

18. Taking the Laplace transform of the ODE, we obtain
$$s^4 Y(s) - s^3 y(0) - s^2 y'(0) - s y''(0) - y'''(0) - Y(s) = 0.$$

Applying the initial conditions,
$$s^4 Y(s) - Y(s) - s^3 - s = 0.$$

Solving for the transform of the solution,
$$Y(s) = \frac{s}{s^2-1}.$$

By inspection, it follows that $y(t) = \mathcal{L}^{-1}[Y(s)] = \cosh t$.

19. Taking the Laplace transform of the ODE, we obtain
$$s^4 Y(s) - s^3 y(0) - s^2 y'(0) - s y''(0) - y'''(0) - 4Y(s) = 0.$$

Applying the initial conditions,
$$s^4 Y(s) - 4Y(s) - s^3 + 2s = 0.$$

Solving for the transform of the solution,
$$Y(s) = \frac{s}{s^2+2}.$$

It follows that $y(t) = \mathcal{L}^{-1}[Y(s)] = \cos\sqrt{2}\, t$.

21. Taking the Laplace transform of both sides of the ODE, we obtain
$$s^2 Y(s) - s y(0) - y'(0) - 2[s Y(s) - y(0)] + 2Y(s) = \frac{s}{s^2+1}.$$

Applying the initial conditions,
$$s^2 Y(s) - 2s Y(s) + 2Y(s) - s + 2 = \frac{s}{s^2+1}.$$

Solving for $Y(s)$, the transform of the solution is
$$Y(s) = \frac{s}{(s^2-2s+2)(s^2+1)} + \frac{s-2}{s^2-2s+2}.$$

Using partial fractions on the first term,

$$\frac{s}{(s^2 - 2s + 2)(s^2 + 1)} = \frac{1}{5}\left[\frac{s - 2}{s^2 + 1} - \frac{s - 4}{s^2 - 2s + 2}\right].$$

Thus we can write

$$Y(s) = \frac{1}{5}\frac{s}{s^2 + 1} - \frac{2}{5}\frac{1}{s^2 + 1} + \frac{2}{5}\frac{2s - 3}{s^2 - 2s + 2}.$$

For the last term, we note that $s^2 - 2s + 2 = (s - 1)^2 + 1$. So that

$$\frac{2s - 3}{s^2 - 2s + 2} = \frac{2(s - 1) - 1}{(s - 1)^2 + 1}.$$

We know that

$$\mathcal{L}^{-1}\left[\frac{2\xi}{\xi^2 + 1} - \frac{1}{\xi^2 + 1}\right] = 2\cos t - \sin t.$$

Based on the translation property of the Laplace transform,

$$\mathcal{L}^{-1}\left[\frac{2s - 3}{s^2 - 2s + 2}\right] = e^t(2\cos t - \sin t).$$

Combining the above, the solution of the IVP is

$$y(t) = \frac{1}{5}\cos t - \frac{2}{5}\sin t + \frac{2}{5}e^t(2\cos t - \sin t).$$

23. Taking the Laplace transform of both sides of the ODE, we obtain

$$s^2 Y(s) - s\,y(0) - y'(0) + 2[s\,Y(s) - y(0)] + Y(s) = \frac{4}{s + 1}.$$

Applying the initial conditions,

$$s^2 Y(s) + 2s\,Y(s) + Y(s) - 2s - 3 = \frac{4}{s + 1}.$$

Solving for $Y(s)$, the transform of the solution is

$$Y(s) = \frac{4}{(s + 1)^3} + \frac{2s + 3}{(s + 1)^2}.$$

First write

$$\frac{2s + 3}{(s + 1)^2} = \frac{2(s + 1) + 1}{(s + 1)^2} = \frac{2}{s + 1} + \frac{1}{(s + 1)^2}.$$

We note that

$$\mathcal{L}^{-1}\left[\frac{4}{\xi^3} + \frac{2}{\xi} + \frac{1}{\xi^2}\right] = 2t^2 + 2 + t.$$

So based on the translation property of the Laplace transform, the solution of the IVP is

$$y(t) = 2t^2 e^{-t} + t e^{-t} + 2 e^{-t}.$$

25. Let $f(t)$ be the forcing function on the right-hand-side. Taking the Laplace transform of both sides of the ODE, we obtain

$$s^2 Y(s) - s\,y(0) - y'(0) + Y(s) = \mathcal{L}\left[f(t)\right].$$

Applying the initial conditions,

$$s^2 Y(s) + Y(s) = \mathcal{L}\left[f(t)\right].$$

Based on the definition of the Laplace transform,

$$\mathcal{L}\left[f(t)\right] = \int_0^\infty f(t)\,e^{-st} dt = \int_0^1 t\,e^{-st} dt = \frac{1}{s^2} - \frac{e^{-s}}{s} - \frac{e^{-s}}{s^2}.$$

Solving for the transform,

$$Y(s) = \frac{1}{s^2(s^2+1)} - e^{-s}\,\frac{s+1}{s^2(s^2+1)}.$$

26. Let $f(t)$ be the forcing function on the right-hand-side. Taking the Laplace transform of both sides of the ODE, we obtain

$$s^2 Y(s) - s\,y(0) - y'(0) + 4\,Y(s) = \mathcal{L}\left[f(t)\right].$$

Applying the initial conditions,

$$s^2 Y(s) + 4\,Y(s) = \mathcal{L}\left[f(t)\right].$$

Based on the definition of the Laplace transform,

$$\mathcal{L}\left[f(t)\right] = \int_0^\infty f(t)\,e^{-st} dt = \int_0^1 t\,e^{-st} dt + \int_1^\infty e^{-st} dt = \frac{1}{s^2} - \frac{e^{-s}}{s^2}.$$

Solving for the transform,

$$Y(s) = \frac{1}{s^2(s^2+4)} - e^{-s}\,\frac{1}{s^2(s^2+4)}.$$

29.(a) Assuming that the conditions of Theorem 6.2.1 are satisfied,

$$F'(s) = \frac{d}{ds}\int_0^\infty e^{-st} f(t) dt = \int_0^\infty \frac{\partial}{\partial s}\left[e^{-st} f(t)\right] dt =$$

$$= \int_0^\infty \left[-t\,e^{-st} f(t)\right] dt = \int_0^\infty e^{-st}\left[-t f(t)\right] dt.$$

(b) Using mathematical induction, suppose that for some $k \geq 1$,

$$F^{(k)}(s) = \int_0^\infty e^{-st}\left[(-t)^k f(t)\right] dt.$$

Differentiating both sides,

$$F^{(k+1)}(s) = \frac{d}{ds}\int_0^\infty e^{-st}\left[(-t)^k f(t)\right] dt = \int_0^\infty \frac{\partial}{\partial s}\left[e^{-st}(-t)^k f(t)\right] dt =$$

$$= \int_0^\infty \left[-t\,e^{-st}(-t)^k f(t)\right] dt = \int_0^\infty e^{-st}\left[(-t)^{k+1} f(t)\right] dt.$$

30. We know that
$$\mathcal{L}\left[e^{at}\right] = \frac{1}{s-a}.$$
Based on Problem 29,
$$\mathcal{L}\left[-t\, e^{at}\right] = \frac{d}{ds}\left[\frac{1}{s-a}\right].$$
Therefore,
$$\mathcal{L}\left[t\, e^{at}\right] = \frac{1}{(s-a)^2}.$$

32. Based on Problem 29,
$$\mathcal{L}\left[(-t)^n\right] = \frac{d^n}{ds^n}\mathcal{L}\left[1\right] = \frac{d^n}{ds^n}\left[\frac{1}{s}\right].$$
Therefore,
$$\mathcal{L}\left[t^n\right] = (-1)^n \frac{(-1)^n n!}{s^{n+1}} = \frac{n!}{s^{n+1}}.$$

34. Using the translation property of the Laplace transform,
$$\mathcal{L}\left[e^{at}\sin bt\right] = \frac{b}{(s-a)^2 + b^2}.$$
Therefore,
$$\mathcal{L}\left[t\, e^{at}\sin bt\right] = -\frac{d}{ds}\left[\frac{b}{(s-a)^2+b^2}\right] = \frac{2b(s-a)}{((s-a)^2+b^2)^2}.$$

35. Using the translation property of the Laplace transform,
$$\mathcal{L}\left[e^{at}\cos bt\right] = \frac{s-a}{(s-a)^2+b^2}.$$
Therefore,
$$\mathcal{L}\left[t\, e^{at}\cos bt\right] = -\frac{d}{ds}\left[\frac{s-a}{(s-a)^2+b^2}\right] = \frac{(s-a)^2 - b^2}{((s-a)^2+b^2)^2}.$$

36.(a) Taking the Laplace transform of the given Bessel equation,
$$\mathcal{L}[t\,y''] + \mathcal{L}[\,y'\,] + \mathcal{L}[t\,y] = 0.$$
Using the differentiation property of the transform,
$$-\frac{d}{ds}\mathcal{L}[y''] + \mathcal{L}[\,y'\,] - \frac{d}{ds}\mathcal{L}[y] = 0.$$
That is,
$$-\frac{d}{ds}\left[s^2 Y(s) - s\,y(0) - y'(0)\right] + s\,Y(s) - y(0) - \frac{d}{ds}Y(s) = 0.$$
It follows that
$$(1+s^2)Y'(s) + s\,Y(s) = 0.$$

(b) We obtain a first-order linear ODE in $Y(s)$:

$$Y'(s) + \frac{s}{s^2+1}Y(s) = 0,$$

with integrating factor

$$\mu(s) = e^{\int \frac{s}{s^2+1}ds} = \sqrt{s^2+1}.$$

The first-order ODE can be written as

$$\frac{d}{ds}\left[\sqrt{s^2+1} \cdot Y(s)\right] = 0,$$

with solution

$$Y(s) = \frac{c}{\sqrt{s^2+1}}.$$

(c) In order to obtain negative powers of s, first write

$$\frac{1}{\sqrt{s^2+1}} = \frac{1}{s}\left[1 + \frac{1}{s^2}\right]^{-1/2}.$$

Expanding $(1+1/s^2)^{-1/2}$ in a binomial series,

$$\frac{1}{\sqrt{1+(1/s^2)}} = 1 - \frac{1}{2}s^{-2} + \frac{1\cdot 3}{2\cdot 4}s^{-4} - \frac{1\cdot 3\cdot 5}{2\cdot 4\cdot 6}s^{-6} + \cdots,$$

valid for $s^{-2} < 1$. Hence, we can formally express $Y(s)$ as

$$Y(s) = c\left[\frac{1}{s} - \frac{1}{2}\frac{1}{s^3} + \frac{1\cdot 3}{2\cdot 4}\frac{1}{s^5} - \frac{1\cdot 3\cdot 5}{2\cdot 4\cdot 6}\frac{1}{s^7} + \cdots\right].$$

Assuming that term-by-term inversion is valid,

$$y(t) = c\left[1 - \frac{1}{2}\frac{t^2}{2!} + \frac{1\cdot 3}{2\cdot 4}\frac{t^4}{4!} - \frac{1\cdot 3\cdot 5}{2\cdot 4\cdot 6}\frac{t^6}{6!} + \cdots\right]$$

$$= c\left[1 - \frac{2!}{2^2}\frac{t^2}{2!} + \frac{4!}{2^2\cdot 4^2}\frac{t^4}{4!} - \frac{6!}{2^2\cdot 4^2\cdot 6^2}\frac{t^6}{6!} + \cdots\right].$$

It follows that

$$y(t) = c\left[1 - \frac{1}{2^2}t^2 + \frac{1}{2^2\cdot 4^2}t^4 - \frac{1}{2^2\cdot 4^2\cdot 6^2}t^6 + \cdots\right] = c\sum_{n=0}^{\infty}\frac{(-1)^n}{2^{2n}(n!)^2}t^{2n}.$$

The series is evidently the expansion, about $x = 0$, of $J_0(t)$.

38. By definition of the Laplace transform, given the appropriate conditions,

$$\mathcal{L}[g(t)] = \int_0^{\infty} e^{-st}\left[\int_0^t f(\tau)d\tau\right] dt = \int_0^{\infty}\int_0^t e^{-st}f(\tau)d\tau dt.$$

Assuming that the order of integration can be exchanged,

$$\mathcal{L}[g(t)] = \int_0^{\infty} f(\tau)\left[\int_{\tau}^{\infty} e^{-st}dt\right] d\tau = \int_0^{\infty} f(\tau)\left[\frac{e^{-s\tau}}{s}\right] d\tau.$$

(Note the region of integration is the area between the lines $\tau(t) = t$ and $\tau(t) = 0$.) Hence
$$\mathcal{L}[g(t)] = \frac{1}{s}\int_0^\infty f(\tau)e^{-s\tau}d\tau = \frac{1}{s}\mathcal{L}[f(t)].$$

6.3

1.

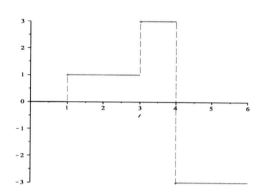

3.

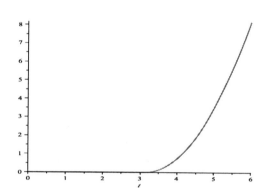

5.

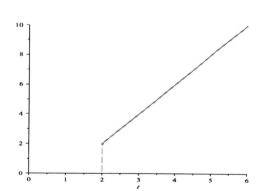

6.

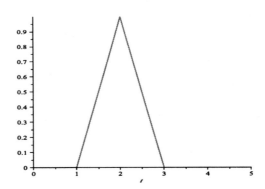

8.(a)

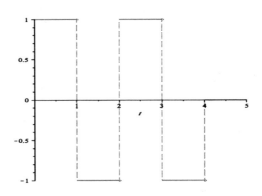

(b) $f(t) = 1 - 2u_1(t) + 2u_2(t) - 2u_3(t) + u_4(t)$.

9.(a)

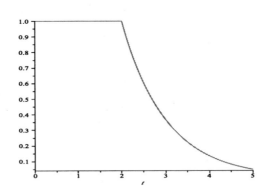

(b) $f(t) = 1 + (e^{-(t-2)} - 1)u_2(t)$.

11.(a)

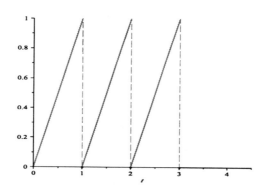

(b) $f(t) = t - u_1(t) - u_2(t) + (2-t)u_3(t)$.

12.(a)

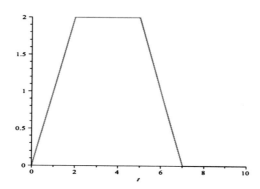

(b) $f(t) = t + (2-t)u_2(t) + (5-t)u_5(t) + (t-7)u_7(t)$.

13. Using the Heaviside function, we can write $f(t) = (t-2)^2 u_2(t)$. The Laplace transform has the property that $\mathcal{L}[u_c(t)f(t-c)] = e^{-cs}\mathcal{L}[f(t)]$. Hence

$$\mathcal{L}\left[(t-2)^2 u_2(t)\right] = \frac{2e^{-2s}}{s^3}.$$

15. The function can be expressed as $f(t) = (t-\pi)[u_\pi(t) - u_{2\pi}(t)]$. Before invoking the translation property of the transform, write the function as

$$f(t) = (t-\pi)u_\pi(t) - (t-2\pi)u_{2\pi}(t) - \pi u_{2\pi}(t).$$

It follows that

$$\mathcal{L}[f(t)] = \frac{e^{-\pi s}}{s^2} - \frac{e^{-2\pi s}}{s^2} - \frac{\pi e^{-2\pi s}}{s}.$$

16. It follows directly from the translation property of the transform that

$$\mathcal{L}[f(t)] = \frac{e^{-s}}{s} + 2\frac{e^{-3s}}{s} - 6\frac{e^{-4s}}{s}.$$

17. Before invoking the translation property of the transform, write the function as
$$f(t) = (t-2)\,u_2(t) - u_2(t) - (t-3)\,u_3(t) - u_3(t).$$
It follows that
$$\mathcal{L}[f(t)] = \frac{e^{-2s}}{s^2} - \frac{e^{-2s}}{s} - \frac{e^{-3s}}{s^2} - \frac{e^{-3s}}{s}.$$

18. It follows directly from the translation property of the transform that
$$\mathcal{L}[f(t)] = \frac{1}{s^2} - \frac{e^{-s}}{s^2}.$$

19. Using the fact that $\mathcal{L}[e^{at} f(t)] = \mathcal{L}[f(t)]_{s \to s-a}$,
$$\mathcal{L}^{-1}\left[\frac{3!}{(s-2)^4}\right] = t^3 e^{2t}.$$

22. The inverse transform of the function $2/(s^2 - 4)$ is $f(t) = \sinh 2t$. Using the translation property of the transform,
$$\mathcal{L}^{-1}\left[\frac{2 e^{-2s}}{s^2 - 4}\right] = \sinh\left(2(t-2)\right) \cdot u_2(t).$$

23. First consider the function
$$G(s) = \frac{(s-2)}{s^2 - 4s + 3}.$$
Completing the square in the denominator,
$$G(s) = \frac{(s-2)}{(s-2)^2 - 1}.$$
It follows that $\mathcal{L}^{-1}[G(s)] = e^{2t} \cosh t$. Hence
$$\mathcal{L}^{-1}\left[\frac{(s-2)e^{-s}}{s^2 - 4s + 3}\right] = e^{2(t-1)} \cosh(t-1)\, u_1(t).$$

24. Write the function as
$$F(s) = \frac{e^{-s}}{s} + \frac{e^{-2s}}{s} - \frac{e^{-3s}}{s} - \frac{e^{-4s}}{s}.$$
It follows from the translation property of the transform, that
$$\mathcal{L}^{-1}\left[\frac{e^{-s} + e^{-2s} - e^{-3s} - e^{-4s}}{s}\right] = u_1(t) + u_2(t) - u_3(t) - u_4(t).$$

25.(a) By definition of the Laplace transform,
$$\mathcal{L}[f(ct)] = \int_0^\infty e^{-st} f(ct)\,dt.$$

6.3

Making a change of variable, $\tau = ct$, we have

$$\mathcal{L}[f(ct)] = \frac{1}{c}\int_0^\infty e^{-s(\tau/c)}f(\tau)d\tau = \frac{1}{c}\int_0^\infty e^{-(s/c)\tau}f(\tau)d\tau.$$

Hence $\mathcal{L}[f(ct)] = (1/c)F(s/c)$, where $s/c > a$.

(b) Using the result in part (a),

$$\mathcal{L}\left[f\left(\frac{t}{k}\right)\right] = kF(ks).$$

Hence

$$\mathcal{L}^{-1}[F(ks)] = \frac{1}{k}f\left(\frac{t}{k}\right).$$

(c) From part (b), $\mathcal{L}^{-1}[F(as)] = (1/a)f(t/a)$ Note that $as + b = a(s + b/a)$. Using the fact that $\mathcal{L}[e^{ct}f(t)] = \mathcal{L}[f(t)]_{s \to s-c}$,

$$\mathcal{L}^{-1}[F(as+b)] = e^{-bt/a}\frac{1}{a}f\left(\frac{t}{a}\right).$$

26. First write

$$F(s) = \frac{n!}{(\frac{s}{2})^{n+1}}.$$

Let $G(s) = n!/s^{n+1}$. Based on the results in Problem 25,

$$\frac{1}{2}\mathcal{L}^{-1}\left[G\left(\frac{s}{2}\right)\right] = g(2t),$$

in which $g(t) = t^n$. Hence $\mathcal{L}^{-1}[F(s)] = 2(2t)^n = 2^{n+1}t^n$.

29. First write

$$F(s) = \frac{e^{-4(s-1/2)}}{2(s-1/2)}.$$

Now consider

$$G(s) = \frac{e^{-2s}}{s}.$$

Using the result in Problem 25(b),

$$\mathcal{L}^{-1}[G(2s)] = \frac{1}{2}g\left(\frac{t}{2}\right),$$

in which $g(t) = u_2(t)$. Hence $\mathcal{L}^{-1}[G(2s)] = u_2(t/2)/2 = u_4(t)/2$. It follows that

$$\mathcal{L}^{-1}[F(s)] = \frac{1}{2}e^{t/2}u_4(t).$$

30. By definition of the Laplace transform,

$$\mathcal{L}[f(t)] = \int_0^\infty e^{-st}u_1(t)dt.$$

That is,
$$\mathcal{L}[f(t)] = \int_0^1 e^{-st}\,dt = \frac{1-e^{-s}}{s}.$$

31. First write the function as $f(t) = u_0(t) - u_1(t) + u_2(t) - u_3(t)$. It follows that
$$\mathcal{L}[f(t)] = \int_0^1 e^{-st}dt + \int_2^3 e^{-st}dt.$$

That is,
$$\mathcal{L}[f(t)] = \frac{1-e^{-s}}{s} + \frac{e^{-2s}-e^{-3s}}{s} = \frac{1-e^{-s}+e^{-2s}-e^{-3s}}{s}.$$

32. The transform may be computed directly. On the other hand, using the translation property of the transform,
$$\mathcal{L}[f(t)] = \frac{1}{s} + \sum_{k=1}^{2n+1}(-1)^k \frac{e^{-ks}}{s} = \frac{1}{s}\left[\sum_{k=0}^{2n+1}(-e^{-s})^k\right] = \frac{1}{s}\frac{1-(-e^{-s})^{2n+2}}{1+e^{-s}}.$$

That is,
$$\mathcal{L}[f(t)] = \frac{1-(e^{-2s})^{n+1}}{s(1+e^{-s})}.$$

35. The given function is periodic, with $T = 2$. Using the result of Problem 34,
$$\mathcal{L}[f(t)] = \frac{1}{1-e^{-2s}}\int_0^2 e^{-st}f(t)dt = \frac{1}{1-e^{-2s}}\int_0^1 e^{-st}dt.$$

That is,
$$\mathcal{L}[f(t)] = \frac{1-e^{-s}}{s(1-e^{-2s})} = \frac{1}{s(1+e^{-s})}.$$

37. The function is periodic, with $T = 1$. Using the result of Problem 34,
$$\mathcal{L}[f(t)] = \frac{1}{1-e^{-s}}\int_0^1 t\,e^{-st}dt.$$

It follows that
$$\mathcal{L}[f(t)] = \frac{1-e^{-s}(1+s)}{s^2(1-e^{-s})}.$$

38. The function is periodic, with $T = \pi$. Using the result of Problem 34,
$$\mathcal{L}[f(t)] = \frac{1}{1-e^{-\pi s}}\int_0^\pi \sin t \cdot e^{-st}dt.$$

We first calculate
$$\int_0^\pi \sin t \cdot e^{-st}dt = \frac{1+e^{-\pi s}}{1+s^2}.$$

Hence
$$\mathcal{L}[f(t)] = \frac{1+e^{-\pi s}}{(1-e^{-\pi s})(1+s^2)}.$$

39.(a)

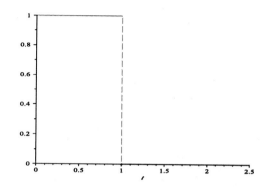

We get that
$$\mathcal{L}[f(t)] = \mathcal{L}[1] - \mathcal{L}[u_1(t)] = \frac{1}{s} - \frac{e^{-s}}{s}.$$

(b)

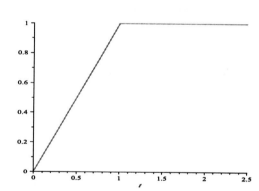

Let $F(s) = \mathcal{L}[1 - u_1(t)]$. Then
$$\mathcal{L}\left[\int_0^t [1 - u_1(\tau)]\, d\tau\right] = \frac{1}{s} F(s) = \frac{1 - e^{-s}}{s^2}.$$

(c)

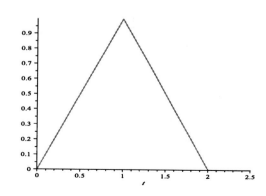

Let $G(s) = \mathcal{L}[g(t)]$. Then

$$\mathcal{L}[h(t)] = G(s) - e^{-s}G(s) = \frac{1-e^{-s}}{s^2} - e^{-s}\frac{1-e^{-s}}{s^2} = \frac{(1-e^{-s})^2}{s^2}.$$

40.(a)

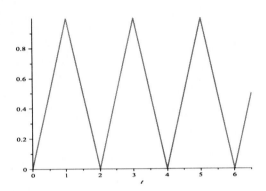

(b) The given function is periodic, with $T = 2$. Using the result of Problem 34,

$$\mathcal{L}[f(t)] = \frac{1}{1-e^{-2s}}\int_0^2 e^{-st}p(t)dt.$$

Based on the piecewise definition of $p(t)$,

$$\int_0^2 e^{-st}p(t)dt = \int_0^1 te^{-st}dt + \int_1^2 (2-t)e^{-st}dt = \frac{1}{s^2}(1-e^{-s})^2.$$

Hence

$$\mathcal{L}[p(t)] = \frac{(1-e^{-s})}{s^2(1+e^{-s})}.$$

(c) Since $p(t)$ satisfies the hypotheses of Theorem 6.2.1, $\mathcal{L}[p'(t)] = s\mathcal{L}[p(t)] - p(0)$. Using the result of Problem 36 from the Student Solutions Manual,

$$\mathcal{L}[p'(t)] = \frac{(1-e^{-s})}{s(1+e^{-s})}.$$

We note the $p(0) = 0$, hence

$$\mathcal{L}[p(t)] = \frac{1}{s}\left[\frac{(1-e^{-s})}{s(1+e^{-s})}\right].$$

6.4

2.(a) Let $h(t)$ be the forcing function on the right-hand-side. Taking the Laplace transform of both sides of the ODE, we obtain

$$s^2Y(s) - sy(0) - y'(0) + 2[sY(s) - y(0)] + 2Y(s) = \mathcal{L}[h(t)].$$

Applying the initial conditions,
$$s^2 Y(s) + 2s Y(s) + 2 Y(s) - 1 = \mathcal{L}[h(t)].$$
The forcing function can be written as $h(t) = u_\pi(t) - u_{2\pi}(t)$. Its transform is
$$\mathcal{L}[h(t)] = \frac{e^{-\pi s} - e^{-2\pi s}}{s}.$$
Solving for $Y(s)$, the transform of the solution is
$$Y(s) = \frac{1}{s^2 + 2s + 2} + \frac{e^{-\pi s} - e^{-2\pi s}}{s(s^2 + 2s + 2)}.$$
First note that
$$\frac{1}{s^2 + 2s + 2} = \frac{1}{(s+1)^2 + 1}.$$
Using partial fractions,
$$\frac{1}{s(s^2 + 2s + 2)} = \frac{1}{2}\frac{1}{s} - \frac{1}{2}\frac{(s+1)+1}{(s+1)^2 + 1}.$$
Taking the inverse transform, term-by-term,
$$\mathcal{L}\left[\frac{1}{s^2 + 2s + 2}\right] = \mathcal{L}\left[\frac{1}{(s+1)^2 + 1}\right] = e^{-t} \sin t.$$
Now let
$$G(s) = \frac{1}{s(s^2 + 2s + 2)}.$$
Then
$$\mathcal{L}^{-1}[G(s)] = \frac{1}{2} - \frac{1}{2}e^{-t} \cos t - \frac{1}{2}e^{-t} \sin t.$$
Using Theorem 6.3.1,
$$\mathcal{L}^{-1}\left[e^{-cs} G(s)\right] = \frac{1}{2}u_c(t) - \frac{1}{2}e^{-(t-c)}\left[\cos(t-c) + \sin(t-c)\right]u_c(t).$$
Hence the solution of the IVP is
$$y(t) = e^{-t} \sin t + \frac{1}{2}u_\pi(t) - \frac{1}{2}e^{-(t-\pi)}\left[\cos(t-\pi) + \sin(t-\pi)\right]u_\pi(t) -$$
$$- \frac{1}{2}u_{2\pi}(t) + \frac{1}{2}e^{-(t-2\pi)}\left[\cos(t-2\pi) + \sin(t-2\pi)\right]u_{2\pi}(t).$$
That is,
$$y(t) = e^{-t} \sin t + \frac{1}{2}\left[u_\pi(t) - u_{2\pi}(t)\right] + \frac{1}{2}e^{-(t-\pi)}\left[\cos t + \sin t\right]u_\pi(t)+$$
$$+ \frac{1}{2}e^{-(t-2\pi)}\left[\cos t + \sin t\right]u_{2\pi}(t).$$

(b)

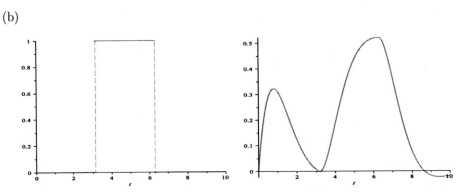

The solution starts out as free oscillation, due to the initial conditions. The amplitude increases, as long as the forcing is present. Thereafter, the solution rapidly decays.

4.(a) Let $h(t)$ be the forcing function on the right-hand-side. Taking the Laplace transform of both sides of the ODE, we obtain

$$s^2 Y(s) - s y(0) - y'(0) + 4 Y(s) = \mathcal{L}\left[h(t)\right].$$

Applying the initial conditions,

$$s^2 Y(s) + 4 Y(s) = \mathcal{L}\left[h(t)\right].$$

The transform of the forcing function is

$$\mathcal{L}\left[h(t)\right] = \frac{1}{s^2+1} + \frac{e^{-\pi s}}{s^2+1}.$$

Solving for $Y(s)$, the transform of the solution is

$$Y(s) = \frac{1}{(s^2+4)(s^2+1)} + \frac{e^{-\pi s}}{(s^2+4)(s^2+1)}.$$

Using partial fractions,

$$\frac{1}{(s^2+4)(s^2+1)} = \frac{1}{3}\left[\frac{1}{s^2+1} - \frac{1}{s^2+4}\right].$$

It follows that

$$\mathcal{L}^{-1}\left[\frac{1}{(s^2+4)(s^2+1)}\right] = \frac{1}{3}\left[\sin t - \frac{1}{2}\sin 2t\right].$$

Based on Theorem 6.3.1,

$$\mathcal{L}^{-1}\left[\frac{e^{-\pi s}}{(s^2+4)(s^2+1)}\right] = \frac{1}{3}\left[\sin(t-\pi) - \frac{1}{2}\sin(2t-2\pi)\right] u_\pi(t).$$

Hence the solution of the IVP is

$$y(t) = \frac{1}{3}\left[\sin t - \frac{1}{2}\sin 2t\right] - \frac{1}{3}\left[\sin t + \frac{1}{2}\sin 2t\right] u_\pi(t).$$

(b)

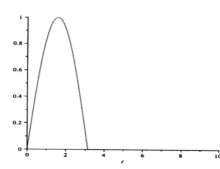
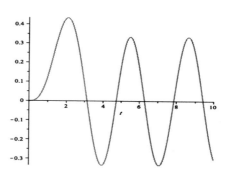

Since there is no damping term, the solution follows the forcing function, after which the response is a steady oscillation about $y = 0$.

5.(a) Let $f(t)$ be the forcing function on the right-hand-side. Taking the Laplace transform of both sides of the ODE, we obtain

$$s^2 Y(s) - s\, y(0) - y'(0) + 3\left[s\, Y(s) - y(0)\right] + 2Y(s) = \mathcal{L}\left[f(t)\right].$$

Applying the initial conditions,

$$s^2 Y(s) + 3s\, Y(s) + 2Y(s) = \mathcal{L}\left[f(t)\right].$$

The transform of the forcing function is

$$\mathcal{L}\left[f(t)\right] = \frac{1}{s} - \frac{e^{-10s}}{s}.$$

Solving for the transform,

$$Y(s) = \frac{1}{s(s^2 + 3s + 2)} - \frac{e^{-10s}}{s(s^2 + 3s + 2)}.$$

Using partial fractions,

$$\frac{1}{s(s^2 + 3s + 2)} = \frac{1}{2}\left[\frac{1}{s} + \frac{1}{s+2} - \frac{2}{s+1}\right].$$

Hence

$$\mathcal{L}^{-1}\left[\frac{1}{s(s^2 + 3s + 2)}\right] = \frac{1}{2} + \frac{e^{-2t}}{2} - e^{-t}.$$

Based on Theorem 6.3.1,

$$\mathcal{L}^{-1}\left[\frac{e^{-10s}}{s(s^2 + 3s + 2)}\right] = \frac{1}{2}\left[1 + e^{-2(t-10)} - 2e^{-(t-10)}\right] u_{10}(t).$$

Hence the solution of the IVP is

$$y(t) = \frac{1}{2}\left[1 - u_{10}(t)\right] + \frac{e^{-2t}}{2} - e^{-t} - \frac{1}{2}\left[e^{-(2t-20)} - 2e^{-(t-10)}\right] u_{10}(t).$$

(b)

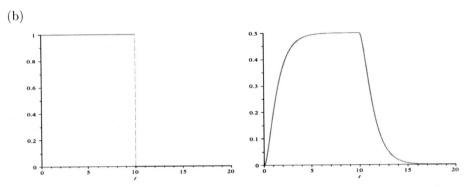

The solution increases to a temporary steady value of $y = 1/2$. After the forcing ceases, the response decays exponentially to $y = 0$.

6.(a) Taking the Laplace transform of both sides of the ODE, we obtain
$$s^2 Y(s) - s y(0) - y'(0) + 3 [s Y(s) - y(0)] + 2 Y(s) = \frac{e^{-2s}}{s}.$$

Applying the initial conditions,
$$s^2 Y(s) + 3s Y(s) + 2 Y(s) - 1 = \frac{e^{-2s}}{s}.$$

Solving for the transform,
$$Y(s) = \frac{1}{s^2 + 3s + 2} + \frac{e^{-2s}}{s(s^2 + 3s + 2)}.$$

Using partial fractions,
$$\frac{1}{s^2 + 3s + 2} = \frac{1}{s+1} - \frac{1}{s+2}$$

and
$$\frac{1}{s(s^2 + 3s + 2)} = \frac{1}{2}\left[\frac{1}{s} + \frac{1}{s+2} - \frac{2}{s+1}\right].$$

Taking the inverse transform. term-by-term, the solution of the IVP is
$$y(t) = e^{-t} - e^{-2t} + \left[\frac{1}{2} - e^{-(t-2)} + \frac{1}{2}e^{-2(t-2)}\right] u_2(t).$$

(b)

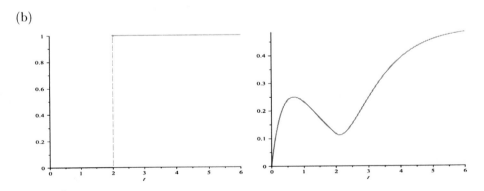

Due to the initial conditions, the response has a transient overshoot, followed by an exponential convergence to a steady value of $y_s = 1/2$.

7.(a) Taking the Laplace transform of both sides of the ODE, we obtain

$$s^2 Y(s) - s\, y(0) - y'(0) + Y(s) = \frac{e^{-3\pi s}}{s}.$$

Applying the initial conditions,

$$s^2 Y(s) + Y(s) - s = \frac{e^{-3\pi s}}{s}.$$

Solving for the transform,

$$Y(s) = \frac{s}{s^2+1} + \frac{e^{-3\pi s}}{s(s^2+1)}.$$

Using partial fractions,

$$\frac{1}{s(s^2+1)} = \frac{1}{s} - \frac{s}{s^2+1}.$$

Hence

$$Y(s) = \frac{s}{s^2+1} + e^{-3\pi s}\left[\frac{1}{s} - \frac{s}{s^2+1}\right].$$

Taking the inverse transform, the solution of the IVP is

$$y(t) = \cos t + [1 - \cos(t - 3\pi)]\, u_{3\pi}(t) = \cos t + [1 + \cos t]\, u_{3\pi}(t).$$

(b)

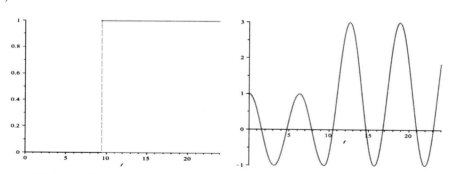

Due to initial conditions, the solution temporarily oscillates about $y = 0$. After the forcing is applied, the response is a steady oscillation about $y_m = 1$.

9.(a) Let $g(t)$ be the forcing function on the right-hand-side. Taking the Laplace transform of both sides of the ODE, we obtain

$$s^2 Y(s) - s\, y(0) - y'(0) + Y(s) = \mathcal{L}\left[g(t)\right].$$

Applying the initial conditions,

$$s^2 Y(s) + Y(s) - 1 = \mathcal{L}\left[g(t)\right].$$

The forcing function can be written as

$$g(t) = \frac{t}{2}[1 - u_6(t)] + 3\, u_6(t) = \frac{t}{2} - \frac{1}{2}(t-6)u_6(t)$$

with Laplace transform

$$\mathcal{L}\left[g(t)\right] = \frac{1}{2s^2} - \frac{e^{-6s}}{2s^2}.$$

Solving for the transform,

$$Y(s) = \frac{1}{s^2+1} + \frac{1}{2s^2(s^2+1)} - \frac{e^{-6s}}{2s^2(s^2+1)}.$$

Using partial fractions,

$$\frac{1}{2s^2(s^2+1)} = \frac{1}{2}\left[\frac{1}{s^2} - \frac{1}{s^2+1}\right].$$

Taking the inverse transform, and using Theorem 6.3.1, the solution of the IVP is

$$y(t) = \sin t + \frac{1}{2}[t - \sin t] - \frac{1}{2}[(t-6) - \sin(t-6)]\, u_6(t)$$
$$= \frac{1}{2}[t + \sin t] - \frac{1}{2}[(t-6) - \sin(t-6)]\, u_6(t).$$

(b)

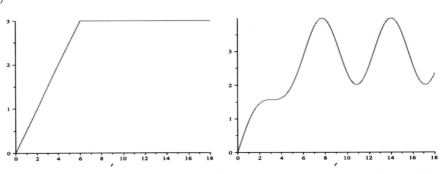

The solution increases, in response to the ramp input, and thereafter oscillates about a mean value of $y_m = 3$.

11.(a) Taking the Laplace transform of both sides of the ODE, we obtain

$$s^2 Y(s) - s\, y(0) - y'(0) + 4\, Y(s) = \frac{e^{-\pi s}}{s} - \frac{e^{-3\pi s}}{s}.$$

Applying the initial conditions,

$$s^2 Y(s) + 4\, Y(s) = \frac{e^{-\pi s}}{s} - \frac{e^{-3\pi s}}{s}.$$

Solving for the transform,

$$Y(s) = \frac{e^{-\pi s}}{s(s^2+4)} - \frac{e^{-3\pi s}}{s(s^2+4)}.$$

Using partial fractions,
$$\frac{1}{s(s^2+4)} = \frac{1}{4}\left[\frac{1}{s} - \frac{s}{s^2+4}\right].$$

Taking the inverse transform, and applying Theorem 6.3.1,
$$y(t) = \frac{1}{4}[1 - \cos(2t - 2\pi)]u_\pi(t) - \frac{1}{4}[1 - \cos(2t - 6\pi)]u_{3\pi}(t)$$
$$= \frac{1}{4}[u_\pi(t) - u_{3\pi}(t)] - \frac{1}{4}\cos 2t \cdot [u_\pi(t) - u_{3\pi}(t)].$$

(b)

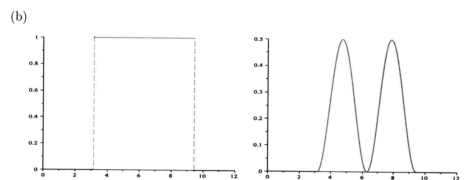

Since there is no damping term, the solution responds immediately to the forcing input. There is a temporary oscillation about $y = 1/4$.

12.(a) Taking the Laplace transform of the ODE, we obtain
$$s^4 Y(s) - s^3 y(0) - s^2 y'(0) - s y''(0) - y'''(0) - Y(s) = \frac{e^{-s}}{s} - \frac{e^{-2s}}{s}.$$

Applying the initial conditions,
$$s^4 Y(s) - Y(s) = \frac{e^{-s}}{s} - \frac{e^{-2s}}{s}.$$

Solving for the transform of the solution,
$$Y(s) = \frac{e^{-s}}{s(s^4-1)} - \frac{e^{-2s}}{s(s^4-1)}.$$

Using partial fractions,
$$\frac{1}{s(s^4-1)} = \frac{1}{4}\left[-\frac{4}{s} + \frac{1}{s+1} + \frac{1}{s-1} + \frac{2s}{s^2+1}\right].$$

It follows that
$$\mathcal{L}^{-1}\left[\frac{1}{s(s^4-1)}\right] = \frac{1}{4}\left[-4 + e^{-t} + e^t + 2\cos t\right].$$

Based on Theorem 6.3.1, the solution of the IVP is
$$y(t) = -[u_1(t) - u_2(t)] + \frac{1}{4}\left[e^{-(t-1)} + e^{(t-1)} + 2\cos(t-1)\right]u_1(t) -$$
$$- \frac{1}{4}\left[e^{-(t-2)} + e^{(t-2)} + 2\cos(t-2)\right]u_2(t).$$

(b)

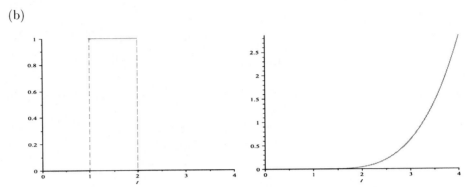

The solution increases without bound, exponentially.

13.(a) Taking the Laplace transform of the ODE, we obtain
$$s^4 Y(s) - s^3 y(0) - s^2 y'(0) - s y''(0) - y'''(0) +$$
$$+ 5 \left[s^2 Y(s) - s y(0) - y'(0) \right] + 4 Y(s) = \frac{1}{s} - \frac{e^{-\pi s}}{s}.$$

Applying the initial conditions,
$$s^4 Y(s) + 5 s^2 Y(s) + 4 Y(s) = \frac{1}{s} - \frac{e^{-\pi s}}{s}.$$

Solving for the transform of the solution,
$$Y(s) = \frac{1}{s(s^4 + 5s^2 + 4)} - \frac{e^{-\pi s}}{s(s^4 + 5s^2 + 4)}.$$

Using partial fractions,
$$\frac{1}{s(s^4 + 5s^2 + 4)} = \frac{1}{12} \left[\frac{3}{s} + \frac{s}{s^2 + 4} - \frac{4s}{s^2 + 1} \right].$$

It follows that
$$\mathcal{L}^{-1} \left[\frac{1}{s(s^4 + 5s^2 + 4)} \right] = \frac{1}{12} \left[3 + \cos 2t - 4 \cos t \right].$$

Based on Theorem 6.3.1, the solution of the IVP is
$$y(t) = \frac{1}{4} \left[1 - u_\pi(t) \right] + \frac{1}{12} \left[\cos 2t - 4 \cos t \right] - \frac{1}{12} \left[\cos 2(t - \pi) - 4 \cos(t - \pi) \right] u_\pi(t).$$

That is,
$$y(t) = \frac{1}{4} \left[1 - u_\pi(t) \right] + \frac{1}{12} \left[\cos 2t - 4 \cos t \right] - \frac{1}{12} \left[\cos 2t + 4 \cos t \right] u_\pi(t).$$

(b)

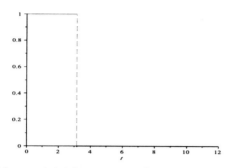
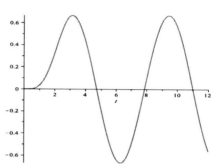

After an initial transient, the solution oscillates about $y_m = 0$.

14. The specified function is defined by

$$f(t) = \begin{cases} 0, & 0 \leq t < t_0 \\ \frac{h}{k}(t - t_0), & t_0 \leq t < t_0 + k \\ h, & t \geq t_0 + k \end{cases}$$

which can conveniently be expressed as

$$f(t) = \frac{h}{k}(t - t_0)\, u_{t_0}(t) - \frac{h}{k}(t - t_0 - k)\, u_{t_0+k}(t).$$

15. The function is defined by

$$g(t) = \begin{cases} 0, & 0 \leq t < t_0 \\ \frac{h}{k}(t - t_0), & t_0 \leq t < t_0 + k \\ -\frac{h}{k}(t - t_0 - 2k), & t_0 + k \leq t < t_0 + 2k \\ 0, & t \geq t_0 + 2k \end{cases}$$

which can also be written as

$$y(t) = \frac{h}{k}(t - t_0)\, u_{t_0}(t) - \frac{2h}{k}(t - t_0 - k)\, u_{t_0+k}(t) + \frac{h}{k}(t - t_0 - 2k)\, u_{t_0+2k}(t).$$

17. We consider the initial value problem

$$y'' + 4y = \frac{1}{k}\left[(t - 5)\, u_5(t) - (t - 5 - k)\, u_{5+k}(t)\right],$$

with $y(0) = y'(0) = 0$.

(a) The specified function is defined by

$$f(t) = \begin{cases} 0, & 0 \leq t < 5 \\ \frac{1}{k}(t - 5), & 5 \leq t < 5 + k \\ 1, & t \geq 5 + k \end{cases}$$

so k controls the point at which $f(t)$ reaches 1. When $k = 5$, $f(t) = g(t)$ in Ex.2.

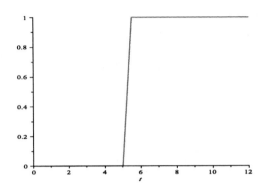

(b) Taking the Laplace transform of both sides of the ODE, we obtain

$$s^2 Y(s) - s\, y(0) - y'(0) + 4Y(s) = \frac{e^{-5s}}{ks^2} - \frac{e^{-(5+k)s}}{ks^2}.$$

Applying the initial conditions,

$$s^2 Y(s) + 4Y(s) = \frac{e^{-5s}}{ks^2} - \frac{e^{-(5+k)s}}{ks^2}.$$

Solving for the transform,

$$Y(s) = \frac{e^{-5s}}{ks^2(s^2+4)} - \frac{e^{-(5+k)s}}{ks^2(s^2+4)}.$$

Using partial fractions,

$$\frac{1}{s^2(s^2+4)} = \frac{1}{4}\left[\frac{1}{s^2} - \frac{1}{s^2+4}\right].$$

It follows that

$$\mathcal{L}^{-1}\left[\frac{1}{s^2(s^2+4)}\right] = \frac{1}{4}t - \frac{1}{8}\sin 2t\,.$$

Using Theorem 6.3.1, the solution of the IVP is

$$y(t) = \frac{1}{k}\left[h(t-5)\, u_5(t) - h(t-5-k)\, u_{5+k}(t)\right],$$

in which $h(t) = t/4 - \sin 2t\,/8$.

(c) Note that for $t > 5 + k$, the solution is given by

$$y(t) = \frac{1}{4} - \frac{1}{8k}\sin(2t-10) + \frac{1}{8k}\sin(2t-10-2k) = \frac{1}{4} - \frac{\sin k}{4k}\cos(2t-10-k).$$

So for $t > 5 + k$, the solution oscillates about $y_m = 1/4$, with an amplitude of

$$A = \frac{|\sin(k)|}{4k}.$$

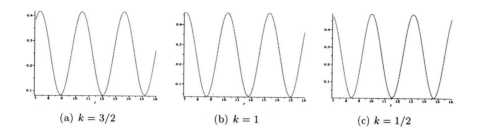

(a) $k = 3/2$ (b) $k = 1$ (c) $k = 1/2$

18.(a) The graph shows f_k for $k = 2$, $k = 1$ and $k = 1/2$.

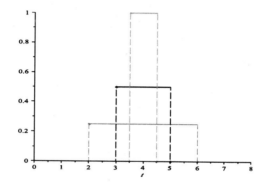

(b) The forcing function can be expressed as

$$f_k(t) = \frac{1}{2k}\left[u_{4-k}(t) - u_{4+k}(t)\right].$$

Taking the Laplace transform of both sides of the ODE, we obtain

$$s^2 Y(s) - s\,y(0) - y'(0) + \frac{1}{3}\left[s\,Y(s) - y(0)\right] + 4\,Y(s) = \frac{e^{-(4-k)s}}{2ks} - \frac{e^{-(4+k)s}}{2ks}.$$

Applying the initial conditions,

$$s^2 Y(s) + \frac{1}{3}s\,Y(s) + 4\,Y(s) = \frac{e^{-(4-k)s}}{2ks} - \frac{e^{-(4+k)s}}{2ks}.$$

Solving for the transform,

$$Y(s) = \frac{3\,e^{-(4-k)s}}{2ks(3s^2+s+12)} - \frac{3\,e^{-(4+k)s}}{2ks(3s^2+s+12)}.$$

Using partial fractions,

$$\frac{1}{s(3s^2+s+12)} = \frac{1}{12}\left[\frac{1}{s} - \frac{1+3s}{3s^2+s+12}\right] = \frac{1}{12}\left[\frac{1}{s} - \frac{1}{6}\frac{1+6(s+\frac{1}{6})}{(s+\frac{1}{6})^2+\frac{143}{36}}\right].$$

Let

$$H(s) = \frac{1}{8k}\left[\frac{1}{s} - \frac{\frac{1}{6}}{(s+\frac{1}{6})^2+\frac{143}{36}} - \frac{s+\frac{1}{6}}{(s+\frac{1}{6})^2+\frac{143}{36}}\right].$$

It follows that
$$h(t) = \mathcal{L}^{-1}[H(s)] = \frac{1}{8k} - \frac{e^{-t/6}}{8k}\left[\frac{1}{\sqrt{143}}\sin\left(\frac{\sqrt{143}\,t}{6}\right) + \cos\left(\frac{\sqrt{143}\,t}{6}\right)\right].$$

Based on Theorem 6.3.1, the solution of the IVP is
$$y(t) = h(t-4+k)\,u_{4-k}(t) - h(t-4-k)\,u_{4+k}(t).$$

(c)

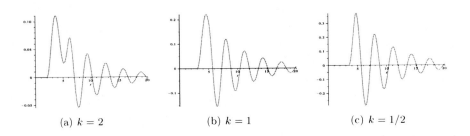

(a) $k=2$ (b) $k=1$ (c) $k=1/2$

As the parameter k decreases, the solution remains null for a longer period of time. Since the magnitude of the impulsive force increases, the initial overshoot of the response also increases. The duration of the impulse decreases. All solutions eventually decay to $y=0$.

21.(a)

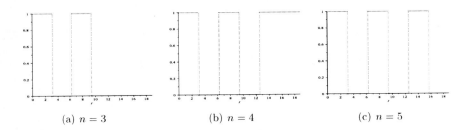

(a) $n=3$ (b) $n=4$ (c) $n=5$

(b) Taking the Laplace transform of both sides of the ODE, we obtain
$$s^2 Y(s) - s\,y(0) - y'(0) + Y(s) = \frac{1}{s} + \sum_{k=1}^{n}\frac{(-1)^k e^{-k\pi s}}{s}.$$

Applying the initial conditions,
$$s^2 Y(s) + Y(s) = \frac{1}{s} + \sum_{k=1}^{n}\frac{(-1)^k e^{-k\pi s}}{s}.$$

Solving for the transform,
$$Y(s) = \frac{1}{s(s^2+1)} + \sum_{k=1}^{n} \frac{(-1)^k e^{-k\pi s}}{s(s^2+1)}.$$

Using partial fractions,
$$\frac{1}{s(s^2+1)} = \frac{1}{s} - \frac{s}{s^2+1}.$$

Let
$$h(t) = \mathcal{L}^{-1}\left[\frac{1}{s(s^2+1)}\right] = 1 - \cos t.$$

Applying Theorem 6.3.1, term-by-term, the solution of the IVP is
$$y(t) = h(t) + \sum_{k=1}^{n} (-1)^k h(t - k\pi)\, u_{k\pi}(t).$$

Note that
$$h(t - k\pi) = u_0(t - k\pi) - \cos(t - k\pi) = u_{k\pi}(t) - (-1)^k \cos t.$$

Hence
$$y(t) = 1 - \cos t + \sum_{k=1}^{n}(-1)^k u_{k\pi}(t) - (\cos t)\sum_{k=1}^{n} u_{k\pi}(t).$$

(c)

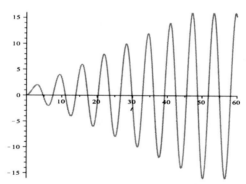

The ODE has no damping term. Each interval of forcing adds to the energy of the system, so the amplitude will increase. For $n = 15$, $g(t) = 0$ when $t > 15\pi$. Therefore the oscillation will eventually become steady, with an amplitude depending on the values of $y(15\pi)$ and $y'(15\pi)$.

(d) As n increases, the interval of forcing also increases. Hence the amplitude of the transient will increase with n. Eventually, the forcing function will be constant. In fact, for large values of t,
$$g(t) = \begin{cases} 1, & n \text{ even} \\ 0, & n \text{ odd} \end{cases}$$

Further, for $t > n\pi$,

$$y(t) = 1 - \cos t - n \cos t - \frac{1-(-1)^n}{2}.$$

Hence the steady state solution will oscillate about 0 or 1, depending on n, with an amplitude of $A = n+1$. In the limit, as $n \to \infty$, the forcing function will be a periodic function, with period 2π. From Problem 33, in Section 6.3,

$$\mathcal{L}[g(t)] = \frac{1}{s(1+e^{-s})}.$$

As n increases, the duration and magnitude of the transient will increase without bound.

22.(a) Taking the initial conditions into consideration, the transform of the ODE is

$$s^2 Y(s) + 0.1\, s\, Y(s) + Y(s) = \frac{1}{s} + \sum_{k=1}^{n} \frac{(-1)^k e^{-k\pi s}}{s}.$$

Solving for the transform,

$$Y(s) = \frac{1}{s(s^2 + 0.1s + 1)} + \sum_{k=1}^{n} \frac{(-1)^k e^{-k\pi s}}{s(s^2 + 0.1s + 1)}.$$

Using partial fractions,

$$\frac{1}{s(s^2+0.1s+1)} = \frac{1}{s} - \frac{s+0.1}{s^2+0.1s+1}.$$

Since the denominator in the second term is irreducible, write

$$\frac{s+0.1}{s^2+0.1s+1} = \frac{(s+0.05)+0.05}{(s+0.05)^2 + (399/400)}.$$

Let

$$h(t) = \mathcal{L}^{-1}\left[\frac{1}{s} - \frac{(s+0.05)}{(s+0.05)^2+(399/400)} - \frac{0.05}{(s+0.05)^2+(399/400)}\right]$$

$$= 1 - e^{-t/20}\left[\cos\left(\frac{\sqrt{399}}{20}t\right) + \frac{1}{\sqrt{399}}\sin\left(\frac{\sqrt{399}}{20}t\right)\right].$$

Applying Theorem 6.3.1, term-by-term, the solution of the IVP is

$$y(t) = h(t) + \sum_{k=1}^{n}(-1)^k h(t-k\pi)\, u_{k\pi}(t).$$

For odd values of n, the solution approaches $y = 0$. (On the next figure, $n = 5$.)

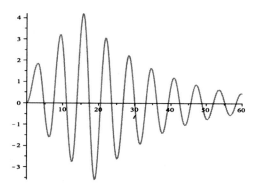

For even values of n, the solution approaches $y = 1$. (On the next figure, $n = 6$.)

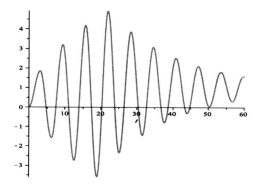

(b) The solution is a sum of damped sinusoids, each of frequency $\omega = \sqrt{399}/20 \approx 1$. Each term has an initial amplitude of approximately 1. For any given n, the solution contains $n+1$ such terms. Although the amplitude will increase with n, the amplitude will also be bounded by $n+1$.

(c) Suppose that the forcing function is replaced by $g(t) = \sin t$. Based on the methods in Chapter 3, the general solution of the differential equation is

$$y(t) = e^{-t/20}\left[c_1 \cos(\frac{\sqrt{399}}{20} t) + c_2 \sin(\frac{\sqrt{399}}{20} t)\right] + y_p(t).$$

Note that $y_p(t) = A \cos t + B \sin t$. Using the method of undetermined coefficients, $A = -10$ and $B = 0$. Based on the initial conditions, the solution of the IVP is

$$y(t) = 10\, e^{-t/20}\left[\cos(\frac{\sqrt{399}}{20} t) + \frac{1}{\sqrt{399}} \sin(\frac{\sqrt{399}}{20} t)\right] - 10 \cos t.$$

Observe that both solutions have the same frequency, $\omega = \sqrt{399}/20 \approx 1$.

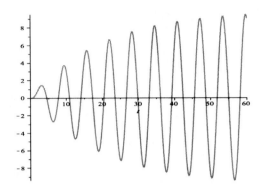

23.(a) Taking the initial conditions into consideration, the transform of the ODE is

$$s^2 Y(s) + Y(s) = \frac{1}{s} + 2 \sum_{k=1}^{n} \frac{(-1)^k e^{-(11k/4)s}}{s}.$$

Solving for the transform,

$$Y(s) = \frac{1}{s(s^2+1)} + 2 \sum_{k=1}^{n} \frac{(-1)^k e^{-(11k/4)s}}{s(s^2+1)}.$$

Using partial fractions,

$$\frac{1}{s(s^2+1)} = \frac{1}{s} - \frac{s}{s^2+1}.$$

Let

$$h(t) = \mathcal{L}^{-1}\left[\frac{1}{s(s^2+1)}\right] = 1 - \cos t.$$

Applying Theorem 6.3.1, term-by-term, the solution of the IVP is

$$y(t) = h(t) + 2 \sum_{k=1}^{n} (-1)^k h(t - \frac{11k}{4}) u_{11k/4}(t).$$

That is,

$$y(t) = 1 - \cos t + 2 \sum_{k=1}^{n} (-1)^k \left[1 - \cos(t - \frac{11k}{4})\right] u_{11k/4}(t).$$

(b) On the figure we see the solution for $n = 35$.

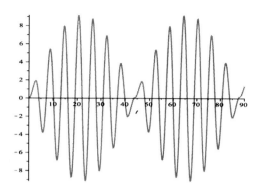

(c) Based on the plot, the slow period appears to be 88. The fast period appears to be about 6. These values correspond to a slow frequency of $\omega_s = 0.0714$ and a fast frequency $\omega_f = 1.0472$.

(d) The natural frequency of the system is $\omega_0 = 1$. The forcing function is initially periodic, with period $T = 11/2 = 5.5$. Hence the corresponding forcing frequency is $w = 1.1424$. Using the results in Section 3.8, the slow frequency is given by

$$\omega_s = \frac{|\omega - \omega_0|}{2} = 0.0712$$

and the fast frequency is given by

$$\omega_f = \frac{|\omega + \omega_0|}{2} = 1.0712.$$

Based on theses values, the slow period is predicted as 88.247 and the fast period is given as 5.8656.

6.5

2.(a) Taking the Laplace transform of both sides of the ODE, we obtain

$$s^2 Y(s) - s\, y(0) - y'(0) + 4Y(s) = e^{-\pi s} - e^{-2\pi s}.$$

Applying the initial conditions,

$$s^2 Y(s) + 4Y(s) = e^{-\pi s} - e^{-2\pi s}.$$

Solving for the transform,

$$Y(s) = \frac{e^{-\pi s} - e^{-2\pi s}}{s^2 + 4} = \frac{e^{-\pi s}}{s^2 + 4} - \frac{e^{-2\pi s}}{s^2 + 4}.$$

Applying Theorem 6.3.1, the solution of the IVP is

$$y(t) = \frac{1}{2}\sin(2t - 2\pi)u_\pi(t) - \frac{1}{2}\sin(2t - 4\pi)u_{2\pi}(t) = \frac{1}{2}\sin(2t)\left[u_\pi(t) - u_{2\pi}(t)\right].$$

(b)

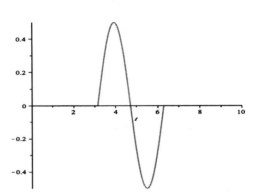

4.(a) Taking the Laplace transform of both sides of the ODE, we obtain
$$s^2 Y(s) - s\, y(0) - y'(0) - Y(s) = -20\, e^{-3s}.$$

Applying the initial conditions,
$$s^2 Y(s) - Y(s) - s = -20\, e^{-3s}.$$

Solving for the transform,
$$Y(s) = \frac{s}{s^2 - 1} - \frac{20\, e^{-3s}}{s^2 - 1}.$$

Using a table of transforms, and Theorem 6.3.1, the solution of the IVP is
$$y(t) = \cosh t - 20\, \sinh(t-3) u_3(t).$$

(b)

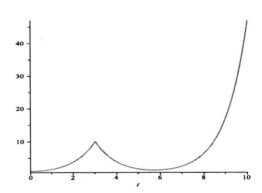

6.(a) Taking the initial conditions into consideration, the transform of the ODE is
$$s^2 Y(s) + 4Y(s) - s/2 = e^{-4\pi s}.$$

Solving for the transform,
$$Y(s) = \frac{s/2}{s^2 + 4} + \frac{e^{-4\pi s}}{s^2 + 4}.$$

Using a table of transforms, and Theorem 6.3.1, the solution of the IVP is

$$y(t) = \frac{1}{2}\cos 2t + \frac{1}{2}\sin(2t - 8\pi)u_{4\pi}(t) = \frac{1}{2}\cos 2t + \frac{1}{2}\sin(2t)\, u_{4\pi}(t).$$

(b)

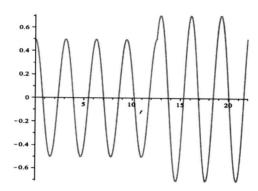

8.(a) Taking the Laplace transform of both sides of the ODE, we obtain

$$s^2 Y(s) - s\, y(0) - y'(0) + 4Y(s) = 2\, e^{-(\pi/4)s}.$$

Applying the initial conditions,

$$s^2 Y(s) + 4Y(s) = 2\, e^{-(\pi/4)s}.$$

Solving for the transform,

$$Y(s) = \frac{2\, e^{-(\pi/4)s}}{s^2 + 4}.$$

Applying Theorem 6.3.1, the solution of the IVP is

$$y(t) = \sin(2t - \frac{\pi}{2})u_{\pi/4}(t) = -\cos(2t)\, u_{\pi/4}(t).$$

(b)

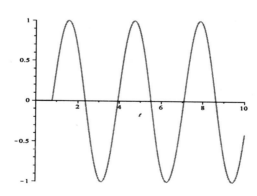

9.(a) Taking the initial conditions into consideration, the transform of the ODE is
$$s^2 Y(s) + Y(s) = \frac{e^{-(\pi/2)s}}{s} + 3e^{-(3\pi/2)s} - \frac{e^{-2\pi s}}{s}.$$

Solving for the transform,
$$Y(s) = \frac{e^{-(\pi/2)s}}{s(s^2+1)} + \frac{3e^{-(3\pi/2)s}}{s^2+1} - \frac{e^{-2\pi s}}{s(s^2+1)}.$$

Using partial fractions,
$$\frac{1}{s(s^2+1)} = \frac{1}{s} - \frac{s}{s^2+1}.$$

Hence
$$Y(s) = \frac{e^{-(\pi/2)s}}{s} - \frac{se^{-(\pi/2)s}}{s^2+1} + \frac{3e^{-(3\pi/2)s}}{s^2+1} - \frac{e^{-2\pi s}}{s} + \frac{se^{-2\pi s}}{s^2+1}.$$

Based on Theorem 6.3.1, the solution of the IVP is
$$y(t) = u_{\pi/2}(t) - \cos(t - \frac{\pi}{2})u_{\pi/2}(t) + 3\sin(t - \frac{3\pi}{2})u_{3\pi/2}(t) - u_{2\pi}(t) + \cos(t - 2\pi)u_{2\pi}(t).$$

That is,
$$y(t) = [1 - \sin(t)]\, u_{\pi/2}(t) + 3\cos(t)\, u_{3\pi/2}(t) - [1 - \cos(t)]\, u_{2\pi}(t).$$

(b)

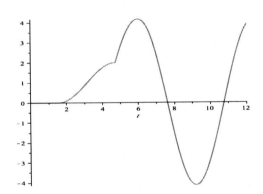

10.(a) Taking the transform of both sides of the ODE,
$$2s^2 Y(s) + sY(s) + 4Y(s) = \int_0^\infty e^{-st}\delta(t - \frac{\pi}{6})\sin t\, dt = \frac{1}{2}e^{-(\pi/6)s}.$$

Solving for the transform,
$$Y(s) = \frac{e^{-(\pi/6)s}}{2(2s^2 + s + 4)}.$$

First write
$$\frac{1}{2(2s^2 + s + 4)} = \frac{\frac{1}{4}}{(s + \frac{1}{4})^2 + \frac{31}{16}}.$$

It follows that
$$y(t) = \mathcal{L}^{-1}[Y(s)] = \frac{1}{\sqrt{31}} e^{-(t-\pi/6)/4} \cdot \sin\frac{\sqrt{31}}{4}(t - \frac{\pi}{6}) u_{\pi/6}(t).$$

(b)

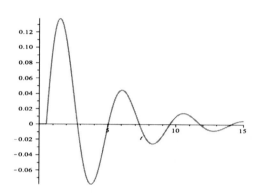

11.(a) Taking the initial conditions into consideration, the transform of the ODE is
$$s^2 Y(s) + 2s Y(s) + 2 Y(s) = \frac{s}{s^2+1} + e^{-(\pi/2)s}.$$

Solving for the transform,
$$Y(s) = \frac{s}{(s^2+1)(s^2+2s+2)} + \frac{e^{-(\pi/2)s}}{s^2+2s+2}.$$

Using partial fractions,
$$\frac{s}{(s^2+1)(s^2+2s+2)} = \frac{1}{5}\left[\frac{s}{s^2+1} + \frac{2}{s^2+1} - \frac{s+4}{s^2+2s+2}\right].$$

We can also write
$$\frac{s+4}{s^2+2s+2} = \frac{(s+1)+3}{(s+1)^2+1}.$$

Let
$$Y_1(s) = \frac{s}{(s^2+1)(s^2+2s+2)}.$$

Then
$$\mathcal{L}^{-1}[Y_1(s)] = \frac{1}{5} \cos t + \frac{2}{5} \sin t - \frac{1}{5} e^{-t} [\cos t + 3 \sin t].$$

Applying Theorem 6.3.1,
$$\mathcal{L}^{-1}\left[\frac{e^{-(\pi/2)s}}{s^2+2s+2}\right] = e^{-(t-\frac{\pi}{2})} \sin\left(t - \frac{\pi}{2}\right) u_{\pi/2}(t).$$

Hence the solution of the IVP is
$$y(t) = \frac{1}{5} \cos t + \frac{2}{5} \sin t - \frac{1}{5} e^{-t} [\cos t + 3 \sin t] - e^{-(t-\frac{\pi}{2})} \cos(t) \, u_{\pi/2}(t).$$

(b)

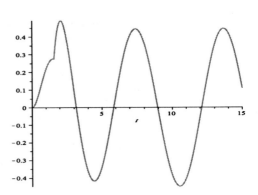

12.(a) Taking the initial conditions into consideration, the transform of the ODE is

$$s^4 Y(s) - Y(s) = e^{-s}.$$

Solving for the transform,

$$Y(s) = \frac{e^{-s}}{s^4 - 1}.$$

Using partial fractions,

$$\frac{1}{s^4 - 1} = \frac{1}{2}\left[\frac{1}{s^2 - 1} - \frac{1}{s^2 + 1}\right].$$

It follows that

$$\mathcal{L}^{-1}\left[\frac{1}{s^4 - 1}\right] = \frac{1}{2}\sinh t - \frac{1}{2}\sin t.$$

Applying Theorem 6.3.1, the solution of the IVP is

$$y(t) = \frac{1}{2}\left[\sinh(t - 1) - \sin(t - 1)\right] u_1(t).$$

(b)

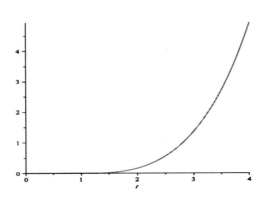

14.(a) The Laplace transform of the ODE is

$$s^2 Y(s) + \frac{1}{2} s Y(s) + Y(s) = e^{-s}.$$

Solving for the transform of the solution,

$$Y(s) = \frac{e^{-s}}{s^2 + s/2 + 1}.$$

First write

$$\frac{1}{s^2 + s/2 + 1} = \frac{1}{(s + \frac{1}{4})^2 + \frac{15}{16}}.$$

Taking the inverse transform and applying both shifting theorems,

$$y(t) = \frac{4}{\sqrt{15}} e^{-(t-1)/4} \sin \frac{\sqrt{15}}{4}(t-1) \, u_1(t).$$

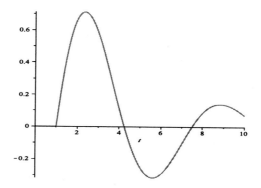

(b) As shown on the graph, the maximum is attained at some $t_1 > 2$. Note that for $t > 2$,

$$y(t) = \frac{4}{\sqrt{15}} e^{-(t-1)/4} \sin \frac{\sqrt{15}}{4}(t-1).$$

Setting $y'(t) = 0$, we find that $t_1 \approx 2.3613$. The maximum value is calculated as $y(2.3613) \approx 0.71153$.

(c) Setting $\gamma = 1/4$, the transform of the solution is

$$Y(s) = \frac{e^{-s}}{s^2 + s/4 + 1}.$$

Following the same steps, it follows that

$$y(t) = \frac{8}{3\sqrt{7}} e^{-(t-1)/8} \sin \frac{3\sqrt{7}}{8}(t-1) \, u_1(t).$$

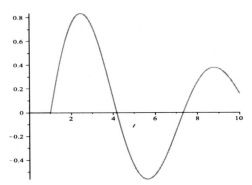

Once again, the maximum is attained at some $t_1 > 2$. Setting $y'(t) = 0$, we find that $t_1 \approx 2.4569$, with $y(t_1) \approx 0.8335$.

(d) Now suppose that $0 < \gamma < 1$. Then the transform of the solution is

$$Y(s) = \frac{e^{-s}}{s^2 + \gamma s + 1}.$$

First write

$$\frac{1}{s^2 + \gamma s + 1} = \frac{1}{(s + \gamma/2)^2 + (1 - \gamma^2/4)}.$$

It follows that

$$h(t) = \mathcal{L}^{-1}\left[\frac{1}{s^2 + \gamma s + 1}\right] = \frac{2}{\sqrt{4 - \gamma^2}} e^{-\gamma t/2} \sin(\sqrt{1 - \gamma^2/4} \cdot t).$$

Hence the solution is

$$y(t) = h(t - 1) u_1(t).$$

The solution is nonzero only if $t > 1$, in which case $y(t) = h(t - 1)$. Setting $y'(t) = 0$, we obtain

$$\tan\left[\sqrt{1 - \gamma^2/4} \cdot (t - 1)\right] = \frac{1}{\gamma}\sqrt{4 - \gamma^2},$$

that is,

$$\frac{\tan\left[\sqrt{1 - \gamma^2/4} \cdot (t - 1)\right]}{\sqrt{1 - \gamma^2/4}} = \frac{2}{\gamma}.$$

As $\gamma \to 0$, we obtain the formal equation $\tan(t - 1) = \infty$. Hence $t_1 \to 1 + \frac{\pi}{2}$. Setting $t = \pi/2$ in $h(t)$, and letting $\gamma \to 0$, we find that $y_1 \to 1$. These conclusions agree with the case $\gamma = 0$, for which it is easy to show that the solution is

$$y(t) = \sin(t - 1) u_1(t).$$

15.(a) See Problem 14. It follows that the solution of the IVP is

$$y(t) = \frac{4k}{\sqrt{15}} e^{-(t-1)/4} \sin \frac{\sqrt{15}}{4}(t - 1) u_1(t).$$

This function is a multiple of the answer in Problem 14(a). Hence the peak value occurs at $t_1 \approx 2.3613$. The maximum value is calculated as $y(2.3613) \approx 0.71153\,k$. We find that the appropriate value of k is $k_1 = 2/0.71153 \approx 2.8108$.

(b) Based on Problem 14(c), the solution is

$$y(t) = \frac{8k}{3\sqrt{7}} e^{-(t-1)/8} \sin \frac{3\sqrt{7}}{8}(t-1)\, u_1(t).$$

Since this function is a multiple of the solution in Problem 14(c), we have that $t_1 \approx 2.4569$, with $y(t_1) \approx 0.8335\,k$. The solution attains a value of $y = 2$ when $k_1 = 2/0.8335$, that is, $k_1 \approx 2.3995$.

(c) Similar to Problem 14(d), for $0 < \gamma < 1$, the solution is

$$y(t) = h(t-1)\, u_1(t),$$

in which

$$h(t) = \frac{2k}{\sqrt{4-\gamma^2}} e^{-\gamma t/2} \sin(\sqrt{1-\gamma^2/4}\cdot t).$$

It follows that $t_1 - 1 \to \pi/2$. Setting $t = \pi/2$ in $h(t)$, and letting $\gamma \to 0$, we find that $y_1 \to k$. Requiring that the peak value remains at $y = 2$, the limiting value of k is $k_1 = 2$. These conclusions agree with the case $\gamma = 0$, for which it is easy to show that the solution is

$$y(t) = k\, \sin(t-1)\, u_1(t).$$

16.(a) Taking the initial conditions into consideration, the transformation of the ODE is

$$s^2 Y(s) + Y(s) = \frac{1}{2k}\left[\frac{e^{-(4-k)s}}{s} - \frac{e^{-(4+k)s}}{s}\right].$$

Solving for the transform of the solution,

$$Y(s) = \frac{1}{2k}\left[\frac{e^{-(4-k)s}}{s(s^2+1)} - \frac{e^{-(4+k)s}}{s(s^2+1)}\right].$$

Using partial fractions,

$$\frac{1}{s(s^2+1)} = \frac{1}{s} - \frac{s}{s^2+1}.$$

Now let

$$h(t) = \mathcal{L}^{-1}\left[\frac{1}{s(s^2+1)}\right] = 1 - \cos t.$$

Applying Theorem 6.3.1, the solution is

$$\phi(t,k) = \frac{1}{2k}\left[h(t-4+k)\, u_{4-k}(t) - h(t-4-k)\, u_{4+k}(t)\right].$$

That is,

$$\phi(t,k) = \frac{1}{2k}[u_{4-k}(t) - u_{4+k}(t)] -$$
$$- \frac{1}{2k}[\cos(t-4+k)\,u_{4-k}(t) - \cos(t-4-k)\,u_{4+k}(t)].$$

(b) Consider various values of t. For any fixed $t < 4$, $\phi(t,k) = 0$, as long as $4 - k > t$. If $t \geq 4$, then for $4 + k < t$,

$$\phi(t,k) = -\frac{1}{2k}[\cos(t-4+k) - \cos(t-4-k)].$$

It follows that

$$\lim_{k \to 0} \phi(t,k) = \lim_{k \to 0} -\frac{\cos(t-4+k) - \cos(t-4-k)}{2k} = \sin(t-4).$$

Hence

$$\lim_{k \to 0} \phi(t,k) = \sin(t-4)\,u_4(t).$$

(c) The Laplace transform of the differential equation

$$y'' + y = \delta(t-4),$$

with $y(0) = y'(0) = 0$, is

$$s^2 Y(s) + Y(s) = e^{-4s}.$$

Solving for the transform of the solution,

$$Y(s) = \frac{e^{-4s}}{s^2 + 1}.$$

It follows that the solution is $\phi_0(t) = \sin(t-4)\,u_4(t)$, and this means that

$$\lim_{k \to 0} \phi(t,k) = \phi_0(t).$$

(d) We can see the convergence on the graphs.

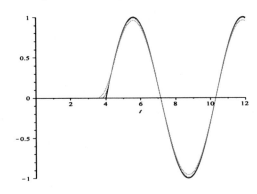

18.(b) The transform of the ODE (given the specified initial conditions) is
$$s^2 Y(s) + Y(s) = \sum_{k=1}^{20} (-1)^{k+1} e^{-k\pi s}.$$
Solving for the transform of the solution,
$$Y(s) = \frac{1}{s^2+1} \sum_{k=1}^{20} (-1)^{k+1} e^{-k\pi s}.$$
Applying Theorem 6.3.1, term-by-term,
$$y(t) = \sum_{k=1}^{20} (-1)^{k+1} \sin(t - k\pi)\, u_{k\pi}(t) = -\sin(t) \cdot \sum_{k=1}^{20} u_{k\pi}(t).$$

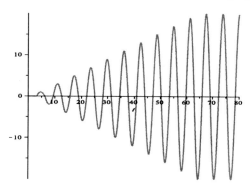

(c) For $t > 20\pi$, $y(t) = -20 \sin t$.

19.(b) Taking the initial conditions into consideration, the transform of the ODE is
$$s^2 Y(s) + Y(s) = \sum_{k=1}^{20} e^{-(k\pi/2)s}.$$
Solving for the transform of the solution,
$$Y(s) = \frac{1}{s^2+1} \sum_{k=1}^{20} e^{-(k\pi/2)s}.$$
Applying Theorem 6.3.1, term-by-term,
$$y(t) = \sum_{k=1}^{20} \sin\left(t - \frac{k\pi}{2}\right) u_{k\pi/2}(t).$$

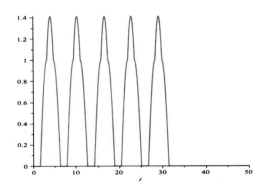

(c) For $t > 10\pi$, $y(t) = 0$.

20.(b) The transform of the ODE (given the specified initial conditions) is

$$s^2 Y(s) + Y(s) = \sum_{k=1}^{20} (-1)^{k+1} e^{-(k\pi/2)s}.$$

Solving for the transform of the solution,

$$Y(s) = \sum_{k=1}^{20} (-1)^{k+1} \frac{e^{-(k\pi/2)s}}{s^2 + 1}.$$

Applying Theorem 6.3.1, term-by-term,

$$y(t) = \sum_{k=1}^{20} (-1)^{k+1} \sin(t - \frac{k\pi}{2}) u_{k\pi/2}(t).$$

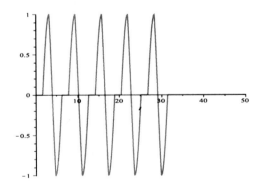

(c) For $t > 10\pi$, $y(t) = 0$.

22.(b) Taking the initial conditions into consideration, the transform of the ODE is

$$s^2 Y(s) + Y(s) = \sum_{k=1}^{40} (-1)^{k+1} e^{-(11k/4)s}.$$

Solving for the transform of the solution,

$$Y(s) = \sum_{k=1}^{40} (-1)^{k+1} \frac{e^{-(11k/4)s}}{s^2+1}.$$

Applying Theorem 6.3.1, term-by-term,

$$y(t) = \sum_{k=1}^{40} (-1)^{k+1} \sin(t - \frac{11k}{4}) u_{11k/4}(t).$$

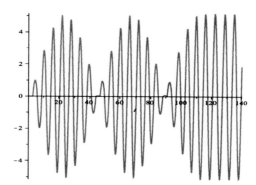

(c) For $t > 110$,

$$y(t) = \sum_{k=1}^{40}(-1)^{k+1}\sin\left(t - \frac{11k}{4}\right) \approx -5.13887\cos(56.375 - t).$$

23.(b) The transform of the ODE (given the specified initial conditions) is

$$s^2 Y(s) + 0.1s\, Y(s) + Y(s) = \sum_{k=1}^{20} (-1)^{k+1} e^{-k\pi s}.$$

Solving for the transform of the solution,

$$Y(s) = \sum_{k=1}^{20} \frac{e^{-k\pi s}}{s^2 + 0.1s + 1}.$$

First write

$$\frac{1}{s^2 + 0.1s + 1} = \frac{1}{(s+\frac{1}{20})^2 + \frac{399}{400}}.$$

It follows that

$$\mathcal{L}^{-1}\left[\frac{1}{s^2+0.1s+1}\right] = \frac{20}{\sqrt{399}} e^{-t/20} \sin(\frac{\sqrt{399}}{20} t).$$

Applying Theorem 6.3.1, term-by-term,

$$y(t) = \sum_{k=1}^{20} (-1)^{k+1} h(t - k\pi)\, u_{k\pi}(t),$$

in which
$$h(t) = \frac{20}{\sqrt{399}} e^{-t/20} \sin(\frac{\sqrt{399}}{20} t).$$

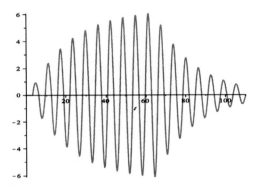

(c) For $t > 20\pi$, each term in the sum for $y(t)$ will contain a decaying exponential term multiplied by a bounded term. Thus $y(t) \to 0$.

24.(b) Taking the initial conditions into consideration, the transform of the ODE is
$$s^2 Y(s) + 0.1s\, Y(s) + Y(s) = \sum_{k=1}^{15} e^{-(2k-1)\pi s}.$$

Solving for the transform of the solution,
$$Y(s) = \sum_{k=1}^{15} \frac{e^{-(2k-1)\pi s}}{s^2 + 0.1s + 1}.$$

As shown in Problem 23,
$$\mathcal{L}^{-1}\left[\frac{1}{s^2 + 0.1s + 1}\right] = \frac{20}{\sqrt{399}} e^{-t/20} \sin(\frac{\sqrt{399}}{20} t).$$

Applying Theorem 6.3.1, term-by-term,
$$y(t) = \sum_{k=1}^{15} h\left[t - (2k-1)\pi\right] u_{(2k-1)\pi}(t),$$

in which
$$h(t) = \frac{20}{\sqrt{399}} e^{-t/20} \sin(\frac{\sqrt{399}}{20} t).$$

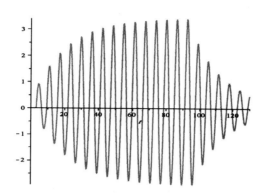

(c) For $t > 29\pi$, each term in the sum for $y(t)$ will contain a decaying exponential term multiplied by a bounded term. Thus $y(t) \to 0$.

6.6

2. Let $f(t) = e^t$. Then

$$(f * 1)(t) = \int_0^t e^{t-\tau} \cdot 1 \, d\tau = e^t \int_0^t e^{-\tau} d\tau = e^t - 1.$$

3. It follows directly that

$$(f * f)(t) = \int_0^t \sin(t-\tau) \sin(\tau) \, d\tau = \frac{1}{2} \int_0^t [\cos(t - 2\tau) - \cos(t)] d\tau = \frac{1}{2}(\sin t - t \cos t).$$

The range of the resulting function is $\mathbb{R}$.

5. We have $\mathcal{L}[e^{-t}] = 1/(s+1)$ and $\mathcal{L}[\sin t] = 1/(s^2+1)$. Based on Theorem 6.6.1,

$$\mathcal{L}\left[\int_0^t e^{-(t-\tau)} \sin(\tau) \, d\tau\right] = \frac{1}{s+1} \cdot \frac{1}{s^2+1} = \frac{1}{(s+1)(s^2+1)}.$$

6. Let $g(t) = t$ and $h(t) = e^t$. Then $f(t) = (g * h)(t)$. Applying Theorem 6.6.1,

$$\mathcal{L}\left[\int_0^t g(t-\tau)h(\tau) \, d\tau\right] = \frac{1}{s^2} \cdot \frac{1}{s-1} = \frac{1}{s^2(s-1)}.$$

7. We have $f(t) = (g * h)(t)$, in which $g(t) = \sin t$ and $h(t) = \cos t$. The transform of the convolution integral is

$$\mathcal{L}\left[\int_0^t g(t-\tau)h(\tau) \, d\tau\right] = \frac{1}{s^2+1} \cdot \frac{s}{s^2+1} = \frac{s}{(s^2+1)^2}.$$

9. It is easy to see that

$$\mathcal{L}^{-1}\left[\frac{1}{s+1}\right] = e^{-t} \quad \text{and} \quad \mathcal{L}^{-1}\left[\frac{s}{s^2+4}\right] = \cos 2t.$$

Applying Theorem 6.6.1,

$$\mathcal{L}^{-1}\left[\frac{s}{(s+1)(s^2+4)}\right] = \int_0^t e^{-(t-\tau)} \cos 2\tau \, d\tau \, .$$

10. We first note that

$$\mathcal{L}^{-1}\left[\frac{1}{(s+1)^2}\right] = t e^{-t} \quad \text{and} \quad \mathcal{L}^{-1}\left[\frac{1}{s^2+4}\right] = \frac{1}{2}\sin 2t \, .$$

Based on the convolution theorem,

$$\mathcal{L}^{-1}\left[\frac{1}{(s+1)^2(s^2+4)}\right] = \frac{1}{2}\int_0^t (t-\tau)e^{-(t-\tau)} \sin 2\tau \, d\tau$$

$$= \frac{1}{2}\int_0^t \tau e^{-\tau} \sin(2t-2\tau) \, d\tau \, .$$

11. Let $g(t) = \mathcal{L}^{-1}[G(s)]$. Since $\mathcal{L}^{-1}[1/(s^2+1)] = \sin t$, the inverse transform of the product is

$$\mathcal{L}^{-1}\left[\frac{G(s)}{s^2+1}\right] = \int_0^t g(t-\tau) \sin \tau \, d\tau = \int_0^t \sin(t-\tau) g(\tau) \, d\tau \, .$$

12.(a) By definition,

$$f*g = \int_0^t (t-\tau)^m \tau^n \, d\tau \, .$$

Set $\tau = t - tu$, $d\tau = -t \, du$, so that

$$\int_0^t (t-\tau)^m \tau^n \, d\tau = -\int_1^0 (tu)^m (t-tu)^n t \, du = t^{m+n+1} \int_0^1 u^m (1-u)^n \, du \, .$$

(b) The Convolution Theorem states that $\mathcal{L}[f*g] = \mathcal{L}[f] \cdot \mathcal{L}[g]$. Noting that

$$\mathcal{L}[t^k] = \frac{k!}{s^{k+1}},$$

it follows that

$$\frac{m!}{s^{m+1}} \frac{n!}{s^{n+1}} = \frac{(m+n+1)!}{s^{m+n+2}} \int_0^1 u^m (1-u)^n \, du \, .$$

Therefore

$$\int_0^1 u^m (1-u)^n \, du = \frac{m! \, n!}{(m+n+1)!} \, .$$

(c) If k is not an integer, we can write

$$\mathcal{L}[t^k] = \frac{\Gamma(k+1)}{s^{k+1}}$$

and

$$\frac{\Gamma(m+1)}{s^{m+1}} \cdot \frac{\Gamma(n+1)}{s^{n+1}} = \frac{\Gamma(m+1)}{s^{m+n+2}} \int_0^1 u^m (1-u)^n \, du,$$

so
$$\int_0^1 u^m(1-u)^n \, du = \frac{\Gamma(m+1)\Gamma(n+1)}{\Gamma(m+n+2)}.$$

13. Taking the initial conditions into consideration, the transform of the ODE is
$$s^2 Y(s) - 1 + \omega^2 Y(s) = G(s).$$
Solving for the transform of the solution,
$$Y(s) = \frac{1}{s^2 + \omega^2} + \frac{G(s)}{s^2 + \omega^2}.$$
As shown in a related situation, Problem 11,
$$\mathcal{L}^{-1}\left[\frac{G(s)}{s^2 + \omega^2}\right] = \frac{1}{\omega} \int_0^t \sin(\omega(t-\tau)) g(\tau) \, d\tau.$$
Hence the solution of the IVP is
$$y(t) = \frac{1}{\omega} \sin(\omega t) + \frac{1}{\omega} \int_0^t \sin(\omega(t-\tau)) g(\tau) \, d\tau.$$

15. The transform of the ODE (given the specified initial conditions) is
$$4s^2 Y(s) + 4s Y(s) + 17 Y(s) = G(s).$$
Solving for the transform of the solution,
$$Y(s) = \frac{G(s)}{4s^2 + 4s + 17}.$$
First write
$$\frac{1}{4s^2 + 4s + 17} = \frac{\frac{1}{4}}{(s + \frac{1}{2})^2 + 4}.$$
Based on the elementary properties of the Laplace transform,
$$\mathcal{L}^{-1}\left[\frac{1}{4s^2 + 4s + 17}\right] = \frac{1}{8} e^{-t/2} \sin 2t.$$
Applying the convolution theorem, the solution of the IVP is
$$y(t) = \frac{1}{8} \int_0^t e^{-(t-\tau)/2} \sin 2(t-\tau) g(\tau) \, d\tau.$$

17. Taking the initial conditions into consideration, the transform of the ODE is
$$s^2 Y(s) - 2s + 3 + 4[s Y(s) - 2] + 4 Y(s) = G(s).$$
Solving for the transform of the solution,
$$Y(s) = \frac{2s + 5}{(s+2)^2} + \frac{G(s)}{(s+2)^2}.$$
We can write
$$\frac{2s + 5}{(s+2)^2} = \frac{2}{s+2} + \frac{1}{(s+2)^2}.$$

It follows that
$$\mathcal{L}^{-1}\left[\frac{2}{s+2}\right] = 2e^{-2t} \quad \text{and} \quad \mathcal{L}^{-1}\left[\frac{1}{(s+2)^2}\right] = te^{-2t}.$$
Based on the convolution theorem, the solution of the IVP is
$$y(t) = 2e^{-2t} + te^{-2t} + \int_0^t (t-\tau)e^{-2(t-\tau)}g(\tau)\,d\tau.$$

19. The transform of the ODE (given the specified initial conditions) is
$$s^4 Y(s) - Y(s) = G(s).$$
Solving for the transform of the solution,
$$Y(s) = \frac{G(s)}{s^4 - 1}.$$
First write
$$\frac{1}{s^4 - 1} = \frac{1}{2}\left[\frac{1}{s^2 - 1} - \frac{1}{s^2 + 1}\right].$$
It follows that
$$\mathcal{L}^{-1}\left[\frac{1}{s^4 - 1}\right] = \frac{1}{2}\left[\sinh t - \sin t\right].$$
Based on the convolution theorem, the solution of the IVP is
$$y(t) = \frac{1}{2}\int_0^t \left[\sinh(t-\tau) - \sin(t-\tau)\right]g(\tau)\,d\tau.$$

20. Taking the initial conditions into consideration, the transform of the ODE is
$$s^4 Y(s) - s^3 + 5s^2 Y(s) - 5s + 4Y(s) = G(s).$$
Solving for the transform of the solution,
$$Y(s) = \frac{s^3 + 5s}{(s^2 + 1)(s^2 + 4)} + \frac{G(s)}{(s^2 + 1)(s^2 + 4)}.$$
Using partial fractions, we find that
$$\frac{s^3 + 5s}{(s^2 + 1)(s^2 + 4)} = \frac{1}{3}\left[\frac{4s}{s^2 + 1} - \frac{s}{s^2 + 4}\right],$$
and
$$\frac{1}{(s^2 + 1)(s^2 + 4)} = \frac{1}{3}\left[\frac{1}{s^2 + 1} - \frac{1}{s^2 + 4}\right].$$
It follows that
$$\mathcal{L}^{-1}\left[\frac{s(s^2 + 5)}{(s^2 + 1)(s^2 + 4)}\right] = \frac{4}{3}\cos t - \frac{1}{3}\cos 2t,$$
and
$$\mathcal{L}^{-1}\left[\frac{1}{(s^2 + 1)(s^2 + 4)}\right] = \frac{1}{3}\sin t - \frac{1}{6}\sin 2t.$$

Based on the convolution theorem, the solution of the IVP is

$$y(t) = \frac{4}{3}\cos t - \frac{1}{3}\cos 2t + \frac{1}{6}\int_0^t [2\sin(t-\tau) - \sin 2(t-\tau)]g(\tau)\,d\tau.$$

22.(a) Taking the Laplace transform of the integral equation, with $\Phi(s) = \mathcal{L}[\phi(t)]$,

$$\Phi(s) + \frac{1}{s^2}\cdot\Phi(s) = \frac{2}{s^2+4}.$$

Note that the convolution theorem was applied. Solving for the transform $\Phi(s)$,

$$\Phi(s) = \frac{2s^2}{(s^2+1)(s^2+4)}.$$

Using partial fractions, we can write

$$\frac{2s^2}{(s^2+1)(s^2+4)} = \frac{2}{3}\left[\frac{4}{s^2+4} - \frac{1}{s^2+1}\right].$$

Therefore the solution of the integral equation is

$$\phi(t) = \frac{4}{3}\sin 2t - \frac{2}{3}\sin t.$$

(b) Differentiate both sides of the equation, we get

$$\phi'(t) + (t-t)\phi(t) + \int_0^t \phi(\xi)d\xi = 2\cos 2t.$$

Clearly, $t - t = 0$, so differentiating this equation again we obtain

$$\phi''(t) + \phi(t) = -4\sin 2t.$$

Plugging $t = 0$ into the original equation gives us $\phi(0) = 0$. Also, $t = 0$ in the first equation here in part (b) gives $\phi'(0) = 2$.

(c) Taking the Laplace transform of the ODE, with $\Phi(s) = \mathcal{L}[\phi(t)]$,

$$s^2\Phi(s) - 2 + \Phi(s) = -\frac{8}{s^2+4}.$$

Solving for the transform of the solution,

$$\Phi(s) = \frac{2s^2}{(s^2+1)(s^2+4)}.$$

This is identical to the Laplace transform we obtained in part (a), so the solution will be the same.

23.(a) Taking the Laplace transform of both sides of the integral equation (using the Convolution Theorem)

$$\Phi(s) + \frac{1}{s^2}\Phi(s) = \frac{1}{s}.$$

It follows that

$$\Phi(s) = \frac{s}{s^2+1}$$

and
$$\phi(t) = \mathcal{L}^{-1}\left[\frac{s}{s^2+1}\right] = \cos t.$$

(b) Differentiating both sides of the equation twice, we get
$$\phi'(t) + \int_0^t \phi(\xi)\,d\xi = 0,$$
and then $\phi''(t) + \phi(t) = 0$. Plugging $t = 0$ into the original equation and the first equation above gives $\phi(0) = 1$ and $\phi'(0) = 0$.

(c) The function $\phi(t) = \cos t$ clearly solves the initial value problem in part (b).

25.(a) The Laplace transform of both sides of the integral equation (using the Convolution Theorem) is
$$\Phi(s) + \frac{2s}{s^2+1}\Phi(s) = \frac{1}{s+1}.$$

Solving for $\Phi(s)$:
$$\Phi(s) = \frac{s^2+1}{(s+1)^3}.$$

Rewriting,
$$\Phi(s) = \frac{(s+1)^2 - 2(s+1) + 2}{(s+1)^3} = \frac{1}{s+1} - \frac{2}{(s+1)^2} + \frac{2}{(s+1)^3}.$$

The solution of the integral equation is
$$\phi(t) = \mathcal{L}^{-1}\left[\frac{1}{s+1} - \frac{2}{(s+1)^2} + \frac{2}{(s+1)^3}\right] = e^{-t} - 2t\,e^{-t} + t^2\,e^{-t}.$$

(b) Differentiating both sides of the equation twice, we get
$$\phi'(t) + 2\phi(t) - 2\int_0^t \sin(t-\xi)\phi(\xi)\,d\xi = -e^{-t},$$
and then
$$\phi''(t) + 2\phi'(t) - 2\int_0^t \cos(t-\xi)\phi(\xi)\,d\xi = e^{-t}.$$

Using the original equation, we can convert the second equation to
$$\phi''(t) + 2\phi'(t) + \phi(t) = 2e^{-t}.$$

Plugging $t = 0$ into the original equation and the first equation above gives $\phi(0) = 1$ and $\phi'(0) = -3$.

(c) It is easily confirmed that the function $\phi(t) = e^{-t} - 2te^{-t} + t^2e^{-t}$ solves the initial value problem in part (b).

27.(a) Taking the Laplace transform of both sides of the integro-differential equation

$$s\,\Phi(s) - 1 - \frac{1}{s^3}\,\Phi(s) = -\frac{1}{s^2}.$$

Solving for $\Phi(s)$:

$$\Phi(s) = \frac{s}{s^2+1}.$$

Taking the inverse Laplace transform,

$$\phi(t) = \cos t.$$

(b) Differentiating both sides of the equation three times, we get

$$\phi''(t) - \int_0^t (t-\xi)\phi(\xi)\,d\xi = -1,$$

then

$$\phi'''(t) - \int_0^t \phi(\xi)\,d\xi = 0,$$

and then $\phi^{(4)}(t) - \phi(t) = 0$. Plugging $t = 0$ into the original equation and the first two equations above gives $\phi'(0) = 0$, $\phi''(0) = -1$ and $\phi'''(0) = 0$.

(c) It is easily confirmed that the function $\phi(t) = \cos t$ solves the initial value problem in part (b).

28.(a) The Laplace transform of both sides of the integro-differential equation is

$$s\,\Phi(s) - 1 + \Phi(s) = \frac{1}{s^2+1}\,\Phi(s).$$

Solving for $\Phi(s)$:

$$\Phi(s) = \frac{s^2+1}{s(s^2+s+1)} = \frac{1}{s} - \frac{1}{(s^2+s+1)}.$$

Note further that

$$\frac{1}{(s^2+s+1)} = \frac{1}{(s+1/2)^2 + 3/4} = \frac{2}{\sqrt{3}}\,\frac{\sqrt{3}/2}{(s+1/2)^2 + 3/4}.$$

Taking the inverse Laplace transform,

$$\phi(t) = 1 - \frac{2}{\sqrt{3}}e^{-t/2}\sin\left(\frac{\sqrt{3}}{2}t\right).$$

(b) Differentiating both sides of the equation twice, we get

$$\phi''(t) + \phi'(t) - \int_0^t \cos(t-\xi)\phi(\xi)\,d\xi = 0,$$

and then

$$\phi'''(t) + \phi''(t) - \phi(t) + \int_0^t \sin(t-\xi)\phi(\xi)\,d\xi = 0.$$

Using the original equation, we can convert the second equation to
$$\phi'''(t) + \phi''(t) + \phi'(t) = 0.$$
Plugging $t = 0$ into the original equation and the first equation above gives $\phi'(0) = -1$ and $\phi''(0) = 1$.

(c) It is easily confirmed that the function $\phi(t) = 1 - (2/\sqrt{3})e^{-t/2}\sin(\sqrt{3}\,t/2)$ solves the initial value problem in part (b).

29.(a) First note that
$$\int_0^b \frac{f(y)}{\sqrt{b-y}}\,dy = \left(\frac{1}{\sqrt{y}} * f\right)(b).$$
Take the Laplace transformation of both sides of the equation. Using the convolution theorem, with $F(s) = \mathcal{L}[f(y)]$,
$$\frac{T_0}{s} = \frac{1}{\sqrt{2g}} F(s) \cdot \mathcal{L}\left[\frac{1}{\sqrt{y}}\right].$$
It was shown in Problem 31(c), Section 6.1, that
$$\mathcal{L}\left[\frac{1}{\sqrt{y}}\right] = \sqrt{\frac{\pi}{s}}.$$
Hence
$$\frac{T_0}{s} = \frac{1}{\sqrt{2g}} F(s) \cdot \sqrt{\frac{\pi}{s}},$$
and
$$F(s) = \sqrt{\frac{2g}{\pi}} \cdot \frac{T_0}{\sqrt{s}}.$$
Taking the inverse transform, we obtain
$$f(y) = \frac{T_0}{\pi}\sqrt{\frac{2g}{y}}.$$

(b) Combining equations (i) and (iv),
$$\frac{2g\,T_0^2}{\pi^2\,y} = 1 + \left(\frac{dx}{dy}\right)^2.$$
Solving for the derivative dx/dy,
$$\frac{dx}{dy} = \sqrt{\frac{2\alpha - y}{y}},$$
in which $\alpha = gT_0^2/\pi^2$.

(c) Consider the change of variable $y = 2\alpha \sin^2(\theta/2)$. Using the chain rule,
$$\frac{dy}{dx} = 2\alpha \sin(\theta/2)\cos(\theta/2) \cdot \frac{d\theta}{dx}$$

and
$$\frac{dx}{dy} = \frac{1}{2\alpha \sin(\theta/2)\cos(\theta/2)} \cdot \frac{dx}{d\theta}.$$

It follows that
$$\frac{dx}{d\theta} = 2\alpha \sin(\theta/2)\cos(\theta/2)\sqrt{\frac{\cos^2(\theta/2)}{\sin^2(\theta/2)}} = 2\alpha \cos^2(\theta/2) = \alpha + \alpha \cos\theta.$$

Direct integration results in
$$x(\theta) = \alpha\theta + \alpha \sin\theta + C.$$

Since the curve passes through the origin, we require $y(0) = x(0) = 0$. Hence $C = 0$, and $x(\theta) = \alpha\theta + \alpha \sin\theta$. We also have
$$y(\theta) = 2\alpha \sin^2(\theta/2) = \alpha - \alpha \cos\theta.$$

CHAPTER 7

Systems of First Order Linear Equations

7.1

1. Introduce the variables $x_1 = u$ and $x_2 = u'$. It follows that $x_1' = x_2$ and
$$x_2' = u'' = -2u - 0.5\,u'.$$
In terms of the new variables, we obtain the system of two first order ODEs
$$x_1' = x_2$$
$$x_2' = -2x_1 - 0.5\,x_2.$$

3. First divide both sides of the equation by t^2, and write
$$u'' = -\frac{1}{t}u' - (1 - \frac{1}{4t^2})u.$$
Set $x_1 = u$ and $x_2 = u'$. It follows that $x_1' = x_2$ and
$$x_2' = u'' = -\frac{1}{t}u' - (1 - \frac{1}{4t^2})u.$$
We obtain the system of equations
$$x_1' = x_2$$
$$x_2' = -(1 - \frac{1}{4t^2})x_1 - \frac{1}{t}x_2.$$

5. Let $x_1 = u$ and $x_2 = u'$; then $u'' = x_2'$. In terms of the new variables, we have
$$x_2' + 0.25 x_2 + 4 x_1 = 2 \cos 3t$$
with the initial conditions $x_1(0) = 1$ and $x_2(0) = -2$. The equivalent first order system is
$$x_1' = x_2$$
$$x_2' = -4 x_1 - 0.25 x_2 + 2 \cos 3t$$
with the above initial conditions.

7.(a) Solving the first equation for x_2, we have $x_2 = x_1' + 2x_1$. Substitution into the second equation results in $(x_1' + 2x_1)' = x_1 - 2(x_1' + 2x_1)$. That is, $x_1'' + 4 x_1' + 3 x_1 = 0$. The resulting equation is a second order differential equation with constant coefficients. The general solution is $x_1(t) = c_1 e^{-t} + c_2 e^{-3t}$. With x_2 given in terms of x_1, it follows that $x_2(t) = c_1 e^{-t} - c_2 e^{-3t}$.

(b) Imposing the specified initial conditions, we obtain
$$c_1 + c_2 = 2, \quad c_1 - c_2 = 3,$$
with solution $c_1 = 5/2$ and $c_2 = -1/2$. Hence
$$x_1(t) = \frac{5}{2} e^{-t} - \frac{1}{2} e^{-3t} \text{ and } x_2(t) = \frac{5}{2} e^{-t} + \frac{1}{2} e^{-3t}.$$

(c)

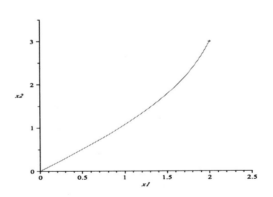

10.(a) Solving the first equation for x_2, we obtain $x_2 = (x_1 - x_1')/2$. Substitution into the second equation results in $(x_1 - x_1')'/2 = 3 x_1 - 2(x_1 - x_1')$. Rearranging the terms, the single differential equation for x_1 is $x_1'' + 3 x_1' + 2 x_1 = 0$.

(b) The general solution is $x_1(t) = c_1 e^{-t} + c_2 e^{-2t}$. With x_2 given in terms of x_1, it follows that $x_2(t) = c_1 e^{-t} + 3 c_2 e^{-2t}/2$. Invoking the specified initial conditions, $c_1 = -7$ and $c_2 = 6$. Hence
$$x_1(t) = -7 e^{-t} + 6 e^{-2t} \text{ and } x_2(t) = -7 e^{-t} + 9 e^{-2t}.$$

(c)

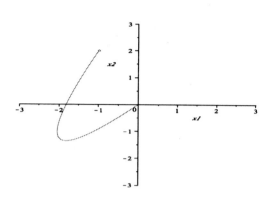

11.(a) Solving the first equation for x_2, we have $x_2 = x_1'/2$. Substitution into the second equation results in $x_1''/2 = -2\,x_1$. The resulting equation is $x_1'' + 4\,x_1 = 0$.

(b) The general solution is $x_1(t) = c_1 \cos 2t + c_2 \sin 2t$. With x_2 given in terms of x_1, it follows that $x_2(t) = -c_1 \sin 2t + c_2 \cos 2t$. Imposing the specified initial conditions, we obtain $c_1 = 3$ and $c_2 = 4$. Hence

$$x_1(t) = 3\cos 2t + 4\sin 2t \text{ and } x_2(t) = -3\sin 2t + 4\cos 2t\,.$$

(c)

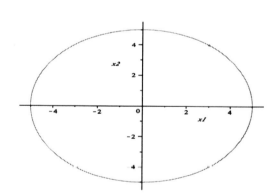

13. Solving the first equation for V, we obtain $V = L \cdot I'$. Substitution into the second equation results in

$$L \cdot I'' = -\frac{I}{C} - \frac{L}{RC}I'.$$

Rearranging the terms, the single differential equation for I is

$$LRC \cdot I'' + L \cdot I' + R \cdot I = 0\,.$$

15. Let $x = c_1 x_1(t) + c_2 x_2(t)$ and $y = c_1 y_1(t) + c_2 y_2(t)$. Then

$$x' = c_1 x_1'(t) + c_2 x_2'(t)$$
$$y' = c_1 y_1'(t) + c_2 y_2'(t).$$

Since $x_1(t)$, $y_1(t)$ and $x_2(t)$, $y_2(t)$ are solutions for the original system,

$$x' = c_1(p_{11}x_1(t) + p_{12}y_1(t)) + c_2(p_{11}x_2(t) + p_{12}y_2(t))$$
$$y' = c_1(p_{21}x_1(t) + p_{22}y_1(t)) + c_2(p_{21}x_2(t) + p_{22}y_2(t)).$$

Rearranging terms gives

$$x' = p_{11}(c_1x_1(t) + c_2x_2(t)) + p_{12}(c_1y_1(t) + c_2y_2(t))$$
$$y' = p_{21}(c_1x_1(t) + c_2x_2(t)) + p_{22}(c_1y_1(t) + c_2y_2(t)),$$

and so x and y solve the original system.

16. Based on the hypothesis,

$$x_1'(t) = p_{11}(t)x_1(t) + p_{12}(t)y_1(t) + g_1(t)$$
$$x_2'(t) = p_{11}(t)x_2(t) + p_{12}(t)y_2(t) + g_1(t).$$

Subtracting the two equations,

$$x_1'(t) - x_2'(t) = p_{11}(t)\left[x_1'(t) - x_2'(t)\right] + p_{12}(t)\left[y_1'(t) - y_2'(t)\right].$$

Similarly,

$$y_1'(t) - y_2'(t) = p_{21}(t)\left[x_1'(t) - x_2'(t)\right] + p_{22}(t)\left[y_1'(t) - y_2'(t)\right].$$

Hence the difference of the two solutions satisfies the homogeneous ODE.

17. For rectilinear motion in one dimension, Newton's second law can be stated as

$$\sum F = m\,x''.$$

The resisting force exerted by a linear spring is given by $F_s = k\,\delta$, in which δ is the displacement of the end of a spring from its equilibrium configuration. Hence, with $0 < x_1 < x_2$, the first two springs are in tension, and the last spring is in compression. The sum of the spring forces on m_1 is

$$F_s^1 = -k_1 x_1 - k_2(x_2 - x_1).$$

The total force on m_1 is

$$\sum F^1 = -k_1 x_1 + k_2(x_2 - x_1) + F_1(t).$$

Similarly, the total force on m_2 is

$$\sum F^2 = -k_2(x_2 - x_1) - k_3 x_2 + F_2(t).$$

18. One of the ways to transform the system is to assign the variables

$$y_1 = x_1, \qquad y_2 = x_2, \qquad y_3 = x_1', \qquad y_4 = x_2'.$$

Before proceeding, note that

$$x_1'' = \frac{1}{m_1}\left[-(k_1 + k_2)x_1 + k_2 x_2 + F_1(t)\right]$$
$$x_2'' = \frac{1}{m_2}\left[k_2 x_1 - (k_2 + k_3)x_2 + F_2(t)\right].$$

Differentiating the new variables, we obtain the system of four first order equations

$$y_1' = y_3$$
$$y_2' = y_4$$
$$y_3' = \frac{1}{m_1}(-(k_1+k_2)y_1 + k_2 y_2 + F_1(t))$$
$$y_4' = \frac{1}{m_2}(k_2 y_1 - (k_2+k_3)y_2 + F_2(t)).$$

19.(a) Taking a clockwise loop around each of the paths, it is easy to see that voltage drops are given by $V_1 - V_2 = 0$, and $V_2 - V_3 = 0$.

(b) Consider the right node. The current in is given by $I_1 + I_2$. The current leaving the node is $-I_3$. Hence the current passing through the node is $(I_1 + I_2) - (-I_3)$. Based on Kirchhoff's first law, $I_1 + I_2 + I_3 = 0$.

(c) In the capacitor,
$$CV_1' = I_1.$$
In the resistor,
$$V_2 = RI_2.$$
In the inductor,
$$LI_3' = V_3.$$

(d) Based on part (a), $V_3 = V_2 = V_1$. Based on part (b),
$$CV_1' + \frac{1}{R}V_2 + I_3 = 0.$$
It follows that
$$CV_1' = -\frac{1}{R}V_1 - I_3 \quad \text{and} \quad LI_3' = V_1.$$

21. Let $I_1, I_2, I_3,$ and I_4 be the current through the resistors, inductor, and capacitor, respectively. Assign $V_1, V_2, V_3,$ and V_4 as the respective voltage drops. Based on Kirchhoff's second law, the net voltage drops, around each loop, satisfy

$$V_1 + V_3 + V_4 = 0, \quad V_1 + V_3 + V_2 = 0 \quad \text{and} \quad V_4 - V_2 = 0.$$

Applying Kirchhoff's first law to the upper-right node,
$$I_3 - (I_2 + I_4) = 0.$$
Likewise, in the remaining nodes,
$$I_1 - I_3 = 0 \quad \text{and} \quad I_2 + I_4 - I_1 = 0.$$
That is,
$$V_4 - V_2 = 0, \quad V_1 + V_3 + V_4 = 0 \quad \text{and} \quad I_2 + I_4 - I_3 = 0.$$
Using the current-voltage relations,
$$V_1 = R_1 I_1, \quad V_2 = R_2 I_2, \quad LI_3' = V_3, \quad CV_4' = I_4.$$

Combining these equations,

$$R_1 I_3 + L I_3' + V_4 = 0 \quad \text{and} \quad C V_4' = I_3 - \frac{V_4}{R_2}.$$

Now set $I_3 = I$ and $V_4 = V$, to obtain the system of equations

$$L I' = -R_1 I - V \quad \text{and} \quad C V' = I - \frac{V}{R_2}.$$

23.(a)

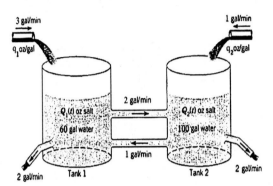

Let $Q_1(t)$ and $Q_2(t)$ be the amount of salt in the respective tanks at time t. Note that the volume of each tank remains constant. Based on conservation of mass, the rate of increase of salt, in any given tank, is given by

$$\text{rate of increase} = \text{rate in} - \text{rate out}.$$

The rate of salt flowing into Tank 1 is

$$r_{in} = \left[q_1 \frac{\text{oz}}{\text{gal}} \right] \left[3 \frac{\text{gal}}{\text{min}} \right] + \left[\frac{Q_2}{100} \frac{\text{oz}}{\text{gal}} \right] \left[1 \frac{\text{gal}}{\text{min}} \right] = 3 q_1 + \frac{Q_2}{100} \frac{\text{oz}}{\text{min}}.$$

The rate at which salt flows out of Tank 1 is

$$r_{out} = \left[\frac{Q_1}{60} \frac{\text{oz}}{\text{gal}} \right] \left[4 \frac{\text{gal}}{\text{min}} \right] = \frac{Q_1}{15} \frac{\text{oz}}{\text{min}}.$$

Hence

$$\frac{dQ_1}{dt} = 3 q_1 + \frac{Q_2}{100} - \frac{Q_1}{15}.$$

Similarly, for Tank 2,

$$\frac{dQ_2}{dt} = q_2 + \frac{Q_1}{30} - \frac{3 Q_2}{100}.$$

The process is modeled by the system of equations

$$Q_1' = -\frac{Q_1}{15} + \frac{Q_2}{100} + 3 q_1$$

$$Q_2' = \frac{Q_1}{30} - \frac{3 Q_2}{100} + q_2.$$

The initial conditions are $Q_1(0) = Q_1^0$ and $Q_2(0) = Q_2^0$.

(b) The equilibrium values are obtained by solving the system

$$-\frac{Q_1}{15} + \frac{Q_2}{100} + 3q_1 = 0$$
$$\frac{Q_1}{30} - \frac{3Q_2}{100} + q_2 = 0.$$

Its solution leads to $Q_1^E = 54\,q_1 + 6\,q_2$ and $Q_2^E = 60\,q_1 + 40\,q_2$.

(c) The question refers to a possible solution of the system

$$54\,q_1 + 6\,q_2 = 60$$
$$60\,q_1 + 40\,q_2 = 50.$$

It is possible to formally solve the system of equations, but the unique solution gives

$$q_1 = \frac{7}{6}\frac{\text{oz}}{\text{gal}} \quad \text{and} \quad q_2 = -\frac{1}{2}\frac{\text{oz}}{\text{gal}},$$

which is not physically possible.

(d) We can write

$$q_2 = -9\,q_1 + \frac{Q_1^E}{6}$$
$$q_2 = -\frac{3}{2}\,q_1 + \frac{Q_2^E}{40},$$

which are the equations of two lines in the q_1-q_2-plane:

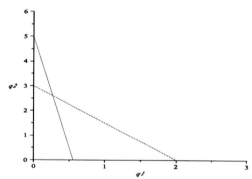

The intercepts of the first line are $Q_1^E/54$ and $Q_1^E/6$. The intercepts of the second line are $Q_2^E/60$ and $Q_2^E/40$. Therefore the system will have a unique solution, in the first quadrant, as long as $Q_1^E/54 \leq Q_2^E/60$ or $Q_2^E/40 \leq Q_1^E/6$. That is,

$$\frac{10}{9} \leq \frac{Q_2^E}{Q_1^E} \leq \frac{20}{3}.$$

7.2

2.(a)
$$\mathbf{A} - 2\mathbf{B} = \begin{pmatrix} 1+i-2i & -1+2i-6 \\ 3+2i-4 & 2-i+4i \end{pmatrix} = \begin{pmatrix} 1-i & -7+2i \\ -1+2i & 2+3i \end{pmatrix}.$$

(b)
$$3\mathbf{A} + \mathbf{B} = \begin{pmatrix} 3+3i+i & -3+6i+3 \\ 9+6i+2 & 6-3i-2i \end{pmatrix} = \begin{pmatrix} 3+4i & 6i \\ 11+6i & 6-5i \end{pmatrix}.$$

(c)
$$\mathbf{AB} = \begin{pmatrix} (1+i)i+2(-1+2i) & 3(1+i)+(-1+2i)(-2i) \\ (3+2i)i+2(2-i) & 3(3+2i)+(2-i)(-2i) \end{pmatrix}$$
$$= \begin{pmatrix} -3+5i & 7+5i \\ 2+i & 7+2i \end{pmatrix}.$$

(d)
$$\mathbf{BA} = \begin{pmatrix} (1+i)i+3(3+2i) & (-1+2i)i+3(2-i) \\ 2(1+i)+(-2i)(3+2i) & 2(-1+2i)+(-2i)(2-i) \end{pmatrix}$$
$$= \begin{pmatrix} 8+7i & 4-4i \\ 6-4i & -4 \end{pmatrix}.$$

3.(c,d)
$$\mathbf{A}^T + \mathbf{B}^T = \begin{pmatrix} -2 & 1 & 2 \\ 1 & 0 & -1 \\ 2 & -3 & 1 \end{pmatrix} + \begin{pmatrix} 1 & 3 & -2 \\ 2 & -1 & 1 \\ 3 & -1 & 0 \end{pmatrix}$$
$$= \begin{pmatrix} -1 & 4 & 0 \\ 3 & -1 & 0 \\ 5 & -4 & 1 \end{pmatrix} = (\mathbf{A}+\mathbf{B})^T.$$

4.(b)
$$\overline{\mathbf{A}} = \begin{pmatrix} 3+2i & 1-i \\ 2+i & -2-3i \end{pmatrix}.$$

(c) By definition,
$$\mathbf{A}^* = \overline{\mathbf{A}^T} = (\overline{\mathbf{A}})^T = \begin{pmatrix} 3+2i & 2+i \\ 1-i & -2-3i \end{pmatrix}.$$

5.
$$2(\mathbf{A}+\mathbf{B}) = 2\begin{pmatrix} 5 & 3 & -2 \\ 0 & 2 & 5 \\ 2 & 2 & 3 \end{pmatrix} = \begin{pmatrix} 10 & 6 & -4 \\ 0 & 4 & 10 \\ 4 & 4 & 6 \end{pmatrix}.$$

7. Let $\mathbf{A} = (a_{ij})$ and $\mathbf{B} = (b_{ij})$. The given operations in (a)-(d) are performed elementwise. That is,

(a) $a_{ij} + b_{ij} = b_{ij} + a_{ij}$.

(b) $a_{ij} + (b_{ij} + c_{ij}) = (a_{ij} + b_{ij}) + c_{ij}$.

(c) $\alpha(a_{ij} + b_{ij}) = \alpha\, a_{ij} + \alpha\, b_{ij}$.

(d) $(\alpha + \beta)\, a_{ij} = \alpha\, a_{ij} + \beta\, a_{ij}$.

In the following, let $\mathbf{A} = (a_{ij})$, $\mathbf{B} = (b_{ij})$ and $\mathbf{C} = (c_{ij})$.

(e) Calculating the generic element,

$$(\mathbf{BC})_{ij} = \sum_{k=1}^{n} b_{ik}\, c_{kj}.$$

Therefore

$$[\mathbf{A}(\mathbf{BC})]_{ij} = \sum_{r=1}^{n} a_{ir}\left(\sum_{k=1}^{n} b_{rk}\, c_{kj}\right) = \sum_{r=1}^{n}\sum_{k=1}^{n} a_{ir}\, b_{rk}\, c_{kj} = \sum_{k=1}^{n}\left(\sum_{r=1}^{n} a_{ir}\, b_{rk}\right) c_{kj}.$$

The inner summation is recognized as

$$\sum_{r=1}^{n} a_{ir}\, b_{rk} = (\mathbf{AB})_{ik},$$

which is the ik-th element of the matrix $\mathbf{AB}$. Thus $[\mathbf{A}(\mathbf{BC})]_{ij} = [(\mathbf{AB})\mathbf{C}]_{ij}$.

(f) Likewise,

$$[\mathbf{A}(\mathbf{B} + \mathbf{C})]_{ij} = \sum_{k=1}^{n} a_{ik}(b_{kj} + c_{kj}) = \sum_{k=1}^{n} a_{ik}\, b_{kj} + \sum_{k=1}^{n} a_{ik}\, c_{kj} = (\mathbf{AB})_{ij} + (\mathbf{AC})_{ij}.$$

8.(a) $\mathbf{x}^T \mathbf{y} = 2(-1 + i) + 2(3i) + (1 - i)(3 - i) = 4i$.

(b) $\mathbf{y}^T \mathbf{y} = (-1 + i)^2 + 2^2 + (3 - i)^2 = 12 - 8i$.

(c) $(\mathbf{x}, \mathbf{y}) = 2(-1 - i) + 2(3i) + (1 - i)(3 + i) = 2 + 2i$.

(d) $(\mathbf{y}, \mathbf{y}) = (-1 + i)(-1 - i) + 2^2 + (3 - i)(3 + i) = 16$.

9. Indeed,

$$5 + 3i = \mathbf{x}^T \mathbf{y} = \sum_{j=1}^{n} x_j\, y_j = \mathbf{y}^T \mathbf{x},$$

and

$$3 - 5i = (\mathbf{x}, \mathbf{y}) = \sum_{j=1}^{n} x_j\, \overline{y}_j = \sum_{j=1}^{n} \overline{y}_j\, x_j = \overline{\sum_{j=1}^{n} y_j\, \overline{x}_j} = \overline{(\mathbf{y}, \mathbf{x})}.$$

11. First augment the given matrix by the identity matrix:

$$[\mathbf{A} \,|\, \mathbf{I}] = \begin{pmatrix} 3 & -1 & 1 & 0 \\ 6 & 2 & 0 & 1 \end{pmatrix}.$$

Divide the first row by 3, to obtain

$$\begin{pmatrix} 1 & -1/3 & 1/3 & 0 \\ 6 & 2 & 0 & 1 \end{pmatrix}.$$

Adding -6 times the first row to the second row results in

$$\begin{pmatrix} 1 & -1/3 & 1/3 & 0 \\ 0 & 4 & -2 & 1 \end{pmatrix}.$$

Divide the second row by 4, to obtain

$$\begin{pmatrix} 1 & -1/3 & 1/3 & 0 \\ 0 & 1 & -1/2 & 1/4 \end{pmatrix}.$$

Finally, adding $1/3$ times the second row to the first row results in

$$\begin{pmatrix} 1 & 0 & 1/6 & 1/12 \\ 0 & 1 & -1/2 & 1/4 \end{pmatrix}.$$

Hence

$$\begin{pmatrix} 3 & -1 \\ 6 & 2 \end{pmatrix}^{-1} = \frac{1}{12}\begin{pmatrix} 2 & 1 \\ -6 & 3 \end{pmatrix}.$$

13. The augmented matrix is

$$\begin{pmatrix} 1 & 1 & -1 & 1 & 0 & 0 \\ 2 & -1 & 1 & 0 & 1 & 0 \\ 1 & 1 & 2 & 0 & 0 & 1 \end{pmatrix}.$$

Combining the elements of the first row with the elements of the second and third rows results in

$$\begin{pmatrix} 1 & 1 & -1 & 1 & 0 & 0 \\ 0 & -3 & 3 & -2 & 1 & 0 \\ 0 & 0 & 3 & -1 & 0 & 1 \end{pmatrix}.$$

Divide the elements of the second row by -3, and the elements of the third row by 3. Now subtracting the new second row from the first row yields

$$\begin{pmatrix} 1 & 0 & 0 & 1/3 & 1/3 & 0 \\ 0 & 1 & -1 & 2/3 & -1/3 & 0 \\ 0 & 0 & 1 & -1/3 & 0 & 1/3 \end{pmatrix}.$$

Finally, combine the third row with the second row to obtain

$$\begin{pmatrix} 1 & 0 & 0 & 1/3 & 1/3 & 0 \\ 0 & 1 & 0 & 1/3 & -1/3 & 1/3 \\ 0 & 0 & 1 & -1/3 & 0 & 1/3 \end{pmatrix}.$$

Hence
$$\begin{pmatrix} 1 & 1 & -1 \\ 2 & -1 & 1 \\ 1 & 1 & 2 \end{pmatrix}^{-1} = \frac{1}{3}\begin{pmatrix} 1 & 1 & 0 \\ 1 & -1 & 1 \\ -1 & 0 & 1 \end{pmatrix}.$$

15. Elementary row operations yield

$$\begin{pmatrix} 2 & 1 & 0 & 1 & 0 & 0 \\ 0 & 2 & 1 & 0 & 1 & 0 \\ 0 & 0 & 2 & 0 & 0 & 1 \end{pmatrix} \to \begin{pmatrix} 1 & 1/2 & 0 & 1/2 & 0 & 0 \\ 0 & 1 & 1/2 & 0 & 1/2 & 0 \\ 0 & 0 & 1 & 0 & 0 & 1/2 \end{pmatrix} \to$$
$$\begin{pmatrix} 1 & 0 & -1/4 & 1/2 & -1/4 & 0 \\ 0 & 1 & 0 & 0 & 1/2 & -1/4 \\ 0 & 0 & 1 & 0 & 0 & 1/2 \end{pmatrix} \to \begin{pmatrix} 1 & 0 & -1/4 & 1/2 & -1/4 & 0 \\ 0 & 1 & 0 & 0 & 1/2 & -1/4 \\ 0 & 0 & 1 & 0 & 0 & 1/2 \end{pmatrix}.$$

Finally, combining the first and third rows results in

$$\begin{pmatrix} 1 & 0 & 0 & 1/2 & -1/4 & 1/8 \\ 0 & 1 & 0 & 0 & 1/2 & -1/4 \\ 0 & 0 & 1 & 0 & 0 & 1/2 \end{pmatrix}, \text{ so } A^{-1} = \begin{pmatrix} 1/2 & -1/4 & 1/8 \\ 0 & 1/2 & -1/4 \\ 0 & 0 & 1/2 \end{pmatrix}.$$

16. Elementary row operations yield

$$\begin{pmatrix} 1 & -1 & -1 & 1 & 0 & 0 \\ 2 & 1 & 0 & 0 & 1 & 0 \\ 3 & -2 & 1 & 0 & 0 & 1 \end{pmatrix} \to \begin{pmatrix} 1 & -1 & -1 & 1 & 0 & 0 \\ 0 & 3 & 2 & -2 & 1 & 0 \\ 0 & 1 & 4 & -3 & 0 & 1 \end{pmatrix} \to$$
$$\begin{pmatrix} 1 & 0 & -1/3 & 1/3 & 1/3 & 0 \\ 0 & 1 & 2/3 & -2/3 & 1/3 & 0 \\ 0 & 0 & 10/3 & -7/3 & -1/3 & 1 \end{pmatrix} \to \begin{pmatrix} 1 & 0 & 0 & 1/10 & 3/10 & 1/10 \\ 0 & 1 & 0 & -1/5 & 2/5 & -1/5 \\ 0 & 0 & 10/3 & -7/3 & -1/3 & 1 \end{pmatrix}.$$

Finally, normalizing the last row results in

$$\begin{pmatrix} 1 & 0 & 0 & 1/10 & 3/10 & 1/10 \\ 0 & 1 & 0 & -1/5 & 2/5 & -1/5 \\ 0 & 0 & 1 & -7/10 & -1/10 & 3/10 \end{pmatrix}, \text{ so } A^{-1} = \begin{pmatrix} 1/10 & 3/10 & 1/10 \\ -1/5 & 2/5 & -1/5 \\ -7/10 & -1/10 & 3/10 \end{pmatrix}.$$

17. Elementary row operations on the augmented matrix yield the row-reduced form of the augmented matrix

$$\begin{pmatrix} 1 & 0 & -1/7 & 0 & 1/7 & 2/7 \\ 0 & 1 & 3/7 & 0 & 4/7 & 1/7 \\ 0 & 0 & 0 & 1 & -2 & -1 \end{pmatrix}.$$

The left submatrix cannot be converted to the identity matrix. Hence the given matrix is singular.

18. Elementary row operations on the augmented matrix yield

$$\begin{pmatrix} 1 & 0 & 0 & -1 & 1 & 0 & 0 & 0 \\ 0 & -1 & 1 & 0 & 0 & 1 & 0 & 0 \\ -1 & 0 & 1 & 0 & 0 & 0 & 1 & 0 \\ 0 & 1 & -1 & 1 & 0 & 0 & 0 & 1 \end{pmatrix} \to \begin{pmatrix} 1 & 0 & 0 & -1 & 1 & 0 & 0 & 0 \\ 0 & -1 & 1 & 0 & 0 & 1 & 0 & 0 \\ 0 & 0 & 1 & -1 & 1 & 0 & 1 & 0 \\ 0 & 1 & -1 & 1 & 0 & 0 & 0 & 1 \end{pmatrix} \to$$

$$\begin{pmatrix} 1 & 0 & 0 & -1 & 1 & 0 & 0 & 0 \\ 0 & 1 & -1 & 0 & 0 & -1 & 0 & 0 \\ 0 & 0 & 1 & -1 & 1 & 0 & 1 & 0 \\ 0 & 0 & 0 & 1 & 0 & 1 & 0 & 1 \end{pmatrix} \to \begin{pmatrix} 1 & 0 & 0 & 0 & 1 & 1 & 0 & 1 \\ 0 & 1 & 0 & 0 & 1 & 0 & 1 & 1 \\ 0 & 0 & 1 & 0 & 1 & 1 & 1 & 1 \\ 0 & 0 & 0 & 1 & 0 & 1 & 0 & 1 \end{pmatrix},$$

so

$$A^{-1} = \begin{pmatrix} 1 & 1 & 0 & 1 \\ 1 & 0 & 1 & 1 \\ 1 & 1 & 1 & 1 \\ 0 & 1 & 0 & 1 \end{pmatrix}.$$

19. Elementary row operations on the augmented matrix yield

$$\begin{pmatrix} 1 & -1 & 2 & 0 & 1 & 0 & 0 & 0 \\ -1 & 2 & -4 & 2 & 0 & 1 & 0 & 0 \\ 1 & 0 & 1 & 3 & 0 & 0 & 1 & 0 \\ -2 & 2 & 0 & -1 & 0 & 0 & 0 & 1 \end{pmatrix} \to \begin{pmatrix} 1 & -1 & 2 & 0 & 1 & 0 & 0 & 0 \\ 0 & 1 & -2 & 2 & 1 & 1 & 0 & 0 \\ 0 & 1 & -1 & 3 & -1 & 0 & 1 & 0 \\ 0 & 0 & 4 & -1 & 2 & 0 & 0 & 1 \end{pmatrix} \to$$

$$\begin{pmatrix} 1 & 0 & 0 & 2 & 2 & 1 & 0 & 0 \\ 0 & 1 & -2 & 2 & 1 & 1 & 0 & 0 \\ 0 & 0 & 1 & 1 & -2 & -1 & 1 & 0 \\ 0 & 0 & 4 & -1 & 2 & 0 & 0 & 1 \end{pmatrix} \to \begin{pmatrix} 1 & 0 & 0 & 2 & 2 & 1 & 0 & 0 \\ 0 & 1 & 0 & 4 & -3 & -1 & 2 & 0 \\ 0 & 0 & 1 & 1 & -2 & -1 & 1 & 0 \\ 0 & 0 & 0 & -5 & 10 & 4 & -4 & 1 \end{pmatrix}.$$

Normalizing the-last row and combining it with the others results in

$$\begin{pmatrix} 1 & 0 & 0 & 2 & 2 & 1 & 0 & 0 \\ 0 & 1 & 0 & 4 & -3 & -1 & 2 & 0 \\ 0 & 0 & 1 & 1 & -2 & -1 & 1 & 0 \\ 0 & 0 & 0 & 1 & -2 & -4/5 & 4/5 & -1/5 \end{pmatrix} \to \begin{pmatrix} 1 & 0 & 0 & 0 & 6 & 13/5 & -8/5 & 2/5 \\ 0 & 1 & 0 & 0 & 5 & 11/5 & -6/5 & 4/5 \\ 0 & 0 & 1 & 0 & 0 & -1/5 & 1/5 & 1/5 \\ 0 & 0 & 0 & 1 & -2 & -4/5 & 4/5 & -1/5 \end{pmatrix}.$$

so

$$A^{-1} = \begin{pmatrix} 6 & 13/5 & -8/5 & 2/5 \\ 5 & 11/5 & -6/5 & 4/5 \\ 0 & -1/5 & 1/5 & 1/5 \\ -2 & -4/5 & 4/5 & -1/5 \end{pmatrix}.$$

20. Suppose that there exist matrices **B** and **C**, such that $\mathbf{AB} = \mathbf{I}$ and $\mathbf{CA} = \mathbf{I}$. Then $\mathbf{CAB} = \mathbf{IB} = \mathbf{B}$, also, $\mathbf{CAB} = \mathbf{CI} = \mathbf{C}$. This shows that $\mathbf{B} = \mathbf{C}$.

23. First note that

$$\mathbf{x}' = \begin{pmatrix} 1 \\ 0 \end{pmatrix} e^t + 2 \begin{pmatrix} 1 \\ 1 \end{pmatrix} (e^t + t e^t) = \begin{pmatrix} 3e^t + 2t\, e^t \\ 2e^t + 2t\, e^t \end{pmatrix}.$$

We also have
$$\begin{pmatrix} 2 & -1 \\ 3 & -2 \end{pmatrix} \mathbf{x} = \begin{pmatrix} 2 & -1 \\ 3 & -2 \end{pmatrix} \begin{pmatrix} 1 \\ 0 \end{pmatrix} e^t + \begin{pmatrix} 2 & -1 \\ 3 & -2 \end{pmatrix} \begin{pmatrix} 2 \\ 2 \end{pmatrix} (te^t)$$
$$= \begin{pmatrix} 2 \\ 3 \end{pmatrix} e^t + \begin{pmatrix} 2 \\ 2 \end{pmatrix} (te^t) = \begin{pmatrix} 2e^t + 2t\,e^t \\ 3e^t + 2t\,e^t \end{pmatrix}.$$

It follows that
$$\begin{pmatrix} 2 & -1 \\ 3 & -2 \end{pmatrix} \mathbf{x} + \begin{pmatrix} 1 \\ -1 \end{pmatrix} e^t = \begin{pmatrix} 3e^t + 2t\,e^t \\ 2e^t + 2t\,e^t \end{pmatrix}.$$

24. It is easy to see that
$$\mathbf{x}' = \begin{pmatrix} -6 \\ 8 \\ 4 \end{pmatrix} e^{-t} + \begin{pmatrix} 0 \\ 4 \\ -4 \end{pmatrix} e^{2t} = \begin{pmatrix} -6e^{-t} \\ 8e^{-t} + 4e^{2t} \\ 4e^{-t} - 4e^{-2t} \end{pmatrix}.$$

On the other hand,
$$\begin{pmatrix} 1 & 1 & 1 \\ 2 & 1 & -1 \\ 0 & -1 & 1 \end{pmatrix} \mathbf{x} = \begin{pmatrix} 1 & 1 & 1 \\ 2 & 1 & -1 \\ 0 & -1 & 1 \end{pmatrix} \begin{pmatrix} 6 \\ -8 \\ -4 \end{pmatrix} e^{-t} + \begin{pmatrix} 1 & 1 & 1 \\ 2 & 1 & -1 \\ 0 & -1 & 1 \end{pmatrix} \begin{pmatrix} 0 \\ 2 \\ -2 \end{pmatrix} e^{2t}$$
$$= \begin{pmatrix} -6 \\ 8 \\ 4 \end{pmatrix} e^{-t} + \begin{pmatrix} 0 \\ 4 \\ -4 \end{pmatrix} e^{2t}.$$

26. Differentiation, elementwise, results in
$$\boldsymbol{\Psi}' = \begin{pmatrix} e^t & -2e^{-2t} & 3e^{3t} \\ -4e^t & 2e^{-2t} & 6e^{3t} \\ -e^t & 2e^{-2t} & 3e^{3t} \end{pmatrix}.$$

On the other hand,
$$\begin{pmatrix} 1 & -1 & 4 \\ 3 & 2 & -1 \\ 2 & 1 & -1 \end{pmatrix} \boldsymbol{\Psi} = \begin{pmatrix} 1 & -1 & 4 \\ 3 & 2 & -1 \\ 2 & 1 & -1 \end{pmatrix} \begin{pmatrix} e^t & e^{-2t} & e^{3t} \\ -4e^t & -e^{-2t} & 2e^{3t} \\ -e^t & -e^{-2t} & e^{3t} \end{pmatrix}$$
$$= \begin{pmatrix} e^t & -2e^{-2t} & 3e^{3t} \\ -4e^t & 2e^{-2t} & 6e^{3t} \\ -e^t & 2e^{-2t} & 3e^{3t} \end{pmatrix}.$$

7.3

4. The augmented matrix is
$$\begin{pmatrix} 1 & 2 & -1 & | & 0 \\ 2 & 1 & 1 & | & 0 \\ 1 & -1 & 2 & | & 0 \end{pmatrix}.$$

Adding -2 times the first row to the second row and subtracting the first row from the third row results in
$$\begin{pmatrix} 1 & 2 & -1 & | & 0 \\ 0 & -3 & 3 & | & 0 \\ 0 & -3 & 3 & | & 0 \end{pmatrix}.$$
Adding the negative of the second row to the third row results in
$$\begin{pmatrix} 1 & 2 & -1 & | & 0 \\ 0 & -3 & 3 & | & 0 \\ 0 & 0 & 0 & | & 0 \end{pmatrix}.$$
We evidently end up with an equivalent system of equations
$$x_1 + 2x_2 - x_3 = 0$$
$$-x_2 + x_3 = 0.$$
Since there is no unique solution, let $x_3 = \alpha$, where α is arbitrary. It follows that $x_2 = \alpha$, and $x_1 = -\alpha$. Hence all solutions have the form
$$x = \alpha \begin{pmatrix} -1 \\ 1 \\ 1 \end{pmatrix}.$$

5. The augmented matrix is
$$\begin{pmatrix} 1 & 0 & -1 & | & 0 \\ 3 & 1 & 1 & | & 0 \\ -1 & 1 & 2 & | & 0 \end{pmatrix}.$$
Adding -3 times the first row to the second row and adding the first row to the last row yields
$$\begin{pmatrix} 1 & 0 & -1 & | & 0 \\ 0 & 1 & 3 & | & 0 \\ 0 & 1 & 1 & | & 0 \end{pmatrix}.$$
Now add the negative of the second row to the third row to obtain
$$\begin{pmatrix} 1 & 0 & -1 & | & 0 \\ 0 & 1 & 3 & | & 0 \\ 0 & 0 & -2 & | & 0 \end{pmatrix}.$$
We end up with an equivalent linear system
$$x_1 - x_3 = 0$$
$$x_2 + 3x_3 = 0$$
$$x_3 = 0.$$
Hence the unique solution of the given system of equations is $x_1 = x_2 = x_3 = 0$.

6. The augmented matrix is
$$\begin{pmatrix} 1 & 2 & -1 & | & -2 \\ -2 & -4 & 2 & | & 4 \\ 2 & 4 & -2 & | & -4 \end{pmatrix}.$$

Adding 2 times the first row to the second row and subtracting 2 times the first row from the third row results in

$$\begin{pmatrix} 1 & 2 & -1 & | & -2 \\ 0 & 0 & 0 & | & 0 \\ 0 & 0 & 0 & | & 0 \end{pmatrix}.$$

We evidently end up with an equivalent system of equations

$$x_1 + 2x_2 - x_3 = -2.$$

Since there is no unique solution, let $x_2 = \alpha$, and $x_3 = \beta$, where α, β are arbitrary. It follows that $x_1 = -2 - 2\alpha + \beta$. Hence all solutions have the form

$$\mathbf{x} = \begin{pmatrix} -2 - 2\alpha + \beta \\ \alpha \\ \beta \end{pmatrix}.$$

8. Write the given vectors as columns of the matrix

$$\mathbf{X} = \begin{pmatrix} 2 & 0 & -1 \\ 1 & 1 & 2 \\ 0 & 0 & 0 \end{pmatrix}.$$

It is evident that $\det(\mathbf{X}) = 0$. Hence the vectors are linearly dependent. In order to find a linear relationship between them, write $c_1 \mathbf{x}^{(1)} + c_2 \mathbf{x}^{(2)} + c_3 \mathbf{x}^{(3)} = \mathbf{0}$. The latter equation is equivalent to

$$\begin{pmatrix} 2 & 0 & -1 \\ 1 & 1 & 2 \\ 0 & 0 & 0 \end{pmatrix} \begin{pmatrix} c_1 \\ c_2 \\ c_3 \end{pmatrix} = \begin{pmatrix} 0 \\ 0 \\ 0 \end{pmatrix}.$$

Performing elementary row operations,

$$\begin{pmatrix} 2 & 0 & -1 & | & 0 \\ 1 & 1 & 2 & | & 0 \\ 0 & 0 & 0 & | & 0 \end{pmatrix} \to \begin{pmatrix} 1 & 0 & -1/2 & | & 0 \\ 0 & 1 & 5/2 & | & 0 \\ 0 & 0 & 0 & | & 0 \end{pmatrix}.$$

We obtain the system of equations

$$c_1 - c_3/2 = 0$$
$$c_2 + 5c_3/2 = 0.$$

Setting $c_3 = 2$, it follows that $c_1 = 1$ and $c_3 = -5$. Hence

$$\mathbf{x}^{(1)} - 5\mathbf{x}^{(2)} + 2\mathbf{x}^{(3)} = \mathbf{0}.$$

10. The matrix containing the given vectors as columns is

$$\mathbf{X} = \begin{pmatrix} 1 & 2 & -1 & 3 \\ 2 & 3 & 0 & -1 \\ -1 & 1 & 2 & 1 \\ 0 & -1 & 2 & 3 \end{pmatrix}.$$

We find that $\det(\mathbf{X}) = -70$. Hence the given vectors are linearly independent.

11. Write the given vectors as columns of the matrix

$$\mathbf{X} = \begin{pmatrix} 1 & 3 & 2 & 4 \\ 2 & 1 & -1 & 3 \\ -2 & 0 & 1 & -2 \end{pmatrix}.$$

The four vectors are necessarily linearly dependent. Hence there are nonzero scalars such that $c_1\mathbf{x}^{(1)} + c_2\mathbf{x}^{(2)} + c_3\mathbf{x}^{(3)} + c_4\mathbf{x}^{(4)} = \mathbf{0}$. The latter equation is equivalent to

$$\begin{pmatrix} 1 & 3 & 2 & 4 \\ 2 & 1 & -1 & 3 \\ -2 & 0 & 1 & -2 \end{pmatrix} \begin{pmatrix} c_1 \\ c_2 \\ c_3 \\ c_4 \end{pmatrix} = \begin{pmatrix} 0 \\ 0 \\ 0 \end{pmatrix}.$$

Performing elementary row operations,

$$\begin{pmatrix} 1 & 3 & 2 & 4 & | & 0 \\ 2 & 1 & -1 & 3 & | & 0 \\ -2 & 0 & 1 & -2 & | & 0 \end{pmatrix} \to \begin{pmatrix} 1 & 0 & 0 & 1 & | & 0 \\ 0 & 1 & 0 & 1 & | & 0 \\ 0 & 0 & 1 & 0 & | & 0 \end{pmatrix}.$$

We end up with an equivalent linear system

$$c_1 + c_4 = 0$$
$$c_2 + c_4 = 0$$
$$c_3 = 0.$$

Let $c_4 = -1$. Then $c_1 = 1$ and $c_2 = 1$. Therefore we find that

$$\mathbf{x}^{(1)} + \mathbf{x}^{(2)} - \mathbf{x}^{(4)} = \mathbf{0}.$$

12. The matrix containing the given vectors as columns, $\mathbf{X}$, is of size $n \times m$. Since $n < m$, we can augment the matrix with $m - n$ rows of zeros. The resulting matrix, $\tilde{\mathbf{X}}$, is of size $m \times m$. Since $\tilde{\mathbf{X}}$ is a square matrix, with at least one row of zeros, it follows that $\det(\tilde{\mathbf{X}}) = 0$. Hence the column vectors of $\tilde{\mathbf{X}}$ are linearly dependent. That is, there is a nonzero vector, $\mathbf{c}$, such that $\tilde{\mathbf{X}}\mathbf{c} = \mathbf{0}_{m \times 1}$. If we write only the first n rows of the latter equation, we have $\mathbf{X}\mathbf{c} = \mathbf{0}_{n \times 1}$. Therefore the column vectors of $\mathbf{X}$ are linearly dependent.

13. By inspection, we find that

$$\mathbf{x}^{(1)}(t) - 2\mathbf{x}^{(2)}(t) = \begin{pmatrix} -e^{-t} \\ 0 \end{pmatrix}.$$

Hence $3\mathbf{x}^{(1)}(t) - 6\mathbf{x}^{(2)}(t) + \mathbf{x}^{(3)}(t) = \mathbf{0}$, and the vectors are linearly dependent.

17. The eigenvalues λ and eigenvectors $\mathbf{x}$ satisfy the equation

$$\begin{pmatrix} 3 - \lambda & -2 \\ 4 & -1 - \lambda \end{pmatrix} \begin{pmatrix} x_1 \\ x_2 \end{pmatrix} = \begin{pmatrix} 0 \\ 0 \end{pmatrix}.$$

For a nonzero solution, we must have $(3 - \lambda)(-1 - \lambda) + 8 = 0$, that is,

$$\lambda^2 - 2\lambda + 5 = 0.$$

The eigenvalues are $\lambda_1 = 1 - 2i$ and $\lambda_2 = 1 + 2i$. The components of the eigenvector $\mathbf{x}^{(1)}$ are solutions of the system

$$\begin{pmatrix} 2+2i & -2 \\ 4 & -2+2i \end{pmatrix} \begin{pmatrix} x_1 \\ x_2 \end{pmatrix} = \begin{pmatrix} 0 \\ 0 \end{pmatrix}.$$

The two equations reduce to $(1+i)x_1 = x_2$. Hence $\mathbf{x}^{(1)} = (1, 1+i)^T$. Now setting $\lambda = \lambda_2 = 1 + 2i$, we have

$$\begin{pmatrix} 2-2i & -2 \\ 4 & -2-2i \end{pmatrix} \begin{pmatrix} x_1 \\ x_2 \end{pmatrix} = \begin{pmatrix} 0 \\ 0 \end{pmatrix},$$

with solution given by $\mathbf{x}^{(2)} = (1, 1-i)^T$.

18. The eigenvalues λ and eigenvectors $\mathbf{x}$ satisfy the equation

$$\begin{pmatrix} -2-\lambda & 1 \\ 1 & -2-\lambda \end{pmatrix} \begin{pmatrix} x_1 \\ x_2 \end{pmatrix} = \begin{pmatrix} 0 \\ 0 \end{pmatrix}.$$

For a nonzero solution, we must have $(-2-\lambda)(-2-\lambda) - 1 = 0$, that is,

$$\lambda^2 + 4\lambda + 3 = 0.$$

The eigenvalues are $\lambda_1 = -3$ and $\lambda_2 = -1$. For $\lambda_1 = -3$, the system of equations becomes

$$\begin{pmatrix} 1 & 1 \\ 1 & 1 \end{pmatrix} \begin{pmatrix} x_1 \\ x_2 \end{pmatrix} = \begin{pmatrix} 0 \\ 0 \end{pmatrix},$$

which reduces to $x_1 + x_2 = 0$. A solution vector is given by $\mathbf{x}^{(1)} = (1, -1)^T$. Substituting $\lambda = \lambda_2 = -1$, we have

$$\begin{pmatrix} -1 & 1 \\ 1 & -1 \end{pmatrix} \begin{pmatrix} x_1 \\ x_2 \end{pmatrix} = \begin{pmatrix} 0 \\ 0 \end{pmatrix}.$$

The equations reduce to $x_1 = x_2$. Hence a solution vector is given by $\mathbf{x}^{(2)} = (1, 1)^T$.

20. The eigensystem is obtained from analysis of the equation

$$\begin{pmatrix} 1-\lambda & \sqrt{3} \\ \sqrt{3} & -1-\lambda \end{pmatrix} \begin{pmatrix} x_1 \\ x_2 \end{pmatrix} = \begin{pmatrix} 0 \\ 0 \end{pmatrix}.$$

For a nonzero solution, the determinant of the coefficient matrix must be zero. That is,

$$\lambda^2 - 4 = 0.$$

Hence the eigenvalues are $\lambda_1 = -2$ and $\lambda_2 = 2$. Substituting the first eigenvalue, $\lambda = -2$, yields

$$\begin{pmatrix} 3 & \sqrt{3} \\ \sqrt{3} & 1 \end{pmatrix} \begin{pmatrix} x_1 \\ x_2 \end{pmatrix} = \begin{pmatrix} 0 \\ 0 \end{pmatrix}.$$

The system is equivalent to the equation $\sqrt{3}\, x_1 + x_2 = 0$. A solution vector is given by $\mathbf{x}^{(1)} = (1, -\sqrt{3})^T$. Substitution of $\lambda = 2$ results in

$$\begin{pmatrix} -1 & \sqrt{3} \\ \sqrt{3} & -3 \end{pmatrix} \begin{pmatrix} x_1 \\ x_2 \end{pmatrix} = \begin{pmatrix} 0 \\ 0 \end{pmatrix},$$

which reduces to $x_1 = \sqrt{3}\, x_2$. A corresponding solution vector is $\mathbf{x}^{(2)} = (\sqrt{3}, 1)^T$.

21. The eigenvalues λ and eigenvectors x satisfy the equation
$$\begin{pmatrix} -3-\lambda & 3/4 \\ -5 & 1-\lambda \end{pmatrix} \begin{pmatrix} x_1 \\ x_2 \end{pmatrix} = \begin{pmatrix} 0 \\ 0 \end{pmatrix}.$$
For a nonzero solution, we must have $(-3-\lambda)(1-\lambda) + 15/4 = 0$, that is,
$$\lambda^2 + 2\lambda + 3/4 = 0.$$
Hence the eigenvalues are $\lambda_1 = -3/2$ and $\lambda_2 = -1/2$. In order to determine the eigenvector corresponding to λ_1, set $\lambda = -3/2$. The system of equations becomes
$$\begin{pmatrix} -3/2 & 3/4 \\ -5 & 5/2 \end{pmatrix} \begin{pmatrix} x_1 \\ x_2 \end{pmatrix} = \begin{pmatrix} 0 \\ 0 \end{pmatrix},$$
which reduces to $-2\,x_1 + x_2 = 0$. A solution vector is given by $\mathbf{x}^{(1)} = (1, 2)^T$. Substitution of $\lambda = \lambda_2 = -1/2$ results in
$$\begin{pmatrix} -5/2 & 3/4 \\ -5 & 3/2 \end{pmatrix} \begin{pmatrix} x_1 \\ x_2 \end{pmatrix} = \begin{pmatrix} 0 \\ 0 \end{pmatrix},$$
which reduces to $10\,x_1 = 3\,x_2$. A corresponding solution vector is $\mathbf{x}^{(2)} = (3, 10)^T$.

23. The eigensystem is obtained from analysis of the equation
$$\begin{pmatrix} 3-\lambda & 2 & 2 \\ 1 & 4-\lambda & 1 \\ -2 & -4 & -1-\lambda \end{pmatrix} \begin{pmatrix} x_1 \\ x_2 \\ x_3 \end{pmatrix} = \begin{pmatrix} 0 \\ 0 \\ 0 \end{pmatrix}.$$
The characteristic equation of the coefficient matrix is $\lambda^3 - 6\lambda^2 + 11\lambda - 6 = 0$, with roots $\lambda_1 = 1$, $\lambda_2 = 2$ and $\lambda_3 = 3$. Setting $\lambda = \lambda_1 = 1$, we have
$$\begin{pmatrix} 2 & 2 & 2 \\ 1 & 3 & 1 \\ -2 & -4 & -2 \end{pmatrix} \begin{pmatrix} x_1 \\ x_2 \\ x_3 \end{pmatrix} = \begin{pmatrix} 0 \\ 0 \\ 0 \end{pmatrix}.$$
This system is reduces to the equations
$$x_1 + x_3 = 0$$
$$x_2 = 0.$$
A corresponding solution vector is given by $\mathbf{x}^{(1)} = (1, 0, -1)^T$. Setting $\lambda = \lambda_2 = 2$, the reduced system of equations is
$$x_1 + 2\,x_2 = 0$$
$$x_3 = 0.$$
A corresponding solution vector is given by $\mathbf{x}^{(2)} = (-2, 1, 0)^T$. Finally, setting $\lambda = \lambda_3 = 3$, the reduced system of equations is
$$x_1 = 0$$
$$x_2 + x_3 = 0.$$
A corresponding solution vector is given by $\mathbf{x}^{(3)} = (0, 1, -1)^T$.

24. For computational purposes, note that if λ is an eigenvalue of $\mathbf{B}$, then $c\lambda$ is an eigenvalue of the matrix $\mathbf{A} = c\mathbf{B}$. Eigenvectors are unaffected, since they are only determined up to a scalar multiple. So with

$$\mathbf{B} = \begin{pmatrix} 11 & -2 & 8 \\ -2 & 2 & 10 \\ 8 & 10 & 5 \end{pmatrix},$$

the associated characteristic equation is $\mu^3 - 18\mu^2 - 81\mu + 1458 = 0$, with roots $\mu_1 = -9$, $\mu_2 = 9$ and $\mu_3 = 18$. Hence the eigenvalues of the given matrix, $\mathbf{A}$, are $\lambda_1 = -1$, $\lambda_2 = 1$ and $\lambda_3 = 2$. Setting $\lambda = \lambda_1 = -1$, (which corresponds to using $\mu_1 = -9$ in the modified problem) the reduced system of equations is

$$2x_1 + x_3 = 0$$
$$x_2 + x_3 = 0.$$

A corresponding solution vector is given by $\mathbf{x}^{(1)} = (1, 2, -2)^T$. Setting $\lambda = \lambda_2 = 1$, the reduced system of equations is

$$x_1 + 2x_3 = 0$$
$$x_2 - 2x_3 = 0.$$

A corresponding solution vector is given by $\mathbf{x}^{(2)} = (2, -2, -1)^T$. Finally, setting $\lambda = \lambda_2 = 1$, the reduced system of equations is

$$x_1 - x_3 = 0$$
$$2x_2 - x_3 = 0.$$

A corresponding solution vector is given by $\mathbf{x}^{(3)} = (2, 1, 2)^T$.

26.(b) By definition,

$$(\mathbf{Ax}, \mathbf{y}) = \sum_{i=0}^{n} (\mathbf{Ax})_i \, \overline{y_i} = \sum_{i=0}^{n} \sum_{j=0}^{n} a_{ij} x_j \, \overline{y_i}.$$

Let $b_{ij} = \overline{a_{ji}}$, so that $a_{ij} = \overline{b_{ji}}$. Now interchanging the order or summation,

$$(\mathbf{Ax}, \mathbf{y}) = \sum_{j=0}^{n} x_j \sum_{i=0}^{n} a_{ij} \, \overline{y_i} = \sum_{j=0}^{n} x_j \sum_{i=0}^{n} \overline{b_{ji}} \, \overline{y_i}.$$

Now note that

$$\sum_{i=0}^{n} \overline{b_{ji}} \, \overline{y_i} = \overline{\sum_{i=0}^{n} b_{ji} \, y_i} = \overline{(\mathbf{A}^*\mathbf{y})_j}.$$

Therefore

$$(\mathbf{Ax}, \mathbf{y}) = \sum_{j=0}^{n} x_j \, \overline{(\mathbf{A}^*\mathbf{y})_j} = (\mathbf{x}, \mathbf{A}^*\mathbf{y}).$$

(c) By definition of a Hermitian matrix, $\mathbf{A} = \mathbf{A}^*$.

27. Suppose that $\mathbf{Ax}= \mathbf{0}$, but that $\mathbf{x}\neq \mathbf{0}$. Let $\mathbf{A}= (a_{ij})$. Using elementary row operations, it is possible to transform the matrix into one that is not upper triangular. If it were upper triangular, backsubstitution would imply that $\mathbf{x}= \mathbf{0}$. Hence a linear combination of all the rows results in a row containing only zeros. That is, there are n scalars, β_i, one for each row and not all zero, such that for each for column j,

$$\sum_{i=1}^{n} \beta_i\, a_{ij} = 0.$$

Now consider $\mathbf{A}^* = (b_{ij})$. By definition, $b_{ij} = \overline{a_{ji}}$, or $a_{ij} = \overline{b_{ji}}$. It follows that for each j,

$$\sum_{i=1}^{n} \beta_i\, \overline{b_{ji}} = \sum_{k=1}^{n} \overline{b_{jk}}\, \beta_k = \overline{\sum_{k=1}^{n} b_{jk}\, \overline{\beta_k}} = 0.$$

Let $\mathbf{y}= (\overline{\beta_1}, \overline{\beta_2}, \cdots, \overline{\beta_n})^T$. Hence we have a nonzero vector, $\mathbf{y}$, such that $\mathbf{A}^*\mathbf{y}= \mathbf{0}$.

29. By linearity,

$$\mathbf{A}(\mathbf{x}^{(0)} + \alpha\, \boldsymbol{\xi}) = \mathbf{Ax}^{(0)} + \alpha\, \mathbf{A}\boldsymbol{\xi} = \mathbf{b} + \mathbf{0} = \mathbf{b}.$$

30. Let $c_{ij} = \overline{a_{ji}}$. By the hypothesis, there is a nonzero vector, $\mathbf{y}$, such that

$$\sum_{j=1}^{n} c_{ij}\, y_j = \sum_{j=1}^{n} \overline{a_{ji}}\, y_j = 0,\ i = 1, 2, \cdots, n.$$

Taking the conjugate of both sides, and interchanging the indices, we have

$$\sum_{i=1}^{n} a_{ij}\, \overline{y_i} = 0.$$

This implies that a linear combination of each row of $\mathbf{A}$ is equal to zero. Now consider the augmented matrix $[\mathbf{A}\,|\,\mathbf{B}]$. Replace the last row by

$$\sum_{i=1}^{n} \overline{y_i}\, [a_{i1}, a_{i2}, \cdots, a_{in}, b_i] = \left[0, 0, \cdots, 0, \sum_{i=1}^{n} \overline{y_i}\, b_i\right].$$

We find that if $(\mathbf{B}, \mathbf{y}) = 0$, then the last row of the augmented matrix contains only zeros. Hence there are $n - 1$ remaining equations. We can now set $x_n = \alpha$, some parameter, and solve for the other variables in terms of α. Therefore the system of equations $\mathbf{Ax}=\mathbf{b}$ has a solution.

31. If $\lambda = 0$ is an eigenvalue of $\mathbf{A}$, then there is a nonzero vector, $\mathbf{x}$, such that

$$\mathbf{Ax} = \lambda \mathbf{x} = \mathbf{0}.$$

That is, $\mathbf{Ax}= \mathbf{0}$ has a nonzero solution. This implies that the mapping defined by $\mathbf{A}$ is not 1-to-1, and hence not invertible. On the other hand, if $\mathbf{A}$ is singular, then $\det(\mathbf{A}) = 0$. Thus, $\mathbf{Ax}= \mathbf{0}$ has a nonzero solution. The latter equation can be written as $\mathbf{Ax}= 0\,\mathbf{x}$.

32.(a) Based on Problem 26, $(\mathbf{Ax},\mathbf{x}) = (\mathbf{x},\mathbf{Ax})$.

(b) Let $\mathbf{x}$ be an eigenvector corresponding to an eigenvalue λ. It then follows that $(\mathbf{Ax},\mathbf{x}) = (\lambda\mathbf{x},\mathbf{x})$ and $(\mathbf{x},\mathbf{Ax}) = (\mathbf{x},\lambda\mathbf{x})$. Based on the properties of the inner product, $(\lambda\mathbf{x},\mathbf{x}) = \lambda(\mathbf{x},\mathbf{x})$ and $(\mathbf{x},\lambda\mathbf{x}) = \overline{\lambda}(\mathbf{x},\mathbf{x})$. Then from part (a),
$$\lambda(\mathbf{x},\mathbf{x}) = \overline{\lambda}(\mathbf{x},\mathbf{x}).$$

(c) From part (b),
$$(\lambda - \overline{\lambda})(\mathbf{x},\mathbf{x}) = 0.$$
Based on the definition of an eigenvector, $(\mathbf{x},\mathbf{x}) = \|\mathbf{x}\|^2 > 0$. Hence we must have $\lambda - \overline{\lambda} = 0$, which implies that λ is real.

33. From Problem 26(c),
$$(\mathbf{Ax}^{(1)},\mathbf{x}^{(2)}) = (\mathbf{x}^{(1)},\mathbf{Ax}^{(2)}).$$
Hence
$$\lambda_1(\mathbf{x}^{(1)},\mathbf{x}^{(2)}) = \overline{\lambda_2}(\mathbf{x}^{(1)},\mathbf{x}^{(2)}) = \lambda_2(\mathbf{x}^{(1)},\mathbf{x}^{(2)}),$$
since the eigenvalues are real. Therefore
$$(\lambda_1 - \lambda_2)(\mathbf{x}^{(1)},\mathbf{x}^{(2)}) = 0.$$
Given that $\lambda_1 \neq \lambda_2$, we must have $(\mathbf{x}^{(1)},\mathbf{x}^{(2)}) = 0$.

7.4

3. Equation (14) states that the Wronskian satisfies the first order linear ODE
$$\frac{dW}{dt} = (p_{11} + p_{22} + \cdots + p_{nn})W.$$
The general solution of this is given by Equation (15):
$$W(t) = C e^{\int (p_{11}+p_{22}+\cdots+p_{nn})dt},$$
in which C is an arbitrary constant. Let $\mathbf{X}_1$ and $\mathbf{X}_2$ be matrices representing two sets of fundamental solutions. It follows that
$$\det(\mathbf{X}_1) = W_1(t) = C_1 e^{\int (p_{11}+p_{22}+\cdots+p_{nn})dt}$$
$$\det(\mathbf{X}_2) = W_2(t) = C_2 e^{\int (p_{11}+p_{22}+\cdots+p_{nn})dt}.$$
Hence $\det(\mathbf{X}_1)/\det(\mathbf{X}_2) = C_1/C_2$. Note that $C_2 \neq 0$.

4. First note that $p_{11} + p_{22} = -p(t)$. As shown in Problem 3,
$$W\left[\mathbf{x}^{(1)},\mathbf{x}^{(2)}\right] = c e^{-\int p(t)dt}.$$

For second order linear ODE, the Wronskian (as defined in Chapter 3) satisfies the first order differential equation $W' + p(t)W = 0$. It follows that

$$W\left[\mathbf{y}^{(1)}, \mathbf{y}^{(2)}\right] = c_1 e^{-\int p(t)dt}.$$

Alternatively, based on the hypothesis,

$$\mathbf{y}^{(1)} = \alpha_{11} \mathbf{x}_{11} + \alpha_{12} \mathbf{x}_{12}$$
$$\mathbf{y}^{(2)} = \alpha_{21} \mathbf{x}_{11} + \alpha_{22} \mathbf{x}_{12}.$$

Direct calculation shows that

$$W\left[\mathbf{y}^{(1)}, \mathbf{y}^{(2)}\right] = \begin{vmatrix} \alpha_{11} x_{11} + \alpha_{12} x_{12} & \alpha_{21} x_{11} + \alpha_{22} x_{12} \\ \alpha_{11} x'_{11} + \alpha_{12} x'_{12} & \alpha_{21} x'_{11} + \alpha_{22} x'_{12} \end{vmatrix}$$
$$= (\alpha_{11}\alpha_{22} - \alpha_{12}\alpha_{21})x_{11}x'_{12} - (\alpha_{11}\alpha_{22} - \alpha_{12}\alpha_{21})x_{12}x'_{11}$$
$$= (\alpha_{11}\alpha_{22} - \alpha_{12}\alpha_{21})x_{11}x_{22} - (\alpha_{11}\alpha_{22} - \alpha_{12}\alpha_{21})x_{12}x_{21}.$$

Here we used the fact that $\mathbf{x}'_1 = \mathbf{x}_2$. Hence

$$W\left[\mathbf{y}^{(1)}, \mathbf{y}^{(2)}\right] = (\alpha_{11}\alpha_{22} - \alpha_{12}\alpha_{21})W\left[\mathbf{x}^{(1)}, \mathbf{x}^{(2)}\right].$$

5. The particular solution satisfies the ODE $(\mathbf{x}^{(p)})' = \mathbf{P}(t)\mathbf{x}^{(p)} + \mathbf{g}(t)$. Now let $\mathbf{x}$ be any solution of the homogeneous equation, $\mathbf{x}' = \mathbf{P}(t)\mathbf{x}$. We know that $\mathbf{x} = \mathbf{x}^{(c)}$, in which $\mathbf{x}^{(c)}$ is a linear combination of some fundamental solution. By linearity of the differential equation, it follows that $\mathbf{x} = \mathbf{x}^{(p)} + \mathbf{x}^{(c)}$ is a solution of the ODE. Based on the uniqueness theorem, all solutions must have this form.

7.(a) By definition,

$$W\left[\mathbf{x}^{(1)}, \mathbf{x}^{(2)}\right] = \begin{vmatrix} t^2 & e^t \\ 2t & e^t \end{vmatrix} = (t^2 - 2t)e^t.$$

(b) The Wronskian vanishes at $t_0 = 0$ and $t_0 = 2$. Hence the vectors are linearly independent on $\mathcal{D} = (-\infty, 0) \cup (0, 2) \cup (2, \infty)$.

(c) It follows from Theorem 7.4.3 that one or more of the coefficients of the ODE must be discontinuous at $t_0 = 0$ and $t_0 = 2$. If not, the Wronskian would not vanish.

(d) Let

$$\mathbf{x} = c_1 \begin{pmatrix} t^2 \\ 2t \end{pmatrix} + c_2 \begin{pmatrix} e^t \\ e^t \end{pmatrix}.$$

Then

$$\mathbf{x}' = c_1 \begin{pmatrix} 2t \\ 2 \end{pmatrix} + c_2 \begin{pmatrix} e^t \\ e^t \end{pmatrix}.$$

On the other hand,

$$\begin{pmatrix} p_{11} & p_{12} \\ p_{21} & p_{22} \end{pmatrix} \mathbf{x} = c_1 \begin{pmatrix} p_{11} & p_{12} \\ p_{21} & p_{22} \end{pmatrix} \begin{pmatrix} t^2 \\ 2t \end{pmatrix} + c_2 \begin{pmatrix} p_{11} & p_{12} \\ p_{21} & p_{22} \end{pmatrix} \begin{pmatrix} e^t \\ e^t \end{pmatrix}$$

$$= \begin{pmatrix} c_1 \left[p_{11}t^2 + 2p_{12}t \right] + c_2 \left[p_{11} + p_{12} \right] e^t \\ c_1 \left[p_{21}t^2 + 2p_{22}t \right] + c_2 \left[p_{21} + p_{22} \right] e^t \end{pmatrix}.$$

Comparing coefficients, we find that

$$p_{11}t^2 + 2p_{12}t = 2t$$
$$p_{11} + p_{12} = 1$$
$$p_{21}t^2 + 2p_{22}t = 2$$
$$p_{21} + p_{22} = 1.$$

Solution of this system of equations results in

$$p_{11}(t) = 0, \quad p_{12}(t) = 1, \quad p_{21}(t) = \frac{2 - 2t}{t^2 - 2t}, \quad p_{22}(t) = \frac{t^2 - 2}{t^2 - 2t}.$$

Hence the vectors are solutions of the ODE

$$\mathbf{x}' = \frac{1}{t^2 - 2t} \begin{pmatrix} 0 & t^2 - 2t \\ 2 - 2t & t^2 - 2 \end{pmatrix} \mathbf{x}.$$

8. Suppose that the solutions $\mathbf{x}^{(1)}, \mathbf{x}^{(2)}, \cdots, \mathbf{x}^{(m)}$ are linearly dependent at $t = t_0$. Then there are constants $c_1, c_2, \cdots, c_m$ (not all zero) such that

$$c_1 \mathbf{x}^{(1)}(t_0) + c_2 \mathbf{x}^{(2)}(t_0) + \cdots + c_m \mathbf{x}^{(m)}(t_0) = \mathbf{0}.$$

Now let $\mathbf{z}(t) = c_1 \mathbf{x}^{(1)}(t) + c_2 \mathbf{x}^{(2)}(t) + \cdots + c_m \mathbf{x}^{(m)}(t)$. Then clearly, $\mathbf{z}(t)$ is a solution of $\mathbf{x}' = \mathbf{P}(t)\mathbf{x}$, with $\mathbf{z}(t_0) = \mathbf{0}$. Furthermore, $\mathbf{y}(t) \equiv \mathbf{0}$ is also a solution, with $\mathbf{y}(t_0) = \mathbf{0}$. By the uniqueness theorem, $\mathbf{z}(t) = \mathbf{y}(t) = \mathbf{0}$. Hence

$$c_1 \mathbf{x}^{(1)}(t) + c_2 \mathbf{x}^{(2)}(t) + \cdots + c_m \mathbf{x}^{(m)}(t) = \mathbf{0}$$

on the entire interval $\alpha < t < \beta$. Going in the other direction is trivial.

9.(a) Let $\mathbf{y}(t)$ be any solution of $\mathbf{x}' = \mathbf{P}(t)\mathbf{x}$. It follows that

$$\mathbf{z}(t) + \mathbf{y}(t) = c_1 \mathbf{x}^{(1)}(t) + c_2 \mathbf{x}^{(2)}(t) + \cdots + c_n \mathbf{x}^{(n)}(t) + \mathbf{y}(t)$$

is also a solution. Now let $t_0 \in (\alpha, \beta)$. Then the collection of vectors

$$\mathbf{x}^{(1)}(t_0), \mathbf{x}^{(2)}(t_0), \ldots, \mathbf{x}^{(n)}(t_0), \mathbf{y}(t_0)$$

constitutes $n + 1$ vectors, each with n components. Based on the assertion in Problem 12, Section 7.3, these vectors are necessarily linearly dependent. That is, there are $n + 1$ constants $b_1, b_2, \ldots, b_n, b_{n+1}$ (not all zero) such that

$$b_1 \mathbf{x}^{(1)}(t_0) + b_2 \mathbf{x}^{(2)}(t_0) + \cdots + b_n \mathbf{x}^{(n)}(t_0) + b_{n+1} \mathbf{y}(t_0) = \mathbf{0}.$$

From Problem 8, we have

$$b_1 \mathbf{x}^{(1)}(t) + b_2 \mathbf{x}^{(2)}(t) + \cdots + b_n \mathbf{x}^{(n)}(t) + b_{n+1} \mathbf{y}(t) = \mathbf{0}$$

for all $t \in (\alpha, \beta)$. Now $b_{n+1} \neq 0$, otherwise that would contradict the fact that the first n vectors are linearly independent. Hence

$$\mathbf{y}(t) = -\frac{1}{b_{n+1}}(b_1 \mathbf{x}^{(1)}(t) + b_2 \mathbf{x}^{(2)}(t) + \cdots + b_n \mathbf{x}^{(n)}(t)),$$

and the assertion is true.

(b) Consider $\mathbf{z}(t) = c_1 \mathbf{x}^{(1)}(t) + c_2\, \mathbf{x}^{(2)}(t) + \cdots + c_n\, \mathbf{x}^{(n)}(t)$, and suppose that we also have

$$\mathbf{z}(t) = k_1 \mathbf{x}^{(1)}(t) + k_2 \mathbf{x}^{(2)}(t) + \cdots + k_n \mathbf{x}^{(n)}(t).$$

Based on the assumption,

$$(k_1 - c_1)\mathbf{x}^{(1)}(t) + (k_2 - c_2)\mathbf{x}^{(2)}(t) + \cdots + (k_n - c_n)\mathbf{x}^{(n)}(t) = \mathbf{0}.$$

The collection of vectors

$$\mathbf{x}^{(1)}(t), \mathbf{x}^{(2)}(t), \ldots, \mathbf{x}^{(n)}(t)$$

is linearly independent on $\alpha < t < \beta$. It follows that $k_i - c_i = 0$, for $i = 1, 2, \cdots, n$.

7.5

2.(a) Setting $\mathbf{x} = \boldsymbol{\xi} e^{rt}$, and substituting into the ODE, we obtain the algebraic equations

$$\begin{pmatrix} 1-r & -2 \\ 3 & -4-r \end{pmatrix} \begin{pmatrix} \xi_1 \\ \xi_2 \end{pmatrix} = \begin{pmatrix} 0 \\ 0 \end{pmatrix}.$$

For a nonzero solution, we must have $\det(\mathbf{A} - r\mathbf{I}) = r^2 + 3r + 2 = 0$. The roots of the characteristic equation are $r_1 = -1$ and $r_2 = -2$. For $r = -1$, the two equations reduce to $\xi_1 = \xi_2$. The corresponding eigenvector is $\boldsymbol{\xi}^{(1)} = (1, 1)^T$. Substitution of $r = -2$ results in the single equation $3\xi_1 = 2\xi_2$. A corresponding eigenvector is $\boldsymbol{\xi}^{(2)} = (2, 3)^T$. Since the eigenvalues are distinct, the general solution is

$$\mathbf{x} = c_1 \begin{pmatrix} 1 \\ 1 \end{pmatrix} e^{-t} + c_2 \begin{pmatrix} 2 \\ 3 \end{pmatrix} e^{-2t}.$$

(b)

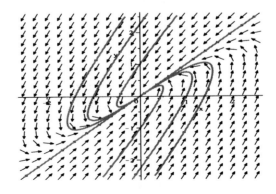

3.(a) Setting $\mathbf{x} = \boldsymbol{\xi} e^{rt}$ results in the algebraic equations
$$\begin{pmatrix} 2-r & -1 \\ 3 & -2-r \end{pmatrix} \begin{pmatrix} \xi_1 \\ \xi_2 \end{pmatrix} = \begin{pmatrix} 0 \\ 0 \end{pmatrix}.$$

For a nonzero solution, we must have $\det(\mathbf{A} - r\mathbf{I}) = r^2 - 1 = 0$. The roots of the characteristic equation are $r_1 = 1$ and $r_2 = -1$. For $r = 1$, the system of equations reduces to $\xi_1 = \xi_2$. The corresponding eigenvector is $\boldsymbol{\xi}^{(1)} = (1, 1)^T$. Substitution of $r = -1$ results in the single equation $3\xi_1 = \xi_2$. A corresponding eigenvector is $\boldsymbol{\xi}^{(2)} = (1, 3)^T$. Since the eigenvalues are distinct, the general solution is
$$\mathbf{x} = c_1 \begin{pmatrix} 1 \\ 1 \end{pmatrix} e^t + c_2 \begin{pmatrix} 1 \\ 3 \end{pmatrix} e^{-t}.$$

(b)

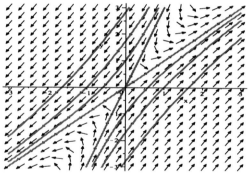

The system has an unstable eigendirection along $\boldsymbol{\xi}^{(1)} = (1, 1)^T$. Unless $c_1 = 0$, all solutions will diverge.

4.(a) Solution of the ODE requires analysis of the algebraic equations
$$\begin{pmatrix} 1-r & 1 \\ 4 & -2-r \end{pmatrix} \begin{pmatrix} \xi_1 \\ \xi_2 \end{pmatrix} = \begin{pmatrix} 0 \\ 0 \end{pmatrix}.$$

For a nonzero solution, we must have $\det(\mathbf{A} - r\mathbf{I}) = r^2 + r - 6 = 0$. The roots of the characteristic equation are $r_1 = 2$ and $r_2 = -3$. For $r = 2$, the system of equations reduces to $\xi_1 = \xi_2$. The corresponding eigenvector is $\boldsymbol{\xi}^{(1)} = (1, 1)^T$. Substitution of $r = -3$ results in the single equation $4\xi_1 + \xi_2 = 0$. A corresponding eigenvector is $\boldsymbol{\xi}^{(2)} = (1, -4)^T$. Since the eigenvalues are distinct, the general solution is
$$\mathbf{x} = c_1 \begin{pmatrix} 1 \\ 1 \end{pmatrix} e^{2t} + c_2 \begin{pmatrix} 1 \\ -4 \end{pmatrix} e^{-3t}.$$

(b)

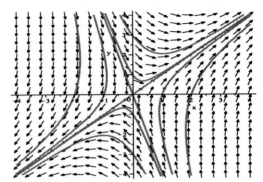

The system has an unstable eigendirection along $\boldsymbol{\xi}^{(1)} = (1,1)^T$. Unless $c_1 = 0$, all solutions will diverge.

8.(a) Setting $\mathbf{x} = \boldsymbol{\xi} e^{rt}$ results in the algebraic equations

$$\begin{pmatrix} 3-r & 6 \\ -1 & -2-r \end{pmatrix} \begin{pmatrix} \xi_1 \\ \xi_2 \end{pmatrix} = \begin{pmatrix} 0 \\ 0 \end{pmatrix}.$$

For a nonzero solution, we must have $\det(\mathbf{A} - r\mathbf{I}) = r^2 - r = 0$. The roots of the characteristic equation are $r_1 = 1$ and $r_2 = 0$. With $r = 1$, the system of equations reduces to $\xi_1 + 3\xi_2 = 0$. The corresponding eigenvector is $\boldsymbol{\xi}^{(1)} = (3, -1)^T$. For the case $r = 0$, the system is equivalent to the equation $\xi_1 + 2\xi_2 = 0$. An eigenvector is $\boldsymbol{\xi}^{(2)} = (2, -1)^T$. Since the eigenvalues are distinct, the general solution is

$$\mathbf{x} = c_1 \begin{pmatrix} 3 \\ -1 \end{pmatrix} e^t + c_2 \begin{pmatrix} 2 \\ -1 \end{pmatrix}.$$

(b)

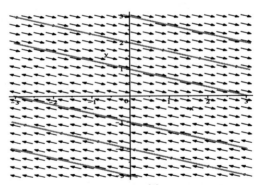

The entire line along the eigendirection $\boldsymbol{\xi}^{(2)} = (2, -1)^T$ consists of equilibrium points. All other solutions diverge. The direction field changes across the line $x_1 + 2x_2 = 0$. Eliminating the exponential terms in the solution, the trajectories are given by $x_1 + 3x_2 = -c_2$.

10. The characteristic equation is given by

$$\begin{vmatrix} 2-r & 2+i \\ -1 & -1-i-r \end{vmatrix} = r^2 - (1-i)r - i = 0.$$

The equation has complex roots $r_1 = 1$ and $r_2 = -i$. For $r = 1$, the components of the solution vector must satisfy $\xi_1 + (2+i)\xi_2 = 0$. Thus the corresponding eigenvector is $\boldsymbol{\xi}^{(1)} = (2+i, -1)^T$. Substitution of $r = -i$ results in the single equation $\xi_1 + \xi_2 = 0$. A corresponding eigenvector is $\boldsymbol{\xi}^{(2)} = (1, -1)^T$. Since the eigenvalues are distinct, the general solution is

$$\mathbf{x} = c_1 \begin{pmatrix} 2+i \\ -1 \end{pmatrix} e^t + c_2 \begin{pmatrix} 1 \\ -1 \end{pmatrix} e^{-it}.$$

11. Setting $\mathbf{x} = \boldsymbol{\xi} e^{rt}$ results in the algebraic equations

$$\begin{pmatrix} 1-r & 1 & 2 \\ 1 & 2-r & 1 \\ 2 & 1 & 1-r \end{pmatrix} \begin{pmatrix} \xi_1 \\ \xi_2 \\ \xi_3 \end{pmatrix} = \begin{pmatrix} 0 \\ 0 \\ 0 \end{pmatrix}.$$

For a nonzero solution, we must have $\det(\mathbf{A} - r\mathbf{I}) = r^3 - 4r^2 - r + 4 = 0$. The roots of the characteristic equation are $r_1 = 4, r_2 = 1$ and $r_3 = -1$. Setting $r = 4$, we have

$$\begin{pmatrix} -3 & 1 & 2 \\ 1 & -2 & 1 \\ 2 & 1 & -3 \end{pmatrix} \begin{pmatrix} \xi_1 \\ \xi_2 \\ \xi_3 \end{pmatrix} = \begin{pmatrix} 0 \\ 0 \\ 0 \end{pmatrix}.$$

This system is reduces to the equations

$$\xi_1 - \xi_3 = 0$$
$$\xi_2 - \xi_3 = 0.$$

A corresponding solution vector is given by $\boldsymbol{\xi}^{(1)} = (1, 1, 1)^T$. Setting $\lambda = 1$, the reduced system of equations is

$$\xi_1 - \xi_3 = 0$$
$$\xi_2 + 2\xi_3 = 0.$$

A corresponding solution vector is given by $\boldsymbol{\xi}^{(2)} = (1, -2, 1)^T$. Finally, setting $\lambda = -1$, the reduced system of equations is

$$\xi_1 + \xi_3 = 0$$
$$\xi_2 = 0.$$

A corresponding solution vector is given by $\boldsymbol{\xi}^{(3)} = (1, 0, -1)^T$. Since the eigenvalues are distinct, the general solution is

$$\mathbf{x} = c_1 \begin{pmatrix} 1 \\ 1 \\ 1 \end{pmatrix} e^{4t} + c_2 \begin{pmatrix} 1 \\ -2 \\ 1 \end{pmatrix} e^t + c_3 \begin{pmatrix} 1 \\ 0 \\ -1 \end{pmatrix} e^{-t}.$$

12. The eigensystem is obtained from analysis of the equation
$$\begin{pmatrix} 3-r & 2 & 4 \\ 2 & -r & 2 \\ 4 & 2 & 3-r \end{pmatrix} \begin{pmatrix} \xi_1 \\ \xi_2 \\ \xi_3 \end{pmatrix} = \begin{pmatrix} 0 \\ 0 \\ 0 \end{pmatrix}.$$

The characteristic equation of the coefficient matrix is $r^3 - 6r^2 - 15r - 8 = 0$, with roots $r_1 = 8$, $r_2 = -1$ and $r_3 = -1$. Setting $r = r_1 = 8$, we have
$$\begin{pmatrix} -5 & 2 & 4 \\ 2 & -8 & 2 \\ 4 & 2 & -5 \end{pmatrix} \begin{pmatrix} \xi_1 \\ \xi_2 \\ \xi_3 \end{pmatrix} = \begin{pmatrix} 0 \\ 0 \\ 0 \end{pmatrix}.$$

This system is reduced to the equations
$$\xi_1 - \xi_3 = 0$$
$$2\xi_2 - \xi_3 = 0.$$

A corresponding solution vector is given by $\boldsymbol{\xi}^{(1)} = (2, 1, 2)^T$. Setting $r = -1$, the system of equations is reduced to the single equation
$$2\xi_1 + \xi_2 + 2\xi_3 = 0.$$

Two independent solutions are obtained as
$$\boldsymbol{\xi}^{(2)} = (1, -2, 0)^T \text{ and } \boldsymbol{\xi}^{(3)} = (0, -2, 1)^T.$$

Hence the general solution is
$$\mathbf{x} = c_1 \begin{pmatrix} 2 \\ 1 \\ 2 \end{pmatrix} e^{8t} + c_2 \begin{pmatrix} 1 \\ -2 \\ 0 \end{pmatrix} e^{-t} + c_3 \begin{pmatrix} 0 \\ -2 \\ 1 \end{pmatrix} e^{-t}.$$

13. Setting $\mathbf{x} = \boldsymbol{\xi} e^{rt}$ results in the algebraic equations
$$\begin{pmatrix} 1-r & 1 & 1 \\ 2 & 1-r & -1 \\ -8 & -5 & -3-r \end{pmatrix} \begin{pmatrix} \xi_1 \\ \xi_2 \\ \xi_3 \end{pmatrix} = \begin{pmatrix} 0 \\ 0 \\ 0 \end{pmatrix}.$$

For a nonzero solution, we must have $\det(\mathbf{A} - r\mathbf{I}) = r^3 + r^2 - 4r - 4 = 0$. The roots of the characteristic equation are $r_1 = 2$, $r_2 = -2$ and $r_3 = -1$. Setting $r = 2$, we have
$$\begin{pmatrix} -1 & 1 & 1 \\ 2 & -1 & -1 \\ -8 & -5 & -5 \end{pmatrix} \begin{pmatrix} \xi_1 \\ \xi_2 \\ \xi_3 \end{pmatrix} = \begin{pmatrix} 0 \\ 0 \\ 0 \end{pmatrix}.$$

This system is reduces to the equations
$$\xi_1 = 0$$
$$\xi_2 + \xi_3 = 0.$$

A corresponding solution vector is given by $\boldsymbol{\xi}^{(1)} = (0, 1, -1)^T$. Setting $\lambda = -1$, the reduced system of equations is
$$2\xi_1 + 3\xi_3 = 0$$
$$\xi_2 - 2\xi_3 = 0.$$

A corresponding solution vector is given by $\boldsymbol{\xi}^{(2)} = (3, -4, -2)^T$. Finally, setting $\lambda = -2$, the reduced system of equations is

$$7\xi_1 + 4\xi_3 = 0$$
$$7\xi_2 - 5\xi_3 = 0.$$

A corresponding solution vector is given by $\boldsymbol{\xi}^{(3)} = (4, -5, -7)^T$. Since the eigenvalues are distinct, the general solution is

$$\mathbf{x} = c_1 \begin{pmatrix} 0 \\ 1 \\ -1 \end{pmatrix} e^{2t} + c_2 \begin{pmatrix} 3 \\ -4 \\ -2 \end{pmatrix} e^{-t} + c_3 \begin{pmatrix} 4 \\ -5 \\ -7 \end{pmatrix} e^{-2t}.$$

15. Setting $\mathbf{x} = \boldsymbol{\xi} e^{rt}$ results in the algebraic equations

$$\begin{pmatrix} 5 - r & -1 \\ 3 & 1 - r \end{pmatrix} \begin{pmatrix} \xi_1 \\ \xi_2 \end{pmatrix} = \begin{pmatrix} 0 \\ 0 \end{pmatrix}.$$

For a nonzero solution, we must have $\det(\mathbf{A} - r\mathbf{I}) = r^2 - 6r + 8 = 0$. The roots of the characteristic equation are $r_1 = 4$ and $r_2 = 2$. With $r = 4$, the system of equations reduces to $\xi_1 - \xi_2 = 0$. The corresponding eigenvector is $\boldsymbol{\xi}^{(1)} = (1, 1)^T$. For the case $r = 2$, the system is equivalent to the equation $3\xi_1 - \xi_2 = 0$. An eigenvector is $\boldsymbol{\xi}^{(2)} = (1, 3)^T$. Since the eigenvalues are distinct, the general solution is

$$\mathbf{x} = c_1 \begin{pmatrix} 1 \\ 1 \end{pmatrix} e^{4t} + c_2 \begin{pmatrix} 1 \\ 3 \end{pmatrix} e^{2t}.$$

Invoking the initial conditions, we obtain the system of equations

$$c_1 + c_2 = 2$$
$$c_1 + 3c_2 = -1.$$

Hence $c_1 = 7/2$ and $c_2 = -3/2$, and the solution of the IVP is

$$\mathbf{x} = \frac{7}{2}\begin{pmatrix} 1 \\ 1 \end{pmatrix} e^{4t} - \frac{3}{2}\begin{pmatrix} 1 \\ 3 \end{pmatrix} e^{2t}.$$

17. Setting $\mathbf{x} = \boldsymbol{\xi} e^{rt}$ results in the algebraic equations

$$\begin{pmatrix} 1 - r & 1 & 2 \\ 0 & 2 - r & 2 \\ -1 & 1 & 3 - r \end{pmatrix} \begin{pmatrix} \xi_1 \\ \xi_2 \\ \xi_3 \end{pmatrix} = \begin{pmatrix} 0 \\ 0 \\ 0 \end{pmatrix}.$$

For a nonzero solution, we must have $\det(\mathbf{A} - r\mathbf{I}) = r^3 - 6r^2 + 11r - 6 = 0$. The roots of the characteristic equation are $r_1 = 1$, $r_2 = 2$ and $r_3 = 3$. Setting $r = 1$, we have

$$\begin{pmatrix} 0 & 1 & 2 \\ 0 & 1 & 2 \\ -1 & 1 & 2 \end{pmatrix} \begin{pmatrix} \xi_1 \\ \xi_2 \\ \xi_3 \end{pmatrix} = \begin{pmatrix} 0 \\ 0 \\ 0 \end{pmatrix}.$$

This system reduces to the equations

$$\xi_1 = 0$$
$$\xi_2 + 2\xi_3 = 0.$$

A corresponding solution vector is given by $\boldsymbol{\xi}^{(1)} = (0, -2, 1)^T$. Setting $\lambda = 2$, the reduced system of equations is

$$\xi_1 - \xi_2 = 0$$
$$\xi_3 = 0.$$

A corresponding solution vector is given by $\boldsymbol{\xi}^{(2)} = (1, 1, 0)^T$. Finally, upon setting $\lambda = 3$, the reduced system of equations is

$$\xi_1 - 2\xi_3 = 0$$
$$\xi_2 - 2\xi_3 = 0.$$

A corresponding solution vector is given by $\boldsymbol{\xi}^{(3)} = (2, 2, 1)^T$. Since the eigenvalues are distinct, the general solution is

$$\mathbf{x} = c_1 \begin{pmatrix} 0 \\ -2 \\ 1 \end{pmatrix} e^t + c_2 \begin{pmatrix} 1 \\ 1 \\ 0 \end{pmatrix} e^{2t} + c_3 \begin{pmatrix} 2 \\ 2 \\ 1 \end{pmatrix} e^{3t}.$$

Invoking the initial conditions, the coefficients must satisfy the equations

$$c_2 + 2c_3 = 2$$
$$-2c_1 + c_2 + 2c_3 = 0$$
$$c_1 + c_3 = 1.$$

It follows that $c_1 = 1$, $c_2 = 2$ and $c_3 = 0$. Hence the solution of the IVP is

$$\mathbf{x} = \begin{pmatrix} 0 \\ -2 \\ 1 \end{pmatrix} e^t + 2 \begin{pmatrix} 1 \\ 1 \\ 0 \end{pmatrix} e^{2t}.$$

18. The eigensystem is obtained from analysis of the equation

$$\begin{pmatrix} -r & 0 & -1 \\ 2 & -r & 0 \\ -1 & 2 & 4-r \end{pmatrix} \begin{pmatrix} \xi_1 \\ \xi_2 \\ \xi_3 \end{pmatrix} = \begin{pmatrix} 0 \\ 0 \\ 0 \end{pmatrix}.$$

The characteristic equation of the coefficient matrix is $r^3 - 4r^2 - r + 4 = 0$, with roots $r_1 = -1$, $r_2 = 1$ and $r_3 = 4$. Setting $r = r_1 = -1$, we have

$$\begin{pmatrix} -1 & 0 & -1 \\ 2 & -1 & 0 \\ -1 & 2 & 3 \end{pmatrix} \begin{pmatrix} \xi_1 \\ \xi_2 \\ \xi_3 \end{pmatrix} = \begin{pmatrix} 0 \\ 0 \\ 0 \end{pmatrix}.$$

This system is reduced to the equations

$$\xi_1 - \xi_3 = 0$$
$$\xi_2 + 2\xi_3 = 0.$$

A corresponding solution vector is given by $\boldsymbol{\xi}^{(1)} = (1, -2, 1)^T$. Setting $r = 1$, the system reduces to the equations

$$\xi_1 + \xi_3 = 0$$
$$\xi_2 + 2\xi_3 = 0.$$

The corresponding eigenvector is $\boldsymbol{\xi}^{(2)} = (1,2,-1)^T$. Finally, upon setting $r=4$, the system is equivalent to the equations

$$4\xi_1 + \xi_3 = 0$$
$$8\xi_2 + \xi_3 = 0.$$

The corresponding eigenvector is $\boldsymbol{\xi}^{(3)} = (2,1,-8)^T$. Hence the general solution is

$$\mathbf{x} = c_1 \begin{pmatrix} 1 \\ -2 \\ 1 \end{pmatrix} e^{-t} + c_2 \begin{pmatrix} 1 \\ 2 \\ -1 \end{pmatrix} e^{t} + c_3 \begin{pmatrix} 2 \\ 1 \\ -8 \end{pmatrix} e^{4t}.$$

Invoking the initial conditions,

$$c_1 + c_2 + 2c_3 = 7$$
$$-2c_1 + 2c_2 + c_3 = 5$$
$$c_1 - c_2 - 8c_3 = 5.$$

It follows that $c_1 = 3$, $c_2 = 6$ and $c_3 = -1$. Hence the solution of the IVP is

$$\mathbf{x} = 3 \begin{pmatrix} 1 \\ -2 \\ 1 \end{pmatrix} e^{-t} + 6 \begin{pmatrix} 1 \\ 2 \\ -1 \end{pmatrix} e^{t} - \begin{pmatrix} 2 \\ 1 \\ -8 \end{pmatrix} e^{4t}.$$

19. Set $\mathbf{x} = \boldsymbol{\xi} t^r$. Substitution into the system of differential equations results in

$$t \cdot rt^{r-1}\boldsymbol{\xi} = \mathbf{A}\boldsymbol{\xi} t^r,$$

which upon simplification yields is, $\mathbf{A}\boldsymbol{\xi} - r\boldsymbol{\xi} = \mathbf{0}$. Hence the vector $\boldsymbol{\xi}$ and constant r must satisfy $(\mathbf{A} - r\mathbf{I})\boldsymbol{\xi} = \mathbf{0}$.

21. Setting $\mathbf{x} = \boldsymbol{\xi} t^r$ results in the algebraic equations

$$\begin{pmatrix} 5-r & -1 \\ 3 & 1-r \end{pmatrix} \begin{pmatrix} \xi_1 \\ \xi_2 \end{pmatrix} = \begin{pmatrix} 0 \\ 0 \end{pmatrix}.$$

For a nonzero solution, we must have $\det(\mathbf{A} - r\mathbf{I}) = r^2 - 6r + 8 = 0$. The roots of the characteristic equation are $r_1 = 4$ and $r_2 = 2$. With $r = 4$, the system of equations reduces to $\xi_1 - \xi_2 = 0$. The corresponding eigenvector is $\boldsymbol{\xi}^{(1)} = (1,1)^T$. For the case $r = 2$, the system is equivalent to the equation $3\xi_1 - \xi_2 = 0$. An eigenvector is $\boldsymbol{\xi}^{(2)} = (1,3)^T$. It follows that

$$\mathbf{x}^{(1)} = \begin{pmatrix} 1 \\ 1 \end{pmatrix} t^4 \text{ and } \mathbf{x}^{(2)} = \begin{pmatrix} 1 \\ 3 \end{pmatrix} t^2.$$

The Wronskian of this solution set is $W\left[\mathbf{x}^{(1)}, \mathbf{x}^{(2)}\right] = 2t^6$. Thus the solutions are linearly independent for $t > 0$. Hence the general solution is

$$\mathbf{x} = c_1 \begin{pmatrix} 1 \\ 1 \end{pmatrix} t^4 + c_2 \begin{pmatrix} 1 \\ 3 \end{pmatrix} t^2.$$

22. As shown in Problem 19, solution of the ODE requires analysis of the equations

$$\begin{pmatrix} 4-r & -3 \\ 8 & -6-r \end{pmatrix} \begin{pmatrix} \xi_1 \\ \xi_2 \end{pmatrix} = \begin{pmatrix} 0 \\ 0 \end{pmatrix}.$$

For a nonzero solution, we must have $\det(\mathbf{A} - r\mathbf{I}) = r^2 + 2r = 0$. The roots of the characteristic equation are $r_1 = 0$ and $r_2 = -2$. For $r = 0$, the system of equations reduces to $4\xi_1 = 3\xi_2$. The corresponding eigenvector is $\boldsymbol{\xi}^{(1)} = (3,4)^T$. Setting $r = -2$ results in the single equation $2\xi_1 - \xi_2 = 0$. A corresponding eigenvector is $\boldsymbol{\xi}^{(2)} = (1,2)^T$. It follows that

$$\mathbf{x}^{(1)} = \begin{pmatrix} 3 \\ 4 \end{pmatrix} \text{ and } \mathbf{x}^{(2)} = \begin{pmatrix} 1 \\ 2 \end{pmatrix} t^{-2}.$$

The Wronskian of this solution set is $W\left[\mathbf{x}^{(1)}, \mathbf{x}^{(2)}\right] = 2t^{-2}$. These solutions are linearly independent for $t > 0$. Hence the general solution is

$$\mathbf{x} = c_1 \begin{pmatrix} 3 \\ 4 \end{pmatrix} + c_2 \begin{pmatrix} 1 \\ 2 \end{pmatrix} t^{-2}.$$

23. Setting $\mathbf{x} = \boldsymbol{\xi} t^r$ results in the algebraic equations

$$\begin{pmatrix} 3-r & -2 \\ 2 & -2-r \end{pmatrix} \begin{pmatrix} \xi_1 \\ \xi_2 \end{pmatrix} = \begin{pmatrix} 0 \\ 0 \end{pmatrix}.$$

For a nonzero solution, we must have $\det(\mathbf{A} - r\mathbf{I}) = r^2 - r - 2 = 0$. The roots of the characteristic equation are $r_1 = 2$ and $r_2 = -1$. Setting $r = 2$, the system of equations reduces to $\xi_1 - 2\xi_2 = 0$. The corresponding eigenvector is $\boldsymbol{\xi}^{(1)} = (2,1)^T$. With $r = -1$, the system is equivalent to the equation $2\xi_1 - \xi_2 = 0$. An eigenvector is $\boldsymbol{\xi}^{(2)} = (1,2)^T$. It follows that

$$\mathbf{x}^{(1)} = \begin{pmatrix} 2 \\ 1 \end{pmatrix} t^2 \text{ and } \mathbf{x}^{(2)} = \begin{pmatrix} 1 \\ 2 \end{pmatrix} t^{-1}.$$

The Wronskian of this solution set is $W\left[\mathbf{x}^{(1)}, \mathbf{x}^{(2)}\right] = 3t$. Thus the solutions are linearly independent for $t > 0$. Hence the general solution is

$$\mathbf{x} = c_1 \begin{pmatrix} 2 \\ 1 \end{pmatrix} t^2 + c_2 \begin{pmatrix} 1 \\ 2 \end{pmatrix} t^{-1}.$$

24.(a) The general solution is

$$x = c_1 \begin{pmatrix} -1 \\ 2 \end{pmatrix} e^{-t} + c_2 \begin{pmatrix} 1 \\ 2 \end{pmatrix} e^{-2t}.$$

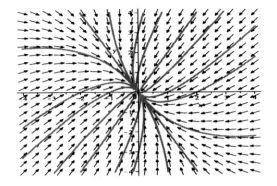

(b)

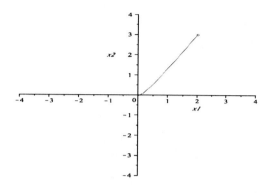

(c)

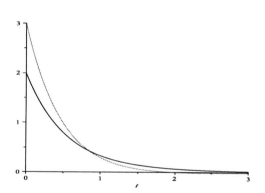

26.(a) The general solution is

$$x = c_1 \begin{pmatrix} -1 \\ 2 \end{pmatrix} e^{-t} + c_2 \begin{pmatrix} 1 \\ 2 \end{pmatrix} e^{2t}.$$

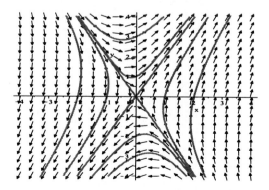

(b)

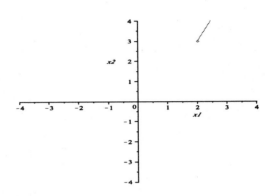

(c)

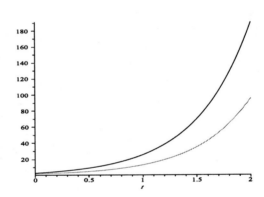

28.(a) We note that $(\mathbf{A} - r_i\mathbf{I})\boldsymbol{\xi}^{(i)} = \mathbf{0}$, for $i = 1, 2$.

(b) It follows that $(\mathbf{A} - r_2\mathbf{I})\boldsymbol{\xi}^{(1)} = \mathbf{A}\boldsymbol{\xi}^{(1)} - r_2\boldsymbol{\xi}^{(1)} = r_1\boldsymbol{\xi}^{(1)} - r_2\boldsymbol{\xi}^{(1)}$.

(c) Suppose that $\boldsymbol{\xi}^{(1)}$ and $\boldsymbol{\xi}^{(2)}$ are linearly dependent. Then there exist constants c_1 and c_2, not both zero, such that $c_1\boldsymbol{\xi}^{(1)} + c_2\boldsymbol{\xi}^{(2)} = \mathbf{0}$. Assume that $c_1 \neq 0$. It is clear that $(\mathbf{A} - r_2\mathbf{I})(c_1\boldsymbol{\xi}^{(1)} + c_2\boldsymbol{\xi}^{(2)}) = \mathbf{0}$. On the other hand,

$$(\mathbf{A} - r_2\mathbf{I})(c_1\boldsymbol{\xi}^{(1)} + c_2\boldsymbol{\xi}^{(2)}) = c_1(r_1 - r_2)\boldsymbol{\xi}^{(1)} + \mathbf{0} = c_1(r_1 - r_2)\boldsymbol{\xi}^{(1)}.$$

Since $r_1 \neq r_2$, we must have $c_1 = 0$, which leads to a contradiction.

(d) Note that $(\mathbf{A} - r_1\mathbf{I})\boldsymbol{\xi}^{(2)} = (r_2 - r_1)\boldsymbol{\xi}^{(2)}$.

(e) Let $n = 3$, with $r_1 \neq r_2 \neq r_3$. Suppose that $\boldsymbol{\xi}^{(1)}$, $\boldsymbol{\xi}^{(2)}$ and $\boldsymbol{\xi}^{(3)}$ are indeed linearly dependent. Then there exist constants c_1, c_2 and c_3, not all zero, such that

$$c_1\boldsymbol{\xi}^{(1)} + c_2\boldsymbol{\xi}^{(2)} + c_3\boldsymbol{\xi}^{(3)} = \mathbf{0}.$$

Assume that $c_1 \neq 0$. It is clear that $(\mathbf{A} - r_2\mathbf{I})(c_1\boldsymbol{\xi}^{(1)} + c_2\boldsymbol{\xi}^{(2)} + c_3\boldsymbol{\xi}^{(3)}) = \mathbf{0}$. On the other hand,

$$(\mathbf{A} - r_2\mathbf{I})(c_1\boldsymbol{\xi}^{(1)} + c_2\boldsymbol{\xi}^{(2)} + c_3\boldsymbol{\xi}^{(3)}) = c_1(r_1 - r_2)\boldsymbol{\xi}^{(1)} + c_3(r_3 - r_2)\boldsymbol{\xi}^{(3)}.$$

It follows that $c_1(r_1 - r_2)\boldsymbol{\xi}^{(1)} + c_3(r_3 - r_2)\boldsymbol{\xi}^{(3)} = \mathbf{0}$. Based on the result of part (a), which is actually not dependent on the value of n, the vectors $\boldsymbol{\xi}^{(1)}$ and $\boldsymbol{\xi}^{(3)}$ are linearly independent. Hence we must have $c_1(r_1 - r_2) = c_3(r_3 - r_2) = 0$, which leads to a contradiction.

29.(a) Let $x_1 = y$ and $x_2 = y'$. It follows that $x_1' = x_2$ and

$$x_2' = y'' = -\frac{1}{a}(cy + by').$$

In terms of the new variables, we obtain the system of two first order ODEs

$$x_1' = x_2$$
$$x_2' = -\frac{1}{a}(cx_1 + bx_2).$$

(b) The coefficient matrix is given by

$$\mathbf{A} = \begin{pmatrix} 0 & 1 \\ -\frac{c}{a} & -\frac{b}{a} \end{pmatrix}.$$

Setting $\mathbf{x} = \boldsymbol{\xi} e^{rt}$ results in the algebraic equations

$$\begin{pmatrix} -r & 1 \\ -\frac{c}{a} & -\frac{b}{a} - r \end{pmatrix} \begin{pmatrix} \xi_1 \\ \xi_2 \end{pmatrix} = \begin{pmatrix} 0 \\ 0 \end{pmatrix}.$$

For a nonzero solution, we must have

$$\det(\mathbf{A} - r\mathbf{I}) = r^2 + \frac{b}{a}r + \frac{c}{a} = 0.$$

Multiplying both sides of the equation by a, we obtain $ar^2 + br + c = 0$.

30.(a) Solution of the ODE requires analysis of the algebraic equations

$$\begin{pmatrix} -1/10 - r & 3/40 \\ 1/10 & -1/5 - r \end{pmatrix} \begin{pmatrix} \xi_1 \\ \xi_2 \end{pmatrix} = \begin{pmatrix} 0 \\ 0 \end{pmatrix}.$$

For a nonzero solution, we must have $\det(\mathbf{A} - r\mathbf{I}) = 0$. The characteristic equation is $80r^2 + 24r + 1 = 0$, with roots $r_1 = -1/4$ and $r_2 = -1/20$. With $r = -1/4$, the system of equations reduces to $2\xi_1 + \xi_2 = 0$. The corresponding eigenvector is $\boldsymbol{\xi}^{(1)} = (1, -2)^T$. Substitution of $r = -1/20$ results in the equation $2\xi_1 - 3\xi_2 = 0$. A corresponding eigenvector is $\boldsymbol{\xi}^{(2)} = (3, 2)^T$. Since the eigenvalues are distinct, the general solution is

$$\mathbf{x} = c_1 \begin{pmatrix} 1 \\ -2 \end{pmatrix} e^{-t/4} + c_2 \begin{pmatrix} 3 \\ 2 \end{pmatrix} e^{-t/20}.$$

Invoking the initial conditions, we obtain the system of equations

$$c_1 + 3c_2 = -17$$
$$-2c_1 + 2c_2 = -21.$$

Hence $c_1 = 29/8$ and $c_2 = -55/8$, and the solution of the IVP is

$$\mathbf{x} = \frac{29}{8}\begin{pmatrix}1\\-2\end{pmatrix}e^{-t/4} - \frac{55}{8}\begin{pmatrix}3\\2\end{pmatrix}e^{-t/20}.$$

(b)

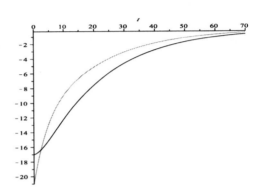

(c) Both functions are monotone increasing. It is easy to show that $-0.5 \le x_1(t) < 0$ and $-0.5 \le x_2(t) < 0$ provided that $t > T \approx 74.39$.

32.(a) The system of differential equations is

$$\frac{d}{dt}\begin{pmatrix}I\\V\end{pmatrix} = \begin{pmatrix}-1/2 & -1/2\\3/2 & -5/2\end{pmatrix}\begin{pmatrix}I\\V\end{pmatrix}.$$

Solution of the system requires analysis of the eigenvalue problem

$$\begin{pmatrix}-1/2-r & -1/2\\3/2 & -5/2-r\end{pmatrix}\begin{pmatrix}\xi_1\\\xi_2\end{pmatrix} = \begin{pmatrix}0\\0\end{pmatrix}.$$

The characteristic equation is $r^2 + 3r + 2 = 0$, with roots $r_1 = -1$ and $r_2 = -2$. With $r = -1$, the equations reduce to $\xi_1 - \xi_2 = 0$. A corresponding eigenvector is given by $\boldsymbol{\xi}^{(1)} = (1,1)^T$. Setting $r = -2$, the system reduces to the equation $3\xi_1 - \xi_2 = 0$. An eigenvector is $\boldsymbol{\xi}^{(2)} = (1,3)^T$. Hence the general solution is

$$\begin{pmatrix}I\\V\end{pmatrix} = c_1\begin{pmatrix}1\\1\end{pmatrix}e^{-t} + c_2\begin{pmatrix}1\\3\end{pmatrix}e^{-2t}.$$

(b) The eigenvalues are distinct and both negative. We find that the equilibrium point $(0,0)$ is a stable node. Hence all solutions converge to $(0,0)$.

33.(a) Solution of the ODE requires analysis of the algebraic equations

$$\begin{pmatrix}-\frac{R_1}{L}-r & -\frac{1}{L}\\ \frac{1}{C} & -\frac{1}{CR_2}-r\end{pmatrix}\begin{pmatrix}\xi_1\\\xi_2\end{pmatrix} = \begin{pmatrix}0\\0\end{pmatrix}.$$

The characteristic equation is

$$r^2 + \left(\frac{L+CR_1R_2}{LCR_2}\right)r + \frac{R_1+R_2}{LCR_2} = 0.$$

The eigenvectors are real and distinct, provided that the discriminant is positive. That is,
$$(\frac{L+CR_1R_2}{LCR_2})^2 - 4(\frac{R_1+R_2}{LCR_2}) > 0,$$
which simplifies to the condition
$$(\frac{1}{CR_2} - \frac{R_1}{L})^2 - \frac{4}{LC} > 0.$$

(b) The parameters in the ODE are all positive. Observe that the sum of the roots is
$$-\frac{L+CR_1R_2}{LCR_2} < 0.$$
Also, the product of the roots is
$$\frac{R_1+R_2}{LCR_2} > 0.$$
It follows that both roots are negative. Hence the equilibrium solution $I=0, V=0$ represents a stable node, which attracts all solutions.

(c) If the condition in part (a) is not satisfied, that is,
$$(\frac{1}{CR_2} - \frac{R_1}{L})^2 - \frac{4}{LC} \leq 0,$$
then the real part of the eigenvalues is
$$\text{Re}(r_{1,2}) = -\frac{L+CR_1R_2}{2LCR_2}.$$
As long as the parameters are all positive, then the solutions will still converge to the equilibrium point $(0,0)$.

7.6

2.(a) Setting $\mathbf{x} = \boldsymbol{\xi} e^{rt}$ results in the algebraic equations
$$\begin{pmatrix} -1-r & -4 \\ 1 & -1-r \end{pmatrix} \begin{pmatrix} \xi_1 \\ \xi_2 \end{pmatrix} = \begin{pmatrix} 0 \\ 0 \end{pmatrix}.$$
For a nonzero solution, we require that $\det(\mathbf{A} - r\mathbf{I}) = r^2 + 2r + 5 = 0$. The roots of the characteristic equation are $r = -1 \pm 2i$. Substituting $r = -1 - 2i$, the two equations reduce to $\xi_1 + 2i\,\xi_2 = 0$. The two eigenvectors are $\boldsymbol{\xi}^{(1)} = (-2i, 1)^T$ and $\boldsymbol{\xi}^{(2)} = (2i, 1)^T$. Hence one of the complex-valued solutions is given by
$$\mathbf{x}^{(1)} = \begin{pmatrix} -2i \\ 1 \end{pmatrix} e^{-(1+2i)t} = \begin{pmatrix} -2i \\ 1 \end{pmatrix} e^{-t}(\cos 2t - i\sin 2t) =$$
$$= e^{-t} \begin{pmatrix} -2\sin 2t \\ \cos 2t \end{pmatrix} + i e^{-t} \begin{pmatrix} -2\cos 2t \\ -\sin 2t \end{pmatrix}.$$

Based on the real and imaginary parts of this solution, the general solution is
$$\mathbf{x} = c_1 e^{-t}\begin{pmatrix} -2\sin 2t \\ \cos 2t \end{pmatrix} + c_2 e^{-t}\begin{pmatrix} 2\cos 2t \\ \sin 2t \end{pmatrix}.$$

(b)

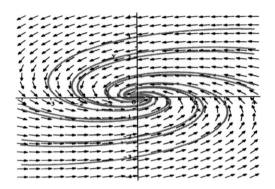

3.(a) Solution of the ODEs is based on the analysis of the algebraic equations
$$\begin{pmatrix} 2-r & -5 \\ 1 & -2-r \end{pmatrix}\begin{pmatrix} \xi_1 \\ \xi_2 \end{pmatrix} = \begin{pmatrix} 0 \\ 0 \end{pmatrix}.$$
For a nonzero solution, we require that $\det(\mathbf{A}-r\mathbf{I}) = r^2 + 1 = 0$. The roots of the characteristic equation are $r = \pm i$. Setting $r = i$, the equations are equivalent to $\xi_1 - (2+i)\xi_2 = 0$. The eigenvectors are $\boldsymbol{\xi}^{(1)} = (2+i,1)^T$ and $\boldsymbol{\xi}^{(2)} = (2-i,1)^T$. Hence one of the complex-valued solutions is given by
$$\mathbf{x}^{(1)} = \begin{pmatrix} 2+i \\ 1 \end{pmatrix}e^{it} = \begin{pmatrix} 2+i \\ 1 \end{pmatrix}(\cos t + i\sin t) =$$
$$= \begin{pmatrix} 2\cos t - \sin t \\ \cos t \end{pmatrix} + i\begin{pmatrix} \cos t + 2\sin t \\ \sin t \end{pmatrix}.$$
Therefore the general solution is
$$\mathbf{x} = c_1\begin{pmatrix} 2\cos t - \sin t \\ \cos t \end{pmatrix} + c_2\begin{pmatrix} \cos t + 2\sin t \\ \sin t \end{pmatrix}.$$
The solution may also be written as
$$\mathbf{x} = c_1\begin{pmatrix} 5\cos t \\ 2\cos t + \sin t \end{pmatrix} + c_2\begin{pmatrix} 5\sin t \\ -\cos t + 2\sin t \end{pmatrix}.$$

(b)

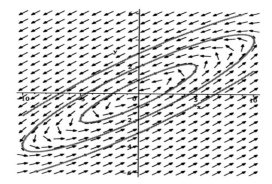

4.(a) Setting $\mathbf{x} = \boldsymbol{\xi}\, e^{rt}$ results in the algebraic equations

$$\begin{pmatrix} 2-r & -5/2 \\ 9/5 & -1-r \end{pmatrix} \begin{pmatrix} \xi_1 \\ \xi_2 \end{pmatrix} = \begin{pmatrix} 0 \\ 0 \end{pmatrix}.$$

For a nonzero solution, we require that $\det(\mathbf{A} - r\mathbf{I}) = r^2 - r + \frac{5}{2} = 0$. The roots of the characteristic equation are $r = (1 \pm 3i)/2$. With $r = (1 + 3i)/2$, the equations reduce to the single equation $(3 - 3i)\xi_1 - 5\xi_2 = 0$. The corresponding eigenvector is given by $\boldsymbol{\xi}^{(1)} = (5, 3 - 3i)^T$. Hence one of the complex-valued solutions is

$$\mathbf{x}^{(1)} = \begin{pmatrix} 5 \\ 3-3i \end{pmatrix} e^{(1+3i)t/2} = \begin{pmatrix} 2+i \\ 1 \end{pmatrix} e^{t/2}(\cos \frac{3}{2}t + i \sin \frac{3}{2}t) =$$

$$= e^{t/2} \begin{pmatrix} 2\cos \frac{3}{2}t - \sin \frac{3}{2}t \\ \cos \frac{3}{2}t \end{pmatrix} + i e^{t/2} \begin{pmatrix} \cos \frac{3}{2}t + 2\sin \frac{3}{2}t \\ \sin \frac{3}{2}t \end{pmatrix}.$$

The general solution is

$$\mathbf{x} = c_1 e^{t/2} \begin{pmatrix} 2\cos \frac{3}{2}t - \sin \frac{3}{2}t \\ \cos \frac{3}{2}t \end{pmatrix} + c_2 e^{t/2} \begin{pmatrix} \cos \frac{3}{2}t + 2\sin \frac{3}{2}t \\ \sin \frac{3}{2}t \end{pmatrix}.$$

The solution may also be written as

$$\mathbf{x} = c_1 e^{t/2} \begin{pmatrix} 5\cos \frac{3}{2}t \\ 3\cos \frac{3}{2}t + 3\sin \frac{3}{2}t \end{pmatrix} + c_2 e^{t/2} \begin{pmatrix} 5\sin \frac{3}{2}t \\ -3\cos \frac{3}{2}t + 3\sin \frac{3}{2}t \end{pmatrix}.$$

(b)

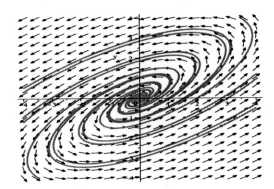

5.(a) Setting $\mathbf{x} = \boldsymbol{\xi} t^r$ results in the algebraic equations

$$\begin{pmatrix} 1-r & -1 \\ 5 & -3-r \end{pmatrix} \begin{pmatrix} \xi_1 \\ \xi_2 \end{pmatrix} = \begin{pmatrix} 0 \\ 0 \end{pmatrix}.$$

The characteristic equation is $r^2 + 2r + 2 = 0$, with roots $r = -1 \pm i$. Substituting $r = -1 - i$ reduces the system of equations to $(2+i)\xi_1 - \xi_2 = 0$. The eigenvectors are $\boldsymbol{\xi}^{(1)} = (1, 2+i)^T$ and $\boldsymbol{\xi}^{(2)} = (1, 2-i)^T$. Hence one of the complex-valued solutions is given by

$$\mathbf{x}^{(1)} = \begin{pmatrix} 1 \\ 2+i \end{pmatrix} e^{-(1+i)t} = \begin{pmatrix} 1 \\ 2+i \end{pmatrix} e^{-t}(\cos t - i \sin t) =$$

$$= e^{-t} \begin{pmatrix} \cos t \\ 2\cos t + \sin t \end{pmatrix} + ie^{-t} \begin{pmatrix} -\sin t \\ \cos t - 2\sin t \end{pmatrix}.$$

The general solution is

$$\mathbf{x} = c_1 e^{-t} \begin{pmatrix} \cos t \\ 2\cos t + \sin t \end{pmatrix} + c_2 e^{-t} \begin{pmatrix} \sin t \\ -\cos t + 2\sin t \end{pmatrix}.$$

(b)

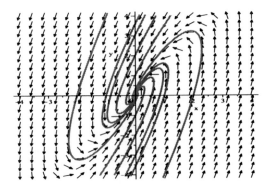

6.(a) Solution of the ODEs is based on the analysis of the algebraic equations
$$\begin{pmatrix} 1-r & 2 \\ -5 & -1-r \end{pmatrix} \begin{pmatrix} \xi_1 \\ \xi_2 \end{pmatrix} = \begin{pmatrix} 0 \\ 0 \end{pmatrix}.$$
For a nonzero solution, we require that $\det(\mathbf{A} - r\mathbf{I}) = r^2 + 9 = 0$. The roots of the characteristic equation are $r = \pm 3i$. Setting $r = 3i$, the two equations reduce to $(1 - 3i)\xi_1 + 2\xi_2 = 0$. The corresponding eigenvector is $\boldsymbol{\xi}^{(1)} = (-2, 1 - 3i)^T$. Hence one of the complex-valued solutions is given by
$$\mathbf{x}^{(1)} = \begin{pmatrix} -2 \\ 1 - 3i \end{pmatrix} e^{3it} = \begin{pmatrix} -2 \\ 1 - 3i \end{pmatrix}(\cos 3t + i \sin 3t) =$$
$$= \begin{pmatrix} -2\cos 3t \\ \cos 3t + 3\sin 3t \end{pmatrix} + i\begin{pmatrix} -2\sin 3t \\ -3\cos 3t + \sin 3t \end{pmatrix}.$$
The general solution is
$$\mathbf{x} = c_1 \begin{pmatrix} -2\cos 3t \\ \cos 3t + 3\sin 3t \end{pmatrix} + c_2 \begin{pmatrix} 2\sin 3t \\ 3\cos 3t - \sin 3t \end{pmatrix}.$$

(b)

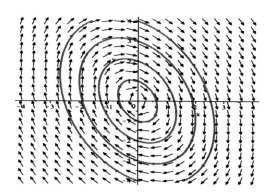

8. The eigensystem is obtained from analysis of the equation
$$\begin{pmatrix} -3-r & 0 & 2 \\ 1 & -1-r & 0 \\ -2 & -1 & -r \end{pmatrix} \begin{pmatrix} \xi_1 \\ \xi_2 \\ \xi_3 \end{pmatrix} = \begin{pmatrix} 0 \\ 0 \\ 0 \end{pmatrix}.$$

The characteristic equation of the coefficient matrix is $r^3 + 4r^2 + 7r + 6 = 0$, with roots $r_1 = -2$, $r_2 = -1 - \sqrt{2}\,i$ and $r_3 = -1 + \sqrt{2}\,i$. Setting $r = -2$, the equations reduce to
$$-\xi_1 + 2\xi_3 = 0$$
$$\xi_1 + \xi_2 = 0.$$
The corresponding eigenvector is $\boldsymbol{\xi}^{(1)} = (2, -2, 1)^T$. With $r = -1 - \sqrt{2}\,i$, the system of equations is equivalent to
$$(2 - i\sqrt{2})\xi_1 - 2\xi_3 = 0$$
$$\xi_1 + i\sqrt{2}\,\xi_2 = 0.$$

An eigenvector is given by $\boldsymbol{\xi}^{(2)} = (-i\sqrt{2}, 1, -1 - i\sqrt{2})^T$. Hence one of the complex-valued solutions is given by

$$\mathbf{x}^{(2)} = \begin{pmatrix} -i\sqrt{2} \\ 1 \\ -1 - i\sqrt{2} \end{pmatrix} e^{-(1+i\sqrt{2})t} = \begin{pmatrix} -i\sqrt{2} \\ 1 \\ -1 - i\sqrt{2} \end{pmatrix} e^{-t}(\cos\sqrt{2}t - i\sin\sqrt{2}t) =$$

$$= e^{-t} \begin{pmatrix} -\sqrt{2}\sin\sqrt{2}t \\ \cos\sqrt{2}t \\ -\cos\sqrt{2}t - \sqrt{2}\sin\sqrt{2}t \end{pmatrix} + ie^{-t} \begin{pmatrix} -\sqrt{2}\cos\sqrt{2}t \\ -\sin\sqrt{2}t \\ -\sqrt{2}\cos\sqrt{2}t + \sin\sqrt{2}t \end{pmatrix}.$$

The other complex-valued solution is $\mathbf{x}^{(3)} = \overline{\boldsymbol{\xi}^{(2)}} e^{r_3 t}$. The general solution is

$$\mathbf{x} = c_1 \begin{pmatrix} 2 \\ -2 \\ 1 \end{pmatrix} e^{-2t} +$$

$$+ c_2 e^{-t} \begin{pmatrix} \sqrt{2}\sin\sqrt{2}t \\ -\cos\sqrt{2}t \\ \cos\sqrt{2}t + \sqrt{2}\sin\sqrt{2}t \end{pmatrix} + c_3 e^{-t} \begin{pmatrix} \sqrt{2}\cos\sqrt{2}t \\ \sin\sqrt{2}t \\ \sqrt{2}\cos\sqrt{2}t - \sin\sqrt{2}t \end{pmatrix}.$$

It is easy to see that all solutions converge to the equilibrium point $(0,0,0)$.

10. Solution of the system of ODEs requires that

$$\begin{pmatrix} -3 - r & 2 \\ -1 & -1 - r \end{pmatrix} \begin{pmatrix} \xi_1 \\ \xi_2 \end{pmatrix} = \begin{pmatrix} 0 \\ 0 \end{pmatrix}.$$

The characteristic equation is $r^2 + 4r + 5 = 0$, with roots $r = -2 \pm i$. Substituting $r = -2 + i$, the equations are equivalent to $\xi_1 - (1 - i)\xi_2 = 0$. The corresponding eigenvector is $\boldsymbol{\xi}^{(1)} = (1 - i, 1)^T$. One of the complex-valued solutions is given by

$$\mathbf{x}^{(1)} = \begin{pmatrix} 1 - i \\ 1 \end{pmatrix} e^{(-2+i)t} = \begin{pmatrix} 1 - i \\ 1 \end{pmatrix} e^{-2t}(\cos t + i\sin t) =$$

$$= e^{-2t} \begin{pmatrix} \cos t + \sin t \\ \cos t \end{pmatrix} + ie^{-2t} \begin{pmatrix} -\cos t + \sin t \\ \sin t \end{pmatrix}.$$

Hence the general solution is

$$\mathbf{x} = c_1 e^{-2t} \begin{pmatrix} \cos t + \sin t \\ \cos t \end{pmatrix} + c_2 e^{-2t} \begin{pmatrix} -\cos t + \sin t \\ \sin t \end{pmatrix}.$$

Invoking the initial conditions, we obtain the system of equations

$$c_1 - c_2 = 1$$
$$c_1 = -2.$$

Solving for the coefficients, the solution of the initial value problem is

$$\mathbf{x} = -2e^{-2t} \begin{pmatrix} \cos t + \sin t \\ \cos t \end{pmatrix} - 3e^{-2t} \begin{pmatrix} -\cos t + \sin t \\ \sin t \end{pmatrix}$$

$$= e^{-2t} \begin{pmatrix} \cos t - 5\sin t \\ -2\cos t - 3\sin t \end{pmatrix}.$$

The solution converges to $(0,0)$ as $t \to \infty$.

12. Solution of the ODEs is based on the analysis of the algebraic equations
$$\begin{pmatrix} -\frac{4}{5} - r & 2 \\ -1 & \frac{6}{5} - r \end{pmatrix} \begin{pmatrix} \xi_1 \\ \xi_2 \end{pmatrix} = \begin{pmatrix} 0 \\ 0 \end{pmatrix}.$$

The characteristic equation is $25 r^2 - 10 r + 26 = 0$, with roots $r = 1/5 \pm i$. Setting $r = 1/5 + i$, the two equations reduce to $\xi_1 - (1 - i)\xi_2 = 0$. The corresponding eigenvector is $\boldsymbol{\xi}^{(1)} = (1 - i, 1)^T$. One of the complex-valued solutions is given by

$$\mathbf{x}^{(1)} = \begin{pmatrix} 1 - i \\ 1 \end{pmatrix} e^{(\frac{1}{5}+i)t} = \begin{pmatrix} 1 - i \\ 1 \end{pmatrix} e^{t/5}(\cos t + i \sin t) =$$

$$= e^{t/5} \begin{pmatrix} \cos t + \sin t \\ \cos t \end{pmatrix} + i e^{t/5} \begin{pmatrix} -\cos t + \sin t \\ \sin t \end{pmatrix}.$$

Hence the general solution is

$$\mathbf{x} = c_1 e^{t/5} \begin{pmatrix} \cos t + \sin t \\ \cos t \end{pmatrix} + c_2 e^{t/5} \begin{pmatrix} -\cos t + \sin t \\ \sin t \end{pmatrix}.$$

(b) Let $\mathbf{x}(0) = (x_1^0, x_2^0)^T$. The solution of the initial value problem is

$$\mathbf{x} = x_2^0 \, e^{t/5} \begin{pmatrix} \cos t + \sin t \\ \cos t \end{pmatrix} + (x_2^0 - x_1^0) e^{t/5} \begin{pmatrix} -\cos t + \sin t \\ \sin t \end{pmatrix}$$

$$= e^{t/5} \begin{pmatrix} x_1^0 \cos t + (2 x_2^0 - x_1^0) \sin t \\ x_2^0 \cos t + (x_2^0 - x_1^0) \sin t \end{pmatrix}.$$

With $\mathbf{x}(0) = (1, 2)^T$, the solution is

$$\mathbf{x} = e^{t/5} \begin{pmatrix} \cos t + 3 \sin t \\ 2 \cos t + \sin t \end{pmatrix}.$$

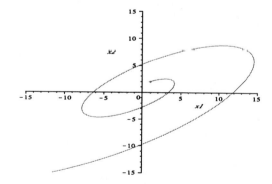

(c)

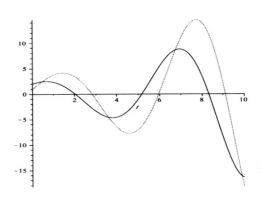

(d)

13.(a) The characteristic equation is $r^2 - 2\alpha r + 1 + \alpha^2 = 0$, with roots $r = \alpha \pm i$.

(b) When $\alpha < 0$ and $\alpha > 0$, the equilibrium point $(0,0)$ is a stable spiral and an unstable spiral, respectively. The equilibrium point is a center when $\alpha = 0$.

(c)

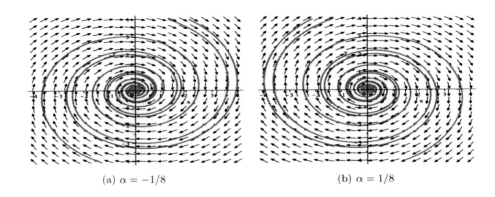

(a) $\alpha = -1/8$ (b) $\alpha = 1/8$

14.(a) The roots of the characteristic equation, $r^2 - \alpha r + 5 = 0$, are
$$r_{1,2} = \frac{\alpha}{2} \pm \frac{1}{2}\sqrt{\alpha^2 - 20}\,.$$

(b) Note that the roots are complex when $-\sqrt{20} < \alpha < \sqrt{20}$. For the case when $\alpha \in (-\sqrt{20},0)$, the equilibrium point $(0,0)$ is a stable spiral. On the other hand, when $\alpha \in (0,\sqrt{20})$, the equilibrium point is an unstable spiral. For the case $\alpha = 0$, the roots are purely imaginary, so the equilibrium point is a center. When $\alpha^2 > 20$, the roots are real and distinct. The equilibrium point becomes a node, with its stability dependent on the sign of α. Finally, the case $\alpha^2 = 20$ marks the transition from spirals to nodes.

(c)

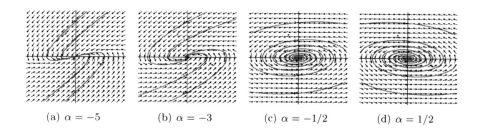

(a) $\alpha = -5$ (b) $\alpha = -3$ (c) $\alpha = -1/2$ (d) $\alpha = 1/2$

17. The characteristic equation of the coefficient matrix is $r^2 + 2r + 1 + \alpha = 0$, with roots given formally as $r_{1,2} = -1 \pm \sqrt{-\alpha}$. The roots are real provided that $\alpha \leq 0$. First note that the sum of the roots is -2 and the product of the roots is $1 + \alpha$. For negative values of α, the roots are distinct, with one always negative. When $\alpha < -1$, the roots have opposite signs. Hence the equilibrium point is a saddle. For the case $-1 < \alpha < 0$, the roots are both negative, and the equilibrium point is a stable node. $\alpha = -1$ represents a transition from saddle to node. When $\alpha = 0$, both roots are equal. For the case $\alpha > 0$, the roots are complex conjugates, with negative real part. Hence the equilibrium point is a stable spiral.

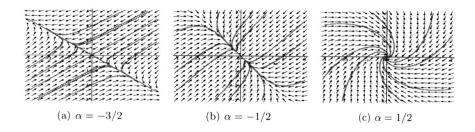

(a) $\alpha = -3/2$ (b) $\alpha = -1/2$ (c) $\alpha = 1/2$

19. The characteristic equation for the system is given by
$$r^2 + (4 - \alpha)r + 10 - 4\alpha = 0.$$
The roots are
$$r_{1,2} = -2 + \frac{\alpha}{2} \pm \sqrt{\alpha^2 + 8\alpha - 24}.$$

First note that the roots are complex when $-4 - 2\sqrt{10} < \alpha < -4 + 2\sqrt{10}$. We also find that when $-4 - 2\sqrt{10} < \alpha < -4 + 2\sqrt{10}$, the equilibrium point is a stable spiral. For $\alpha > -4 + 2\sqrt{10}$, the roots are real. When $\alpha > 2.5$, the roots have opposite signs, with the equilibrium point being a saddle. For the case $-4 + 2\sqrt{10} < \alpha < 2.5$, the roots are both negative, and the equilibrium point is a stable node. Finally, when $\alpha < -4 - 2\sqrt{10}$, both roots are negative, with the equilibrium point being a stable node.

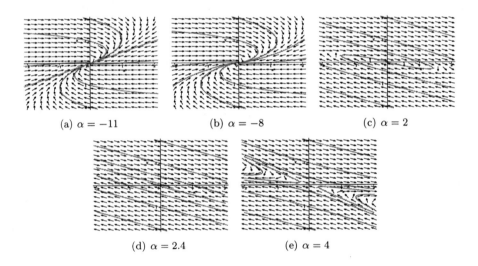

(a) $\alpha = -11$ (b) $\alpha = -8$ (c) $\alpha = 2$

(d) $\alpha = 2.4$ (e) $\alpha = 4$

20. The characteristic equation is $r^2 + 2r - (24 + 8\alpha) = 0$, with roots
$$r_{1,2} = -1 \pm \sqrt{25 + 8\alpha}.$$

The roots are complex when $\alpha < -25/8$. Since the real part is negative, the origin is a stable spiral. Otherwise the roots are real. When $-25/8 < \alpha < -3$, both roots are negative, and hence the equilibrium point is a stable node. For $\alpha > -3$, the roots are of opposite sign and the origin is a saddle.

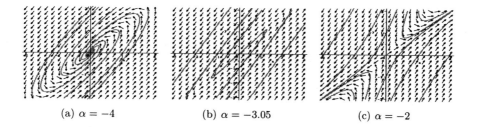

(a) $\alpha = -4$ (b) $\alpha = -3.05$ (c) $\alpha = -2$

22. Based on the method in Problem 19 of Section 7.5, setting $\mathbf{x} = \boldsymbol{\xi} t^r$ results in the algebraic equations

$$\begin{pmatrix} 2-r & -5 \\ 1 & -2-r \end{pmatrix} \begin{pmatrix} \xi_1 \\ \xi_2 \end{pmatrix} = \begin{pmatrix} 0 \\ 0 \end{pmatrix}.$$

The characteristic equation for the system is $r^2 + 1 = 0$, with roots $r_{1,2} = \pm i$. With $r = i$, the equations reduce to the single equation $\xi_1 - (2+i)\xi_2 = 0$. A corresponding eigenvector is $\boldsymbol{\xi}^{(1)} = (2+i, 1)^T$. One complex-valued solution is

$$\mathbf{x}^{(1)} = \begin{pmatrix} 2+i \\ 1 \end{pmatrix} t^i.$$

We can write $t^i = e^{i \ln t}$. Hence

$$\mathbf{x}^{(1)} = \begin{pmatrix} 2+i \\ 1 \end{pmatrix} e^{i \ln t} = \begin{pmatrix} 2+i \\ 1 \end{pmatrix} [\cos(\ln t) + i \sin(\ln t)] =$$

$$= \begin{pmatrix} 2\cos(\ln t) - \sin(\ln t) \\ \cos(\ln t) \end{pmatrix} + i \begin{pmatrix} \cos(\ln t) + 2\sin(\ln t) \\ \sin(\ln t) \end{pmatrix}.$$

Therefore the general solution is

$$\mathbf{x} = c_1 \begin{pmatrix} 2\cos(\ln t) - \sin(\ln t) \\ \cos(\ln t) \end{pmatrix} + c_2 \begin{pmatrix} \cos(\ln t) + 2\sin(\ln t) \\ \sin(\ln t) \end{pmatrix}.$$

Other combinations are also possible.

24.(a) The characteristic equation of the system is

$$r^3 + \frac{2}{5}r^2 + \frac{81}{80}r - \frac{17}{160} = 0,$$

with eigenvalues $r_1 = 1/10$, and $r_{2,3} = -1/4 \pm i$. For $r = 1/10$, simple calculations reveal that a corresponding eigenvector is $\boldsymbol{\xi}^{(1)} = (0, 0, 1)^T$. Setting $r = -1/4 - i$, we obtain the system of equations

$$\xi_1 - i\,\xi_2 = 0$$
$$\xi_3 = 0.$$

A corresponding eigenvector is $\boldsymbol{\xi}^{(2)} = (i, 1, 0)^T$. Hence one solution is

$$\mathbf{x}^{(1)} = \begin{pmatrix} 0 \\ 0 \\ 1 \end{pmatrix} e^{t/10}.$$

Another solution, which is complex-valued, is given by

$$\mathbf{x}^{(2)} = \begin{pmatrix} i \\ 1 \\ 0 \end{pmatrix} e^{-(\frac{1}{4}+i)t} = \begin{pmatrix} i \\ 1 \\ 0 \end{pmatrix} e^{-t/4}(\cos t - i \sin t) =$$

$$= e^{-t/4} \begin{pmatrix} \sin t \\ \cos t \\ 0 \end{pmatrix} + i e^{-t/4} \begin{pmatrix} \cos t \\ -\sin t \\ 0 \end{pmatrix}.$$

Using the real and imaginary parts of $\mathbf{x}^{(2)}$, the general solution is constructed as

$$\mathbf{x} = c_1 \begin{pmatrix} 0 \\ 0 \\ 1 \end{pmatrix} e^{t/10} + c_2 e^{-t/4} \begin{pmatrix} \sin t \\ \cos t \\ 0 \end{pmatrix} + c_3 e^{-t/4} \begin{pmatrix} \cos t \\ -\sin t \\ 0 \end{pmatrix}.$$

(b) Let $\mathbf{x}(0) = (x_1^0, x_2^0, x_3^0)$. The solution can be written as

$$\mathbf{x} = \begin{pmatrix} 0 \\ 0 \\ x_3^0 e^{t/10} \end{pmatrix} + e^{-t/4} \begin{pmatrix} x_2^0 \sin t + x_1^0 \cos t \\ x_2^0 \cos t - x_1^0 \sin t \\ 0 \end{pmatrix}.$$

With $\mathbf{x}(0) = (1, 1, 1)$, the solution of the initial value problem is

$$\mathbf{x} = \begin{pmatrix} 0 \\ 0 \\ e^{t/10} \end{pmatrix} + e^{-t/4} \begin{pmatrix} \sin t + \cos t \\ \cos t - \sin t \\ 0 \end{pmatrix}.$$

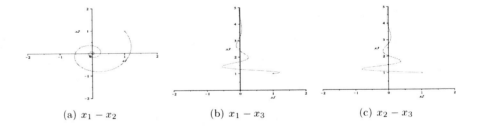

(a) $x_1 - x_2$ (b) $x_1 - x_3$ (c) $x_2 - x_3$

(c)

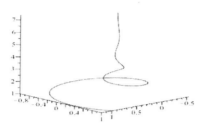

25.(a) Based on Problems 19-21 of Section 7.1, the system of differential equations is
$$\frac{d}{dt}\begin{pmatrix} I \\ V \end{pmatrix} = \begin{pmatrix} -\frac{R_1}{L} & -\frac{1}{L} \\ \frac{1}{C} & -\frac{1}{CR_2} \end{pmatrix}\begin{pmatrix} I \\ V \end{pmatrix} = \begin{pmatrix} -\frac{1}{2} & -\frac{1}{8} \\ 2 & -\frac{1}{2} \end{pmatrix}\begin{pmatrix} I \\ V \end{pmatrix},$$
since $R_1 = R_2 = 4$ ohms, $C = 1/2$ farads and $L = 8$ henrys.

(b) The eigenvalue problem is
$$\begin{pmatrix} -\frac{1}{2}-r & -\frac{1}{8} \\ 2 & -\frac{1}{2}-r \end{pmatrix}\begin{pmatrix} \xi_1 \\ \xi_2 \end{pmatrix} = \begin{pmatrix} 0 \\ 0 \end{pmatrix}.$$
The characteristic equation of the system is $r^2 + r + \frac{1}{2} = 0$, with eigenvalues
$$r_{1,2} = -\frac{1}{2} \pm \frac{1}{2}i.$$
Setting $r = -1/2 + i/2$, the algebraic equations reduce to $4i\xi_1 + \xi_2 = 0$. It follows that $\boldsymbol{\xi}^{(1)} = (1, -4i)^T$. Hence one complex-valued solution is
$$\begin{pmatrix} I \\ V \end{pmatrix}^{(1)} = \begin{pmatrix} 1 \\ -4i \end{pmatrix}e^{(-1+i)t/2} = \begin{pmatrix} 1 \\ -4i \end{pmatrix}e^{-t/2}[\cos(t/2) + i\sin(t/2)] =$$
$$= e^{-t/2}\begin{pmatrix} \cos(t/2) \\ 4\sin(t/2) \end{pmatrix} + ie^{-t/2}\begin{pmatrix} \sin(t/2) \\ -4\cos(t/2) \end{pmatrix}.$$
Therefore the general solution is
$$\begin{pmatrix} I \\ V \end{pmatrix} = c_1 e^{-t/2}\begin{pmatrix} \cos(t/2) \\ 4\sin(t/2) \end{pmatrix} + c_2 e^{-t/2}\begin{pmatrix} \sin(t/2) \\ -4\cos(t/2) \end{pmatrix}.$$

(c) Imposing the initial conditions, we arrive at the equations $c_1 = 2$ and $c_2 = -3/4$, and
$$\begin{pmatrix} I \\ V \end{pmatrix} = e^{-t/2}\begin{pmatrix} 2\cos(t/2) - \frac{3}{4}\sin(t/2) \\ 8\sin(t/2) + 3\cos(t/2) \end{pmatrix}.$$

(d) Since the eigenvalues have negative real parts, all solutions converge to the origin.

26.(a) The characteristic equation of the system is
$$r^2 + \frac{1}{RC}r + \frac{1}{CL} = 0,$$
with eigenvalues
$$r_{1,2} = -\frac{1}{2RC} \pm \frac{1}{2RC}\sqrt{1 - \frac{4R^2C}{L}}.$$
The eigenvalues are real and different provided that
$$1 - \frac{4R^2C}{L} > 0.$$
The eigenvalues are complex conjugates as long as
$$1 - \frac{4R^2C}{L} < 0.$$

(b) With the specified values, the eigenvalues are $r_{1,2} = -1 \pm i$. The eigenvector corresponding to $r = -1 + i$ is $\boldsymbol{\xi}^{(1)} = (1, -4i)^T$. Hence one complex-valued solution is
$$\begin{pmatrix} I \\ V \end{pmatrix}^{(1)} = \begin{pmatrix} 1 \\ -1+i \end{pmatrix} e^{(-1+i)t} = \begin{pmatrix} 1 \\ -1+i \end{pmatrix} e^{-t}(\cos t + i \sin t) =$$
$$= e^{-t}\begin{pmatrix} \cos t \\ -\cos t - \sin t \end{pmatrix} + ie^{-t}\begin{pmatrix} \sin t \\ \cos t - \sin t \end{pmatrix}.$$
Therefore the general solution is
$$\begin{pmatrix} I \\ V \end{pmatrix} = c_1 e^{-t}\begin{pmatrix} \cos t \\ -\cos t - \sin t \end{pmatrix} + c_2 e^{-t}\begin{pmatrix} \sin t \\ \cos t - \sin t \end{pmatrix}.$$

(c) Imposing the initial conditions, we arrive at the equations
$$c_1 = 2$$
$$-c_1 + c_2 = 1,$$
with $c_1 = 2$ and $c_2 = 3$. Therefore the solution of the IVP is
$$\begin{pmatrix} I \\ V \end{pmatrix} = e^{-t}\begin{pmatrix} 2\cos t + 3\sin t \\ \cos t - 5\sin t \end{pmatrix}.$$

(d) Since $\operatorname{Re}(r_{1,2}) = -1$, all solutions converge to the origin.

27.(a) Suppose that $c_1\mathbf{a} + c_2\mathbf{b} = \mathbf{0}$. Since $\mathbf{a}$ and $\mathbf{b}$ are the real and imaginary parts of the vector $\boldsymbol{\xi}^{(1)}$, respectively, $\mathbf{a} = (\boldsymbol{\xi}^{(1)} + \overline{\boldsymbol{\xi}^{(1)}})/2$ and $\mathbf{b} = (\boldsymbol{\xi}^{(1)} - \overline{\boldsymbol{\xi}^{(1)}})/2i$. Hence
$$c_1(\boldsymbol{\xi}^{(1)} + \overline{\boldsymbol{\xi}^{(1)}}) - ic_2(\boldsymbol{\xi}^{(1)} - \overline{\boldsymbol{\xi}^{(1)}}) = \mathbf{0},$$
which leads to
$$(c_1 - ic_2)\boldsymbol{\xi}^{(1)} + (c_1 + ic_2)\overline{\boldsymbol{\xi}^{(1)}} = \mathbf{0}.$$

(b) Now since $\boldsymbol{\xi}^{(1)}$ and $\overline{\boldsymbol{\xi}^{(1)}}$ are linearly independent, we must have

$$c_1 - ic_2 = 0$$
$$c_1 + ic_2 = 0.$$

It follows that $c_1 = c_2 = 0$.

(c) Recall that
$$\mathbf{u}(t) = e^{\lambda t}(\mathbf{a}\cos\mu t - \mathbf{b}\sin\mu t)$$
$$\mathbf{v}(t) = e^{\lambda t}(\mathbf{a}\cos\mu t + \mathbf{b}\sin\mu t).$$

Consider the equation $c_1\mathbf{u}(t_0) + c_2\mathbf{v}(t_0) = \mathbf{0}$, for some t_0. We can then write
$$c_1 e^{\lambda t_0}(\mathbf{a}\cos\mu t_0 - \mathbf{b}\sin\mu t_0) + c_2 e^{\lambda t_0}(\mathbf{a}\cos\mu t_0 + \mathbf{b}\sin\mu t_0) = \mathbf{0}. \quad (*)$$

Rearranging the terms, and dividing by the exponential,
$$(c_1 + c_2)\cos\mu t_0 \,\mathbf{a} + (c_2 - c_1)\sin\mu t_0 \,\mathbf{b} = \mathbf{0}.$$

From part (b), since $\mathbf{a}$ and $\mathbf{b}$ are linearly independent, it follows that
$$(c_1 + c_2)\cos\mu t_0 = (c_2 - c_1)\sin\mu t_0 = 0.$$

Without loss of generality, assume that the trigonometric factors are nonzero. Otherwise proceed again from Equation $(*)$, above. We then conclude that
$$c_1 + c_2 = 0 \text{ and } c_2 - c_1 = 0,$$
which leads to $c_1 = c_2 = 0$. Thus $\mathbf{u}(t_0)$ and $\mathbf{v}(t_0)$ are linearly independent for some t_0, and hence the functions are linearly independent at every point.

28.(a) Let $x_1 = u$ and $x_2 = u'$. It follows that $x_1' = x_2$ and
$$x_2' = u'' = -\frac{k}{m}u.$$

In terms of the new variables, we obtain the system of two first order ODEs
$$x_1' = x_2$$
$$x_2' = -\frac{k}{m}x_1.$$

(b) The associated eigenvalue problem is
$$\begin{pmatrix} -r & 1 \\ -k/m & -r \end{pmatrix}\begin{pmatrix} \xi_1 \\ \xi_2 \end{pmatrix} = \begin{pmatrix} 0 \\ 0 \end{pmatrix}.$$

The characteristic equation is $r^2 + k/m = 0$, with roots $r_{1,2} = \pm i\sqrt{k/m}$.

(c) Since the eigenvalues are purely imaginary, the origin is a center. Hence the phase curves are ellipses, with a clockwise flow. For computational purposes, let $k = 1$ and $m = 2$.

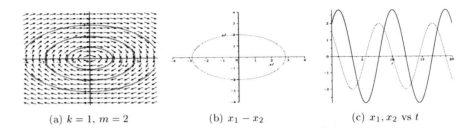

(a) $k = 1$, $m = 2$ (b) $x_1 - x_2$ (c) x_1, x_2 vs t

(d) The general solution of the second order equation is

$$u(t) = c_1 \cos\sqrt{\frac{k}{m}}\,t + c_2 \sin\sqrt{\frac{k}{m}}\,t.$$

The general solution of the system of ODEs is given by

$$\mathbf{x} = c_1 \begin{pmatrix} \sqrt{\frac{m}{k}} \sin\sqrt{\frac{k}{m}}\,t \\ \cos\sqrt{\frac{k}{m}}\,t \end{pmatrix} + c_2 \begin{pmatrix} \sqrt{\frac{m}{k}} \cos\sqrt{\frac{k}{m}}\,t \\ -\sin\sqrt{\frac{k}{m}}\,t \end{pmatrix}.$$

It is evident that the natural frequency of the system is equal to $|r_1| = |r_2|$.

29.(a) Set $\mathbf{x} = (x_1, x_2)^T$. We can rewrite Equation (22) in the form

$$\begin{pmatrix} 2 & 0 \\ 0 & 9/4 \end{pmatrix} \begin{pmatrix} \frac{d^2 x_1}{dt^2} \\ \frac{d^2 x_2}{dt^2} \end{pmatrix} = \begin{pmatrix} -4 & 3 \\ 3 & -\frac{27}{4} \end{pmatrix} \begin{pmatrix} x_1 \\ x_2 \end{pmatrix}.$$

Multiplying both sides of this equation by the inverse of the diagonal matrix, we obtain

$$\begin{pmatrix} \frac{d^2 x_1}{dt^2} \\ \frac{d^2 x_2}{dt^2} \end{pmatrix} = \begin{pmatrix} -2 & 3/2 \\ 4/3 & -3 \end{pmatrix} \begin{pmatrix} x_1 \\ x_2 \end{pmatrix}.$$

(b) Substituting $\mathbf{x} = \boldsymbol{\xi}\, e^{rt}$,

$$r^2 \begin{pmatrix} \xi_1 \\ \xi_2 \end{pmatrix} e^{rt} = \begin{pmatrix} -2 & 3/2 \\ 4/3 & -3 \end{pmatrix} \begin{pmatrix} \xi_1 \\ \xi_2 \end{pmatrix} e^{rt},$$

which can be written as

$$(\mathbf{A} - r^2 \mathbf{I})\boldsymbol{\xi} = \mathbf{0}.$$

(c) The eigenvalues are $r_1^2 = -1$ and $r_2^2 = -4$, with corresponding eigenvectors

$$\boldsymbol{\xi}^{(1)} = \begin{pmatrix} 3 \\ 2 \end{pmatrix} \text{ and } \boldsymbol{\xi}^{(2)} = \begin{pmatrix} 3 \\ -4 \end{pmatrix}.$$

(d) The linearly independent solutions are

$$\mathbf{x}^{(1)} = \tilde{C}_1 \begin{pmatrix} 3 \\ 2 \end{pmatrix} e^{it} \text{ and } \mathbf{x}^{(2)} = \tilde{C}_2 \begin{pmatrix} 3 \\ -4 \end{pmatrix} e^{2it}.$$

in which $\tilde{C}_1$ and $\tilde{C}_2$ are arbitrary complex coefficients. In scalar form,
$$x_1 = 3c_1 \cos t + 3c_2 \sin t + 3c_3 \cos 2t + 3c_4 \sin 2t$$
$$x_2 = 2c_1 \cos t + 2c_2 \sin t - 4c_3 \cos 2t - 4c_4 \sin 2t$$

(e) Differentiating the above expressions,
$$x_1' = -3c_1 \sin t + 3c_2 \cos t - 6c_3 \sin 2t + 6c_4 \cos 2t$$
$$x_2' = -2c_1 \sin t + 2c_2 \cos t + 8c_3 \sin 2t - 8c_4 \cos 2t$$

It is evident that $\mathbf{y} = (x_1, x_2, x_1', x_2')^T$ as in Equation (31).

31.(a) The second order system is given by
$$\frac{d^2 x_1}{dt^2} = -2x_1 + x_2$$
$$\frac{d^2 x_2}{dt^2} = x_1 - 2x_2$$

Let $y_1 = x_1$, $y_2 = x_2$, $y_3 = x_1'$ and $y_4 = x_2'$. In terms of the new variables, we have
$$y_1' = y_3$$
$$y_2' = y_4$$
$$y_3' = -2y_1 + y_2$$
$$y_4' = y_1 - 2y_2$$

hence the coefficient matrix is
$$\mathbf{A} = \begin{pmatrix} 0 & 0 & 1 & 0 \\ 0 & 0 & 0 & 1 \\ -2 & 1 & 0 & 0 \\ 1 & -2 & 0 & 0 \end{pmatrix}.$$

(b) The eigenvalues and corresponding eigenvectors of $\mathbf{A}$ are:
$$r_1 = i, \quad \boldsymbol{\xi}^{(1)} = (1, 1, i, i)^T$$
$$r_2 = -i, \quad \boldsymbol{\xi}^{(2)} = (1, 1, -i, -i)^T$$
$$r_3 = \sqrt{3}\, i, \quad \boldsymbol{\xi}^{(3)} = (1, -1, \sqrt{3}\, i, -\sqrt{3}\, i)^T$$
$$r_4 = -\sqrt{3}\, i, \quad \boldsymbol{\xi}^{(4)} = (1, -1, -\sqrt{3}\, i, \sqrt{3}\, i)^T$$

(c) Note that
$$\boldsymbol{\xi}^{(1)} e^{it} = \begin{pmatrix} 1 \\ 1 \\ i \\ i \end{pmatrix} (\cos t + i \sin t)$$

and
$$\boldsymbol{\xi}^{(3)} e^{\sqrt{3}\, it} = \begin{pmatrix} 1 \\ -1 \\ \sqrt{3}\, i \\ -\sqrt{3}\, i \end{pmatrix} (\cos \sqrt{3}\, t + i \sin \sqrt{3}\, t).$$

Hence the general solution is

$$\mathbf{y} = c_1 \begin{pmatrix} \cos t \\ \cos t \\ -\sin t \\ -\sin t \end{pmatrix} + c_2 \begin{pmatrix} \sin t \\ \sin t \\ \cos t \\ \cos t \end{pmatrix} + c_3 \begin{pmatrix} \cos \sqrt{3}\,t \\ -\cos \sqrt{3}\,t \\ -\sqrt{3} \sin \sqrt{3}\,t \\ \sqrt{3} \sin \sqrt{3}\,t \end{pmatrix} + c_4 \begin{pmatrix} \sin \sqrt{3}\,t \\ -\sin \sqrt{3}\,t \\ \sqrt{3} \cos \sqrt{3}\,t \\ -\sqrt{3} \cos \sqrt{3}\,t \end{pmatrix}.$$

(d) The two modes have natural frequencies of $\omega_1 = 1$ rad/sec and $\omega_2 = \sqrt{3}$ rad/sec.

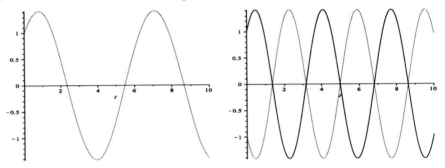

(e) For the initial condition $\mathbf{y}(0) = (-1, 3, 0, 0)^T$, it is necessary that

$$\begin{pmatrix} -1 \\ 3 \\ 0 \\ 0 \end{pmatrix} = c_1 \begin{pmatrix} 1 \\ 1 \\ 0 \\ 0 \end{pmatrix} + c_2 \begin{pmatrix} 0 \\ 0 \\ 1 \\ 1 \end{pmatrix} + c_3 \begin{pmatrix} 1 \\ -1 \\ 0 \\ 0 \end{pmatrix} + c_4 \begin{pmatrix} 0 \\ 0 \\ \sqrt{3} \\ -\sqrt{3} \end{pmatrix},$$

resulting in the coefficients $c_1 = 1$, $c_2 = 0$, $c_3 = -2$ and $c_4 = 0$.

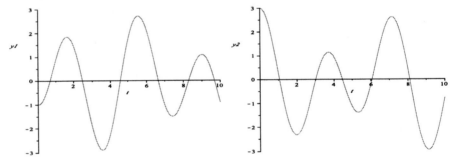

The solutions are not periodic, since the two natural frequencies are incommensurate.

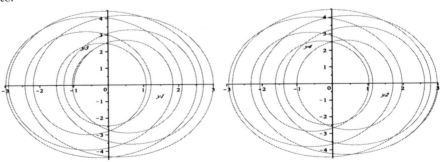

7.7

1.(a) The eigenvalues and eigenvectors were found in Problem 1, Section 7.5.

$$r_1 = -1, \quad \boldsymbol{\xi}^{(1)} = \begin{pmatrix} 1 \\ 2 \end{pmatrix}; \quad r_2 = 2, \quad \boldsymbol{\xi}^{(2)} = \begin{pmatrix} 2 \\ 1 \end{pmatrix}.$$

The general solution is

$$\mathbf{x} = c_1 \begin{pmatrix} e^{-t} \\ 2e^{-t} \end{pmatrix} + c_2 \begin{pmatrix} 2e^{2t} \\ e^{2t} \end{pmatrix}.$$

Hence a fundamental matrix is given by

$$\boldsymbol{\Psi}(t) = \begin{pmatrix} e^{-t} & 2e^{2t} \\ 2e^{-t} & e^{2t} \end{pmatrix}.$$

(b) We now have

$$\boldsymbol{\Psi}(0) = \begin{pmatrix} 1 & 2 \\ 2 & 1 \end{pmatrix} \text{ and } \boldsymbol{\Psi}^{-1}(0) = \frac{1}{3}\begin{pmatrix} -1 & 2 \\ 2 & -1 \end{pmatrix},$$

So that

$$\boldsymbol{\Phi}(t) = \boldsymbol{\Psi}(t)\boldsymbol{\Psi}^{-1}(0) = \frac{1}{3}\begin{pmatrix} -e^{-t} + 4e^{2t} & 2e^{-t} - 2e^{2t} \\ -2e^{-t} + 2e^{2t} & 4e^{-t} - e^{2t} \end{pmatrix}.$$

3.(a) The eigenvalues and eigenvectors were found in Problem 3, Section 7.5. The general solution of the system is

$$\mathbf{x} = c_1 \begin{pmatrix} e^{t} \\ e^{t} \end{pmatrix} + c_2 \begin{pmatrix} e^{-t} \\ 3e^{-t} \end{pmatrix}.$$

Hence a fundamental matrix is given by

$$\boldsymbol{\Psi}(t) = \begin{pmatrix} e^{t} & e^{-t} \\ e^{t} & 3e^{-t} \end{pmatrix}.$$

(b) Given the initial conditions $\mathbf{x}(0) = \mathbf{e}^{(1)}$, we solve the equations

$$c_1 + c_2 = 1$$
$$c_1 + 3c_2 = 0,$$

to obtain $c_1 = 3/2$, $c_2 = -1/2$. The corresponding solution is

$$\mathbf{x} = \begin{pmatrix} \frac{3}{2}e^{t} - \frac{1}{2}e^{-t} \\ \frac{3}{2}e^{t} - \frac{3}{2}e^{-t} \end{pmatrix}.$$

Given the initial conditions $\mathbf{x}(0) = \mathbf{e}^{(2)}$, we solve the equations

$$c_1 + c_2 = 0$$
$$c_1 + 3c_2 = 1,$$

to obtain $c_1 = -1/2$, $c_2 = 1/2$. The corresponding solution is
$$\mathbf{x} = \begin{pmatrix} -\frac{1}{2}e^t + \frac{1}{2}e^{-t} \\ -\frac{1}{2}e^t + \frac{3}{2}e^{-t} \end{pmatrix}.$$
Therefore the fundamental matrix is
$$\Phi(t) = \frac{1}{2}\begin{pmatrix} 3e^t - e^{-t} & -e^t + e^{-t} \\ 3e^t - 3e^{-t} & -e^t + 3e^{-t} \end{pmatrix}.$$

5.(a) The general solution, found in Problem 3, Section 7.6, is given by
$$\mathbf{x} = c_1 \begin{pmatrix} 5\cos t \\ 2\cos t + \sin t \end{pmatrix} + c_2 \begin{pmatrix} 5\sin t \\ -\cos t + 2\sin t \end{pmatrix}.$$
Hence a fundamental matrix is given by
$$\Psi(t) = \begin{pmatrix} 5\cos t & 5\sin t \\ 2\cos t + \sin t & -\cos t + 2\sin t \end{pmatrix}.$$

(b) Given the initial conditions $\mathbf{x}(0) = \mathbf{e}^{(1)}$, we solve the equations
$$5c_1 = 1$$
$$2c_1 - c_2 = 0,$$
resulting in $c_1 = 1/5$, $c_2 = 2/5$. The corresponding solution is
$$\mathbf{x} = \begin{pmatrix} \cos t + 2\sin t \\ \sin t \end{pmatrix}.$$
Given the initial conditions $\mathbf{x}(0) = \mathbf{e}^{(2)}$, we solve the equations
$$5c_1 = 0$$
$$2c_1 - c_2 = 1,$$
resulting in $c_1 = 0$, $c_2 = -1$. The corresponding solution is
$$\mathbf{x} = \begin{pmatrix} -5\sin t \\ \cos t - 2\sin t \end{pmatrix}.$$
Therefore the fundamental matrix is
$$\Phi(t) = \begin{pmatrix} \cos t + 2\sin t & -5\sin t \\ \sin t & \cos t - 2\sin t \end{pmatrix}.$$

7.(a) The general solution, found in Problem 15, Section 7.5, is given by
$$\mathbf{x} = c_1 \begin{pmatrix} e^{2t} \\ 3e^{2t} \end{pmatrix} + c_2 \begin{pmatrix} e^{4t} \\ e^{4t} \end{pmatrix}.$$
Hence a fundamental matrix is given by
$$\Psi(t) = \begin{pmatrix} e^{2t} & e^{4t} \\ 3e^{2t} & e^{4t} \end{pmatrix}.$$

(b) Given the initial conditions $\mathbf{x}(0) = \mathbf{e}^{(1)}$, we solve the equations

$$c_1 + c_2 = 1$$
$$3c_1 + c_2 = 0,$$

resulting in $c_1 = -1/2$, $c_2 = 3/2$. The corresponding solution is

$$\mathbf{x} = \frac{1}{2}\begin{pmatrix} -e^{2t} + 3e^{4t} \\ -3e^{2t} + 3e^{4t} \end{pmatrix}.$$

The initial conditions $\mathbf{x}(0) = \mathbf{e}^{(2)}$ require that

$$c_1 + c_2 = 0$$
$$3c_1 + c_2 = 1,$$

resulting in $c_1 = 1/2$, $c_2 = -1/2$. The corresponding solution is

$$\mathbf{x} = \frac{1}{2}\begin{pmatrix} e^{2t} - e^{4t} \\ 3e^{2t} - e^{4t} \end{pmatrix}.$$

Therefore the fundamental matrix is

$$\Phi(t) = \frac{1}{2}\begin{pmatrix} -e^{2t} + 3e^{4t} & e^{2t} - e^{4t} \\ -3e^{2t} + 3e^{4t} & 3e^{2t} - e^{4t} \end{pmatrix}.$$

8.(a) The general solution, found in Problem 5, Section 7.6, is given by

$$\mathbf{x} = c_1 e^{-t}\begin{pmatrix} \cos t \\ 2\cos t + \sin t \end{pmatrix} + c_2 e^{-t}\begin{pmatrix} \sin t \\ -\cos t + 2\sin t \end{pmatrix}.$$

Hence a fundamental matrix is given by

$$\Psi(t) = \begin{pmatrix} e^{-t}\cos t & e^{-t}\sin t \\ 2e^{-t}\cos t + e^{-t}\sin t & -e^{-t}\cos t + 2e^{-t}\sin t \end{pmatrix}.$$

(b) The specific solution corresponding to the initial conditions $\mathbf{x}(0) = \mathbf{e}^{(1)}$ is

$$\mathbf{x} = e^{-t}\begin{pmatrix} \cos t + 2\sin t \\ 5\sin t \end{pmatrix}.$$

For the initial conditions $\mathbf{x}(0) = \mathbf{e}^{(2)}$, the solution is

$$\mathbf{x} = e^{-t}\begin{pmatrix} -\sin t \\ \cos t - 2\sin t \end{pmatrix}.$$

Therefore the fundamental matrix is

$$\Phi(t) = e^{-t}\begin{pmatrix} \cos t + 2\sin t & -\sin t \\ 5\sin t & \cos t - 2\sin t \end{pmatrix}.$$

9.(a) The general solution, found in Problem 13, Section 7.5, is given by

$$\mathbf{x} = c_1 \begin{pmatrix} 4e^{-2t} \\ -5e^{-2t} \\ -7e^{-2t} \end{pmatrix} + c_2 \begin{pmatrix} 3e^{-t} \\ -4e^{-t} \\ -2e^{-t} \end{pmatrix} + c_3 \begin{pmatrix} 0 \\ e^{2t} \\ -e^{2t} \end{pmatrix}.$$

Hence a fundamental matrix is given by
$$\Psi(t) = \begin{pmatrix} 4e^{-2t} & 3e^{-t} & 0 \\ -5e^{-2t} & -4e^{-t} & e^{2t} \\ -7e^{-2t} & -2e^{-t} & -e^{2t} \end{pmatrix}.$$

(b) Given the initial conditions $\mathbf{x}(0) = \mathbf{e}^{(1)}$, we solve the equations
$$4c_1 + 3c_2 = 1$$
$$-5c_1 - 4c_2 + c_3 = 0$$
$$-7c_1 - 2c_2 - c_3 = 0,$$
resulting in $c_1 = -1/2$, $c_2 = 1$, $c_3 = 3/2$. The corresponding solution is
$$\mathbf{x} = \begin{pmatrix} -2e^{-2t} + 3e^{-t} \\ 5e^{-2t}/2 - 4e^{-t} + 3e^{2t}/2 \\ 7e^{-2t}/2 - 2e^{-t} - 3e^{2t}/2 \end{pmatrix}.$$

The initial conditions $\mathbf{x}(0) = \mathbf{e}^{(2)}$, we solve the equations
$$4c_1 + 3c_2 = 0$$
$$-5c_1 - 4c_2 + c_3 = 1$$
$$-7c_1 - 2c_2 - c_3 = 0,$$
resulting in $c_1 = -1/4$, $c_2 = 1/3$, $c_3 = 13/12$. The corresponding solution is
$$\mathbf{x} = \begin{pmatrix} -e^{-2t} + e^{-t} \\ 5e^{-2t}/4 - 4e^{-t}/3 + 13e^{2t}/12 \\ 7e^{-2t}/4 - 2e^{-t}/3 - 13e^{2t}/12 \end{pmatrix}.$$

The initial conditions $\mathbf{x}(0) = \mathbf{e}^{(3)}$, we solve the equations
$$4c_1 + 3c_2 = 0$$
$$-5c_1 - 4c_2 + c_3 = 0$$
$$-7c_1 - 2c_2 - c_3 = 1,$$
resulting in $c_1 = -1/4$, $c_2 = 1/3$, $c_3 = 1/12$. The corresponding solution is
$$\mathbf{x} = \begin{pmatrix} -e^{-2t} + e^{-t} \\ 5e^{-2t}/4 - 4e^{-t}/3 + e^{2t}/12 \\ 7e^{-2t}/4 - 2e^{-t}/3 - e^{2t}/12 \end{pmatrix}.$$

Therefore the fundamental matrix is
$$\Phi(t) = \frac{1}{12}\begin{pmatrix} -24e^{-2t} + 36e^{-t} & -12e^{-2t} + 12e^{-t} & -12e^{-2t} + 12e^{-t} \\ 30e^{-2t} - 48e^{-t} + 18e^{2t} & 15e^{-2t} - 16e^{-t} + 13e^{2t} & 15e^{-2t} - 16e^{-t} + e^{2t} \\ 42e^{-2t} - 24e^{-t} - 18e^{2t} & 21e^{-2t} - 8e^{-t} - 13e^{2t} & 21e^{-2t} - 8e^{-t} - e^{2t} \end{pmatrix}.$$

12. The solution of the initial value problem is given by
$$\mathbf{x} = \Phi(t)\mathbf{x}(0) = \begin{pmatrix} e^{-t}\cos 2t & -2e^{-t}\sin 2t \\ \frac{1}{2}e^{-t}\sin 2t & e^{-t}\cos 2t \end{pmatrix}\begin{pmatrix} 3 \\ 1 \end{pmatrix} =$$
$$= e^{-t}\begin{pmatrix} 3\cos 2t - 2\sin 2t \\ \frac{3}{2}\sin 2t + \cos 2t \end{pmatrix}.$$

13. Let
$$\Psi(t) = \begin{pmatrix} x_1^{(1)}(t) & \cdots & x_1^{(n)}(t) \\ \vdots & & \vdots \\ x_n^{(1)}(t) & \cdots & x_n^{(n)}(t) \end{pmatrix}.$$

It follows that
$$\Psi(t_0) = \begin{pmatrix} x_1^{(1)}(t_0) & \cdots & x_1^{(n)}(t_0) \\ \vdots & & \vdots \\ x_n^{(1)}(t_0) & \cdots & x_n^{(n)}(t_0) \end{pmatrix}$$

is a scalar matrix, which is invertible, since the solutions are linearly independent. Let $\Psi^{-1}(t_0) = (c_{ij})$. Then

$$\Psi(t)\Psi^{-1}(t_0) = \begin{pmatrix} x_1^{(1)}(t) & \cdots & x_1^{(n)}(t) \\ \vdots & & \vdots \\ x_n^{(1)}(t) & \cdots & x_n^{(n)}(t) \end{pmatrix} \begin{pmatrix} c_{11} & \cdots & c_{1n} \\ \vdots & & \vdots \\ c_{n1} & \cdots & c_{nn} \end{pmatrix}.$$

The j-th column of the product matrix is

$$\left[\Psi(t)\Psi^{-1}(t_0)\right]^{(j)} = \sum_{k=1}^{n} c_{kj}\,\mathbf{x}^{(k)},$$

which is a solution vector, since it is a linear combination of solutions. Furthermore, the columns are all linearly independent, since the vectors $\mathbf{x}^{(k)}$ are. Hence the product is a fundamental matrix. Finally, setting $t = t_0$, $\Psi(t_0)\Psi^{-1}(t_0) = \mathbf{I}$. This is precisely the definition of $\Phi(t)$.

14. The fundamental matrix $\Phi(t)$ for the system
$$\mathbf{x}' = \begin{pmatrix} 1 & 1 \\ 4 & 1 \end{pmatrix} \mathbf{x}$$

is given by
$$\Phi(t) = \frac{1}{4}\begin{pmatrix} 2e^{3t} + 2e^{-t} & e^{3t} - e^{-t} \\ 4e^{3t} - 4e^{-t} & 2e^{3t} + 2e^{-t} \end{pmatrix}.$$

Direct multiplication results in

$$\Phi(t)\Phi(s) = \frac{1}{16}\begin{pmatrix} 2e^{3t} + 2e^{-t} & e^{3t} - e^{-t} \\ 4e^{3t} - 4e^{-t} & 2e^{3t} + 2e^{-t} \end{pmatrix}\begin{pmatrix} 2e^{3s} + 2e^{-s} & e^{3s} - e^{-s} \\ 4e^{3s} - 4e^{-s} & 2e^{3s} + 2e^{-s} \end{pmatrix}$$
$$= \frac{1}{16}\begin{pmatrix} 8(e^{3t+3s} + e^{-t-s}) & 4(e^{3t+3s} - e^{-t-s}) \\ 16(e^{3t+3s} - e^{-t-s}) & 8(e^{3t+3s} + e^{-t-s}) \end{pmatrix}.$$

Hence
$$\Phi(t)\Phi(s) = \frac{1}{4}\begin{pmatrix} 2e^{3(t+s)} + 2e^{-(t+s)} & e^{3(t+s)} - e^{-(t+s)} \\ 4e^{3(t+s)} - 4e^{-(t+s)} & 2e^{3(t+s)} + 2e^{-(t+s)} \end{pmatrix} = \Phi(t+s).$$

15.(a) Let s be arbitrary, but fixed, and t variable. Similar to the argument in Problem 13, the columns of the matrix $\mathbf{\Phi}(t)\mathbf{\Phi}(s)$ are linear combinations of fundamental solutions. Hence the columns of $\mathbf{\Phi}(t)\mathbf{\Phi}(s)$ are also solution of the system of equations. Further, setting $t = 0$, $\mathbf{\Phi}(0)\mathbf{\Phi}(s) = \mathbf{I}\,\mathbf{\Phi}(s) = \mathbf{\Phi}(s)$. That is, $\mathbf{\Phi}(t)\mathbf{\Phi}(s)$ is a solution of the initial value problem $\mathbf{Z}' = \mathbf{AZ}$, with $\mathbf{Z}(0) = \mathbf{\Phi}(s)$. Now consider the change of variable $\tau = t + s$. Let $\mathbf{W}(\tau) = \mathbf{Z}(\tau - s)$. The given initial value problem can be reformulated as

$$\frac{d}{d\tau}\mathbf{W} = \mathbf{AW}, \text{ with } \mathbf{W}(s) = \mathbf{\Phi}(s).$$

Since $\mathbf{\Phi}(t)$ is a fundamental matrix satisfying $\mathbf{\Phi}' = \mathbf{A}\mathbf{\Phi}$, with $\mathbf{\Phi}(0) = \mathbf{I}$, it follows that

$$\mathbf{W}(\tau) = \left[\mathbf{\Phi}(\tau)\mathbf{\Phi}^{-1}(s)\right]\mathbf{\Phi}(s) = \mathbf{\Phi}(\tau).$$

That is, $\mathbf{\Phi}(t + s) = \mathbf{\Phi}(\tau) = \mathbf{W}(\tau) = \mathbf{Z}(t) = \mathbf{\Phi}(t)\mathbf{\Phi}(s)$.

(b) Based on part (a), $\mathbf{\Phi}(t)\mathbf{\Phi}(-t) = \mathbf{\Phi}(t + (-t)) = \mathbf{\Phi}(0) = \mathbf{I}$. Hence $\mathbf{\Phi}(-t) = \mathbf{\Phi}^{-1}(t)$.

(c) It also follows that $\mathbf{\Phi}(t - s) = \mathbf{\Phi}(t + (-s)) = \mathbf{\Phi}(t)\mathbf{\Phi}(-s) = \mathbf{\Phi}(t)\mathbf{\Phi}^{-1}(s)$.

16. Let $\mathbf{A}$ be a diagonal matrix, with $\mathbf{A} = \left[a_1\mathbf{e}^{(1)}, a_2\mathbf{e}^{(2)}, \cdots, a_n\mathbf{e}^{(n)}\right]$. Note that for any positive integer k,

$$\mathbf{A}^k = \left[a_1^k\,\mathbf{e}^{(1)}, a_2^k\,\mathbf{e}^{(2)}, \cdots, a_n^k\,\mathbf{e}^{(n)}\right].$$

It follows, from basic matrix algebra, that

$$\mathbf{I} + \sum_{k=1}^{m} \mathbf{A}^k \frac{t^k}{k!} = \begin{pmatrix} \sum_{k=0}^{m} a_1^k \frac{t^k}{k!} & 0 & \cdots & 0 \\ 0 & \sum_{k=0}^{m} a_2^k \frac{t^k}{k!} & \cdots & 0 \\ \vdots & \vdots & & \vdots \\ 0 & 0 & \cdots & \sum_{k=0}^{m} a_n^k \frac{t^k}{k!} \end{pmatrix}.$$

It can be shown that the partial sums on the left hand side converge for all t. Taking the limit as $m \to \infty$ on both sides of the equation, we obtain

$$e^{\mathbf{A}t} = \begin{pmatrix} e^{a_1 t} & 0 & \cdots & 0 \\ 0 & e^{a_2 t} & \cdots & 0 \\ \vdots & \vdots & & \vdots \\ 0 & 0 & \cdots & e^{a_n t} \end{pmatrix}.$$

Alternatively, consider the system $\mathbf{x}' = \mathbf{Ax}$. Since the ODEs are uncoupled, the vectors $\mathbf{x}^{(j)} = e^{a_j t}\mathbf{e}^{(j)}$, $j = 1, 2, \cdots n$, are a set of linearly independent solutions. Hence the matrix

$$\mathbf{x} = \left[e^{a_1 t}\,\mathbf{e}^{(1)}, e^{a_2 t}\,\mathbf{e}^{(2)}, \cdots, e^{a_n t}\,\mathbf{e}^{(n)}\right]$$

is a fundamental matrix. Finally, since $\mathbf{X}(0) = \mathbf{I}$, it follows that

$$\left[e^{a_1 t}\,\mathbf{e}^{(1)}, e^{a_2 t}\,\mathbf{e}^{(2)}, \cdots, e^{a_n t}\,\mathbf{e}^{(n)}\right] = \mathbf{\Phi}(t) = e^{\mathbf{A}t}.$$

17.(a) Let $x_1 = u$ and $x_2 = u'$; then $u'' = x_2'$. In terms of the new variables, we have
$$x_2' + \omega^2 x_1 = 0$$
with the initial conditions $x_1(0) = u_0$ and $x_2(0) = v_0$. The equivalent first order system is
$$x_1' = x_2$$
$$x_2' = -\omega^2 x_1$$
which can be expressed in the form
$$\begin{pmatrix} x_1 \\ x_2 \end{pmatrix}' = \begin{pmatrix} 0 & 1 \\ -\omega^2 & 0 \end{pmatrix} \begin{pmatrix} x_1 \\ x_2 \end{pmatrix}; \quad \begin{pmatrix} x_1(0) \\ x_2(0) \end{pmatrix} = \begin{pmatrix} u_0 \\ v_0 \end{pmatrix}.$$

(b) Setting
$$\mathbf{A} = \begin{pmatrix} 0 & 1 \\ -\omega^2 & 0 \end{pmatrix},$$
it is easy to show that
$$\mathbf{A}^2 = -\omega^2 \mathbf{I}, \ \mathbf{A}^3 = -\omega^2 \mathbf{A} \ \text{and} \ \mathbf{A}^4 = \omega^4 \mathbf{I}.$$
It follows inductively that
$$\mathbf{A}^{2k} = (-1)^k \omega^{2k} \mathbf{I}$$
and
$$\mathbf{A}^{2k+1} = (-1)^k \omega^{2k} \mathbf{A}.$$
Hence
$$e^{\mathbf{A}t} = \sum_{k=0}^{\infty} \left[(-1)^k \frac{\omega^{2k} t^{2k}}{(2k)!} \mathbf{I} + (-1)^k \frac{\omega^{2k} t^{2k+1}}{(2k+1)!} \mathbf{A} \right]$$
$$= \left[\sum_{k=0}^{\infty} (-1)^k \frac{\omega^{2k} t^{2k}}{(2k)!} \right] \mathbf{I} + \frac{1}{\omega} \left[\sum_{k=0}^{\infty} (-1)^k \frac{\omega^{2k+1} t^{2k+1}}{(2k+1)!} \right] \mathbf{A}$$
and therefore
$$e^{\mathbf{A}t} = \cos \omega t \, \mathbf{I} + \frac{1}{\omega} \sin \omega t \, \mathbf{A}.$$

(c) From Equation (28),
$$\begin{pmatrix} x_1 \\ x_2 \end{pmatrix} = \left[\cos \omega t \, \mathbf{I} + \frac{1}{\omega} \sin \omega t \, \mathbf{A} \right] \begin{pmatrix} u_0 \\ v_0 \end{pmatrix}$$
$$= \cos \omega t \begin{pmatrix} u_0 \\ v_0 \end{pmatrix} + \frac{1}{\omega} \sin \omega t \begin{pmatrix} v_0 \\ -\omega^2 u_0 \end{pmatrix}.$$

18.(a) Assuming that $\mathbf{x} = \phi(t)$ is a solution, then $\phi' = \mathbf{A}\phi$, with $\phi(0) = \mathbf{x}^0$. Integrate both sides of the equation to obtain
$$\phi(t) - \phi(0) = \int_0^t \mathbf{A}\phi(s) ds.$$

Hence
$$\phi(t) = \mathbf{x}^0 + \int_0^t \mathbf{A}\phi(s)ds.$$

(b) Proceed with the iteration
$$\phi^{(i+1)}(t) = \mathbf{x}^0 + \int_0^t \mathbf{A}\phi^{(i)}(s)ds.$$

With $\phi^{(0)}(t) = \mathbf{x}^0$, and noting that $\mathbf{A}$ is a constant matrix,
$$\phi^{(1)}(t) = \mathbf{x}^0 + \int_0^t \mathbf{A}\mathbf{x}^0 ds = \mathbf{x}^0 + \mathbf{A}\mathbf{x}^0 t.$$

That is, $\phi^{(1)}(t) = (\mathbf{I} + \mathbf{A}t)\mathbf{x}^0$.

(c) We then have
$$\phi^{(2)}(t) = \mathbf{x}^0 + \int_0^t \mathbf{A}(\mathbf{I}+\mathbf{A}t)\mathbf{x}^0 ds = \mathbf{x}^0 + \mathbf{A}\mathbf{x}^0 t + \mathbf{A}^2 \mathbf{x}^0 \frac{t^2}{2} = (\mathbf{I} + \mathbf{A}t + \mathbf{A}^2\frac{t^2}{2})\mathbf{x}^0.$$

Now suppose that
$$\phi^{(n)}(t) = (\mathbf{I} + \mathbf{A}t + \mathbf{A}^2\frac{t^2}{2} + \cdots + \mathbf{A}^n\frac{t^n}{n!})\mathbf{x}^0.$$

It follows that
$$\int_0^t \mathbf{A}(\mathbf{I} + \mathbf{A}t + \mathbf{A}^2\frac{t^2}{2} + \cdots + \mathbf{A}^n\frac{t^n}{n!})\mathbf{x}^0 ds =$$
$$= \mathbf{A}(\mathbf{I}t + \mathbf{A}\frac{t^2}{2} + \mathbf{A}^2\frac{t^3}{3!} + \cdots + \mathbf{A}^n\frac{t^{n+1}}{(n+1)!})\mathbf{x}^0$$
$$= (\mathbf{A}t + \mathbf{A}^2\frac{t^2}{2} + \mathbf{A}^3\frac{t^3}{3!} + \cdots + \mathbf{A}^{n+1}\frac{t^n}{n!})\mathbf{x}^0.$$

Therefore
$$\phi^{(n+1)}(t) = (\mathbf{I} + \mathbf{A}t + \mathbf{A}^2\frac{t^2}{2} + \cdots + \mathbf{A}^{n+1}\frac{t^{n+1}}{(n+1)!})\mathbf{x}^0.$$

By induction, the asserted form of $\phi^{(n)}(t)$ is valid for all $n \geq 0$.

(d) Define $\phi^{(\infty)}(t) = \lim_{n \to \infty} \phi^{(n)}(t)$. It can be shown that the limit does exist. In fact,
$$\phi^{(\infty)}(t) = e^{\mathbf{A}t}\mathbf{x}^0.$$

Term-by-term differentiation results in
$$\frac{d}{dt}\phi^{(\infty)}(t) = \frac{d}{dt}(\mathbf{I} + \mathbf{A}t + \mathbf{A}^2\frac{t^2}{2} + \cdots + \mathbf{A}^n\frac{t^n}{n!} + \cdots)\mathbf{x}^0$$
$$= (\mathbf{A} + \mathbf{A}^2 t + \cdots + \mathbf{A}^n\frac{t^{n-1}}{(n-1)!} + \cdots)\mathbf{x}^0$$
$$= \mathbf{A}(\mathbf{I} + \mathbf{A}t + \mathbf{A}^2\frac{t^2}{2} + \cdots + \mathbf{A}^{n-1}\frac{t^{n-1}}{(n-1)!} + \cdots)\mathbf{x}^0.$$

That is,
$$\frac{d}{dt}\phi^{(\infty)}(t) = \mathbf{A}\phi^{(\infty)}(t).$$

Furthermore, $\phi^{(\infty)}(0) = \mathbf{x}^0$. Based on uniqueness of solutions, $\phi(t) = \phi^{(\infty)}(t)$.

7.8

2.(a)

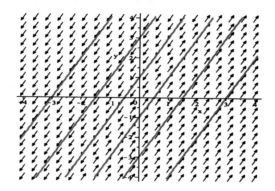

(b) All of the points on the line $x_2 = 2x_1$ are equilibrium points. Solutions starting at all other points become unbounded.

(c) Setting $\mathbf{x} = \boldsymbol{\xi} t^r$ results in the algebraic equations
$$\begin{pmatrix} 4-r & -2 \\ 8 & -4-r \end{pmatrix} \begin{pmatrix} \xi_1 \\ \xi_2 \end{pmatrix} = \begin{pmatrix} 0 \\ 0 \end{pmatrix}.$$

The characteristic equation is $r^2 = 0$, with the single root $r = 0$. Substituting $r = 0$ reduces the system of equations to $2\xi_1 - \xi_2 = 0$. Therefore the only eigenvector is $\boldsymbol{\xi} = (1, 2)^T$. One solution is
$$\mathbf{x}^{(1)} = \begin{pmatrix} 1 \\ 2 \end{pmatrix},$$

which is a constant vector. In order to generate a second linearly independent solution, we must search for a generalized eigenvector. This leads to the system of equations
$$\begin{pmatrix} 4 & -2 \\ 8 & -4 \end{pmatrix} \begin{pmatrix} \eta_1 \\ \eta_2 \end{pmatrix} = \begin{pmatrix} 1 \\ 2 \end{pmatrix}.$$

This system also reduces to a single equation, $2\eta_1 - \eta_2 = 1/2$. Setting $\eta_1 = k$, some arbitrary constant, we obtain $\eta_2 = 2k - 1/2$. A second solution is
$$\mathbf{x}^{(2)} = \begin{pmatrix} 1 \\ 2 \end{pmatrix} t + \begin{pmatrix} k \\ 2k-1/2 \end{pmatrix} = \begin{pmatrix} 1 \\ 2 \end{pmatrix} t + \begin{pmatrix} 0 \\ -1/2 \end{pmatrix} + k \begin{pmatrix} 1 \\ 2 \end{pmatrix}.$$

Note that the last term is a multiple of $\mathbf{x}^{(1)}$ and may be dropped. Hence
$$\mathbf{x}^{(2)} = \begin{pmatrix} 1 \\ 2 \end{pmatrix} t + \begin{pmatrix} 0 \\ -1/2 \end{pmatrix}.$$

The general solution is
$$\mathbf{x} = c_1 \begin{pmatrix} 1 \\ 2 \end{pmatrix} + c_2 \left[\begin{pmatrix} 1 \\ 2 \end{pmatrix} t + \begin{pmatrix} 0 \\ -1/2 \end{pmatrix} \right].$$

4.(a)

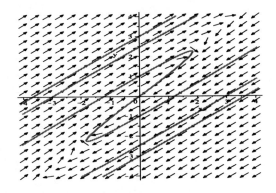

(b) All trajectories converge to the origin.

(c) Solution of the ODE requires analysis of the algebraic equations
$$\begin{pmatrix} -3-r & \frac{5}{2} \\ -\frac{5}{2} & 2-r \end{pmatrix} \begin{pmatrix} \xi_1 \\ \xi_2 \end{pmatrix} = \begin{pmatrix} 0 \\ 0 \end{pmatrix}.$$

For a nonzero solution, we must have $\det(\mathbf{A} - r\mathbf{I}) = r^2 + r + 1/4 = 0$. The only root is $r = -1/2$, which is an eigenvalue of multiplicity two. Setting $r = -1/2$ is the coefficient matrix reduces the system to the single equation $-\xi_1 + \xi_2 = 0$. Hence the corresponding eigenvector is $\boldsymbol{\xi} = (1, 1)^T$. One solution is
$$\mathbf{x}^{(1)} = \begin{pmatrix} 1 \\ 1 \end{pmatrix} e^{-t/2}.$$

In order to obtain a second linearly independent solution, we find a solution of the system
$$\begin{pmatrix} -5/2 & 5/2 \\ -5/2 & 5/2 \end{pmatrix} \begin{pmatrix} \eta_1 \\ \eta_2 \end{pmatrix} = \begin{pmatrix} 1 \\ 1 \end{pmatrix}.$$

There equations reduce to $-5\eta_1 + 5\eta_2 = 2$. Set $\eta_1 = k$, some arbitrary constant. Then $\eta_2 = k + 2/5$. A second solution is
$$\mathbf{x}^{(2)} = \begin{pmatrix} 1 \\ 1 \end{pmatrix} t e^{-t/2} + \begin{pmatrix} k \\ k + 2/5 \end{pmatrix} e^{-t/2} = \begin{pmatrix} 1 \\ 1 \end{pmatrix} t e^{-t/2} + \begin{pmatrix} 0 \\ 2/5 \end{pmatrix} e^{-t/2} + k \begin{pmatrix} 1 \\ 1 \end{pmatrix} e^{-t/2}.$$

Dropping the last term, the general solution is
$$\mathbf{x} = c_1 \begin{pmatrix} 1 \\ 1 \end{pmatrix} e^{-t/2} + c_2 \left[\begin{pmatrix} 1 \\ 1 \end{pmatrix} t e^{-t/2} + \begin{pmatrix} 0 \\ 2/5 \end{pmatrix} e^{-t/2} \right].$$

6. The eigensystem is obtained from analysis of the equation

$$\begin{pmatrix} -r & 1 & 1 \\ 1 & -r & 1 \\ 1 & 1 & -r \end{pmatrix} \begin{pmatrix} \xi_1 \\ \xi_2 \\ \xi_3 \end{pmatrix} = \begin{pmatrix} 0 \\ 0 \\ 0 \end{pmatrix}.$$

The characteristic equation of the coefficient matrix is $r^3 - 3r - 2 = 0$, with roots $r_1 = 2$ and $r_{2,3} = -1$. Setting $r = 2$, we have

$$\begin{pmatrix} -2 & 1 & 1 \\ 1 & -2 & 1 \\ 1 & 1 & -2 \end{pmatrix} \begin{pmatrix} \xi_1 \\ \xi_2 \\ \xi_3 \end{pmatrix} = \begin{pmatrix} 0 \\ 0 \\ 0 \end{pmatrix}.$$

This system is reduced to the equations

$$\xi_1 - \xi_3 = 0$$
$$\xi_2 - \xi_3 = 0.$$

A corresponding eigenvector is given by $\boldsymbol{\xi}^{(1)} = (1,1,1)^T$. Setting $r = -1$, the system of equations is reduced to the single equation

$$\xi_1 + \xi_2 + \xi_3 = 0.$$

An eigenvector vector is given by $\boldsymbol{\xi}^{(2)} = (1,0,-1)^T$. Since the last equation has two free variables, a third linearly independent eigenvector (associated with $r = -1$) is $\boldsymbol{\xi}^{(3)} = (0,1,-1)^T$. Therefore the general solution may be written as

$$\mathbf{x} = c_1 \begin{pmatrix} 1 \\ 1 \\ 1 \end{pmatrix} e^{2t} + c_2 \begin{pmatrix} 1 \\ 0 \\ -1 \end{pmatrix} e^{-t} + c_3 \begin{pmatrix} 0 \\ 1 \\ -1 \end{pmatrix} e^{-t}.$$

7.(a) Solution of the ODE requires analysis of the algebraic equations

$$\begin{pmatrix} 1-r & -4 \\ 4 & -7-r \end{pmatrix} \begin{pmatrix} \xi_1 \\ \xi_2 \end{pmatrix} = \begin{pmatrix} 0 \\ 0 \end{pmatrix}.$$

For a nonzero solution, we must have $\det(\mathbf{A} - r\mathbf{I}) = r^2 + 6r + 9 = 0$. The only root is $r = -3$, which is an eigenvalue of multiplicity two. Substituting $r = -3$ into the coefficient matrix, the system reduces to the single equation $\xi_1 - \xi_2 = 0$. Hence the corresponding eigenvector is $\boldsymbol{\xi} = (1,1)^T$. One solution is

$$\mathbf{x}^{(1)} = \begin{pmatrix} 1 \\ 1 \end{pmatrix} e^{-3t}.$$

For a second linearly independent solution, we search for a generalized eigenvector. Its components satisfy

$$\begin{pmatrix} 4 & -4 \\ 4 & -4 \end{pmatrix} \begin{pmatrix} \eta_1 \\ \eta_2 \end{pmatrix} = \begin{pmatrix} 1 \\ 1 \end{pmatrix},$$

that is, $4\eta_1 - 4\eta_2 = 1$. Let $\eta_2 = k$, some arbitrary constant. Then $\eta_1 = k + 1/4$. It follows that a second solution is given by

$$\mathbf{x}^{(2)} = \begin{pmatrix} 1 \\ 1 \end{pmatrix} t e^{-3t} + \begin{pmatrix} k + 1/4 \\ k \end{pmatrix} e^{-3t} = \begin{pmatrix} 1 \\ 1 \end{pmatrix} t e^{-3t} + \begin{pmatrix} 1/4 \\ 0 \end{pmatrix} e^{-3t} + k \begin{pmatrix} 1 \\ 1 \end{pmatrix} e^{-3t}.$$

Dropping the last term, the general solution is

$$\mathbf{x} = c_1 \begin{pmatrix} 1 \\ 1 \end{pmatrix} e^{-3t} + c_2 \left[\begin{pmatrix} 1 \\ 1 \end{pmatrix} t e^{-3t} + \begin{pmatrix} 1/4 \\ 0 \end{pmatrix} e^{-3t} \right].$$

Imposing the initial conditions, we require that $c_1 + c_2/4 = 3$, $c_1 = 2$, which results in $c_1 = 2$ and $c_2 = 4$. Therefore the solution of the IVP is

$$\mathbf{x} = \begin{pmatrix} 3 \\ 2 \end{pmatrix} e^{-3t} + \begin{pmatrix} 4 \\ 4 \end{pmatrix} t e^{-3t}.$$

(b)

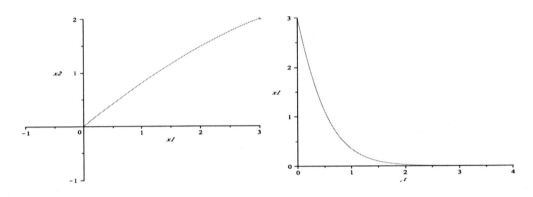

8.(a) Solution of the ODEs is based on the analysis of the algebraic equations

$$\begin{pmatrix} -\frac{5}{2} - r & \frac{3}{2} \\ -\frac{3}{2} & \frac{1}{2} - r \end{pmatrix} \begin{pmatrix} \xi_1 \\ \xi_2 \end{pmatrix} = \begin{pmatrix} 0 \\ 0 \end{pmatrix}.$$

The characteristic equation is $r^2 + 2r + 1 = 0$, with a single root $r = -1$. Setting $r = -1$, the two equations reduce to $-\xi_1 + \xi_2 = 0$. The corresponding eigenvector is $\boldsymbol{\xi} = (1,1)^T$. One solution is

$$\mathbf{x}^{(1)} = \begin{pmatrix} 1 \\ 1 \end{pmatrix} e^{-t}.$$

A second linearly independent solution is obtained by solving the system

$$\begin{pmatrix} -3/2 & 3/2 \\ -3/2 & 3/2 \end{pmatrix} \begin{pmatrix} \eta_1 \\ \eta_2 \end{pmatrix} = \begin{pmatrix} 1 \\ 1 \end{pmatrix}.$$

The equations reduce to the single equation $-3\eta_1 + 3\eta_2 = 2$. Let $\eta_1 = k$. We obtain $\eta_2 = 2/3 + k$, and a second linearly independent solution is

$$\mathbf{x}^{(2)} = \begin{pmatrix} 1 \\ 1 \end{pmatrix} t e^{-t} + \begin{pmatrix} k \\ 2/3 + k \end{pmatrix} e^{-t} = \begin{pmatrix} 1 \\ 1 \end{pmatrix} t e^{-t} + \begin{pmatrix} 0 \\ 2/3 \end{pmatrix} e^{-t} + k \begin{pmatrix} 1 \\ 1 \end{pmatrix} e^{-t}.$$

Dropping the last term, the general solution is

$$\mathbf{x} = c_1 \begin{pmatrix} 1 \\ 1 \end{pmatrix} e^{-t} + c_2 \left[\begin{pmatrix} 1 \\ 1 \end{pmatrix} t e^{-t} + \begin{pmatrix} 0 \\ 2/3 \end{pmatrix} e^{-t} \right].$$

Imposing the initial conditions, we find that $c_1 = 3$, $c_1 + 2c_2/3 = -1$, so that $c_1 = 3$ and $c_2 = -6$. Therefore the solution of the IVP is

$$\mathbf{x} = \begin{pmatrix} 3 \\ -1 \end{pmatrix} e^{-t} - \begin{pmatrix} 6 \\ 6 \end{pmatrix} t e^{-t}.$$

(b)

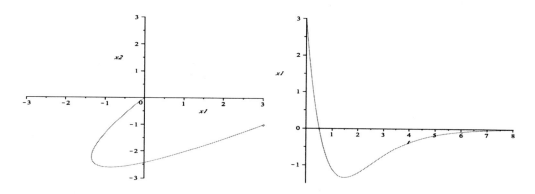

10.(a) The eigensystem is obtained from analysis of the equation

$$\begin{pmatrix} 3-r & 9 \\ -1 & -3-r \end{pmatrix} \begin{pmatrix} \xi_1 \\ \xi_2 \end{pmatrix} = \begin{pmatrix} 0 \\ 0 \end{pmatrix}.$$

The characteristic equation is $r^2 = 0$, with a single root $r = 0$. Setting $r = 0$, the two equations reduce to $\xi_1 + 3\xi_2 = 0$. The corresponding eigenvector is $\boldsymbol{\xi} = (-3, 1)^T$. Hence one solution is

$$\mathbf{x}^{(1)} = \begin{pmatrix} -3 \\ 1 \end{pmatrix},$$

which is a constant vector. A second linearly independent solution is obtained from the system

$$\begin{pmatrix} 3 & 9 \\ -1 & -3 \end{pmatrix} \begin{pmatrix} \eta_1 \\ \eta_2 \end{pmatrix} = \begin{pmatrix} -3 \\ 1 \end{pmatrix}.$$

The equations reduce to the single equation $\eta_1 + 3\eta_2 = -1$. Let $\eta_2 = k$. We obtain $\eta_1 = -1 - 3k$, and a second linearly independent solution is

$$\mathbf{x}^{(2)} = \begin{pmatrix} -3 \\ 1 \end{pmatrix} t + \begin{pmatrix} -1 - 3k \\ k \end{pmatrix} = \begin{pmatrix} -3 \\ 1 \end{pmatrix} t + \begin{pmatrix} -1 \\ 0 \end{pmatrix} + k \begin{pmatrix} -3 \\ 1 \end{pmatrix}.$$

Dropping the last term, the general solution is

$$\mathbf{x} = c_1 \begin{pmatrix} -3 \\ 1 \end{pmatrix} + c_2 \left[\begin{pmatrix} -3 \\ 1 \end{pmatrix} t + \begin{pmatrix} -1 \\ 0 \end{pmatrix} \right].$$

Imposing the initial conditions, we require that $-3c_1 - c_2 = 2$, $c_1 = 4$, which results in $c_1 = 4$ and $c_2 = -14$. Therefore the solution of the IVP is

$$\mathbf{x} = \begin{pmatrix} 2 \\ 4 \end{pmatrix} - 14 \begin{pmatrix} -3 \\ 1 \end{pmatrix} t.$$

(b)

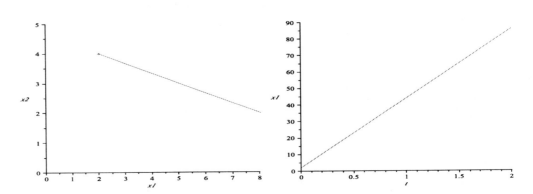

13. Setting $\mathbf{x} = \boldsymbol{\xi} t^r$ results in the algebraic equations
$$\begin{pmatrix} 3-r & -4 \\ 1 & -1-r \end{pmatrix} \begin{pmatrix} \xi_1 \\ \xi_2 \end{pmatrix} = \begin{pmatrix} 0 \\ 0 \end{pmatrix}.$$

The characteristic equation is $r^2 - 2r + 1 = 0$, with a single root of $r_{1,2} = 1$. With $r = 1$, the system reduces to a single equation $\xi_1 - 2\xi_2 = 0$. An eigenvector is given by $\boldsymbol{\xi} = (2, 1)^T$. Hence one solution is
$$\mathbf{x}^{(1)} = \begin{pmatrix} 2 \\ 1 \end{pmatrix} t.$$

In order to find a second linearly independent solution, we search for a generalized eigenvector whose components satisfy
$$\begin{pmatrix} 2 & -4 \\ 1 & -2 \end{pmatrix} \begin{pmatrix} \eta_1 \\ \eta_2 \end{pmatrix} = \begin{pmatrix} 2 \\ 1 \end{pmatrix}.$$

These equations reduce to $\eta_1 - 2\eta_2 = 1$. Let $\eta_2 = k$, some arbitrary constant. Then $\eta_1 = 1 + 2k$. (Before proceeding, note that if we set $u = \ln t$, the original equation is transformed into a constant coefficient equation with independent variable u. Recall that a second solution is obtained by multiplication of the first solution by the factor u. This implies that we must multiply first solution by a factor of $\ln t$.) Hence a second linearly independent solution is
$$\mathbf{x}^{(2)} = \begin{pmatrix} 2 \\ 1 \end{pmatrix} t \ln t + \begin{pmatrix} 1 + 2k \\ k \end{pmatrix} t = \begin{pmatrix} 2 \\ 1 \end{pmatrix} t \ln t + \begin{pmatrix} 1 \\ 0 \end{pmatrix} t + k \begin{pmatrix} 2 \\ 1 \end{pmatrix} t.$$

Dropping the last term, the general solution is
$$\mathbf{x} = c_1 \begin{pmatrix} 2 \\ 1 \end{pmatrix} t + c_2 \left[\begin{pmatrix} 2 \\ 1 \end{pmatrix} t \ln t + \begin{pmatrix} 1 \\ 0 \end{pmatrix} t \right].$$

16.(a) Using the result in Problem 15, the eigenvalues are
$$r_{1,2} = -\frac{1}{2RC} \pm \frac{\sqrt{L^2 - 4R^2CL}}{2RCL}.$$

The discriminant vanishes when $L = 4R^2C$.

(b) The system of differential equations is
$$\frac{d}{dt}\begin{pmatrix}I\\V\end{pmatrix}=\begin{pmatrix}0&\frac{1}{4}\\-1&-1\end{pmatrix}\begin{pmatrix}I\\V\end{pmatrix}.$$

The associated eigenvalue problem is
$$\begin{pmatrix}-r&\frac{1}{4}\\-1&-1-r\end{pmatrix}\begin{pmatrix}\xi_1\\\xi_2\end{pmatrix}=\begin{pmatrix}0\\0\end{pmatrix}.$$

The characteristic equation is $r^2+r+1/4=0$, with a single root of $r_{1,2}=-1/2$. Setting $r=-1/2$, the algebraic equations reduce to $2\xi_1+\xi_2=0$. An eigenvector is given by $\boldsymbol{\xi}=(1,-2)^T$. Hence one solution is

$$\begin{pmatrix}I\\V\end{pmatrix}^{(1)}=\begin{pmatrix}1\\-2\end{pmatrix}e^{-t/2}.$$

A second solution is obtained from a generalized eigenvector whose components satisfy

$$\begin{pmatrix}\frac{1}{2}&\frac{1}{4}\\-1&-\frac{1}{2}\end{pmatrix}\begin{pmatrix}\eta_1\\\eta_2\end{pmatrix}=\begin{pmatrix}1\\-2\end{pmatrix}.$$

It follows that $\eta_1=k$ and $\eta_2=4-2k$. A second linearly independent solution is

$$\begin{pmatrix}I\\V\end{pmatrix}^{(2)}=\begin{pmatrix}1\\-2\end{pmatrix}te^{-t/2}+\begin{pmatrix}k\\4-2k\end{pmatrix}e^{-t/2}=\begin{pmatrix}1\\-2\end{pmatrix}te^{-t/2}+\begin{pmatrix}0\\4\end{pmatrix}e^{-t/2}+k\begin{pmatrix}1\\-2\end{pmatrix}e^{-t/2}.$$

Dropping the last term, the general solution is

$$\begin{pmatrix}I\\V\end{pmatrix}=c_1\begin{pmatrix}1\\-2\end{pmatrix}e^{-t/2}+c_2\left[\begin{pmatrix}1\\-2\end{pmatrix}te^{-t/2}+\begin{pmatrix}0\\4\end{pmatrix}e^{-t/2}\right].$$

Imposing the initial conditions, we require that $c_1=1$, $-2c_1+4c_2=2$, which results in $c_1=1$ and $c_2=1$. Therefore the solution of the IVP is

$$\begin{pmatrix}I\\V\end{pmatrix}=\begin{pmatrix}1\\2\end{pmatrix}e^{-t/2}+\begin{pmatrix}1\\-2\end{pmatrix}te^{-t/2}.$$

19.(a) The eigensystem is obtained from analysis of the equation

$$\begin{pmatrix}5-r&-3&-2\\8&-5-r&-4\\-4&3&3-r\end{pmatrix}\begin{pmatrix}\xi_1\\\xi_2\\\xi_3\end{pmatrix}=\begin{pmatrix}0\\0\\0\end{pmatrix}.$$

The characteristic equation of the coefficient matrix is $r^3-3r^2+3r-1=0$, with a single root of multiplicity three, $r=1$. Setting $r=1$, we have

$$\begin{pmatrix}4&-3&-2\\8&-6&-4\\-4&3&2\end{pmatrix}\begin{pmatrix}\xi_1\\\xi_2\\\xi_3\end{pmatrix}=\begin{pmatrix}0\\0\\0\end{pmatrix}.$$

The system of algebraic equations reduces to a single equation
$$4\xi_1-3\xi_2-2\xi_3=0.$$

An eigenvector vector is given by $\boldsymbol{\xi}^{(1)} = (1,0,2)^T$. Since the last equation has two free variables, a second linearly independent eigenvector (associated with $r = 1$) is $\boldsymbol{\xi}^{(2)} = (0,2,-3)^T$. Therefore two solutions are obtained as

$$\mathbf{x}^{(1)} = \begin{pmatrix} 1 \\ 0 \\ 2 \end{pmatrix} e^t \text{ and } \mathbf{x}^{(2)} = \begin{pmatrix} 0 \\ 2 \\ -3 \end{pmatrix} e^t.$$

(b) It follows directly that $\mathbf{x}' = \boldsymbol{\xi} t e^t + \boldsymbol{\xi} e^t + \boldsymbol{\eta} e^t$. Hence the coefficient vectors must satisfy $\boldsymbol{\xi} t e^t + \boldsymbol{\xi} e^t + \boldsymbol{\eta} e^t = \mathbf{A}\boldsymbol{\xi} t e^t + \mathbf{A}\boldsymbol{\eta} e^t$. Rearranging the terms, we have

$$\boldsymbol{\xi} e^t = (\mathbf{A} - \mathbf{I})\boldsymbol{\xi} t e^t + (\mathbf{A} - \mathbf{I})\boldsymbol{\eta} e^t.$$

Given an eigenvector $\boldsymbol{\xi}$, it follows that $(\mathbf{A} - \mathbf{I})\boldsymbol{\xi} = \mathbf{0}$ and $(\mathbf{A} - \mathbf{I})\boldsymbol{\eta} = \boldsymbol{\xi}$.

(c) Clearly, $(\mathbf{A} - \mathbf{I})^2 \boldsymbol{\eta} = (\mathbf{A} - \mathbf{I})(\mathbf{A} - \mathbf{I})\boldsymbol{\eta} = (\mathbf{A} - \mathbf{I})\boldsymbol{\xi} = \mathbf{0}$. Also,

$$\begin{pmatrix} 4 & -3 & -2 \\ 8 & -6 & -4 \\ -4 & 3 & 2 \end{pmatrix} \begin{pmatrix} 4 & -3 & -2 \\ 8 & -6 & -4 \\ -4 & 3 & 2 \end{pmatrix} = \begin{pmatrix} 0 & 0 & 0 \\ 0 & 0 & 0 \\ 0 & 0 & 0 \end{pmatrix}.$$

(d) We get that

$$\boldsymbol{\xi} = (\mathbf{A} - \mathbf{I})\boldsymbol{\eta} = \begin{pmatrix} 4 & -3 & -2 \\ 8 & -6 & -4 \\ -4 & 3 & 2 \end{pmatrix} \begin{pmatrix} 0 \\ 0 \\ 1 \end{pmatrix} = \begin{pmatrix} -2 \\ -4 \\ 2 \end{pmatrix}.$$

This is an eigenvector:

$$\begin{pmatrix} 5 & -3 & -2 \\ 8 & -5 & -4 \\ -4 & 3 & 3 \end{pmatrix} \begin{pmatrix} -2 \\ -4 \\ 2 \end{pmatrix} = \begin{pmatrix} -2 \\ -4 \\ 2 \end{pmatrix}.$$

(e) Given the three linearly independent solutions, a fundamental matrix is given by

$$\boldsymbol{\Psi}(t) = \begin{pmatrix} e^t & 0 & -2t\, e^t \\ 0 & 2e^t & -4t\, e^t \\ 2e^t & -3e^t & 2t\, e^t + e^t \end{pmatrix}.$$

(f) We construct the transformation matrix

$$\mathbf{T} = \begin{pmatrix} 1 & -2 & 0 \\ 0 & -4 & 0 \\ 2 & 2 & 1 \end{pmatrix},$$

with inverse

$$\mathbf{T}^{-1} = \begin{pmatrix} 1 & -1/2 & 0 \\ 0 & -1/4 & 0 \\ -2 & 3/2 & 1 \end{pmatrix}.$$

The Jordan form of the matrix $\mathbf{A}$ is

$$\mathbf{J} = \mathbf{T}^{-1}\mathbf{A}\mathbf{T} = \begin{pmatrix} 1 & 0 & 0 \\ 0 & 1 & 1 \\ 0 & 0 & 1 \end{pmatrix}.$$

21.(a) Direct multiplication results in

$$\mathbf{J}^2 = \begin{pmatrix} \lambda^2 & 0 & 0 \\ 0 & \lambda^2 & 2\lambda \\ 0 & 0 & \lambda^2 \end{pmatrix}, \mathbf{J}^3 = \begin{pmatrix} \lambda^3 & 0 & 0 \\ 0 & \lambda^3 & 3\lambda^2 \\ 0 & 0 & \lambda^3 \end{pmatrix}, \mathbf{J}^4 = \begin{pmatrix} \lambda^4 & 0 & 0 \\ 0 & \lambda^4 & 4\lambda^3 \\ 0 & 0 & \lambda^4 \end{pmatrix}.$$

(b) Suppose that

$$\mathbf{J}^n = \begin{pmatrix} \lambda^n & 0 & 0 \\ 0 & \lambda^n & n\lambda^{n-1} \\ 0 & 0 & \lambda^n \end{pmatrix}.$$

Then

$$\mathbf{J}^{n+1} = \begin{pmatrix} \lambda^n & 0 & 0 \\ 0 & \lambda^n & n\lambda^{n-1} \\ 0 & 0 & \lambda^n \end{pmatrix} \begin{pmatrix} \lambda & 0 & 0 \\ 0 & \lambda & 1 \\ 0 & 0 & \lambda \end{pmatrix} = \begin{pmatrix} \lambda \cdot \lambda^n & 0 & 0 \\ 0 & \lambda \cdot \lambda^n & \lambda^n + n\lambda \cdot \lambda^{n-1} \\ 0 & 0 & \lambda \cdot \lambda^n \end{pmatrix}.$$

Hence the result follows by mathematical induction.

(c) Note that $\mathbf{J}$ is block diagonal. Hence each block may be exponentiated. Using the result in Problem 20,

$$e^{\mathbf{J}t} = \begin{pmatrix} e^{\lambda t} & 0 & 0 \\ 0 & e^{\lambda t} & t e^{\lambda t} \\ 0 & 0 & e^{\lambda t} \end{pmatrix}.$$

(d) Setting $\lambda = 1$, and using the transformation matrix $\mathbf{T}$ in Problem 19,

$$\mathbf{T} e^{\mathbf{J}t} = \begin{pmatrix} 1 & 2 & 0 \\ 0 & 4 & 0 \\ 2 & -2 & -1 \end{pmatrix} \begin{pmatrix} e^t & 0 & 0 \\ 0 & e^t & t e^t \\ 0 & 0 & e^t \end{pmatrix} = \begin{pmatrix} e^t & 2e^t & 2t e^t \\ 0 & 4e^t & 4t e^t \\ 2e^t & -2e^t & -2t e^t - e^t \end{pmatrix}.$$

Based on the form of $\mathbf{J}$, $e^{\mathbf{J}t}$ is the fundamental matrix associated with the solutions

$$\mathbf{y}^{(1)} = \boldsymbol{\xi}^{(1)} e^t, \mathbf{y}^{(2)} = (2\boldsymbol{\xi}^{(1)} + 2\boldsymbol{\xi}^{(2)}) e^t \text{ and } \mathbf{y}^{(3)} = (2\boldsymbol{\xi}^{(1)} + 2\boldsymbol{\xi}^{(2)}) t e^t + \boldsymbol{\eta} e^t.$$

Hence the resulting matrix is the fundamental matrix associated with the solution set

$$\left\{ \boldsymbol{\xi}^{(1)} e^t, (2\boldsymbol{\xi}^{(1)} + 2\boldsymbol{\xi}^{(2)}) e^t, (2\boldsymbol{\xi}^{(1)} + 2\boldsymbol{\xi}^{(2)}) t e^t + \boldsymbol{\eta} e^t \right\},$$

as opposed to the solution set in Problem 19, given by

$$\left\{ \boldsymbol{\xi}^{(1)} e^t, \boldsymbol{\xi}^{(2)} e^t, (2\boldsymbol{\xi}^{(1)} + 2\boldsymbol{\xi}^{(2)}) t e^t + \boldsymbol{\eta} e^t \right\}.$$

22.(a) Direct multiplication results in

$$\mathbf{J}^2 = \begin{pmatrix} \lambda^2 & 2\lambda & 1 \\ 0 & \lambda^2 & 2\lambda \\ 0 & 0 & \lambda^2 \end{pmatrix}, \mathbf{J}^3 = \begin{pmatrix} \lambda^3 & 3\lambda^2 & 3\lambda \\ 0 & \lambda^3 & 3\lambda^2 \\ 0 & 0 & \lambda^3 \end{pmatrix}, \mathbf{J}^4 = \begin{pmatrix} \lambda^4 & 4\lambda^3 & 6\lambda^2 \\ 0 & \lambda^4 & 4\lambda^3 \\ 0 & 0 & \lambda^4 \end{pmatrix}.$$

(b) Suppose that
$$\mathbf{J}^n = \begin{pmatrix} \lambda^n & n\lambda^{n-1} & \frac{n(n-1)}{2}\lambda^{n-2} \\ 0 & \lambda^n & n\lambda^{n-1} \\ 0 & 0 & \lambda^n \end{pmatrix}.$$

Then
$$\mathbf{J}^{n+1} = \begin{pmatrix} \lambda^n & n\lambda^{n-1} & \frac{n(n-1)}{2}\lambda^{n-2} \\ 0 & \lambda^n & n\lambda^{n-1} \\ 0 & 0 & \lambda^n \end{pmatrix} \begin{pmatrix} \lambda & 1 & 0 \\ 0 & \lambda & 1 \\ 0 & 0 & \lambda \end{pmatrix}$$

$$= \begin{pmatrix} \lambda \cdot \lambda^n & \lambda^n + n\lambda \cdot \lambda^{n-1} & n\lambda^{n-1} + \frac{n(n-1)}{2}\lambda \cdot \lambda^{n-2} \\ 0 & \lambda \cdot \lambda^n & \lambda^n + n\lambda \cdot \lambda^{n-1} \\ 0 & 0 & \lambda \cdot \lambda^n \end{pmatrix}.$$

The result follows by noting that
$$n\lambda^{n-1} + \frac{n(n-1)}{2}\lambda \cdot \lambda^{n-2} = \left[n + \frac{n(n-1)}{2}\right]\lambda^{n-1} = \frac{n^2+n}{2}\lambda^{n-1}.$$

(c) We first observe that
$$\sum_{n=0}^{\infty} \lambda^n \frac{t^n}{n!} = e^{\lambda t}$$

$$\sum_{n=0}^{\infty} n\lambda^{n-1} \frac{t^n}{n!} = t\sum_{n=1}^{\infty} \lambda^{n-1} \frac{t^{n-1}}{(n-1)!} = t e^{\lambda t}$$

$$\sum_{n=0}^{\infty} \frac{n(n-1)}{2}\lambda^{n-2} \frac{t^n}{n!} = \frac{t^2}{2}\sum_{n=2}^{\infty} \lambda^{n-2} \frac{t^{n-2}}{(n-2)!} = \frac{t^2}{2}e^{\lambda t}.$$

Therefore
$$e^{\mathbf{J}t} = \begin{pmatrix} e^{\lambda t} & te^{\lambda t} & \frac{t^2}{2}e^{\lambda t} \\ 0 & e^{\lambda t} & te^{\lambda t} \\ 0 & 0 & e^{\lambda t} \end{pmatrix}.$$

(d) Setting $\lambda = 2$, and using the transformation matrix $\mathbf{T}$ in Problem 18,
$$\mathbf{T}e^{\mathbf{J}t} = \begin{pmatrix} 0 & 1 & 2 \\ 1 & 1 & 0 \\ -1 & 0 & 3 \end{pmatrix} \begin{pmatrix} e^{2t} & te^{2t} & \frac{t^2}{2}e^{2t} \\ 0 & e^{2t} & te^{2t} \\ 0 & 0 & e^{2t} \end{pmatrix} = \begin{pmatrix} 0 & e^{2t} & te^{2t} + 2e^{2t} \\ e^{2t} & te^{2t} + e^{2t} & \frac{t^2}{2}e^{2t} + te^{2t} \\ -e^{2t} & -te^{2t} & -\frac{t^2}{2}e^{2t} + 3e^{2t} \end{pmatrix}.$$

7.9

5. As shown in Problem 2, Section 7.8, the general solution of the homogeneous equation is
$$\mathbf{x}_c = c_1 \begin{pmatrix} 1 \\ 2 \end{pmatrix} + c_2 \begin{pmatrix} t \\ 2t - \frac{1}{2} \end{pmatrix}.$$

An associated fundamental matrix is

$$\Psi(t) = \begin{pmatrix} 1 & t \\ 2 & 2t - \frac{1}{2} \end{pmatrix}.$$

The inverse of the fundamental matrix is easily determined as

$$\Psi^{-1}(t) = \begin{pmatrix} 4t - 3 & -2t + 2 \\ 8t - 8 & -4t + 5 \end{pmatrix}.$$

We can now compute

$$\Psi^{-1}(t)\mathbf{g}(t) = -\frac{1}{t^3}\begin{pmatrix} 2t^2 + 4t - 1 \\ -2t - 4 \end{pmatrix},$$

and

$$\int \Psi^{-1}(t)\mathbf{g}(t)\,dt = \begin{pmatrix} -\frac{1}{2}t^{-2} + 4t^{-1} - 2\ln t \\ -2t^{-2} - 2t^{-1} \end{pmatrix}.$$

Finally,

$$\mathbf{v}(t) = \Psi(t)\int \Psi^{-1}(t)\mathbf{g}(t)\,dt,$$

where

$$v_1(t) = -\frac{1}{2}t^{-2} + 2t^{-1} - 2\ln t - 2, \qquad v_2(t) = 5t^{-1} - 4\ln t - 4.$$

Note that the vector $(2,4)^T$ is a multiple of one of the fundamental solutions. Hence we can write the general solution as

$$\mathbf{x} = c_1\begin{pmatrix} 1 \\ 2 \end{pmatrix} + c_2\begin{pmatrix} t \\ 2t - \frac{1}{2} \end{pmatrix} - \frac{1}{t^2}\begin{pmatrix} 1/2 \\ 0 \end{pmatrix} + \frac{1}{t}\begin{pmatrix} 2 \\ 5 \end{pmatrix} - 2\ln t \begin{pmatrix} 1 \\ 2 \end{pmatrix}.$$

6. The eigenvalues of the coefficient matrix are $r_1 = 0$ and $r_2 = -5$. It follows that the solution of the homogeneous equation is

$$\mathbf{x}_c = c_1\begin{pmatrix} 1 \\ 2 \end{pmatrix} + c_2\begin{pmatrix} -2e^{-5t} \\ e^{-5t} \end{pmatrix}.$$

The coefficient matrix is symmetric. Hence the system is diagonalizable. Using the normalized eigenvectors as columns, the transformation matrix, and its inverse, are

$$\mathbf{T} = \frac{1}{\sqrt{5}}\begin{pmatrix} 1 & -2 \\ 2 & 1 \end{pmatrix} \qquad \mathbf{T}^{-1} = \frac{1}{\sqrt{5}}\begin{pmatrix} 1 & 2 \\ -2 & 1 \end{pmatrix}.$$

Setting $\mathbf{x} = \mathbf{T}\mathbf{y}$, and $\mathbf{h}(t) = \mathbf{T}^{-1}\mathbf{g}(t)$, the transformed system is given, in scalar form, as

$$y_1' = \frac{5 + 8t}{\sqrt{5}\,t}$$

$$y_2' = -5y_2 + \frac{4}{\sqrt{5}}.$$

The solutions are readily obtained as

$$y_1(t) = \sqrt{5}\ln t + \frac{8}{\sqrt{5}}t + c_1 \text{ and } y_2(t) = c_2 e^{-5t} + \frac{4}{5\sqrt{5}}.$$

Transforming back to the original variables, we have $\mathbf{x} = \mathbf{T}\mathbf{y}$, with

$$\mathbf{x} = \frac{1}{\sqrt{5}}\begin{pmatrix} 1 & -2 \\ 2 & 1 \end{pmatrix}\begin{pmatrix} y_1(t) \\ y_2(t) \end{pmatrix} = \frac{1}{\sqrt{5}}\begin{pmatrix} 1 \\ 2 \end{pmatrix}y_1(t) + \frac{1}{\sqrt{5}}\begin{pmatrix} -2 \\ 1 \end{pmatrix}y_2(t).$$

Hence the general solution is

$$\mathbf{x} = k_1\begin{pmatrix} 1 \\ 2 \end{pmatrix} + k_2\begin{pmatrix} -2e^{-5t} \\ e^{-5t} \end{pmatrix} + \begin{pmatrix} 1 \\ 2 \end{pmatrix}\ln t + \frac{8}{5}\begin{pmatrix} 1 \\ 2 \end{pmatrix}t + \frac{4}{25}\begin{pmatrix} -2 \\ 1 \end{pmatrix}.$$

7. The solution of the homogeneous equation is

$$\mathbf{x}_c = c_1\begin{pmatrix} e^{-t} \\ -2e^{-t} \end{pmatrix} + c_2\begin{pmatrix} e^{3t} \\ 2e^{3t} \end{pmatrix}.$$

Based on the simple form of the right hand side, we use the method of undetermined coefficients. Set $\mathbf{v} = \mathbf{a}\, e^t$. Substitution into the ODE yields

$$\begin{pmatrix} a_1 \\ a_2 \end{pmatrix}e^t = \begin{pmatrix} 1 & 1 \\ 4 & 1 \end{pmatrix}\begin{pmatrix} a_1 \\ a_2 \end{pmatrix}e^t + \begin{pmatrix} 2 \\ -1 \end{pmatrix}e^t.$$

In scalar form, after canceling the exponential, we have

$$a_1 = a_1 + a_2 + 2$$
$$a_2 = 4a_1 + a_2 - 1,$$

with $a_1 = 1/4$ and $a_2 = -2$. Hence the particular solution is

$$\mathbf{v} = \begin{pmatrix} 1/4 \\ -2 \end{pmatrix}e^t,$$

so that the general solution is

$$\mathbf{x} = c_1\begin{pmatrix} e^{-t} \\ -2e^{-t} \end{pmatrix} + c_2\begin{pmatrix} e^{3t} \\ 2e^{3t} \end{pmatrix} + \frac{1}{4}\begin{pmatrix} e^t \\ -8e^t \end{pmatrix}.$$

9. Note that the coefficient matrix is symmetric. Hence the system is diagonalizable. The eigenvalues and eigenvectors are given by

$$r_1 = -\frac{1}{2},\ \boldsymbol{\xi}^{(1)} = \begin{pmatrix} 1 \\ 1 \end{pmatrix}\ \text{and}\ r_2 = -2,\ \boldsymbol{\xi}^{(2)} = \begin{pmatrix} 1 \\ -1 \end{pmatrix}.$$

Using the normalized eigenvectors as columns, the transformation matrix, and its inverse, are

$$\mathbf{T} = \frac{1}{\sqrt{2}}\begin{pmatrix} 1 & 1 \\ 1 & -1 \end{pmatrix} \quad \mathbf{T}^{-1} = \frac{1}{\sqrt{2}}\begin{pmatrix} 1 & 1 \\ 1 & -1 \end{pmatrix}.$$

Setting $\mathbf{x} = \mathbf{T}\mathbf{y}$, and $\mathbf{h}(t) = \mathbf{T}^{-1}\mathbf{g}(t)$, the transformed system is given, in scalar form, as

$$y_1' = -\frac{1}{2}y_1 + \sqrt{2}\,t + \frac{1}{\sqrt{2}}e^t$$
$$y_2' = -2y_2 + \sqrt{2}\,t - \frac{1}{\sqrt{2}}e^t.$$

Using any elementary method for first order linear equations, the solutions are

$$y_1(t) = k_1 e^{-t/2} + \frac{\sqrt{2}}{3} e^t - 4\sqrt{2} + 2\sqrt{2}\, t$$

$$y_2(t) = k_2 e^{-2t} - \frac{1}{3\sqrt{2}} e^t - \frac{1}{2\sqrt{2}} + \frac{1}{\sqrt{2}} t.$$

Transforming back to the original variables, $\mathbf{x} = \mathbf{Ty}$, the general solution is

$$\mathbf{x} = c_1 \begin{pmatrix} 1 \\ 1 \end{pmatrix} e^{-t/2} + c_2 \begin{pmatrix} 1 \\ -1 \end{pmatrix} e^{-2t} - \frac{1}{4}\begin{pmatrix} 17 \\ 15 \end{pmatrix} + \frac{1}{2}\begin{pmatrix} 5 \\ 3 \end{pmatrix} t + \frac{1}{6}\begin{pmatrix} 1 \\ 3 \end{pmatrix} e^t.$$

10. Since the coefficient matrix is symmetric, the differential equations can be decoupled. The eigenvalues and eigenvectors are given by

$$r_1 = -4,\ \boldsymbol{\xi}^{(1)} = \begin{pmatrix} \sqrt{2} \\ -1 \end{pmatrix} \text{ and } r_2 = -1,\ \boldsymbol{\xi}^{(2)} = \begin{pmatrix} 1 \\ \sqrt{2} \end{pmatrix}.$$

Using the normalized eigenvectors as columns, the transformation matrix, and its inverse, are

$$\mathbf{T} = \frac{1}{\sqrt{3}} \begin{pmatrix} \sqrt{2} & 1 \\ -1 & \sqrt{2} \end{pmatrix} \qquad \mathbf{T}^{-1} = \frac{1}{\sqrt{3}} \begin{pmatrix} \sqrt{2} & -1 \\ 1 & \sqrt{2} \end{pmatrix}.$$

Setting $\mathbf{x} = \mathbf{Ty}$, and $\mathbf{h}(t) = \mathbf{T}^{-1}\mathbf{g}(t)$, the transformed system is given, in scalar form, as

$$y_1' = -4y_1 + \frac{1}{\sqrt{3}}(1 + \sqrt{2})e^{-t}$$

$$y_2' = -y_2 + \frac{1}{\sqrt{3}}(1 - \sqrt{2})e^{-t}.$$

The solutions are easily obtained as

$$y_1(t) = k_1 e^{-4t} + \frac{1}{3\sqrt{3}}(1 + \sqrt{2})e^{-t}, \qquad y_2(t) = k_2 e^{-t} + \frac{1}{\sqrt{3}}(1 - \sqrt{2})t e^{-t}.$$

Transforming back to the original variables, the general solution is

$$\mathbf{x} = c_1 \begin{pmatrix} \sqrt{2} \\ -1 \end{pmatrix} e^{-4t} + c_2 \begin{pmatrix} 1 \\ \sqrt{2} \end{pmatrix} e^{-t} + \frac{1}{9}\begin{pmatrix} 2 + \sqrt{2} + 3\sqrt{3} \\ 3\sqrt{6} - \sqrt{2} - 1 \end{pmatrix} e^{-t} + \frac{1}{3}\begin{pmatrix} 1 - \sqrt{2} \\ \sqrt{2} - 2 \end{pmatrix} t e^{-t}.$$

Note that

$$\begin{pmatrix} 2 + \sqrt{2} + 3\sqrt{3} \\ 3\sqrt{6} - \sqrt{2} - 1 \end{pmatrix} = \begin{pmatrix} 2 + \sqrt{2} \\ -\sqrt{2} - 1 \end{pmatrix} + 3\sqrt{3}\begin{pmatrix} 1 \\ \sqrt{2} \end{pmatrix}.$$

The second vector is an eigenvector, hence the solution may be written as

$$\mathbf{x} = c_1 \begin{pmatrix} \sqrt{2} \\ -1 \end{pmatrix} e^{-4t} + c_2 \begin{pmatrix} 1 \\ \sqrt{2} \end{pmatrix} e^{-t} + \frac{1}{9}\begin{pmatrix} 2 + \sqrt{2} \\ -\sqrt{2} - 1 \end{pmatrix} e^{-t} + \frac{1}{3}\begin{pmatrix} 1 - \sqrt{2} \\ \sqrt{2} - 2 \end{pmatrix} t e^{-t}.$$

11. Based on the solution of Problem 3 of Section 7.6, a fundamental matrix is given by

$$\boldsymbol{\Psi}(t) = \begin{pmatrix} 5\cos t & 5\sin t \\ 2\cos t + \sin t & -\cos t + 2\sin t \end{pmatrix}.$$

The inverse of the fundamental matrix is easily determined as
$$\Psi^{-1}(t) = \frac{1}{5}\begin{pmatrix} \cos t - 2\sin t & 5\sin t \\ 2\cos t + \sin t & -5\cos t \end{pmatrix}.$$

It follows that
$$\Psi^{-1}(t)\mathbf{g}(t) = \begin{pmatrix} \cos t \sin t \\ -\cos^2 t \end{pmatrix},$$

and
$$\int \Psi^{-1}(t)\mathbf{g}(t)\,dt = \begin{pmatrix} \frac{1}{2}\sin^2 t \\ -\frac{1}{2}\cos t \sin t - \frac{1}{2}t \end{pmatrix}.$$

A particular solution is constructed as
$$\mathbf{v}(t) = \Psi(t)\int \Psi^{-1}(t)\mathbf{g}(t)\,dt,$$

where
$$v_1(t) = \frac{5}{2}\cos t \sin t - \cos^2 t + \frac{5}{2}t + 1, \qquad v_2(t) = \cos t \sin t - \frac{1}{2}\cos^2 t + t + \frac{1}{2}.$$

Hence the general solution is
$$\mathbf{x} = c_1\begin{pmatrix} 5\cos t \\ 2\cos t + \sin t \end{pmatrix} + c_2\begin{pmatrix} 5\sin t \\ -\cos t + 2\sin t \end{pmatrix} - t\sin t \begin{pmatrix} 5/2 \\ 1 \end{pmatrix} + t\cos t \begin{pmatrix} 0 \\ 1/2 \end{pmatrix} - \cos t \begin{pmatrix} 5/2 \\ 1 \end{pmatrix}.$$

13.(a) As shown in Problem 25 of Section 7.6, the solution of the homogeneous system is
$$\begin{pmatrix} x_1^{(c)} \\ x_2^{(c)} \end{pmatrix} = c_1 e^{-t/2}\begin{pmatrix} \cos(t/2) \\ 4\sin(t/2) \end{pmatrix} + c_2 e^{-t/2}\begin{pmatrix} \sin(t/2) \\ -4\cos(t/2) \end{pmatrix}.$$

Therefore the associated fundamental matrix is given by
$$\Psi(t) = e^{-t/2}\begin{pmatrix} \cos(t/2) & \sin(t/2) \\ 4\sin(t/2) & -4\cos(t/2) \end{pmatrix}.$$

(b) The inverse of the fundamental matrix is
$$\Psi^{-1}(t) = \frac{e^{t/2}}{4}\begin{pmatrix} 4\cos(t/2) & \sin(t/2) \\ 4\sin(t/2) & -\cos(t/2) \end{pmatrix}.$$

It follows that
$$\Psi^{-1}(t)\mathbf{g}(t) = \frac{1}{2}\begin{pmatrix} \cos(t/2) \\ \sin(t/2) \end{pmatrix},$$

and
$$\int \Psi^{-1}(t)\mathbf{g}(t)\,dt = \begin{pmatrix} \sin(t/2) \\ -\cos(t/2) \end{pmatrix}.$$

A particular solution is constructed as
$$\mathbf{v}(t) = \Psi(t)\int \Psi^{-1}(t)\mathbf{g}(t)\,dt,$$

where $v_1(t) = 0$, $v_2(t) = 4e^{-t/2}$. Hence the general solution is

$$\mathbf{x} = c_1 e^{-t/2}\begin{pmatrix} \cos(t/2) \\ 4\sin(t/2) \end{pmatrix} + c_2 e^{-t/2}\begin{pmatrix} \sin(t/2) \\ -4\cos(t/2) \end{pmatrix} + 4e^{-t/2}\begin{pmatrix} 0 \\ 1 \end{pmatrix}.$$

Imposing the initial conditions, we require that $c_1 = 0$, $-4c_2 + 4 = 0$, which results in $c_1 = 0$ and $c_2 = 1$. Therefore the solution of the IVP is

$$\mathbf{x} = e^{-t/2}\begin{pmatrix} \sin(t/2) \\ 4 - 4\cos(t/2) \end{pmatrix}.$$

15. The general solution of the homogeneous problem is

$$\begin{pmatrix} x_1^{(c)} \\ x_2^{(c)} \end{pmatrix} = c_1 \begin{pmatrix} 1 \\ 2 \end{pmatrix} t^{-1} + c_2 \begin{pmatrix} 2 \\ 1 \end{pmatrix} t^2,$$

which can be verified by substitution into the system of ODEs. Since the vectors are linearly independent, a fundamental matrix is given by

$$\mathbf{\Psi}(t) = \begin{pmatrix} t^{-1} & 2t^2 \\ 2t^{-1} & t^2 \end{pmatrix}.$$

The inverse of the fundamental matrix is

$$\mathbf{\Psi}^{-1}(t) = \frac{1}{3}\begin{pmatrix} -t & 2t \\ 2t^{-2} & -t^{-2} \end{pmatrix}.$$

Dividing both equations by t, we obtain

$$\mathbf{g}(t) = \begin{pmatrix} -2 \\ t^3 - t^{-1} \end{pmatrix}.$$

Proceeding with the method of variation of parameters,

$$\mathbf{\Psi}^{-1}(t)\mathbf{g}(t) = \begin{pmatrix} \frac{2}{3}t^4 + \frac{2}{3}t - \frac{2}{3} \\ -\frac{1}{3}t - \frac{4}{3}t^{-2} + \frac{1}{3}t^{-3} \end{pmatrix},$$

and

$$\int \mathbf{\Psi}^{-1}(t)\mathbf{g}(t)\,dt = \begin{pmatrix} \frac{2}{15}t^5 + \frac{1}{3}t^2 - \frac{2}{3}t \\ \frac{1}{6}t^2 + \frac{4}{3}t^{-1} - \frac{1}{6}t^{-2} \end{pmatrix}.$$

Hence a particular solution is obtained as

$$\mathbf{v} = \begin{pmatrix} -\frac{1}{5}t^4 + 3t - 1 \\ \frac{1}{10}t^4 + 2t - \frac{3}{2} \end{pmatrix}.$$

The general solution is

$$\mathbf{x} = c_1 \begin{pmatrix} 1 \\ 2 \end{pmatrix} t^{-1} + c_2 \begin{pmatrix} 2 \\ 1 \end{pmatrix} t^2 + \frac{1}{10}\begin{pmatrix} -2 \\ 1 \end{pmatrix} t^4 + \begin{pmatrix} 3 \\ 2 \end{pmatrix} t - \begin{pmatrix} 1 \\ 3/2 \end{pmatrix}.$$

16. Based on the hypotheses,

$$\boldsymbol{\phi}'(t) = \mathbf{P}(t)\boldsymbol{\phi}(t) + \mathbf{g}(t) \text{ and } \mathbf{v}'(t) = \mathbf{P}(t)\mathbf{v}(t) + \mathbf{g}(t).$$

Subtracting the two equations results in

$$\boldsymbol{\phi}'(t) - \mathbf{v}'(t) = \mathbf{P}(t)\boldsymbol{\phi}(t) - \mathbf{P}(t)\mathbf{v}(t),$$

that is,
$$[\phi(t) - \mathbf{v}(t)]' = \mathbf{P}(t)\left[\phi(t) - \mathbf{v}(t)\right].$$

It follows that $\phi(t) - \mathbf{v}(t)$ is a solution of the homogeneous equation. According to Theorem 7.4.2,
$$\phi(t) - \mathbf{v}(t) = c_1 \mathbf{x}^{(1)}(t) + c_2 \mathbf{x}^{(2)}(t) + \cdots + c_n \mathbf{x}^{(n)}(t).$$

Hence
$$\phi(t) = \mathbf{u}(t) + \mathbf{v}(t),$$
in which $\mathbf{u}(t)$ is the general solution of the homogeneous problem.

17.(a) Setting $t_0 = 0$ in Equation (34),
$$\mathbf{x} = \boldsymbol{\Phi}(t)\mathbf{x}^0 + \boldsymbol{\Phi}(t)\int_0^t \boldsymbol{\Phi}^{-1}(s)\mathbf{g}(s)ds = \boldsymbol{\Phi}(t)\mathbf{x}^0 + \int_0^t \boldsymbol{\Phi}(t)\boldsymbol{\Phi}^{-1}(s)\mathbf{g}(s)ds.$$

It was shown in Problem 15(c) in Section 7.7 that $\boldsymbol{\Phi}(t)\boldsymbol{\Phi}^{-1}(s) = \boldsymbol{\Phi}(t-s)$. Therefore
$$\mathbf{x} = \boldsymbol{\Phi}(t)\mathbf{x}^0 + \int_0^t \boldsymbol{\Phi}(t-s)\mathbf{g}(s)ds.$$

(b) The principal fundamental matrix is identified as $\boldsymbol{\Phi}(t) = e^{\mathbf{A}t}$. Hence
$$\mathbf{x} = e^{\mathbf{A}t}\mathbf{x}^0 + \int_0^t e^{\mathbf{A}(t-s)}\mathbf{g}(s)ds.$$

In Problem 27 of Section 3.6, the particular solution is given as
$$y(t) = \int_{t_0}^t K(t-s)g(s)ds,$$
in which the kernel $K(t)$ depends on the nature of the fundamental solutions.

18. Similarly to Eq.(43), here
$$(s\mathbf{I} - \mathbf{A})\mathbf{X}(s) = \mathbf{G}(s) + \begin{pmatrix} \alpha_1 \\ \alpha_2 \end{pmatrix},$$
where
$$\mathbf{G}(s) = \begin{pmatrix} 2/(s+1) \\ 3/s^2 \end{pmatrix} \quad \text{and} \quad s\mathbf{I} - \mathbf{A} = \begin{pmatrix} s+2 & -1 \\ -1 & s+2 \end{pmatrix}.$$

The transfer matrix is given by Eq.(46):
$$(s\mathbf{I} - \mathbf{A})^{-1} = \frac{1}{(s+1)(s+3)}\begin{pmatrix} s+2 & 1 \\ 1 & s+2 \end{pmatrix}.$$

From these equations we obtain that
$$\mathbf{X}(s) = \begin{pmatrix} \frac{2(s+2)}{(s+1)^2(s+3)} + \frac{3}{s^2(s+1)(s+3)} + \frac{\alpha_1(s+2)}{(s+1)(s+3)} + \frac{\alpha_2}{(s+1)(s+3)} \\ \frac{2}{(s+1)^2(s+3)} + \frac{3(s+2)}{s^2(s+1)(s+3)} + \frac{\alpha_1}{(s+1)(s+3)} + \frac{\alpha_2(s+2)}{(s+1)(s+3)} \end{pmatrix}.$$

The inverse Laplace transform gives us that
$$\mathbf{x}(t) = \begin{pmatrix} \frac{4+\alpha_1+\alpha_2}{2}e^{-t} + \frac{-4+3\alpha_1-3\alpha_2}{6}e^{-3t} + t + te^{-t} - \frac{4}{3} \\ \frac{2+\alpha_1+\alpha_2}{2}e^{-t} + \frac{4-3\alpha_1+3\alpha_2}{6}e^{-3t} + 2t + te^{-t} - \frac{5}{3} \end{pmatrix},$$
so α_1 and α_2 should be chosen so that
$$\frac{4+\alpha_1+\alpha_2}{2} = c_2 + \frac{1}{2} \text{ and } \frac{-4+3\alpha_1-3\alpha_2}{6} = c_1.$$
This gives us $\alpha_1 = (-5 + 6c_1 + 6c_2)/6$ and $\alpha_2 = -c_1 + c_2 - 13/6$.

CHAPTER 8

Numerical Methods

8.1

2. The Euler formula for this problem is $y_{n+1} = y_n + h(5t_n - 3\sqrt{y_n})$, in which $t_n = t_0 + nh$. Since $t_0 = 0$, we can also write $y_{n+1} = y_n + 5nh^2 - 3h\sqrt{y_n}$ with $y_0 = 2$.

(a) Euler method with $h = 0.05$:

	$n=2$	$n=4$	$n=6$	$n=8$
t_n	0.1	0.2	0.3	0.4
y_n	1.59980	1.29288	1.07242	0.930175

(b) Euler method with $h = 0.025$:

	$n=4$	$n=8$	$n=12$	$n=16$
t_n	0.1	0.2	0.3	0.4
y_n	1.61124	1.31361	1.10012	0.962552

The backward Euler formula is $y_{n+1} = y_n + h(5t_{n+1} - 3\sqrt{y_{n+1}})$ in which $t_n = t_0 + nh$. Since $t_0 = 0$, we can also write $y_{n+1} = y_n + 5(n+1)h^2 - 3h\sqrt{y_{n+1}}$, with $y_0 = 2$. Solving for y_{n+1}, and choosing the positive root, we find that

$$y_{n+1} = \left[-\frac{3}{2}h + \frac{1}{2}\sqrt{(20n+29)h^2 + 4y_n} \right]^2.$$

(c) Backward Euler method with $h = 0.05$:

	$n=2$	$n=4$	$n=6$	$n=8$
t_n	0.1	0.2	0.3	0.4
y_n	1.64337	1.37164	1.17763	1.05334

(d) Backward Euler method with $h = 0.025$:

	$n=4$	$n=8$	$n=12$	$n=16$
t_n	0.1	0.2	0.3	0.4
y_n	1.63301	1.35295	1.15267	1.02407

3. The Euler formula for this problem is $y_{n+1} = y_n + h(2y_n - 3t_n)$, in which $t_n = t_0 + nh$. Since $t_0 = 0$, $y_{n+1} = y_n + 2hy_n - 3nh^2$, with $y_0 = 1$.

(a) Euler method with $h = 0.05$:

	$n=2$	$n=4$	$n=6$	$n=8$
t_n	0.1	0.2	0.3	0.4
y_n	1.2025	1.41603	1.64289	1.88590

(b) Euler method with $h = 0.025$:

	$n=4$	$n=8$	$n=12$	$n=16$
t_n	0.1	0.2	0.3	0.4
y_n	1.20388	1.41936	1.64896	1.89572

The backward Euler formula is $y_{n+1} = y_n + h(2y_{n+1} - 3t_{n+1})$, in which $t_n = t_0 + nh$. Since $t_0 = 0$, we can also write $y_{n+1} = y_n + 2h\,y_{n+1} - 3(n+1)h^2$, with $y_0 = 1$. Solving for y_{n+1}, we find that

$$y_{n+1} = \frac{y_n - 3(n+1)h^2}{1 - 2h}.$$

(c) Backward Euler method with $h = 0.05$:

	$n=2$	$n=4$	$n=6$	$n=8$
t_n	0.1	0.2	0.3	0.4
y_n	1.20864	1.43104	1.67042	1.93076

(d) Backward Euler method with $h = 0.025$:

	$n=4$	$n=8$	$n=12$	$n=16$
t_n	0.1	0.2	0.3	0.4
y_n	1.20693	1.42683	1.66265	1.91802

4. The Euler formula is $y_{n+1} = y_n + h\left[2t_n + e^{-t_n y_n}\right]$. Since $t_n = t_0 + nh$ and $t_0 = 0$, we can also write $y_{n+1} = y_n + 2nh^2 + h e^{-nh y_n}$, with $y_0 = 1$.

(a) Euler method with $h = 0.05$:

	$n=2$	$n=4$	$n=6$	$n=8$
t_n	0.1	0.2	0.3	0.4
y_n	1.10244	1.21426	1.33484	1.46399

(b) Euler method with $h = 0.025$:

	$n=4$	$n=8$	$n=12$	$n=16$
t_n	0.1	0.2	0.3	0.4
y_n	1.10365	1.21656	1.33817	1.46832

The backward Euler formula is $y_{n+1} = y_n + h\left[2t_{n+1} + e^{-t_{n+1} y_{n+1}}\right]$. Since $t_0 = 0$ and $t_{n+1} = (n+1)h$, we can also write

$$y_{n+1} = y_n + 2h^2(n+1) + h e^{-(n+1)h y_{n+1}},$$

with $y_0 = 1$. This equation cannot be solved explicitly for y_{n+1}. At each step, given the current value of y_n, the equation must be solved numerically for y_{n+1}.

(c) Backward Euler method with $h = 0.05$:

	$n=2$	$n=4$	$n=6$	$n=8$
t_n	0.1	0.2	0.3	0.4
y_n	1.10720	1.22333	1.34797	1.48110

(d) Backward Euler method with $h = 0.025$:

	$n=4$	$n=8$	$n=12$	$n=16$
t_n	0.1	0.2	0.3	0.4
y_n	1.10603	1.22110	1.34473	1.47688

6. The Euler formula for this problem is $y_{n+1} = y_n + h(t_n^2 - y_n^2)\sin y_n$. Here $t_0 = 0$ and $t_n = nh$. So that $y_{n+1} = y_n + h(n^2 h^2 - y_n^2)\sin y_n$, with $y_0 = -1$.

(a) Euler method with $h = 0.05$:

	$n=2$	$n=4$	$n=6$	$n=8$
t_n	0.1	0.2	0.3	0.4
y_n	−0.920498	−0.857538	−0.808030	−0.770038

(b) Euler method with $h = 0.025$:

	$n = 4$	$n = 8$	$n = 12$	$n = 16$
t_n	0.1	0.2	0.3	0.4
y_n	-0.922575	-0.860923	-0.82300	-0.774965

The backward Euler formula is $y_{n+1} = y_n + h(t_{n+1}^2 - y_{n+1}^2)\sin y_{n+1}$. Since $t_0 = 0$ and $t_{n+1} = (n+1)h$, we can also write

$$y_{n+1} = y_n + h\left[(n+1)^2 h^2 - y_{n+1}^2\right] \sin y_{n+1},$$

with $y_0 = -1$. Note that this equation cannot be solved explicitly for y_{n+1}. Given y_n, the transcendental equation

$$y_{n+1} + h\, y_{n+1}^2 \sin y_{n+1} = y_n + h(n+1)^2 h^2$$

must be solved numerically for y_{n+1}.

(c) Backward Euler method with $h = 0.05$:

	$n = 2$	$n = 4$	$n = 6$	$n = 8$
t_n	0.1	0.2	0.3	0.4
y_n	-0.928059	-0.870054	-0.824021	-0.788686

(d) Backward Euler method with $h = 0.025$:

	$n = 4$	$n = 8$	$n = 12$	$n = 16$
t_n	0.1	0.2	0.3	0.4
y_n	-0.926341	-0.867163	-0.820279	-0.784275

8. The Euler formula $y_{n+1} = y_n + h(5 t_n - 3\sqrt{y_n})$, in which $t_n = t_0 + nh$. Since $t_0 = 0$, we can also write $y_{n+1} = y_n + 5nh^2 - 3h\sqrt{y_n}$ with $y_0 = 2$.

(a) Euler method with $h = 0.025$:

	$n = 20$	$n = 40$	$n = 60$	$n = 80$
t_n	0.5	1.0	1.5	2.0
y_n	0.891830	1.25225	2.37818	4.07257

(b) Euler method with $h = 0.0125$:

	$n = 40$	$n = 80$	$n = 120$	$n = 160$
t_n	0.5	1.0	1.5	2.0
y_n	0.908902	1.26872	2.39336	4.08799

The backward Euler formula is $y_{n+1} = y_n + h(5 t_{n+1} - 3\sqrt{y_{n+1}})$, in which $t_n = t_0 + nh$. Since $t_0 = 0$, we can also write $y_{n+1} = y_n + 5(n+1)h^2 - 3h\sqrt{y_{n+1}}$ with

$y_0 = 2$. Solving for y_{n+1}, and choosing the positive root, we find that

$$y_{n+1} = \left[-\frac{3}{2}h + \frac{1}{2}\sqrt{(20n+29)h^2 + 4y_n}\right]^2.$$

(c) Backward Euler method with $h = 0.025$:

	$n = 20$	$n = 40$	$n = 60$	$n = 80$
t_n	0.5	1.0	1.5	2.0
y_n	0.958565	1.31786	2.43924	4.13474

(d) Backward Euler method with $h = 0.0125$:

	$n = 40$	$n = 80$	$n = 120$	$n = 160$
t_n	0.5	1.0	1.5	2.0
y_n	0.942261	1.30153	2.24389	4.11908

10. The Euler formula is $y_{n+1} = y_n + h\left[2t_n + e^{-t_n y_n}\right]$. Since $t_n = t_0 + nh$ and $t_0 = 0$, we can also write $y_{n+1} = y_n + 2nh^2 + h e^{-nh y_n}$, with $y_0 = 1$.

(a) Euler method with $h = 0.025$:

	$n = 20$	$n = 40$	$n = 60$	$n = 80$
t_n	0.5	1.0	1.5	2.0
y_n	1.60729	2.46830	3.72167	5.45963

(b) Euler method with $h = 0.0125$:

	$n = 40$	$n = 80$	$n = 120$	$n = 160$
t_n	0.5	1.0	1.5	2.0
y_n	1.60996	2.47460	3.73356	5.47774

The backward Euler formula is $y_{n+1} = y_n + h\left[2t_{n+1} + e^{-t_{n+1} y_{n+1}}\right]$. Since $t_0 = 0$ and $t_{n+1} = (n+1)h$, we can also write

$$y_{n+1} = y_n + 2h^2(n+1) + h e^{-(n+1)h y_{n+1}},$$

with $y_0 = 1$. This equation cannot be solved explicitly for y_{n+1}. At each step, given the current value of y_n, the equation must be solved numerically for y_{n+1}.

(c) Backward Euler method with $h = 0.025$:

	$n = 20$	$n = 40$	$n = 60$	$n = 80$
t_n	0.5	1.0	1.5	2.0
y_n	1.61792	2.49356	3.76940	5.53223

(d) Backward Euler method with $h = 0.0125$:

	$n = 40$	$n = 80$	$n = 120$	$n = 160$
t_n	0.5	1.0	1.5	2.0
y_n	1.61528	2.48723	3.75742	5.51404

11. The Euler formula is $y_{n+1} = y_n + h(4 - t_n y_n)/(1 + y_n^2)$. Since $t_n = t_0 + nh$ and $t_0 = 0$, we can also write $y_{n+1} = y_n + (4h - nh^2 y_n)/(1 + y_n^2)$, with $y_0 = -2$.

(a) Euler method with $h = 0.025$:

	$n = 20$	$n = 40$	$n = 60$	$n = 80$
t_n	0.5	1.0	1.5	2.0
y_n	-1.45865	-0.217545	1.05715	1.41487

(b) Euler method with $h = 0.0125$:

	$n = 40$	$n = 80$	$n = 120$	$n = 160$
t_n	0.5	1.0	1.5	2.0
y_n	-1.45322	-0.180813	1.05903	1.41244

The backward Euler formula is $y_{n+1} = y_n + h(4 - t_{n+1} y_{n+1})/(1 + y_{n+1}^2)$. Since $t_0 = 0$ and $t_{n+1} = (n+1)h$, we can also write $y_{n+1}(1 + y_{n+1}^2) = y_n(1 + y_{n+1}^2) + [4h - (n+1)h^2 y_{n+1}]$, with $y_0 = -2$. This equation cannot be solved explicitly for y_{n+1}. At each step, given the current value of y_n, the equation must be solved numerically for y_{n+1}.

(c) Backward Euler method with $h = 0.025$:

	$n = 20$	$n = 40$	$n = 60$	$n = 80$
t_n	0.5	1.0	1.5	2.0
y_n	-1.43600	-0.0681657	1.06489	1.40575

(d) Backward Euler method with $h = 0.0125$:

	$n = 40$	$n = 80$	$n = 120$	$n = 160$
t_n	0.5	1.0	1.5	2.0
y_n	-1.44190	-0.105737	1.06290	1.40789

12. The Euler formula is $y_{n+1} = y_n + h(y_n^2 + 2 t_n y_n)/(3 + t_n^2)$. Since $t_n = t_0 + nh$ and $t_0 = 0$, we can also write $y_{n+1} = y_n + (h y_n^2 + 2 nh^2 y_n)/(3 + n^2 h^2)$, with $y_0 = 0.5$.

(a) Euler method with $h = 0.025$:

	$n = 20$	$n = 40$	$n = 60$	$n = 80$
t_n	0.5	1.0	1.5	2.0
y_n	0.587987	0.791589	1.14743	1.70973

(b) Euler method with $h = 0.0125$:

	$n = 40$	$n = 80$	$n = 120$	$n = 160$
t_n	0.5	1.0	1.5	2.0
y_n	0.589440	0.795758	1.15693	1.72955

The backward Euler formula is $y_{n+1} = y_n + h(y_{n+1}^2 + 2 t_{n+1} y_{n+1})/(3 + t_{n+1}^2)$. Since $t_0 = 0$ and $t_{n+1} = (n+1)h$, we can also write

$$y_{n+1}\left[3 + (n+1)^2 h^2\right] - h y_{n+1}^2 = y_n \left[3 + (n+1)^2 h^2\right] + 2(n+1) h^2 \, y_{n+1},$$

with $y_0 = 0.5$. Note that although this equation can be solved explicitly for y_{n+1}, it is also possible to use a numerical equation solver. At each time step, given the current value of y_n, the equation may be solved numerically for y_{n+1}.

(c) Backward Euler method with $h = 0.025$:

	$n = 20$	$n = 40$	$n = 60$	$n = 80$
t_n	0.5	1.0	1.5	2.0
y_n	0.593901	0.808716	1.18687	1.79291

(d) Backward Euler method with $h = 0.0125$:

	$n = 40$	$n = 80$	$n = 120$	$n = 160$
t_n	0.5	1.0	1.5	2.0
y_n	0.592396	0.804319	1.17664	1.77111

13. The Euler formula for this problem is $y_{n+1} = y_n + h(1 - t_n + 4 y_n)$, in which $t_n = t_0 + nh$. Since $t_0 = 0$, we can also write $y_{n+1} = y_n + h - nh^2 + 4h y_n$, with $y_0 = 1$. With $h = 0.01$, a total number of 200 iterations is needed to reach $t = 2$. With $h = 0.001$, a total of 2000 iterations are necessary.

14. The backward Euler formula is $y_{n+1} = y_n + h(1 - t_{n+1} + 4 y_{n+1})$. Since the equation is linear, we can solve for y_{n+1} in terms of y_n :

$$y_{n+1} = \frac{y_n + h - h t_{n+1}}{1 - 4h}.$$

Here $t_0 = 0$ and $y_0 = 1$. With $h = 0.01$, a total number of 200 iterations is needed to reach $t = 2$. With $h = 0.001$, a total of 2000 iterations are necessary.

18. Let $\phi(t)$ be a solution of the initial value problem. The local truncation error for the Euler method, on the interval $t_n \leq t \leq t_{n+1}$, is given by

$$e_{n+1} = \frac{1}{2}\phi''(\bar{t}_n)h^2,$$

where $t_n < \bar{t}_n < t_{n+1}$. Since $\phi'(t) = t^2 + [\phi(t)]^2$, it follows that

$$\phi''(t) = 2t + 2\phi(t)\phi'(t) = 2t + 2t^2\phi(t) + 2[\phi(t)]^3$$

and so

$$e_{n+1} = [\bar{t}_n + \bar{t}_n^2\phi(\bar{t}_n) + \phi^3(\bar{t}_n)]h^2$$

where $t_n < \bar{t}_n < t_{n+1}$.

20. Given that $\phi(t)$ is a solution of the initial value problem, the local truncation error for the Euler method, on the interval $t_n \leq t \leq t_{n+1}$, is

$$e_{n+1} = \frac{1}{2}\phi''(\bar{t}_n)h^2,$$

where $t_n < \bar{t}_n < t_{n+1}$. Based on the ODE, $\phi'(t) = \sqrt{t + \phi(t)}$, and hence

$$\phi''(t) = \frac{1 + \phi'(t)}{2\sqrt{t + \phi(t)}} = \frac{1}{2\sqrt{t + \phi(t)}} + \frac{1}{2}$$

and so

$$e_{n+1} = \frac{1}{4}\left[1 + \frac{1}{\sqrt{\bar{t}_n + \phi(\bar{t}_n)}}\right]h^2$$

where $t_n < \bar{t}_n < t_{n+1}$.

21. Let $\phi(t)$ be a solution of the initial value problem. The local truncation error for the Euler method, on the interval $t_n \leq t \leq t_{n+1}$, is given by

$$e_{n+1} = \frac{1}{2}\phi''(\bar{t}_n)h^2,$$

where $t_n < \bar{t}_n < t_{n+1}$. Since $\phi'(t) = 2t + e^{-t\phi(t)}$, it follows that

$$\phi''(t) = 2 - [\phi(t) + t\phi'(t)] \cdot e^{-t\phi(t)} = 2 - \left[\phi(t) + 2t^2 + te^{-t\phi(t)}\right] \cdot e^{-t\phi(t)}.$$

Hence

$$e_{n+1} = h^2 - \frac{h^2}{2}\left[\phi(\bar{t}_n) + 2\bar{t}_n^2 + \bar{t}_n e^{-\bar{t}_n\phi(\bar{t}_n)}\right] \cdot e^{-\bar{t}_n\phi(\bar{t}_n)}.$$

22.(a) Direct integration yields $\phi(t) = (1/5\pi)\sin 5\pi t + 1$.

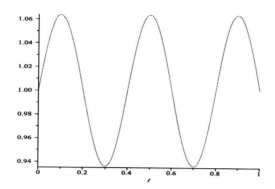

(b)

	$n=0$	$n=1$	$n=2$	$n=3$
t_n	0.0	0.2	0.4	0.6
y_n	1.0	1.2	1.0	1.2

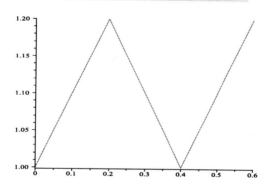

(c)

	$n=0$	$n=1$	$n=2$	$n=3$	$n=4$
t_n	0.0	0.1	0.2	0.3	0.4
y_n	1.0	1.1	1.1	1.0	1.0

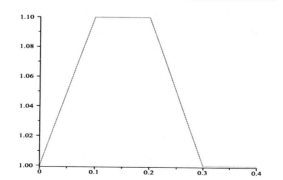

(d) Since $\phi''(t) = -5\pi \sin 5\pi t$, the local truncation error for the Euler method, on the interval $t_n \leq t \leq t_{n+1}$, is given by

$$e_{n+1} = -\frac{5\pi h^2}{2} \sin 5\pi \bar{t}_n.$$

In order to satisfy

$$|e_{n+1}| \leq \frac{5\pi h^2}{2} < 0.05,$$

it is necessary that

$$h < \frac{1}{\sqrt{50\pi}} \approx 0.08.$$

26.(a)
$$1000 \cdot \begin{vmatrix} 6.0 & 18 \\ 2.0 & 6.0 \end{vmatrix} = 1000 \cdot (0) = 0.$$

(b)
$$1000 \cdot \begin{vmatrix} 6.01 & 18.0 \\ 2.00 & 6.00 \end{vmatrix} = 1000(0.06) = 60.$$

(c)
$$1000 \cdot \begin{vmatrix} 6.010 & 18.04 \\ 2.004 & 6.000 \end{vmatrix} = 1000(-0.09216) = -92.16.$$

27. Rounding to three digits, $a(b-c) \approx 0.224$. Likewise, to three digits, $ab \approx 0.702$ and $ac \approx 0.477$. It follows that $ab - ac \approx 0.225$.

8.2

2. The improved Euler formula is

$$y_{n+1} = y_n + \frac{h}{2}(5t_n - 3\sqrt{y_n}) + \frac{h}{2}(5t_{n+1} - 3\sqrt{K_n}),$$

in which $K_n = y_n + h(5t_n - 3\sqrt{y_n})$. Since $t_n = t_0 + nh$ and $t_0 = 0$, we can also write

$$y_{n+1} = y_n + \frac{h}{2}(5nh - 3\sqrt{y_n}) + \frac{h}{2}\left[5(n+1)h - 3\sqrt{K_n}\right],$$

with $y_0 = 2$.

(a) $h = 0.05$:

	$n=2$	$n=4$	$n=6$	$n=8$
t_n	0.1	0.2	0.3	0.4
y_n	1.62283	1.33460	1.12820	0.995445

(b) $h = 0.025$:

	$n = 4$	$n = 8$	$n = 12$	$n = 16$
t_n	0.1	0.2	0.3	0.4
y_n	1.62243	1.33386	1.12718	0.994215

(c) $h = 0.0125$:

	$n = 8$	$n = 16$	$n = 24$	$n = 32$
t_n	0.1	0.2	0.3	0.4
y_n	1.62234	1.33368	1.12693	0.993921

3. The improved Euler formula for this problem is

$$y_{n+1} = y_n + \frac{h}{2}(4y_n - 3t_n - 3t_{n+1}) + h^2(2y_n - 3t_n).$$

Since $t_n = t_0 + nh$ and $t_0 = 0$, we can also write

$$y_{n+1} = y_n + 2h\,y_n + \frac{h^2}{2}(4y_n - 3 - 6n) - 3nh^3,$$

with $y_0 = 1$.

(a) $h = 0.05$:

	$n = 2$	$n = 4$	$n = 6$	$n = 8$
t_n	0.1	0.2	0.3	0.4
y_n	1.20526	1.42273	1.65511	1.90570

(b) $h = 0.025$:

	$n = 4$	$n = 8$	$n = 12$	$n = 16$
t_n	0.1	0.2	0.3	0.4
y_n	1.20533	1.42290	1.65542	1.90621

(c) $h = 0.0125$:

	$n = 8$	$n = 16$	$n = 24$	$n = 32$
t_n	0.1	0.2	0.3	0.4
y_n	1.20534	1.42294	1.65550	1.90634

5. The improved Euler formula is

$$y_{n+1} = y_n + h\frac{y_n^2 + 2t_n y_n}{2(3 + t_n^2)} + h\frac{K_n^2 + 2t_{n+1}K_n}{2(3 + t_{n+1}^2)},$$

in which

$$K_n = y_n + h\frac{y_n^2 + 2t_n y_n}{3 + t_n^2}.$$

Since $t_n = t_0 + nh$ and $t_0 = 0$, we can also write
$$y_{n+1} = y_n + h\frac{y_n^2 + 2nh\,y_n}{2(3 + n^2h^2)} + h\frac{K_n^2 + 2(n+1)hK_n}{2[3 + (n+1)^2h^2]},$$
with $y_0 = 0.5$.

(a) $h = 0.05$:

	$n = 2$	$n = 4$	$n = 6$	$n = 8$
t_n	0.1	0.2	0.3	0.4
y_n	0.510164	0.524126	0.54083	0.564251

(b) $h = 0.025$:

	$n = 4$	$n = 8$	$n = 12$	$n = 16$
t_n	0.1	0.2	0.3	0.4
y_n	0.510168	0.524135	0.542100	0.564277

(c) $h = 0.0125$:

	$n = 8$	$n = 16$	$n = 24$	$n = 32$
t_n	0.1	0.2	0.3	0.4
y_n	0.510169	0.524137	0.542104	0.564284

6. The improved Euler formula for this problem is
$$y_{n+1} = y_n + \frac{h}{2}(t_n^2 - y_n^2)\sin y_n + \frac{h}{2}(t_{n+1}^2 - K_n^2)\sin K_n,$$
in which
$$K_n = y_n + h\,(t_n^2 - y_n^2)\sin y_n\,.$$
Since $t_n = t_0 + nh$ and $t_0 = 0$, we can also write
$$y_{n+1} = y_n + \frac{h}{2}(n^2h^2 - y_n^2)\sin y_n + \frac{h}{2}\left[(n+1)^2h^2 - K_n^2\right]\sin K_n,$$
with $y_0 = -1$.

(a) $h = 0.05$:

	$n = 2$	$n = 4$	$n = 6$	$n = 8$
t_n	0.1	0.2	0.3	0.4
y_n	-0.924650	-0.864338	-0.816642	-0.780008

(b) $h = 0.025$:

	$n = 4$	$n = 8$	$n = 12$	$n = 16$
t_n	0.1	0.2	0.3	0.4
y_n	-0.924550	-0.864177	-0.816442	-0.779781

(c) $h = 0.0125$:

	$n = 8$	$n = 16$	$n = 24$	$n = 32$
t_n	0.1	0.2	0.3	0.4
y_n	-0.924525	-0.864138	-0.816393	-0.779725

7. The improved Euler formula for this problem is

$$y_{n+1} = y_n + \frac{h}{2}(4y_n - t_n - t_{n+1} + 1) + h^2(2y_n - t_n + 0.5).$$

Since $t_n = t_0 + nh$ and $t_0 = 0$, we can also write

$$y_{n+1} = y_n + h(2y_n + 0.5) + h^2(2y_n - n) - nh^3,$$

with $y_0 = 1$.

(a) $h = 0.025$:

	$n = 20$	$n = 40$	$n = 60$	$n = 80$
t_n	0.5	1.0	1.5	2.0
y_n	2.96719	7.88313	20.8114	55.5106

(b) $h = 0.0125$:

	$n = 40$	$n = 80$	$n = 120$	$n = 160$
t_n	0.5	1.0	1.5	2.0
y_n	2.96800	7.88755	20.8294	55.5758

8. The improved Euler formula is

$$y_{n+1} = y_n + \frac{h}{2}(5t_n - 3\sqrt{y_n}) + \frac{h}{2}(5t_{n+1} - 3\sqrt{K_n}),$$

in which $K_n = y_n + h(5t_n - 3\sqrt{y_n})$. Since $t_n = t_0 + nh$ and $t_0 = 0$, we can also write

$$y_{n+1} = y_n + \frac{h}{2}(5nh - 3\sqrt{y_n}) + \frac{h}{2}\left[5(n+1)h - 3\sqrt{K_n}\right],$$

with $y_0 = 2$.

(a) $h = 0.025$:

	$n = 20$	$n = 40$	$n = 60$	$n = 80$
t_n	0.5	1.0	1.5	2.0
y_n	0.926139	1.28558	2.40898	4.10386

(b) $h = 0.0125$:

	$n = 40$	$n = 80$	$n = 120$	$n = 160$
t_n	0.5	1.0	1.5	2.0
y_n	0.925815	1.28525	2.40869	4.10359

9. The improved Euler formula for this problem is

$$y_{n+1} = y_n + \frac{h}{2}\sqrt{t_n + y_n} + \frac{h}{2}\sqrt{t_{n+1} + K_n},$$

in which $K_n = y_n + h\sqrt{t_n + y_n}$. Since $t_n = t_0 + nh$ and $t_0 = 0$, we can also write

$$y_{n+1} = y_n + \frac{h}{2}\sqrt{nh + y_n} + \frac{h}{2}\sqrt{(n+1)h + K_n},$$

with $y_0 = 3$.

(a) $h = 0.025$:

	$n = 20$	$n = 40$	$n = 60$	$n = 80$
t_n	0.5	1.0	1.5	2.0
y_n	3.96217	5.10887	6.43134	7.92332

(b) $h = 0.0125$:

	$n = 40$	$n = 80$	$n = 120$	$n = 160$
t_n	0.5	1.0	1.5	2.0
y_n	3.96218	5.10889	6.43138	7.92337

10. The improved Euler formula is

$$y_{n+1} = y_n + \frac{h}{2}\left[2t_n + e^{-t_n y_n}\right] + \frac{h}{2}\left[2t_{n+1} + e^{-t_{n+1} K_n}\right],$$

in which $K_n = y_n + h\left[2t_n + e^{-t_n y_n}\right]$. Since $t_n = t_0 + nh$ and $t_0 = 0$, we can also write

$$y_{n+1} = y_n + \frac{h}{2}\left[2nh + e^{-nh y_n}\right] + \frac{h}{2}\left[2(n+1)h + e^{-(n+1)h K_n}\right],$$

with $y_0 = 1$.

(a) $h = 0.025$:

	$n = 20$	$n = 40$	$n = 60$	$n = 80$
t_n	0.5	1.0	1.5	2.0
y_n	1.61263	2.48097	3.74556	5.49595

(b) $h = 0.0125$:

	$n=40$	$n=80$	$n=120$	$n=160$
t_n	0.5	1.0	1.5	2.0
y_n	1.61263	2.48092	3.74550	5.49589

12. The improved Euler formula is

$$y_{n+1} = y_n + h\frac{y_n^2 + 2t_n y_n}{2(3+t_n^2)} + h\frac{K_n^2 + 2t_{n+1}K_n}{2(3+t_{n+1}^2)},$$

in which

$$K_n = y_n + h\frac{y_n^2 + 2t_n y_n}{3+t_n^2}.$$

Since $t_n = t_0 + nh$ and $t_0 = 0$, we can also write

$$y_{n+1} = y_n + h\frac{y_n^2 + 2nh\, y_n}{2(3+n^2h^2)} + h\frac{K_n^2 + 2(n+1)hK_n}{2\,[3+(n+1)^2h^2]},$$

with $y_0 = 0.5$.

(a) $h = 0.025$:

	$n=20$	$n=40$	$n=60$	$n=80$
t_n	0.5	1.0	1.5	2.0
y_n	0.590897	0.799950	1.16653	1.74969

(b) $h = 0.0125$:

	$n=40$	$n=80$	$n=120$	$n=160$
t_n	0.5	1.0	1.5	2.0
y_n	0.590906	0.799988	1.16663	1.74992

16. The exact solution of the initial value problem is $\phi(t) = \frac{1}{2} + \frac{1}{2}e^{2t}$. Based on the result in Problem 14(c), the local truncation error for a linear differential equation is

$$e_{n+1} = \frac{1}{6}\phi'''(\bar{t}_n)h^3,$$

where $t_n < \bar{t}_n < t_{n+1}$. Since $\phi'''(t) = 4e^{2t}$, the local truncation error is

$$e_{n+1} = \frac{2}{3}e^{2\bar{t}_n}h^3.$$

Furthermore, with $0 \leq \bar{t}_n \leq 1$,

$$|e_{n+1}| \leq \frac{2}{3}e^2 h^3.$$

It also follows that for $h = 0.1$,

$$|e_1| \leq \frac{2}{3}e^{0.2}(0.1)^3 = \frac{1}{1500}e^{0.2}.$$

Using the improved Euler method, with $h = 0.1$, we have $y_1 \approx 1.11000$. The exact value is given by $\phi(0.1) = 1.1107014$.

17. The exact solution of the initial value problem is given by $\phi(t) = \frac{1}{2}t + e^{2t}$. Using the modified Euler method, the local truncation error for a linear differential equation is
$$e_{n+1} = \frac{1}{6}\phi'''(\bar{t}_n)h^3,$$
where $t_n < \bar{t}_n < t_{n+1}$. Since $\phi'''(t) = 8\,e^{2t}$, the local truncation error is
$$e_{n+1} = \frac{4}{3}e^{2\bar{t}_n}h^3.$$
Furthermore, with $0 \leq \bar{t}_n \leq 1$, the local error is bounded by
$$|e_{n+1}| \leq \frac{4}{3}e^2 h^3.$$
It also follows that for $h = 0.1$,
$$|e_1| \leq \frac{4}{3}e^{0.2}(0.1)^3 = \frac{1}{750}e^{0.2}.$$
Using the improved Euler method, with $h = 0.1$, we have $y_1 \approx 1.27000$. The exact value is given by $\phi(0.1) = 1.271403$.

18. Using the Euler method, $y_1 = 1 + 0.1(0.5 - 0 + 2 \cdot 1) = 1.25$. Using the improved Euler method, $y_1 = 1 + 0.05(0.5 - 0 + 2) + 0.05(0.5 - 0.1 + 2.5) = 1.27$. The estimated error is $e_1 \approx 1.27 - 1.25 = 0.02$. The step size should be adjusted by a factor of $\sqrt{0.0025/0.02} \approx 0.354$. Hence the required step size is estimated as $h \approx (0.1)(0.36) = 0.036$.

20. Using the Euler method, $y_1 = 3 + 0.1\sqrt{0+3} = 3.173205$. Using the improved Euler method,
$$y_1 = 3 + 0.05\sqrt{0+3} + 0.05\sqrt{0.1 + 3.173205} = 3.177063.$$
The estimated error is $e_1 \approx 3.177063 - 3.173205 = 0.003858$. The step size should be adjusted by a factor of $\sqrt{0.0025/0.003858} \approx 0.805$. Hence the required step size is estimated as $h \approx (0.1)(0.805) = 0.0805$.

21. Using the Euler method,
$$y_1 = 0.5 + 0.1\frac{(0.5)^2 + 0}{3 + 0} = 0.508334$$
Using the improved Euler method,
$$y_1 = 0.5 + 0.05\frac{(0.5)^2 + 0}{3+0} + 0.05\frac{(0.508334)^2 + 2(0.1)(0.508334)}{3 + (0.1)^2} = 0.510148.$$
The estimated error is $e_1 \approx 0.510148 - 0.508334 = 0.0018$. The local truncation error is less than the given tolerance. The step size can be adjusted by a factor of $\sqrt{0.0025/0.0018} \approx 1.1785$. Hence it is possible to use a step size of
$$h \approx (0.1)(1.1785) \approx 0.117.$$

22. Assuming that the solution has continuous derivatives at least to the third order,
$$\phi(t_{n+1}) = \phi(t_n) + \phi'(t_n)h + \frac{\phi''(t_n)}{2!}h^2 + \frac{\phi'''(\bar{t}_n)}{3!}h^3,$$
where $t_n < \bar{t}_n < t_{n+1}$. Suppose that $y_n = \phi(t_n)$.

(a) The local truncation error is given by
$$e_{n+1} = \phi(t_{n+1}) - y_{n+1}.$$
The modified Euler formula is defined as
$$y_{n+1} = y_n + h f\left[t_n + \frac{1}{2}h, y_n + \frac{1}{2}h f(t_n, y_n)\right].$$
Observe that $\phi'(t_n) = f(t_n, \phi(t_n)) = f(t_n, y_n)$. It follows that
$$e_{n+1} = \phi(t_{n+1}) - y_{n+1} =$$
$$= h f(t_n, y_n) + \frac{\phi''(t_n)}{2!}h^2 + \frac{\phi'''(\bar{t}_n)}{3!}h^3 - h f\left[t_n + \frac{1}{2}h, y_n + \frac{1}{2}h f(t_n, y_n)\right].$$

(b) As shown in Problem 14(b),
$$\phi''(t_n) = f_t(t_n, y_n) + f_y(t_n, y_n) f(t_n, y_n).$$
Furthermore,
$$f\left[t_n + \frac{1}{2}h, y_n + \frac{1}{2}h f(t_n, y_n)\right] = f(t_n, y_n) + f_t(t_n, y_n)\frac{h}{2} + f_y(t_n, y_n) k +$$
$$+ \frac{1}{2!}\left[\frac{h^2}{4}f_{tt} + hk f_{ty} + k^2 f_{yy}\right]_{t=\xi, y=\eta},$$
in which $k = \frac{1}{2}h f(t_n, y_n)$ and $t_n < \xi < t_n + h/2$, $y_n < \eta < y_n + k$. Therefore
$$e_{n+1} = \frac{\phi'''(\bar{t}_n)}{3!}h^3 - \frac{h}{2!}\left[\frac{h^2}{4}f_{tt} + hk f_{ty} + k^2 f_{yy}\right]_{t=\xi, y=\eta}.$$
Note that each term in the brackets has a factor of h^2. Hence the local truncation error is proportional to h^3.

(c) If $f(t,y)$ is linear, then $f_{tt} = f_{ty} = f_{yy} = 0$, and
$$e_{n+1} = \frac{\phi'''(\bar{t}_n)}{3!}h^3.$$

23. The modified Euler formula for this problem is
$$y_{n+1} = y_n + h\left\{3 + t_n + \frac{1}{2}h - \left[y_n + \frac{1}{2}h(3 + t_n - y_n)\right]\right\}$$
$$= y_n + h(3 + t_n - y_n) + \frac{h^2}{2}(y_n - t_n - 2).$$

Since $t_n = t_0 + nh$ and $t_0 = 0$, we can also write
$$y_{n+1} = y_n + h(3 + nh - y_n) + \frac{h^2}{2}(y_n - nh - 2),$$
with $y_0 = 1$. Setting $h = 0.05$, we obtain the following values:

	$n=2$	$n=4$	$n=6$	$n=8$
t_n	0.1	0.2	0.3	0.4
y_n	1.19512	1.38120	1.55909	1.72956

25. The modified Euler formula is
$$y_{n+1} = y_n + h\left[2y_n - 3t_n - \frac{3}{2}h + h(2y_n - 3t_n)\right]$$
$$= y_n + h(2y_n - 3t_n) + \frac{h^2}{2}(4y_n - 6t_n - 3).$$

Since $t_n = t_0 + nh$ and $t_0 = 0$, we can also write
$$y_{n+1} = y_n + h(2y_n - 3nh) + \frac{h^2}{2}(4y_n - 6nh - 3),$$
with $y_0 = 1$. Setting $h = 0.05$, we obtain:

	$n=2$	$n=4$	$n=6$	$n=8$
t_n	0.1	0.2	0.3	0.4
y_n	1.20526	1.42273	1.65511	1.90570

26. The modified Euler formula for this problem is
$$y_{n+1} = y_n + h\left[2t_n + h + e^{-(t_n + \frac{h}{2})K_n}\right],$$
in which $K_n = y_n + (h/2)[2t_n + e^{-t_n y_n}]$. Now $t_n = t_0 + nh$, with $t_0 = 0$ and $y_0 = 1$. Setting $h = 0.05$, we obtain the following values:

	$n=2$	$n=4$	$n=6$	$n=8$
t_n	0.1	0.2	0.3	0.4
y_n	1.10485	1.21886	1.34149	1.47264

27. Let $f(t, y)$ be linear in both variables. The improved Euler formula is
$$y_{n+1} = y_n + \frac{1}{2}h\left[f(t_n, y_n) + f(t_n + h, y_n + hf(t_n, y_n))\right]$$
$$= y_n + \frac{1}{2}hf(t_n, y_n) + \frac{1}{2}hf(t_n, y_n) + \frac{1}{2}hf[h, hf(t_n, y_n)]$$
$$= y_n + hf(h, y_n) + \frac{1}{2}hf[h, hf(t_n, y_n)].$$

The modified Euler formula is
$$y_{n+1} = y_n + hf\left[t_n + \frac{1}{2}h, y_n + \frac{1}{2}hf(t_n, y_n)\right] = y_n + hf(t_n, y_n) + hf\left[\frac{1}{2}h, \frac{1}{2}hf(t_n, y_n)\right].$$

Since $f(t,y)$ is linear in both variables,
$$f\left[\frac{1}{2}h, \frac{1}{2}hf(t_n, y_n)\right] = \frac{1}{2}f\left[h, hf(t_n, y_n)\right].$$

8.3

1. The ODE is linear, with $f(t,y) = 3 + t - y$. The Runge-Kutta algorithm requires the evaluations

$$k_{n1} = f(t_n, y_n)$$
$$k_{n2} = f(t_n + \frac{1}{2}h, y_n + \frac{1}{2}hk_{n1})$$
$$k_{n3} = f(t_n + \frac{1}{2}h, y_n + \frac{1}{2}hk_{n2})$$
$$k_{n4} = f(t_n + h, y_n + hk_{n3}).$$

The next estimate is given as the weighted average

$$y_{n+1} = y_n + \frac{h}{6}(k_{n1} + 2k_{n2} + 2k_{n3} + k_{n4}).$$

(a) For $h = 0.1$:

	$n=1$	$n=2$	$n=3$	$n=4$
t_n	0.1	0.2	0.3	0.4
y_n	1.19516	1.38127	1.55918	1.72968

(b) For $h = 0.05$:

	$n=2$	$n=4$	$n=6$	$n=8$
t_n	0.1	0.2	0.3	0.4
y_n	1.19516	1.38127	1.55918	1.72968

The exact solution of the IVP is $y(t) = 2 + t - e^{-t}$.

2. In this problem, $f(t,y) = 5t - 3\sqrt{y}$. At each time step, the Runge-Kutta algorithm requires the evaluations

$$k_{n1} = f(t_n, y_n)$$
$$k_{n2} = f(t_n + \frac{1}{2}h, y_n + \frac{1}{2}hk_{n1})$$
$$k_{n3} = f(t_n + \frac{1}{2}h, y_n + \frac{1}{2}hk_{n2})$$
$$k_{n4} = f(t_n + h, y_n + hk_{n3}).$$

The next estimate is given as the weighted average

$$y_{n+1} = y_n + \frac{h}{6}(k_{n1} + 2k_{n2} + 2k_{n3} + k_{n4}).$$

(a) For $h = 0.1$:

	$n = 1$	$n = 2$	$n = 3$	$n = 4$
t_n	0.1	0.2	0.3	0.4
y_n	1.62231	1.33362	1.12686	0.993839

(b) For $h = 0.05$:

	$n = 2$	$n = 4$	$n = 6$	$n = 8$
t_n	0.1	0.2	0.3	0.4
y_n	1.62230	1.33362	1.12685	0.993826

The exact solution of the IVP is given implicitly as

$$\frac{1}{(2\sqrt{y} + 5t)^5(t - \sqrt{y})^2} = \frac{\sqrt{2}}{512}.$$

3. The ODE is linear, with $f(t, y) = 2y - 3t$. The Runge-Kutta algorithm requires the evaluations

$$k_{n1} = f(t_n, y_n)$$
$$k_{n2} = f(t_n + \frac{1}{2}h, y_n + \frac{1}{2}hk_{n1})$$
$$k_{n3} = f(t_n + \frac{1}{2}h, y_n + \frac{1}{2}hk_{n2})$$
$$k_{n4} = f(t_n + h, y_n + hk_{n3}).$$

The next estimate is given as the weighted average

$$y_{n+1} = y_n + \frac{h}{6}(k_{n1} + 2k_{n2} + 2k_{n3} + k_{n4}).$$

(a) For $h = 0.1$:

	$n = 1$	$n = 2$	$n = 3$	$n = 4$
t_n	0.1	0.2	0.3	0.4
y_n	1.20535	1.42295	1.65553	1.90638

(b) For $h = 0.05$:

	$n = 2$	$n = 4$	$n = 6$	$n = 8$
t_n	0.1	0.2	0.3	0.4
y_n	1.20535	1.42296	1.65553	1.90638

The exact solution of the IVP is $y(t) = e^{2t}/4 + 3t/2 + 3/4$.

5. In this problem, $f(t,y) = (y^2 + 2ty)/(3+t^2)$. The Runge-Kutta algorithm requires the evaluations

$$k_{n1} = f(t_n, y_n)$$
$$k_{n2} = f(t_n + \frac{1}{2}h, y_n + \frac{1}{2}hk_{n1})$$
$$k_{n3} = f(t_n + \frac{1}{2}h, y_n + \frac{1}{2}hk_{n2})$$
$$k_{n4} = f(t_n + h, y_n + hk_{n3}).$$

The next estimate is given as the weighted average

$$y_{n+1} = y_n + \frac{h}{6}(k_{n1} + 2k_{n2} + 2k_{n3} + k_{n4}).$$

(a) For $h = 0.1$:

	$n=1$	$n=2$	$n=3$	$n=4$
t_n	0.1	0.2	0.3	0.4
y_n	0.510170	0.524138	0.542105	0.564286

(b) For $h = 0.05$:

	$n=2$	$n=4$	$n=6$	$n=8$
t_n	0.1	0.2	0.3	0.4
y_n	0.520169	0.524138	0.542105	0.564286

The exact solution of the IVP is $y(t) = (3+t^2)/(6-t)$.

6. In this problem, $f(t,y) = (t^2 - y^2)\sin y$. At each time step, the Runge-Kutta algorithm requires the evaluations

$$k_{n1} = f(t_n, y_n)$$
$$k_{n2} = f(t_n + \frac{1}{2}h, y_n + \frac{1}{2}hk_{n1})$$
$$k_{n3} = f(t_n + \frac{1}{2}h, y_n + \frac{1}{2}hk_{n2})$$
$$k_{n4} = f(t_n + h, y_n + hk_{n3}).$$

The next estimate is given as the weighted average

$$y_{n+1} = y_n + \frac{h}{6}(k_{n1} + 2k_{n2} + 2k_{n3} + k_{n4}).$$

(a) For $h = 0.1$:

	$n=1$	$n=2$	$n=3$	$n=4$
t_n	0.1	0.2	0.3	0.4
y_n	-0.924517	-0.864125	-0.816377	-0.779706

(b) For $h = 0.05$:

	$n = 2$	$n = 4$	$n = 6$	$n = 8$
t_n	0.1	0.2	0.3	0.4
y_n	−0.924517	−0.864125	−0.816377	−0.779706

7.(a) For $h = 0.1$:

	$n = 5$	$n = 10$	$n = 15$	$n = 20$
t_n	0.5	1.0	1.5	2.0
y_n	2.96825	7.88889	20.8349	55.5957

(b) For $h = 0.05$:

	$n = 10$	$n = 20$	$n = 30$	$n = 40$
t_n	0.5	1.0	1.5	2.0
y_n	2.96828	7.88904	20.8355	55.5980

The exact solution of the IVP is $y(t) = e^{2t} + t/2$.

8. See Problem 2 for the exact solution.

(a) For $h = 0.1$:

	$n = 5$	$n = 10$	$n = 15$	$n = 20$
t_n	0.5	1.0	1.5	2.0
y_n	0.925725	1.28516	2.40860	4.10350

(b) For $h = 0.05$:

	$n = 10$	$n = 20$	$n = 30$	$n = 40$
t_n	0.5	1.0	1.5	2.0
y_n	0.925711	1.28515	2.40860	4.10350

9.(a) For $h = 0.1$:

	$n = 5$	$n = 10$	$n = 15$	$n = 20$
t_n	0.5	1.0	1.5	2.0
y_n	3.96219	5.10890	6.43139	7.92338

(b) For $h = 0.05$:

	$n = 10$	$n = 20$	$n = 30$	$n = 40$
t_n	0.5	1.0	1.5	2.0
y_n	3.96219	5.10890	6.43139	7.92338

8.3

The exact solution is given implicitly as

$$\ln\left[\frac{2}{y+t-1}\right] + 2\sqrt{t+y} - 2\operatorname{arctanh}\sqrt{t+y} = t + 2\sqrt{3} - 2\operatorname{arctanh}\sqrt{3}.$$

10. See Problem 4.

(a) For $h = 0.1$:

	$n=5$	$n=10$	$n=15$	$n=20$
t_n	0.5	1.0	1.5	2.0
y_n	1.61262	2.48091	3.74548	5.49587

(b) For $h = 0.05$:

	$n=10$	$n=20$	$n=30$	$n=40$
t_n	0.5	1.0	1.5	2.0
y_n	1.61262	2.48091	3.74548	5.49587

12. See Problem 5 for the exact solution.

(a) For $h = 0.1$:

	$n=5$	$n=10$	$n=15$	$n=20$
t_n	0.5	1.0	1.5	2.0
y_n	0.590909	0.800000	1.166667	1.75000

(b) For $h = 0.05$:

	$n=10$	$n=20$	$n=30$	$n=40$
t_n	0.5	1.0	1.5	2.0
y_n	0.590909	0.800000	1.166667	1.75000

13. The ODE is linear, with $f(t,y) = 1 - t + 4y$. The Runge-Kutta algorithm requires the evaluations

$$k_{n1} = f(t_n, y_n)$$
$$k_{n2} = f(t_n + \tfrac{1}{2}h, y_n + \tfrac{1}{2}hk_{n1})$$
$$k_{n3} = f(t_n + \tfrac{1}{2}h, y_n + \tfrac{1}{2}hk_{n2})$$
$$k_{n4} = f(t_n + h, y_n + hk_{n3}).$$

The next estimate is given as the weighted average

$$y_{n+1} = y_n + \frac{h}{6}(k_{n1} + 2k_{n2} + 2k_{n3} + k_{n4}).$$

The exact solution of the IVP is $y(t) = 19e^{4t}/16 + t/4 - 3/16$.

(a) For $h = 0.1$:

	$n = 5$	$n = 10$	$n = 15$	$n = 20$
t_n	0.5	1.0	1.5	2.0
y_n	8.7093175	64.858107	478.81928	3535.8667

(b) For $h = 0.05$:

	$n = 10$	$n = 20$	$n = 30$	$n = 40$
t_n	0.5	1.0	1.5	2.0
y_n	8.7118060	64.894875	479.22674	3539.8804

15.(a)

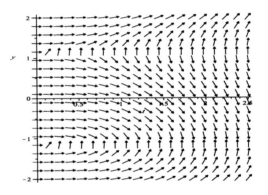

(b) For the integral curve starting at $(0,0)$, the slope turns infinite near $t_M \approx 1.5$. Note that the exact solution of the IVP is defined implicitly as

$$y^3 - 4y = t^3.$$

(c) Using the classic Runge-Kutta algorithm with the given values of h, we obtain the values

$h = 0.1$

	$n=12$	$n=13$	$n=14$	$n=15$	$n=16$
t_n	1.2	1.3	1.4	1.5	1.6
y_n	-0.45566	-0.60448	-0.82806	-1.73807	-1.56618

$h = 0.05$

	$n=26$	$n=28$	$n=30$	$n=32$	$n=34$
t_n	1.3	1.4	1.5	1.6	1.7
y_n	-0.60447	-0.82786	-1.06266	-1.42862	-1.17608

$h = 0.025$

	$n=54$	$n=56$	$n=58$	$n=60$	$n=62$
t_n	1.35	1.4	1.45	1.5	1.55
y_n	-0.70134	-0.82783	-1.05986	-1.49336	-1.30986

$h = 0.01$

	$n=142$	$n=143$	$n=144$	$n=145$	$n=146$
t_n	1.42	1.43	1.44	1.45	1.46
y_n	-0.89513	-0.93617	-0.98653	-1.05951	-0.76554

Based on the direction field, the solution should decrease monotonically to the limiting value $y = -2/\sqrt{3}$. In the following table, the value of t_M corresponds to the approximate time in the iteration process that the calculated values begin to increase.

h	t_M
0.1	1.55
0.05	1.65
0.025	1.525
0.01	1.455

(d) Numerical values will continue to be generated, although they will not be associated with the integral curve starting at $(0,0)$. These values are approximations to nearby integral curves.

(e) We consider the solution associated with the initial condition $y(0) = 1$. The exact solution is given by
$$y^3 - 4y = t^3 - 3.$$

For the integral curve starting at $(0,1)$, the slope becomes infinite near $t_M \approx 2.0$. In the following table, the values of t_M corresponds to the approximate time in the iteration process that the calculated values begin to increase.

h	t_M
0.1	1.85
0.05	1.85
0.025	1.86
0.01	1.835

8.4

1.(a) Using the notation $f_n = f(t_n, y_n)$, the predictor formula is

$$y_{n+1} = y_n + \frac{h}{24}(55 f_n - 59 f_{n-1} + 37 f_{n-2} - 9 f_{n-3}).$$

With $f_{n+1} = f(t_{n+1}, y_{n+1})$, the corrector formula is

$$y_{n+1} = y_n + \frac{h}{24}(9 f_{n+1} + 19 f_n - 5 f_{n-1} + f_{n-2}).$$

We use the starting values generated by the Runge-Kutta method:

	$n=0$	$n=1$	$n=2$	$n=3$
t_n	0.0	0.1	0.2	0.3
y_n	1.0	1.19516	1.38127	1.55918

	$n=4$ (pre)	$n=4$ (cor)	$n=5$ (pre)	$n=5$ (cor)
t_n	0.4	0.4	0.5	0.5
y_n	1.72967690	1.72986801	1.89346436	1.89346973

(b) With $f_{n+1} = f(t_{n+1}, y_{n+1})$, the fourth order Adams-Moulton formula is

$$y_{n+1} = y_n + \frac{h}{24}(9 f_{n+1} + 19 f_n - 5 f_{n-1} + f_{n-2}).$$

In this problem, $f_{n+1} = 3 + t_{n+1} - y_{n+1}$. Since the ODE is linear, we can solve for

$$y_{n+1} = \frac{1}{24 + 9h}[24 y_n + 27 h + 9h\, t_{n+1} + h(19 f_n - 5 f_{n-1} + f_{n-2})].$$

	$n=4$	$n=5$
t_n	0.4	0.5
y_n	1.7296800	1.8934695

(c) The fourth order backward differentiation formula is

$$y_{n+1} = \frac{1}{25}[48 y_n - 36 y_{n-1} + 16 y_{n-2} - 3 y_{n-3} + 12h f_{n+1}].$$

In this problem, $f_{n+1} = 3 + t_{n+1} - y_{n+1}$. Since the ODE is linear, we can solve for

$$y_{n+1} = \frac{1}{25 + 12h}[36h + 12h\,t_{n+1} + 48y_n - 36y_{n-1} + 16y_{n-2} - 3y_{n-3}].$$

	$n = 4$	$n = 5$
t_n	0.4	0.5
y_n	1.7296805	1.8934711

The exact solution of the IVP is $y(t) = 2 + t - e^{-t}$.

2.(a) Using the notation $f_n = f(t_n, y_n)$, the predictor formula is

$$y_{n+1} = y_n + \frac{h}{24}(55\,f_n - 59\,f_{n-1} + 37\,f_{n-2} - 9\,f_{n-3}).$$

With $f_{n+1} = f(t_{n+1}, y_{n+1})$, the corrector formula is

$$y_{n+1} = y_n + \frac{h}{24}(9\,f_{n+1} + 19\,f_n - 5\,f_{n-1} + f_{n-2}).$$

We use the starting values generated by the Runge-Kutta method:

	$n = 0$	$n = 1$	$n = 2$	$n = 3$
t_n	0.0	0.1	0.2	0.3
y_n	2.0	1.62231	1.33362	1.12686

	$n = 4$ (pre)	$n = 4$ (cor)	$n = 5$ (pre)	$n = 5$ (cor)
t_n	0.4	0.4	0.5	0.5
y_n	0.993751	0.993852	0.925469	0.925764

(b) With $f_{n+1} = f(t_{n+1}, y_{n+1})$, the fourth order Adams-Moulton formula is

$$y_{n+1} = y_n + \frac{h}{24}(9\,f_{n+1} + 19\,f_n - 5\,f_{n-1} + f_{n-2}).$$

In this problem, $f_{n+1} = 5t_{n+1} - 3\sqrt{y_{n+1}}$. Since the ODE is nonlinear, an equation solver is needed to approximate the solution of

$$y_{n+1} = y_n + \frac{h}{24}\left[45t_{n+1} - 27\sqrt{y_{n+1}} + 19\,f_n - 5\,f_{n-1} + f_{n-2}\right]$$

at each time step. We obtain the approximate values:

	$n = 4$	$n = 5$
t_n	0.4	0.5
y_n	0.993847	0.925746

(c) The fourth order backward differentiation formula is

$$y_{n+1} = \frac{1}{25}[48\,y_n - 36\,y_{n-1} + 16\,y_{n-2} - 3\,y_{n-3} + 12h f_{n+1}].$$

Since the ODE is nonlinear, an equation solver is used to approximate the solution of

$$y_{n+1} = \frac{1}{25}\left[48\,y_n - 36\,y_{n-1} + 16\,y_{n-2} - 3\,y_{n-3} + 12h(5t_{n+1} - 3\sqrt{y_{n+1}})\right]$$

at each time step.

	$n = 4$	$n = 5$
t_n	0.4	0.5
y_n	0.993869	0.925837

The exact solution of the IVP is given implicitly by

$$\frac{1}{(2\sqrt{y} + 5t)^5 (t - \sqrt{y})^2} = \frac{\sqrt{2}}{512}.$$

3.(a) The predictor formula is

$$y_{n+1} = y_n + \frac{h}{24}(55\,f_n - 59\,f_{n-1} + 37\,f_{n-2} - 9\,f_{n-3}).$$

With $f_{n+1} = f(t_{n+1}, y_{n+1})$, the corrector formula is

$$y_{n+1} = y_n + \frac{h}{24}(9\,f_{n+1} + 19\,f_n - 5\,f_{n-1} + f_{n-2}).$$

Using the starting values generated by the Runge-Kutta method:

	$n = 0$	$n = 1$	$n = 2$	$n = 3$
t_n	0.0	0.1	0.2	0.3
y_n	1.0	1.205350	1.422954	1.655527

	$n = 4$ (pre)	$n = 4$ (cor)	$n = 5$ (pre)	$n = 5$ (cor)
t_n	0.4	0.4	0.5	0.5
y_n	1.906340	1.906382	2.179455	2.179567

(b) With $f_{n+1} = f(t_{n+1}, y_{n+1})$, the fourth order Adams-Moulton formula is

$$y_{n+1} = y_n + \frac{h}{24}(9\,f_{n+1} + 19\,f_n - 5\,f_{n-1} + f_{n-2}).$$

In this problem, $f_{n+1} = 2\,y_{n+1} - 3\,t_{n+1}$. Since the ODE is linear, we can solve for

$$y_{n+1} = \frac{1}{24 - 18h}\left[24\,y_n - 27h\,t_{n+1} + h(19\,f_n - 5\,f_{n-1} + f_{n-2})\right].$$

	$n = 4$	$n = 5$
t_n	0.4	0.5
y_n	1.906385	2.179576

(c) The fourth order backward differentiation formula is

$$y_{n+1} = \frac{1}{25}\left[48\,y_n - 36\,y_{n-1} + 16\,y_{n-2} - 3\,y_{n-3} + 12h f_{n+1}\right].$$

In this problem, $f_{n+1} = 2\,y_{n+1} - 3\,t_{n+1}$. Since the ODE is linear, we can solve for
$$y_{n+1} = \frac{1}{25 - 24h}\left[48\,y_n - 36\,y_{n-1} + 16\,y_{n-2} - 3\,y_{n-3} - 36h\,t_{n+1}\right].$$

	$n = 4$	$n = 5$
t_n	0.4	0.5
y_n	1.906395	2.179611

The exact solution of the IVP is $y(t) = e^{2t}/4 + 3t/2 + 3/4$.

5.(a) The predictor formula is
$$y_{n+1} = y_n + \frac{h}{24}(55\,f_n - 59\,f_{n-1} + 37\,f_{n-2} - 9\,f_{n-3}).$$

With $f_{n+1} = f(t_{n+1}, y_{n+1})$, the corrector formula is
$$y_{n+1} = y_n + \frac{h}{24}(9\,f_{n+1} + 19\,f_n - 5\,f_{n-1} + f_{n-2}).$$

Using the starting values generated by the Runge-Kutta method:

	$n = 0$	$n = 1$	$n = 2$	$n = 3$
t_n	0.0	0.1	0.2	0.3
y_n	0.5	0.51016950	0.52413795	0.54210529

	$n = 4$ (pre)	$n = 4$ (cor)	$n = 5$ (pre)	$n = 5$ (cor)
t_n	0.4	0.4	0.5	0.5
y_n	0.56428532	0.56428577	0.59090816	0.59090918

(b) With $f_{n+1} = f(t_{n+1}, y_{n+1})$, the fourth order Adams-Moulton formula is
$$y_{n+1} = y_n + \frac{h}{24}(9\,f_{n+1} + 19\,f_n - 5\,f_{n-1} + f_{n-2}).$$

In this problem,
$$f_{n+1} = \frac{y_{n+1}^2 + 2\,t_{n+1}\,y_{n+1}}{3 + t_{n+1}^2}.$$

Since the ODE is nonlinear, an equation solver is needed to approximate the solution of
$$y_{n+1} = y_n + \frac{h}{24}\left[9\,\frac{y_{n+1}^2 + 2\,t_{n+1}\,y_{n+1}}{3 + t_{n+1}^2} + 19\,f_n - 5\,f_{n-1} + f_{n-2}\right]$$
at each time step.

	$n = 4$	$n = 5$
t_n	0.4	0.5
y_n	0.56428578	0.59090920

(c) The fourth order backward differentiation formula is
$$y_{n+1} = \frac{1}{25}\left[48\,y_n - 36\,y_{n-1} + 16\,y_{n-2} - 3\,y_{n-3} + 12h f_{n+1}\right].$$

Since the ODE is nonlinear, an equation solver is needed to approximate the solution of

$$y_{n+1} = \frac{1}{25}\left[48\,y_n - 36\,y_{n-1} + 16\,y_{n-2} - 3\,y_{n-3} + 12h\,\frac{y_{n+1}^2 + 2\,t_{n+1}\,y_{n+1}}{3 + t_{n+1}^2}\right]$$

at each time step. We obtain the approximate values:

	$n = 4$	$n = 5$
t_n	0.4	0.5
y_n	0.56428588	0.59090952

The exact solution of the IVP is $y(t) = (3 + t^2)/(6 - t)$.

6.(a) The predictor formula is

$$y_{n+1} = y_n + \frac{h}{24}(55\,f_n - 59\,f_{n-1} + 37\,f_{n-2} - 9\,f_{n-3}).$$

With $f_{n+1} = f(t_{n+1}, y_{n+1})$, the corrector formula is

$$y_{n+1} = y_n + \frac{h}{24}(9\,f_{n+1} + 19\,f_n - 5\,f_{n-1} + f_{n-2}).$$

We use the starting values generated by the Runge-Kutta method:

	$n = 0$	$n = 1$	$n = 2$	$n = 3$
t_n	0.0	0.1	0.2	0.3
y_n	-1.0	-0.924517	-0.864125	-0.816377

	$n = 4$ (pre)	$n = 4$ (cor)	$n = 5$ (pre)	$n = 5$ (cor)
t_n	0.4	0.4	0.5	0.5
y_n	-0.779832	-0.779693	-0.753311	-0.753135

(b) With $f_{n+1} = f(t_{n+1}, y_{n+1})$, the fourth order Adams-Moulton formula is

$$y_{n+1} = y_n + \frac{h}{24}(9\,f_{n+1} + 19\,f_n - 5\,f_{n-1} + f_{n-2}).$$

In this problem, $f_{n+1} = (t_{n+1}^2 - y_{n+1}^2)\sin y_{n+1}$. Since the ODE is nonlinear, we obtain the implicit equation

$$y_{n+1} = y_n + \frac{h}{24}\left[9(t_{n+1}^2 - y_{n+1}^2)\sin y_{n+1} + 19\,f_n - 5\,f_{n-1} + f_{n-2}\right].$$

	$n = 4$	$n = 5$
t_n	0.4	0.5
y_n	-0.779700	-0.753144

(c) The fourth order backward differentiation formula is

$$y_{n+1} = \frac{1}{25}\left[48\,y_n - 36\,y_{n-1} + 16\,y_{n-2} - 3\,y_{n-3} + 12h f_{n+1}\right].$$

Since the ODE is nonlinear, we obtain the implicit equation

$$y_{n+1} = \frac{1}{25}\left[48\,y_n - 36\,y_{n-1} + 16\,y_{n-2} - 3\,y_{n-3} + 12h(t_{n+1}^2 - y_{n+1}^2)\sin y_{n+1}\right].$$

	$n=4$	$n=5$
t_n	0.4	0.5
y_n	-0.779680	-0.753089

8.(a) The predictor formula is

$$y_{n+1} = y_n + \frac{h}{24}(55\,f_n - 59\,f_{n-1} + 37\,f_{n-2} - 9\,f_{n-3}).$$

With $f_{n+1} = f(t_{n+1}, y_{n+1})$, the corrector formula is

$$y_{n+1} = y_n + \frac{h}{24}(9\,f_{n+1} + 19\,f_n - 5\,f_{n-1} + f_{n-2}).$$

We use the starting values generated by the Runge-Kutta method:

	$n=0$	$n=1$	$n=2$	$n=3$
t_n	0.0	0.05	0.1	0.15
y_n	2.0	1.7996296	1.6223042	1.4672503

	$n=10$	$n=20$	$n=30$	$n=40$
t_n	0.5	1.0	1.5	2.0
y_n	0.9257133	1.285148	2.408595	4.103495

(b) Since the ODE is nonlinear, an equation solver is needed to approximate the solution of

$$y_{n+1} = y_n + \frac{h}{24}\left[45t_{n+1} - 27\sqrt{y_{n+1}} + 19\,f_n - 5\,f_{n-1} + f_{n-2}\right]$$

at each time step. We obtain the approximate values:

	$n=10$	$n=20$	$n=30$	$n=40$
t_n	0.5	1.0	1.5	2.0
y_n	0.9257125	1.285148	2.408595	4.103495

(c) The fourth order backward differentiation formula is

$$y_{n+1} = \frac{1}{25}\left[48\,y_n - 36\,y_{n-1} + 16\,y_{n-2} - 3\,y_{n-3} + 12hf_{n+1}\right].$$

Since the ODE is nonlinear, an equation solver is needed to approximate the solution of

$$y_{n+1} = \frac{1}{25}\left[48\,y_n - 36\,y_{n-1} + 16\,y_{n-2} - 3\,y_{n-3} + 12h(5t_{n+1} - 3\sqrt{y_{n+1}})\right]$$

at each time step.

	$n = 10$	$n = 20$	$n = 30$	$n = 40$
t_n	0.5	1.0	1.5	2.0
y_n	0.9257248	1.285158	2.408594	4.103493

The exact solution of the IVP is given implicitly by

$$\frac{1}{(2\sqrt{y} + 5t)^5 (t - \sqrt{y})^2} = \frac{\sqrt{2}}{512}.$$

9.(a) The predictor formula is

$$y_{n+1} = y_n + \frac{h}{24}(55\, f_n - 59\, f_{n-1} + 37\, f_{n-2} - 9\, f_{n-3}).$$

With $f_{n+1} = f(t_{n+1}, y_{n+1})$, the corrector formula is

$$y_{n+1} = y_n + \frac{h}{24}(9\, f_{n+1} + 19\, f_n - 5\, f_{n-1} + f_{n-2}).$$

Using the starting values generated by the Runge-Kutta method:

	$n = 0$	$n = 1$	$n = 2$	$n = 3$
t_n	0.0	0.05	0.1	0.15
y_n	3.0	3.087586	3.177127	3.268609

	$n = 10$	$n = 20$	$n = 30$	$n = 40$
t_n	0.5	1.0	1.5	2.0
y_n	3.962186	5.108903	6.431390	7.923385

(b) With $f_{n+1} = f(t_{n+1}, y_{n+1})$, the fourth order Adams-Moulton formula is

$$y_{n+1} = y_n + \frac{h}{24}(9\, f_{n+1} + 19\, f_n - 5\, f_{n-1} + f_{n-2}).$$

In this problem, $f_{n+1} = \sqrt{t_{n+1} + y_{n+1}}$. Since the ODE is nonlinear, an equation solver must be implemented in order to approximate the solution of

$$y_{n+1} = y_n + \frac{h}{24}\left[9\sqrt{t_{n+1} + y_{n+1}} + 19\, f_n - 5\, f_{n-1} + f_{n-2}\right]$$

at each time step.

	$n = 10$	$n = 20$	$n = 30$	$n = 40$
t_n	0.5	1.0	1.5	2.0
y_n	3.962186	5.108903	6.431390	7.923385

(c) The fourth order backward differentiation formula is

$$y_{n+1} = \frac{1}{25}\left[48\, y_n - 36\, y_{n-1} + 16\, y_{n-2} - 3\, y_{n-3} + 12h f_{n+1}\right].$$

Since the ODE is nonlinear, an equation solver is needed to approximate the solution of
$$y_{n+1} = \frac{1}{25}\left[48\,y_n - 36\,y_{n-1} + 16\,y_{n-2} - 3\,y_{n-3} + 12h\sqrt{t_{n+1} + y_{n+1}}\,\right]$$
at each time step.

	$n = 10$	$n = 20$	$n = 30$	$n = 40$
t_n	0.5	1.0	1.5	2.0
y_n	3.962186	5.108903	6.431390	7.923385

The exact solution is given implicitly by
$$\ln\left[\frac{2}{y+t-1}\right] + 2\sqrt{t+y} - 2\,\text{arctanh}\,\sqrt{t+y} = t + 2\sqrt{3} - 2\,\text{arctanh}\,\sqrt{3}\,.$$

10.(a) The predictor formula is
$$y_{n+1} = y_n + \frac{h}{24}(55\,f_n - 59\,f_{n-1} + 37\,f_{n-2} - 9\,f_{n-3}).$$

With $f_{n+1} = f(t_{n+1}, y_{n+1})$, the corrector formula is
$$y_{n+1} = y_n + \frac{h}{24}(9\,f_{n+1} + 19\,f_n - 5\,f_{n-1} + f_{n-2}).$$

We use the starting values generated by the Runge-Kutta method:

	$n = 0$	$n = 1$	$n = 2$	$n = 3$
t_n	0.0	0.05	0.1	0.15
y_n	1.0	1.051230	1.104843	1.160740

	$n = 10$	$n = 20$	$n = 30$	$n = 40$
t_n	0.5	1.0	1.5	2.0
y_n	1.612622	2.480909	3.7451479	5.495872

(b) With $f_{n+1} = f(t_{n+1}, y_{n+1})$, the fourth order Adams-Moulton formula is
$$y_{n+1} = y_n + \frac{h}{24}(9\,f_{n+1} + 19\,f_n - 5\,f_{n-1} + f_{n-2}).$$

In this problem, $f_{n+1} = 2\,t_{n+1} + e^{-t_{n+1}\,y_{n+1}}$. Since the ODE is nonlinear, an equation solver must be implemented in order to approximate the solution of
$$y_{n+1} = y_n + \frac{h}{24}\left\{9\left[2\,t_{n+1} + e^{-t_{n+1}\,y_{n+1}}\right] + 19\,f_n - 5\,f_{n-1} + f_{n-2}\right\}$$
at each time step.

	$n = 10$	$n = 20$	$n = 30$	$n = 40$
t_n	0.5	1.0	1.5	2.0
y_n	1.612622	2.480909	3.7451479	5.495872

(c) The fourth order backward differentiation formula is

$$y_{n+1} = \frac{1}{25}\left[48\,y_n - 36\,y_{n-1} + 16\,y_{n-2} - 3\,y_{n-3} + 12h f_{n+1}\right].$$

Since the ODE is nonlinear, we obtain the implicit equation

$$y_{n+1} = \frac{1}{25}\left\{48\,y_n - 36\,y_{n-1} + 16\,y_{n-2} - 3\,y_{n-3} + 12h\left[2\,t_{n+1} + e^{-t_{n+1}\,y_{n+1}}\right]\right\}.$$

	$n = 10$	$n = 20$	$n = 30$	$n = 40$
t_n	0.5	1.0	1.5	2.0
y_n	1.612623	2.480905	3.7451473	5.495869

11.(a) The predictor formula is

$$y_{n+1} = y_n + \frac{h}{24}(55\,f_n - 59\,f_{n-1} + 37\,f_{n-2} - 9\,f_{n-3}).$$

With $f_{n+1} = f(t_{n+1}, y_{n+1})$, the corrector formula is

$$y_{n+1} = y_n + \frac{h}{24}(9\,f_{n+1} + 19\,f_n - 5\,f_{n-1} + f_{n-2}).$$

Using the starting values generated by the Runge-Kutta method:

	$n = 0$	$n = 1$	$n = 2$	$n = 3$
t_n	0.0	0.05	0.1	0.15
y_n	-2.0	-1.958833	-1.915221	-1.868975

	$n = 10$	$n = 20$	$n = 30$	$n = 40$
t_n	0.5	1.0	1.5	2.0
y_n	-1.447639	-0.1436281	1.060946	1.410122

(b) With $f_{n+1} = f(t_{n+1}, y_{n+1})$, the fourth order Adams-Moulton formula is

$$y_{n+1} = y_n + \frac{h}{24}(9\,f_{n+1} + 19\,f_n - 5\,f_{n-1} + f_{n-2}).$$

In this problem,

$$f_{n+1} = \frac{4 - t_{n+1}\,y_{n+1}}{1 + y_{n+1}^2}.$$

Since the differential equation is nonlinear, an equation solver is used to approximate the solution of

$$y_{n+1} = y_n + \frac{h}{24}\left[9\,\frac{4 - t_{n+1}\,y_{n+1}}{1 + y_{n+1}^2} + 19\,f_n - 5\,f_{n-1} + f_{n-2}\right]$$

at each time step.

	$n = 10$	$n = 20$	$n = 30$	$n = 40$
t_n	0.5	1.0	1.5	2.0
y_n	-1.447638	-0.1436767	1.060913	1.410103

(c) The fourth order backward differentiation formula is

$$y_{n+1} = \frac{1}{25}[48\,y_n - 36\,y_{n-1} + 16\,y_{n-2} - 3\,y_{n-3} + 12h f_{n+1}].$$

Since the ODE is nonlinear, an equation solver must be implemented in order to approximate the solution of

$$y_{n+1} = \frac{1}{25}\left[48\,y_n - 36\,y_{n-1} + 16\,y_{n-2} - 3\,y_{n-3} + 12h\frac{4 - t_{n+1}\,y_{n+1}}{1 + y_{n+1}^2}\right]$$

at each time step.

	$n = 10$	$n = 20$	$n = 30$	$n = 40$
t_n	0.5	1.0	1.5	2.0
y_n	-1.447621	-0.1447619	1.060717	1.410027

12.(a) The predictor formula is

$$y_{n+1} = y_n + \frac{h}{24}(55\,f_n - 59\,f_{n-1} + 37\,f_{n-2} - 9\,f_{n-3}).$$

With $f_{n+1} = f(t_{n+1}, y_{n+1})$, the corrector formula is

$$y_{n+1} = y_n + \frac{h}{24}(9\,f_{n+1} + 19\,f_n - 5\,f_{n-1} + f_{n-2}).$$

We use the starting values generated by the Runge-Kutta method:

	$n = 0$	$n = 1$	$n = 2$	$n = 3$
t_n	0.0	0.05	0.1	0.15
y_n	0.5	0.5046218	0.5101695	0.5166666

	$n = 10$	$n = 20$	$n = 30$	$n = 40$
t_n	0.5	1.0	1.5	2.0
y_n	0.5909091	0.8000000	1.166667	1.750000

(b) With $f_{n+1} = f(t_{n+1}, y_{n+1})$, the fourth order Adams-Moulton formula is

$$y_{n+1} = y_n + \frac{h}{24}(9\,f_{n+1} + 19\,f_n - 5\,f_{n-1} + f_{n-2}).$$

In this problem,

$$f_{n+1} = \frac{y_{n+1}^2 + 2\,t_{n+1}\,y_{n+1}}{3 + t_{n+1}^2}.$$

Since the ODE is nonlinear, an equation solver is needed to approximate the solution of

$$y_{n+1} = y_n + \frac{h}{24}\left[9\frac{y_{n+1}^2 + 2\,t_{n+1}\,y_{n+1}}{3 + t_{n+1}^2} + 19\,f_n - 5\,f_{n-1} + f_{n-2}\right]$$

at each time step.

	$n=10$	$n=20$	$n=30$	$n=40$
t_n	0.5	1.0	1.5	2.0
y_n	0.5909091	0.8000000	1.166667	1.750000

(c) The fourth order backward differentiation formula is

$$y_{n+1} = \frac{1}{25}\left[48\,y_n - 36\,y_{n-1} + 16\,y_{n-2} - 3\,y_{n-3} + 12h f_{n+1}\right].$$

Since the ODE is nonlinear, we obtain the implicit equation

$$y_{n+1} = \frac{1}{25}\left[48\,y_n - 36\,y_{n-1} + 16\,y_{n-2} - 3\,y_{n-3} + 12h\,\frac{y_{n+1}^2 + 2t_{n+1}y_{n+1}}{3 + t_{n+1}^2}\right].$$

	$n=10$	$n=20$	$n=30$	$n=40$
t_n	0.5	1.0	1.5	2.0
y_n	0.5909092	0.8000002	1.166667	1.750001

The exact solution of the IVP is $y(t) = (3+t^2)/(6-t)$.

13. Both Adams methods entail the approximation of $f(t,y)$, on the interval $[t_n, t_{n+1}]$, by a polynomial. Approximating $\phi'(t) = P_1(t) \equiv A$, which is a constant polynomial, we have

$$\phi(t_{n+1}) - \phi(t_n) = \int_{t_n}^{t_{n+1}} A\,dt = A(t_{n+1} - t_n) = Ah.$$

Setting $A = \lambda f_n + (1-\lambda)f_{n-1}$, where $0 \le \lambda \le 1$, we obtain the approximation

$$y_{n+1} = y_n + h\left[\lambda f_n + (1-\lambda)f_{n-1}\right].$$

An appropriate choice of λ yields the familiar Euler formula. Similarly, setting

$$A = \lambda f_n + (1-\lambda)f_{n+1},$$

where $0 \le \lambda \le 1$, we obtain the approximation

$$y_{n+1} = y_n + h\left[\lambda f_n + (1-\lambda)f_{n+1}\right].$$

14. For a third order Adams-Bashforth formula, we approximate $f(t,y)$, on the interval $[t_n, t_{n+1}]$, by a quadratic polynomial using the points (t_{n-2}, y_{n-2}), (t_{n-1}, y_{n-1}) and (t_n, y_n). Let $P_3(t) = At^2 + Bt + C$. We obtain the system of equations

$$At_{n-2}^2 + Bt_{n-2} + C = f_{n-2}$$
$$At_{n-1}^2 + Bt_{n-1} + C = f_{n-1}$$
$$At_n^2 + Bt_n + C = f_n.$$

For computational purposes, assume that $t_0 = 0$, and $t_n = nh$. It follows that
$$A = \frac{f_n - 2f_{n-1} + f_{n-2}}{2h^2}$$
$$B = \frac{(3-2n)f_n + (4n-4)f_{n-1} + (1-2n)f_{n-2}}{2h}$$
$$C = \frac{n^2 - 3n + 2}{2} f_n + (2n - n^2)f_{n-1} + \frac{n^2 - n}{2} f_{n-2}.$$

We then have
$$y_{n+1} - y_n = \int_{t_n}^{t_{n+1}} \left[At^2 + Bt + C\right] dt = Ah^3\left(n^2 + n + \frac{1}{3}\right) + Bh^2\left(n + \frac{1}{2}\right) + Ch,$$
which yields
$$y_{n+1} - y_n = \frac{h}{12}\left(23 f_n - 16 f_{n-1} + 5 f_{n-2}\right).$$

15. For a third order Adams-Moulton formula, we approximate $f(t,y)$, on the interval $[t_n, t_{n+1}]$, by a quadratic polynomial using the points (t_{n-1}, y_{n-1}), (t_n, y_n) and (t_{n+1}, y_{n+1}). Let $P_3(t) = \alpha t^2 + \beta t + \gamma$. This time we obtain the system of algebraic equations
$$\alpha t_{n-1}^2 + \beta t_{n-1} + \gamma = f_{n-1}$$
$$\alpha t_n^2 + \beta t_n + \gamma = f_n$$
$$\alpha t_{n+1}^2 + \beta t_{n+1} + \gamma = f_{n+1}.$$

For computational purposes, again assume that $t_0 = 0$, and $t_n = nh$. It follows that
$$\alpha = \frac{f_{n-1} - 2f_n + f_{n+1}}{2h^2}$$
$$\beta = \frac{-(2n+1)f_{n-1} + 4nf_n + (1-2n)f_{n+1}}{2h}$$
$$\gamma = \frac{n^2 + n}{2} f_{n-1} + (1 - n^2)f_n + \frac{n^2 - n}{2} f_{n+1}.$$

We then have
$$y_{n+1} - y_n = \int_{t_n}^{t_{n+1}} \left[\alpha t^2 + \beta t + \gamma\right] dt = \alpha h^3\left(n^2 + n + \frac{1}{3}\right) + \beta h^2\left(n + \frac{1}{2}\right) + \gamma h,$$
which results in
$$y_{n+1} - y_n = \frac{h}{12}\left(5 f_{n+1} + 8 f_n - f_{n-1}\right).$$

8.5

1. In vector notation, the initial value problem can be written as
$$\frac{d}{dt}\begin{pmatrix} x \\ y \end{pmatrix} = \begin{pmatrix} x + y + t \\ 4x - 2y \end{pmatrix}, \quad \mathbf{x}(0) = \begin{pmatrix} 1 \\ 0 \end{pmatrix}.$$

(a) The Euler formula is
$$\begin{pmatrix} x_{n+1} \\ y_{n+1} \end{pmatrix} = \begin{pmatrix} x_n \\ y_n \end{pmatrix} + h \begin{pmatrix} x_n + y_n + t_n \\ 4x_n - 2y_n \end{pmatrix}.$$

That is,
$$x_{n+1} = x_n + h(x_n + y_n + t_n)$$
$$y_{n+1} = y_n + h(4x_n - 2y_n).$$

With $h = 0.1$, we obtain the values

	$n=2$	$n=4$	$n=6$	$n=8$	$n=10$
t_n	0.2	0.4	0.6	0.8	1.0
x_n	1.26	1.7714	2.58991	3.82374	5.64246
y_n	0.76	1.4824	2.3703	3.60413	5.38885

(b) The Runge-Kutta method uses the following intermediate calculations:

$$\mathbf{k}_{n1} = (x_n + y_n + t_n, \, 4x_n - 2y_n)^T$$

$$\mathbf{k}_{n2} = \left[x_n + \frac{h}{2}k_{n1}^1 + y_n + \frac{h}{2}k_{n1}^2 + t_n + \frac{h}{2}, \, 4(x_n + \frac{h}{2}k_{n1}^1) - 2(y_n + \frac{h}{2}k_{n1}^2) \right]^T$$

$$\mathbf{k}_{n3} = \left[x_n + \frac{h}{2}k_{n2}^1 + y_n + \frac{h}{2}k_{n2}^2 + t_n + \frac{h}{2}, \, 4(x_n + \frac{h}{2}k_{n2}^1) - 2(y_n + \frac{h}{2}k_{n2}^2) \right]^T$$

$$\mathbf{k}_{n4} = \left[x_n + hk_{n3}^1 + y_n + hk_{n3}^2 + t_n + h, \, 4(x_n + hk_{n3}^1) - 2(y_n + hk_{n3}^2) \right]^T.$$

With $h = 0.2$, we obtain the values:

	$n=1$	$n=2$	$n=3$	$n=4$	$n=5$
t_n	0.2	0.4	0.6	0.8	1.0
x_n	1.32493	1.93679	2.93414	4.48318	6.84236
y_n	0.758933	1.57919	2.66099	4.22639	6.56452

(c) With $h = 0.1$, we obtain

	$n=2$	$n=4$	$n=6$	$n=8$	$n=10$
t_n	0.2	0.4	0.6	0.8	1.0
x_n	1.32489	1.9369	2.93459	4.48422	6.8444
y_n	0.759516	1.57999	2.66201	4.22784	6.56684

The exact solution of the IVP is
$$x(t) = e^{2t} + \frac{2}{9}e^{-3t} - \frac{1}{3}t - \frac{2}{9}$$
$$y(t) = e^{2t} - \frac{8}{9}e^{-3t} - \frac{2}{3}t - \frac{1}{9}.$$

3.(a) The Euler formula is
$$\begin{pmatrix} x_{n+1} \\ y_{n+1} \end{pmatrix} = \begin{pmatrix} x_n \\ y_n \end{pmatrix} + h \begin{pmatrix} -t_n x_n - y_n - 1 \\ x_n \end{pmatrix}.$$

That is,
$$x_{n+1} = x_n + h(-t_n x_n - y_n - 1)$$
$$y_{n+1} = y_n + h(x_n).$$

With $h = 0.1$, we obtain the values

	$n=2$	$n=4$	$n=6$	$n=8$	$n=10$
t_n	0.2	0.4	0.6	0.8	1.0
x_n	0.582	0.117969	-0.336912	-0.730007	-1.02134
y_n	1.18	1.27344	1.27382	1.18572	1.02371

(b) The Runge-Kutta method uses the following intermediate calculations:
$$\mathbf{k}_{n1} = (-t_n x_n - y_n - 1, x_n)^T$$
$$\mathbf{k}_{n2} = \left[-(t_n + \frac{h}{2})(x_n + \frac{h}{2}k_{n1}^1) - (y_n + \frac{h}{2}k_{n1}^2) - 1, x_n + \frac{h}{2}k_{n1}^1\right]^T$$
$$\mathbf{k}_{n3} = \left[-(t_n + \frac{h}{2})(x_n + \frac{h}{2}k_{n2}^1) - (y_n + \frac{h}{2}k_{n2}^2) - 1, x_n + \frac{h}{2}k_{n2}^1\right]^T$$
$$\mathbf{k}_{n4} = \left[-(t_n + h)(x_n + hk_{n3}^1) - (y_n + hk_{n3}^2) - 1, x_n + hk_{n3}^1\right]^T.$$

With $h = 0.2$, we obtain the values:

	$n=1$	$n=2$	$n=3$	$n=4$	$n=5$
t_n	0.2	0.4	0.6	0.8	1.0
x_n	0.568451	0.109776	-0.32208	-0.681296	-0.937852
y_n	1.15775	1.22556	1.20347	1.10162	0.937852

(c) With $h = 0.1$, we obtain

	$n=2$	$n=4$	$n=6$	$n=8$	$n=10$
t_n	0.2	0.4	0.6	0.8	1.0
x_n	0.56845	0.109773	-0.322081	-0.681291	-0.937841
y_n	1.15775	1.22557	1.20347	1.10161	0.93784

4.(a) The Euler formula gives
$$x_{n+1} = x_n + h(x_n - y_n + x_n y_n)$$
$$y_{n+1} = y_n + h(3x_n - 2y_n - x_n y_n).$$

With $h = 0.1$, we obtain the values

	$n=2$	$n=4$	$n=6$	$n=8$	$n=10$
t_n	0.2	0.4	0.6	0.8	1.0
x_n	-0.198	-0.378796	-0.51932	-0.594324	-0.588278
y_n	0.618	0.28329	-0.0321025	-0.326801	-0.57545

(b) Given
$$f(t,x,y) = x - y + xy$$
$$g(t,x,y) = 3x - 2y - xy,$$

the Runge-Kutta method uses the following intermediate calculations:

$$\mathbf{k}_{n1} = [f(t_n, x_n, y_n), g(t_n, x_n, y_n)]^T$$
$$\mathbf{k}_{n2} = \left[f(t_n + \frac{h}{2}, x_n + \frac{h}{2}k_{n1}^1, y_n + \frac{h}{2}k_{n1}^2), g(t_n + \frac{h}{2}, x_n + \frac{h}{2}k_{n1}^1, y_n + \frac{h}{2}k_{n1}^2)\right]^T$$
$$\mathbf{k}_{n3} = \left[f(t_n + \frac{h}{2}, x_n + \frac{h}{2}k_{n2}^1, y_n + \frac{h}{2}k_{n2}^2), g(t_n + \frac{h}{2}, x_n + \frac{h}{2}k_{n2}^1, y_n + \frac{h}{2}k_{n2}^2)\right]^T$$
$$\mathbf{k}_{n4} = \left[f(t_n + h, x_n + hk_{n3}^1, y_n + hk_{n3}^2), g(t_n + h, x_n + hk_{n3}^1, y_n + hk_{n3}^2)\right]^T.$$

With $h = 0.2$, we obtain the values:

	$n=1$	$n=2$	$n=3$	$n=4$	$n=5$
t_n	0.2	0.4	0.6	0.8	1.0
x_n	-0.196904	-0.372643	-0.501302	-0.561270	-0.547053
y_n	0.630936	0.298888	-0.0111429	-0.288943	-0.508303

(c) With $h = 0.1$, we obtain

	$n=2$	$n=4$	$n=6$	$n=8$	$n=10$
t_n	0.2	0.4	0.6	0.8	1.0
x_n	-0.196935	-0.372687	-0.501345	-0.561292	-0.547031
y_n	0.630939	0.298866	-0.0112184	-0.28907	-0.508427

5.(a) The Euler formula gives

$$x_{n+1} = x_n + h\left[x_n(1 - 0.5\,x_n - 0.5\,y_n)\right]$$
$$y_{n+1} = y_n + h\left[y_n(-0.25 + 0.5\,x_n)\right].$$

With $h = 0.1$, we obtain the values

	$n=2$	$n=4$	$n=6$	$n=8$	$n=10$
t_n	0.2	0.4	0.6	0.8	1.0
x_n	2.96225	2.34119	1.90236	1.56602	1.29768
y_n	1.34538	1.67121	1.97158	2.23895	2.46732

(b) Given
$$f(t,x,y) = x(1 - 0.5\,x - 0.5\,y)$$
$$g(t,x,y) = y(-0.25 + 0.5\,x),$$

the Runge-Kutta method uses the following intermediate calculations:

$$\mathbf{k}_{n1} = [f(t_n, x_n, y_n), g(t_n, x_n, y_n)]^T$$

$$\mathbf{k}_{n2} = \left[f(t_n + \frac{h}{2}, x_n + \frac{h}{2}k_{n1}^1, y_n + \frac{h}{2}k_{n1}^2), g(t_n + \frac{h}{2}, x_n + \frac{h}{2}k_{n1}^1, y_n + \frac{h}{2}k_{n1}^2) \right]^T$$

$$\mathbf{k}_{n3} = \left[f(t_n + \frac{h}{2}, x_n + \frac{h}{2}k_{n2}^1, y_n + \frac{h}{2}k_{n2}^2), g(t_n + \frac{h}{2}, x_n + \frac{h}{2}k_{n2}^1, y_n + \frac{h}{2}k_{n2}^2) \right]^T$$

$$\mathbf{k}_{n4} = \left[f(t_n + h, x_n + hk_{n3}^1, y_n + hk_{n3}^2), g(t_n + h, x_n + hk_{n3}^1, y_n + hk_{n3}^2) \right]^T.$$

With $h = 0.2$, we obtain the values:

	$n=1$	$n=2$	$n=3$	$n=4$	$n=5$
t_n	0.2	0.4	0.6	0.8	1.0
x_n	3.06339	2.44497	1.9911	1.63818	1.3555
y_n	1.34858	1.68638	2.00036	2.27981	2.5175

(c) With $h = 0.1$, we obtain

	$n=2$	$n=4$	$n=6$	$n=8$	$n=10$
t_n	0.2	0.4	0.6	0.8	1.0
x_n	3.06314	2.44465	1.99075	1.63781	1.35514
y_n	1.34899	1.68699	2.00107	2.28057	2.51827

6.(a) The Euler formula gives

$$x_{n+1} = x_n + h\left[e^{-x_n + y_n} - \cos x_n \right]$$
$$y_{n+1} = y_n + h\left[\sin(x_n - 3y_n) \right].$$

With $h = 0.1$, we obtain the values

	$n=2$	$n=4$	$n=6$	$n=8$	$n=10$
t_n	0.2	0.4	0.6	0.8	1.0
x_n	1.42386	1.82234	2.21728	2.61118	2.9955
y_n	2.18957	2.36791	2.53329	2.68763	2.83354

(b) The Runge-Kutta method uses the following intermediate calculations:

$$\mathbf{k}_{n1} = [f(t_n, x_n, y_n), g(t_n, x_n, y_n)]^T$$

$$\mathbf{k}_{n2} = \left[f(t_n + \frac{h}{2}, x_n + \frac{h}{2}k_{n1}^1, y_n + \frac{h}{2}k_{n1}^2), g(t_n + \frac{h}{2}, x_n + \frac{h}{2}k_{n1}^1, y_n + \frac{h}{2}k_{n1}^2) \right]^T$$

$$\mathbf{k}_{n3} = \left[f(t_n + \frac{h}{2}, x_n + \frac{h}{2}k_{n2}^1, y_n + \frac{h}{2}k_{n2}^2), g(t_n + \frac{h}{2}, x_n + \frac{h}{2}k_{n2}^1, y_n + \frac{h}{2}k_{n2}^2) \right]^T$$

$$\mathbf{k}_{n4} = \left[f(t_n + h, x_n + hk_{n3}^1, y_n + hk_{n3}^2), g(t_n + h, x_n + hk_{n3}^1, y_n + hk_{n3}^2) \right]^T.$$

With $h = 0.2$, we obtain the values:

	$n=1$	$n=2$	$n=3$	$n=4$	$n=5$
t_n	0.2	0.4	0.6	0.8	1.0
x_n	1.41513	1.81208	2.20635	2.59826	2.97806
y_n	2.18699	2.36233	2.5258	2.6794	2.82487

(c) With $h = 0.1$, we obtain

	$n=2$	$n=4$	$n=6$	$n=8$	$n=10$
t_n	0.2	0.4	0.6	0.8	1.0
x_n	1.41513	1.81209	2.20635	2.59826	2.97806
y_n	2.18699	2.36233	2.52581	2.67941	2.82488

7. The Runge-Kutta method uses the following intermediate calculations:

$$\mathbf{k}_{n1} = [x_n - 4y_n, -x_n + y_n]^T$$

$$\mathbf{k}_{n2} = \left[x_n + \frac{h}{2}k_{n1}^1 - 4(y_n + \frac{h}{2}k_{n1}^2), -(x_n + \frac{h}{2}k_{n1}^1) + y_n + \frac{h}{2}k_{n1}^2\right]^T$$

$$\mathbf{k}_{n3} = \left[x_n + \frac{h}{2}k_{n2}^1 - 4(y_n + \frac{h}{2}k_{n2}^2), -(x_n + \frac{h}{2}k_{n2}^1) + y_n + \frac{h}{2}k_{n2}^2\right]^T$$

$$\mathbf{k}_{n4} = \left[x_n + hk_{n3}^1 - 4(y_n + hk_{n3}^2), -(x_n + hk_{n3}^1) + y_n + hk_{n3}^2\right]^T.$$

Using $h = 0.05$, we obtain the following values:

	$n=4$	$n=8$	$n=12$	$n=16$	$n=20$
t_n	0.2	0.4	0.6	0.8	1.0
x_n	1.3204	1.9952	3.2992	5.7362	10.227
y_n	−0.25085	−0.66245	−1.3752	−2.6434	−4.9294

Using $h = 0.025$, we obtain the following values:

	$n=8$	$n=16$	$n=24$	$n=32$	$n=40$
t_n	0.2	0.4	0.6	0.8	1.0
x_n	1.3204	1.9952	3.2992	5.7362	10.227
y_n	−0.25085	−0.66245	−1.3752	−2.6435	−4.9294

The exact solution is given by

$$\phi(t) = \frac{e^{-t} + e^{3t}}{2}, \quad \psi(t) = \frac{e^{-t} - e^{3t}}{4},$$

and the associated tabulated values:

	$n=5$	$n=10$	$n=15$	$n=20$	$n=25$
t_n	0.2	0.4	0.6	0.8	1.0
$\phi(t_n)$	1.3204	1.9952	3.2992	5.7362	10.227
$\psi(t_n)$	−0.25085	−0.66245	−1.3752	−2.6435	−4.9294

9. The predictor formulas are

$$x_{n+1} = x_n + \frac{h}{24}(55 f_n - 59 f_{n-1} + 37 f_{n-2} - 9 f_{n-3})$$

$$y_{n+1} = y_n + \frac{h}{24}(55 g_n - 59 g_{n-1} + 37 g_{n-2} - 9 g_{n-3}).$$

With $f_{n+1} = x_{n+1} - 4 y_{n+1}$ and $g_{n+1} = -x_{n+1} + y_{n+1}$, the corrector formulas are

$$x_{n+1} = x_n + \frac{h}{24}(9 f_{n+1} + 19 f_n - 5 f_{n-1} + f_{n-2})$$

$$y_{n+1} = y_n + \frac{h}{24}(9 g_{n+1} + 19 g_n - 5 g_{n-1} + g_{n-2}).$$

We use the starting values from the exact solution:

	$n=0$	$n=1$	$n=2$	$n=3$
t_n	0	0.1	0.2	0.3
x_n	1.0	1.12883	1.32042	1.60021
y_n	0.0	−0.11057	−0.250847	−0.429696

One time step using the predictor-corrector method results in the approximate values:

	$n=4$ (pre)	$n=4$ (cor)
t_n	0.4	0.4
x_n	1.99445	1.99521
y_n	−0.662064	−0.662442

8.6

3. The solution of the initial value problem is $\psi(t) = e^{-100 t} + t$.

(a,b) Based on the exact solution, the local truncation error for both of the Euler methods is

$$|e_{loc}| \leq \frac{10^4}{2} e^{-100 \bar{t}_n} h^2.$$

Hence $|e_n| \leq 5000 h^2$, for all $0 < \bar{t}_n < 1$. Furthermore, the local truncation error is greatest near $t=0$. Therefore $|e_1| \leq 5000 h^2 < 0.0005$ for $h < 0.0003$. Now the truncation error accumulates at each time step. Therefore the actual time step should be much smaller than $h \approx 0.0003$. For example, with $h = 0.00025$, we obtain the data

	Euler	B.Euler	$\phi(t)$
$t=0.05$	0.056323	0.057165	0.056738
$t=0.1$	0.100040	0.100051	0.100045

Note that the total number of time steps needed to reach $t = 0.1$ is $N = 400$.

(c) Using the Runge-Kutta method, comparisons are made for several values of h; $h = 0.005$ is sufficient.

$h = 0.1$:

	$\phi(t)$	y_n	$y_n - \phi(t_n)$
$t = 0.05$	0.056738	0.057416	0.000678
$t = 0.1$	0.100045	0.100055	0.000010

$h = 0.005$:

	$\phi(t)$	y_n	$y_n - \phi(t_n)$
$t = 0.05$	0.056738	0.056766	0.000027
$t = 0.1$	0.100045	0.100046	0.0000004

6.(a) Using the method of undetermined coefficients, it is easy to show that the general solution of the ODE is $y(t) = ce^{\lambda t} + t^2$. Imposing the initial condition, it follows that $c = 0$ and hence the solution of the IVP is $\phi(t) = t^2$.

(b) Using the Runge-Kutta method, with $h = 0.01$, numerical solutions are generated for various values of λ :

$\lambda = 1$:

	$\phi(t)$	y_n	$|y_n - \phi(t_n)|$
$t = 0.25$	0.0625	0.0624999	2×10^{-11}
$t = 0.5$	0.25	0.25	0
$t = 0.75$	0.5625	0.5625	0
$t = 1.0$	1.0	1.0	0

$\lambda = 10$:

	$\phi(t)$	y_n	$|y_n - \phi(t_n)|$
$t = 0.25$	0.0625	0.0624998	2.215×10^{-7}
$t = 0.5$	0.25	0.249997	2.920×10^{-6}
$t = 0.75$	0.5625	0.562464	3.579×10^{-5}
$t = 1.0$	1.0	0.999564	4.362×10^{-4}

$\lambda = 20$:

	$\phi(t)$	y_n	$|y_n - \phi(t_n)|$
$t = 0.25$	0.0625	0.062489	1.10×10^{-5}
$t = 0.5$	0.25	0.248342	1.658×10^{-3}
$t = 0.75$	0.5625	0.316455	0.246045
$t = 1.0$	1.0	-35.5143	36.5143

$\lambda = 50$:

	$\phi(t)$	y_n	$|y_n - \phi(t_n)|$
$t = 0.25$	0.0625	-0.044803	0.107303
$t = 0.5$	0.25	-28669.55	28669.804
$t = 0.75$	0.5625	-7.66014×10^9	7.66014×10^9
$t = 1.0$	1.0	-2.04668×10^{15}	2.04668×10^{15}

The following table shows the calculated value, y_1, at the first time step:

$\phi(t)$	$y_1\ (\lambda = 1)$	$y_1\ (\lambda = 10)$	$y_1\ (\lambda = 20)$	$y_1\ (\lambda = 50)$
10^{-2}	9.99999×10^{-5}	9.99979×10^{-5}	9.99833×10^{-5}	9.97396×10^{-5}

(c) Referring back to the exact solution given in part (a), if a nonzero initial condition, say $y(0) = \varepsilon$, is specified, the solution of the IVP becomes

$$\phi_\varepsilon(t) = \varepsilon\, e^{\lambda t} + t^2.$$

We then have $|\phi(t) - \phi_\varepsilon(t)| = |\varepsilon|\, e^{\lambda t}$. It is evident that for any $t > 0$,

$$\lim_{\lambda \to \infty} |\phi(t) - \phi_\varepsilon(t)| = \infty.$$

This implies that virtually any error introduced early in the calculations will be magnified as $\lambda \to \infty$. The initial value problem is inherently unstable.

CHAPTER 9

Nonlinear Differential Equations and Stability

9.1

2.(a) Setting $\mathbf{x} = \boldsymbol{\xi}\, e^{rt}$ results in the algebraic equations

$$\begin{pmatrix} 5-r & -1 \\ 3 & 1-r \end{pmatrix} \begin{pmatrix} \xi_1 \\ \xi_2 \end{pmatrix} = \begin{pmatrix} 0 \\ 0 \end{pmatrix}.$$

For a nonzero solution, we must have $\det(\mathbf{A} - r\,\mathbf{I}) = r^2 - 6r + 8 = 0$. The roots of the characteristic equation are $r_1 = 2$ and $r_2 = 4$. For $r = 2$, the system of equations reduces to $3\xi_1 = \xi_2$. The corresponding eigenvector is $\boldsymbol{\xi}^{(1)} = (1,3)^T$. Substitution of $r = 4$ results in the single equation $\xi_1 = \xi_2$. A corresponding eigenvector is $\boldsymbol{\xi}^{(2)} = (1,1)^T$.

(b) The eigenvalues are real and positive, hence the critical point is an unstable node.

(c,d)

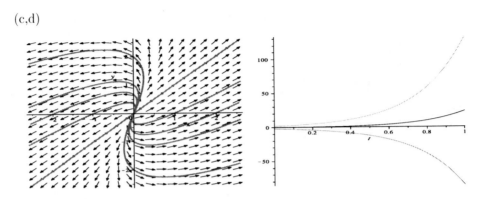

3.(a) Solution of the ODE requires analysis of the algebraic equations

$$\begin{pmatrix} 2-r & -1 \\ 3 & -2-r \end{pmatrix} \begin{pmatrix} \xi_1 \\ \xi_2 \end{pmatrix} = \begin{pmatrix} 0 \\ 0 \end{pmatrix}.$$

For a nonzero solution, we must have $\det(\mathbf{A} - r\mathbf{I}) = r^2 - 1 = 0$. The roots of the characteristic equation are $r_1 = 1$ and $r_2 = -1$. For $r = 1$, the system of equations reduces to $\xi_1 = \xi_2$. The corresponding eigenvector is $\boldsymbol{\xi}^{(1)} = (1,1)^T$. Substitution of $r = -1$ results in the single equation $3\xi_1 - \xi_2 = 0$. A corresponding eigenvector is $\boldsymbol{\xi}^{(2)} = (1,3)^T$.

(b) The eigenvalues are real, with $r_1 r_2 < 0$. Hence the critical point is an unstable saddle point.

(c,d)

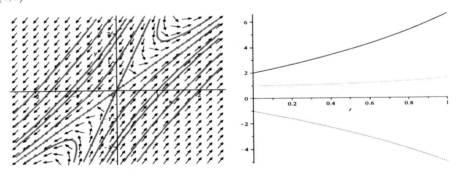

5.(a) The characteristic equation is given by

$$\begin{vmatrix} 1-r & -5 \\ 1 & -3-r \end{vmatrix} = r^2 + 2r + 2 = 0.$$

The equation has complex roots $r_1 = -1 + i$ and $r_2 = -1 - i$. For $r = -1 + i$, the components of the solution vector must satisfy $\xi_1 - (2+i)\xi_2 = 0$. Thus the corresponding eigenvector is $\boldsymbol{\xi}^{(1)} = (2+i, 1)^T$. Substitution of $r = -1 - i$ results in the single equation $\xi_1 - (2-i)\xi_2 = 0$. A corresponding eigenvector is $\boldsymbol{\xi}^{(2)} = (2-i, 1)^T$.

(b) The eigenvalues are complex conjugates, with negative real part. Hence the origin is an asymptotically stable spiral.

(c,d)

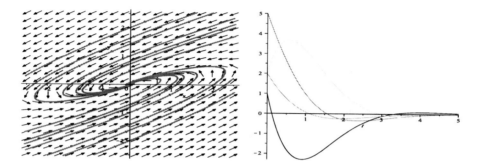

6.(a) Solution of the ODEs is based on the analysis of the algebraic equations

$$\begin{pmatrix} 2-r & -5 \\ 1 & -2-r \end{pmatrix} \begin{pmatrix} \xi_1 \\ \xi_2 \end{pmatrix} = \begin{pmatrix} 0 \\ 0 \end{pmatrix}.$$

For a nonzero solution, we require that $\det(\mathbf{A} - r\mathbf{I}) = r^2 + 1 = 0$. The roots of the characteristic equation are $r = \pm i$. Setting $r = i$, the equations are equivalent to $\xi_1 - (2+i)\xi_2 = 0$. The eigenvectors are $\boldsymbol{\xi}^{(1)} = (2+i, 1)^T$ and $\boldsymbol{\xi}^{(2)} = (2-i, 1)^T$.

(b) The eigenvalues are purely imaginary. Hence the critical point is a stable center.

(c,d)

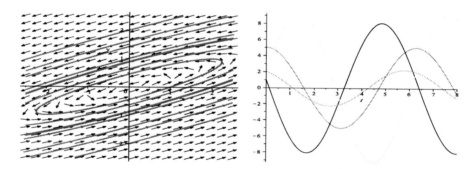

8.(a) The characteristic equation is given by

$$\begin{vmatrix} -1-r & -1 \\ 0 & -1/4-r \end{vmatrix} = (r+1)(r+1/4) = 0,$$

with roots $r_1 = -1$ and $r_2 = -1/4$. For $r = -1$, the components of the solution vector must satisfy $\xi_2 = 0$. Thus the corresponding eigenvector is $\boldsymbol{\xi}^{(1)} = (1, 0)^T$. Substitution of $r = -1/4$ results in the single equation $3\xi_1/4 + \xi_2 = 0$. A corresponding eigenvector is $\boldsymbol{\xi}^{(2)} = (4, -3)^T$.

(b) The eigenvalues are real and both negative. Hence the critical point is an asymptotically stable node.

(c,d)

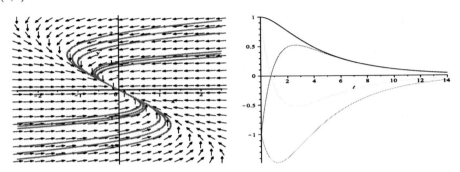

9.(a) Solution of the ODEs is based on the analysis of the algebraic equations

$$\begin{pmatrix} 3-r & -4 \\ 1 & -1-r \end{pmatrix} \begin{pmatrix} \xi_1 \\ \xi_2 \end{pmatrix} = \begin{pmatrix} 0 \\ 0 \end{pmatrix}.$$

For a nonzero solution, we require that $\det(\mathbf{A} - r\mathbf{I}) = r^2 - 2r + 1 = 0$. The single root of the characteristic equation is $r = 1$. Setting $r = 1$, the components of the solution vector must satisfy $\xi_1 - 2\xi_2 = 0$. A corresponding eigenvector is $\boldsymbol{\xi} = (2, 1)^T$.

(b) Since there is only one linearly independent eigenvector, the critical point is an unstable, improper node.

(c,d)

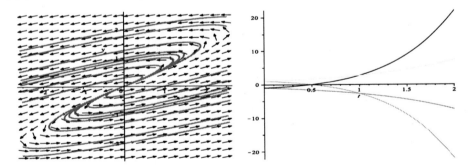

11.(a) The characteristic equation is $(r+1)^2 = 0$, with double root $r = -1$. It is easy to see that the two linearly independent eigenvectors are $\boldsymbol{\xi}^{(1)} = (1, 0)^T$ and $\boldsymbol{\xi}^{(2)} = (0, 1)^T$.

(b) Since there are two linearly independent eigenvectors, the critical point is an asymptotically stable proper node.

(c,d)

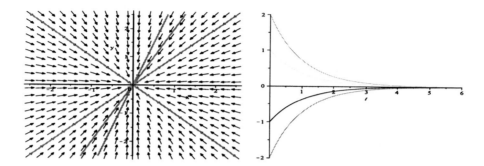

12.(a) Setting $\mathbf{x} = \boldsymbol{\xi}\, e^{rt}$ results in the algebraic equations

$$\begin{pmatrix} 2-r & -5/2 \\ 9/5 & -1-r \end{pmatrix} \begin{pmatrix} \xi_1 \\ \xi_2 \end{pmatrix} = \begin{pmatrix} 0 \\ 0 \end{pmatrix}.$$

For a nonzero solution, we require that $\det(\mathbf{A} - r\mathbf{I}) = r^2 - r + 5/2 = 0$. The roots of the characteristic equation are $r = 1/2 \pm 3i/2$. Substituting $r = 1/2 - 3i/2$, the equations reduce to $(3 + 3i)\xi_1 - 5\,\xi_2 = 0$. Therefore the two eigenvectors are $\boldsymbol{\xi}^{(1)} = (5, 3 + 3i)^T$ and $\boldsymbol{\xi}^{(2)} = (5, 3 - 3i)^T$.

(b) Since the eigenvalues are complex, with positive real part, the critical point is an unstable spiral.

(c,d)

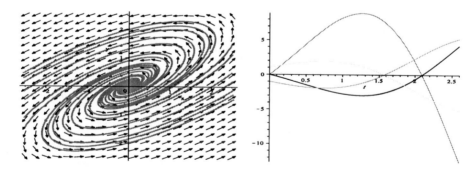

14. Setting $\mathbf{x}' = \mathbf{0}$, that is,

$$\begin{pmatrix} -2 & 1 \\ 1 & -2 \end{pmatrix} \mathbf{x} = \begin{pmatrix} 2 \\ -1 \end{pmatrix},$$

we find that the critical point is $\mathbf{x}^0 = (-1, 0)^T$. With the change of dependent variable, $\mathbf{x} = \mathbf{x}^0 + \mathbf{u}$, the differential equation can be written as

$$\frac{d\mathbf{u}}{dt} = \begin{pmatrix} -2 & 1 \\ 1 & -2 \end{pmatrix} \mathbf{u}.$$

The critical point for the transformed equation is the origin. Setting $\mathbf{u} = \boldsymbol{\xi} e^{rt}$ results in the algebraic equations

$$\begin{pmatrix} -2-r & 1 \\ 1 & -2-r \end{pmatrix} \begin{pmatrix} \xi_1 \\ \xi_2 \end{pmatrix} = \begin{pmatrix} 0 \\ 0 \end{pmatrix}.$$

For a nonzero solution, we require that $\det(\mathbf{A} - r\mathbf{I}) = r^2 + 4r + 3 = 0$. The roots of the characteristic equation are $r = -3, -1$. Hence the critical point is an asymptotically stable node.

15. Setting $\mathbf{x}' = \mathbf{0}$, that is,

$$\begin{pmatrix} -1 & -1 \\ 2 & -1 \end{pmatrix} \mathbf{x} = \begin{pmatrix} 1 \\ -5 \end{pmatrix},$$

we find that the critical point is $\mathbf{x}^0 = (-2, 1)^T$. With the change of dependent variable, $\mathbf{x} = \mathbf{x}^0 + \mathbf{u}$, the differential equation can be written as

$$\frac{d\mathbf{u}}{dt} = \begin{pmatrix} -1 & -1 \\ 2 & -1 \end{pmatrix} \mathbf{u}.$$

The characteristic equation is $\det(\mathbf{A} - r\mathbf{I}) = r^2 + 2r + 3 = 0$, with complex conjugate roots $r = -1 \pm i\sqrt{2}$. Since the real parts of the eigenvalues are negative, the critical point is an asymptotically stable spiral.

16. The critical point $\mathbf{x}^0$ satisfies the system of equations

$$\begin{pmatrix} 0 & -\beta \\ \delta & 0 \end{pmatrix} \mathbf{x} = \begin{pmatrix} -\alpha \\ \gamma \end{pmatrix}.$$

It follows that $x^0 = \gamma/\delta$ and $y^0 = \alpha/\beta$. Using the transformation, $\mathbf{x} = \mathbf{x}^0 + \mathbf{u}$, the differential equation can be written as

$$\frac{d\mathbf{u}}{dt} = \begin{pmatrix} 0 & -\beta \\ \delta & 0 \end{pmatrix} \mathbf{u}.$$

The characteristic equation is $\det(\mathbf{A} - r\mathbf{I}) = r^2 + \beta\delta = 0$. Since $\beta\delta > 0$, the roots are purely imaginary, with $r = \pm i\sqrt{\beta\delta}$. Hence the critical point is a stable center.

21.(a) If $q > 0$ and $p < 0$, then the roots are either complex conjugates with negative real parts, or both real and negative.

(b) If $q > 0$ and $p = 0$, then the roots are purely imaginary.

(c) If $q < 0$, then the roots are real, with $r_1 \cdot r_2 < 0$. If $p > 0$, then either the roots are real, with $r_1 > 0$ or the roots are complex conjugates with positive real parts.

9.2

2. The differential equations can be combined to obtain a related ODE

$$\frac{dy}{dx} = -\frac{2y}{x}.$$

The equation is separable, with

$$\frac{dy}{y} = -\frac{2\,dx}{x}.$$

The solution is given by $y = C\,x^{-2}$. Note that the system is uncoupled, and hence we also have $x = x_0 e^{-t}$ and $y = y_0 e^{2t}$. Matching the initial conditions, for the first case we obtain $x(t) = 4e^{-t}$ and $y(t) = 2e^{2t}$, for the second case we obtain $x(t) = 4e^{-t}$ and $y(t) = 0$.

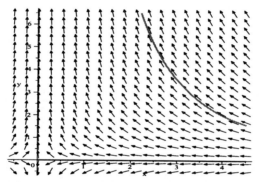

In order to determine the direction of motion along the trajectories, observe that for positive initial conditions, x will decrease, whereas y will increase.

4. The trajectories of the system satisfy the ODE

$$\frac{dy}{dx} = -\frac{bx}{ay}.$$

The equation is separable, with $ay\,dy = -bx\,dx$. Hence the trajectories are given by $b\,x^2 + a\,y^2 = C^2$, in which C is arbitrary. Evidently, the trajectories are ellipses. Invoking the initial condition, we find that $C^2 = ab$. The system of ODEs can also be written as

$$\frac{d\mathbf{x}}{dt} = \begin{pmatrix} 0 & a \\ -b & 0 \end{pmatrix} \mathbf{x}.$$

Using the methods in Chapter 7, it is easy to show that

$$x = \sqrt{a}\,\cos\sqrt{ab}\,t, \qquad y = -\sqrt{b}\,\sin\sqrt{ab}\,t.$$

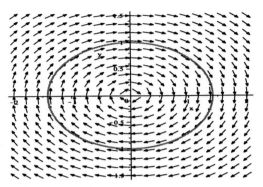

Note that for positive initial conditions, x will increase, whereas y will decrease.

6.(a) The critical points are solutions of the equations
$$1 + 2y = 0$$
$$1 - 3x^2 = 0.$$

There are two critical points, $(-1/\sqrt{3}, -1/2)$ and $(1/\sqrt{3}, -1/2)$.

(b)

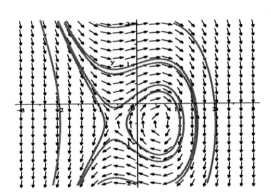

(c) Locally, the trajectories near the point $(-1/\sqrt{3}, -1/2)$ resemble the behavior near a saddle. Hence the critical point is unstable. Near the point $(1/\sqrt{3}, -1/2)$, the solutions are periodic. Therefore the second critical point is stable.

7.(a) The critical points are solutions of the equations
$$2x - x^2 - xy = 0$$
$$3y - 2y^2 - 3xy = 0.$$

There are four critical points, $(0,0)$, $(0, 3/2)$, $(2, 0)$, and $(-1, 3)$.

(b)

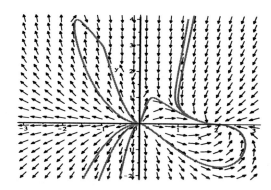

(c) Examining the phase plot we can conclude that $(0,0)$ is an unstable node, $(0, 3/2)$ is a saddle point (hence unstable), $(2, 0)$ is an asymptotically stable node, and $(-1, 3)$ is an asymptotically stable node.

(d) Again, the phase plot shows us that the basin of $(2, 0)$ is the (open) first and fourth quadrants and the basin of $(-1, 3)$ is the (open) second quadrant.

8.(a) The critical points are solutions of the equations
$$-(2+y)(x+y) = 0$$
$$-y(1-x) = 0.$$

There are three critical points, $(0, 0)$, $(1, -1)$, and $(1, -2)$.

(b)

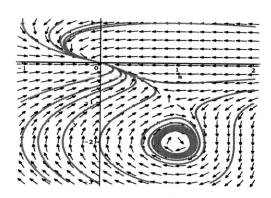

(c) Examining the phase plot we can conclude that $(0, 0)$ is an asymptotically stable node, $(1, -1)$ is a saddle point (hence unstable), and $(1, -2)$ is an asymptotically stable spiral.

(d) The phase plot suggests that the basin of $(0, 0)$ is the whole plane except for a subset of the fourth quadrant that is the basin of $(1, -2)$.

9.(a) The critical points are given by the solution set of the equations
$$y(2 - x - y) = 0$$
$$-x - y - 2xy = 0.$$

Clearly, $(0,0)$ is a critical point. If $x = 2 - y$, then it follows that $y(y - 2) = 1$. The additional critical points are $(1 - \sqrt{2}, 1 + \sqrt{2})$ and $(1 + \sqrt{2}, 1 - \sqrt{2})$.

(b)

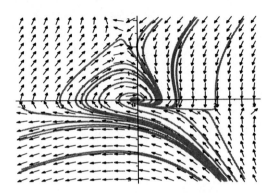

(c) The behavior near the origin is that of a stable spiral. Hence the point $(0,0)$ is asymptotically stable. At the critical point $(1 - \sqrt{2}, 1 + \sqrt{2})$, the trajectories resemble those near a saddle. Hence the critical point is unstable. Near the point $(1 + \sqrt{2}, 1 - \sqrt{2})$, the trajectories resemble those near a saddle. Hence the critical point is also unstable.

(d) Observing the direction field and the trajectories in (b), we can see that the basin of attraction of the origin is a complicated region including portions of all four quadrants.

10.(a) The critical points are solutions of the equations
$$(2 + x)(y - x) = 0$$
$$y(2 + x - x^2) = 0.$$

The origin is evidently a critical point. If $x = -2$, then $y = 0$. If $x = y$, then either $y = 0$ or $x = y = -1$ or $x = y = 2$. Hence the other critical points are $(-2, 0)$, $(-1, -1)$ and $(2, 2)$.

(b)

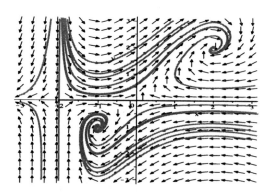

(c) Based on the global phase portrait, the critical points $(0,0)$ and $(-2,0)$ have the characteristics of a saddle. Hence these points are unstable. The behavior near the remaining two critical points resembles those near a stable spiral. Hence the critical points $(-1,-1)$ and $(2,2)$ are asymptotically stable.

(d) The basin of $(2,2)$ is the part of the upper half plane where $x > -2$, the basin of $(-1,-1)$ is the part of the lower half plane where $x > -2$.

11.(a) The critical points are given by the solution set of the equations
$$x(1 - 2y) = 0$$
$$y - x^2 - y^2 = 0.$$

If $x = 0$, then either $y = 0$ or $y = 1$. If $y = 1/2$, then $x = \pm 1/2$. Hence the critical points are at $(0,0)$, $(0,1)$, $(-1/2, 1/2)$ and $(1/2, 1/2)$.

(b)

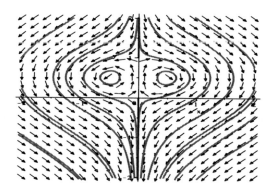

(c) The trajectories near the critical points $(-1/2, 1/2)$ and $(1/2, 1/2)$ are closed curves. Hence the critical points have the characteristics of a center, which is stable. The trajectories near the critical points $(0,0)$ and $(0,1)$ resemble those near a saddle. Hence these critical points are unstable.

(d) As the two stable critical points are centers, they have no basins of attraction. Trajectories near the critical points are ovals around those points.

13.(a) The critical points are solutions of the equations
$$(2+x)(y-x) = 0$$
$$(4-x)(y+x) = 0.$$
If $y = x$, then either $x = y = 0$ or $x = y = 4$. If $x = -2$, then $y = 2$. If $x = -y$, then $y = 2$ or $y = 0$. Hence the critical points are at $(0,0)$, $(4,4)$ and $(-2,2)$.

(b)

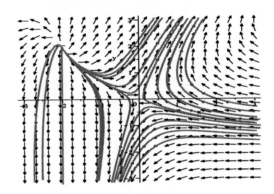

(c) The critical point at $(4,4)$ is evidently a stable spiral, which is asymptotically stable. Closer examination of the critical point at $(0,0)$ reveals that it is a saddle, which is unstable. The trajectories near the critical point $(-2,2)$ resemble those near an unstable node.

(d) The basin of attraction for $(4,4)$ consists of the points for which $x > -2$ and which lie over a curve (inferred from part (b)) passing through the origin and $(-2,2)$.

14.(a) The critical points are given by the solution set of the equations
$$(2-x)(y-x) = 0$$
$$y(2-x-x^2) = 0.$$
If $x = 2$, then $y = 0$. If $y = 0$, then $x = 0$. Also, $x = 1$ and $x = -2$ are roots of the second equation, and then $y = x$ from the first equation. Hence the critical points are at $(2,0)$, $(0,0)$, $(1,1)$ and $(-2,-2)$.

(b)

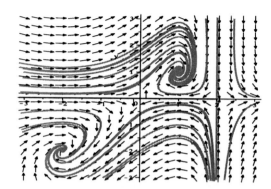

(c) The critical points $(0,0)$ and $(2,0)$ are saddles, hence unstable. The other two critical points are asymptotically stable spirals.

(d) The basin of $(1,1)$ is the part of the upper half plane where $x < 2$ (so all points (x,y) such that $x < 2$ and $y > 0$), the basin of $(-2,-2)$ is all points (x,y) such that $x < 2$ and $y < 0$.

15.(a) The critical points are given by the solution set of the equations
$$x(2 - x - y) = 0$$
$$-x + 3y - 2xy = 0.$$

If $x = 0$, then $y = 0$. The other critical points can be found by setting $y = 2 - x$ and substituting this into the second equation. This gives us $x = 3$ and $x = 1$. Hence the critical points are at $(0,0)$, $(1,1)$ and $(3,-1)$.

(b)

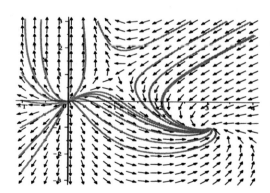

(c) The critical point $(1,1)$ is a saddle, hence unstable. $(0,0)$ is an unstable node and $(3,-1)$ is an asymptotically stable spiral.

(d) The basin of $(3,-1)$ consists of all points (x,y) for which $x > 0$ and $x > y$.

16.(a) The critical points are given by the solution set of the equations
$$x(2-x-y) = 0$$
$$(1-y)(2+x) = 0.$$

If $x = 0$, then $y = 1$. Also, when $x = -2$, then $y = 4$. The last critical point is given by $y = 1$, $x = 1$. Hence the critical points are at $(0,1)$, $(1,1)$ and $(-2,4)$.

(b)

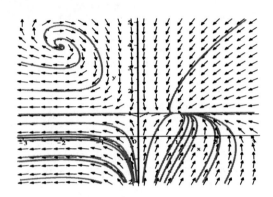

(c) The critical point $(0,1)$ is a saddle, hence unstable. $(-2,4)$ is an unstable spiral and $(1,1)$ is an asymptotically stable node.

(d) The basin of $(1,1)$ is the right half plane, i.e. all the points (x,y) for which $x > 0$.

18.(a) The trajectories are solutions of the differential equation
$$\frac{dy}{dx} = -\frac{4x}{y},$$
which can also be written as $4x\,dx + y\,dy = 0$. Integrating, we obtain
$$4x^2 + y^2 = C^2.$$

Hence the trajectories are ellipses.

(b)

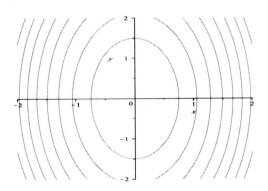

Based on the differential equations, the direction of motion on each trajectory is clockwise.

19.(a) The trajectories of the system satisfy the ODE

$$\frac{dy}{dx} = \frac{2x + y}{y},$$

which can also be written as $(2x + y)dx - ydy = 0$. This differential equation is homogeneous. Setting $y = x\,v(x)$, we obtain

$$v + x\frac{dv}{dx} = \frac{2}{v} + 1,$$

that is,

$$x\frac{dv}{dx} = \frac{2 + v - v^2}{v}.$$

The resulting ODE is separable, with solution $x^3(v + 1)(v - 2)^2 = C$. Reverting back to the original variables, the trajectories are level curves of

$$H(x, y) = (x + y)(y - 2x)^2.$$

(b)

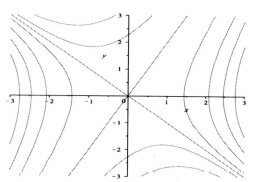

The origin is a saddle. Along the line $y = 2x$, solutions increase without bound. Along the line $y = -x$, solutions converge toward the origin.

20.(a) The trajectories are solutions of the differential equation

$$\frac{dy}{dx} = \frac{x + y}{x - y},$$

which is homogeneous. Setting $y = x\,v(x)$, we obtain

$$v + x\frac{dv}{dx} = \frac{x + xv}{x - xv},$$

that is,

$$x\frac{dv}{dx} = \frac{1 + v^2}{1 - v}.$$

The resulting ODE is separable, with solution

$$\arctan(v) = \ln|x|\sqrt{1+v^2}.$$

Reverting back to the original variables, the trajectories are level curves of

$$H(x,y) = \arctan(y/x) - \ln\sqrt{x^2+y^2}.$$

(b)

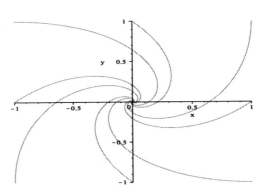

The origin is a stable spiral.

22.(a) The trajectories are solutions of the differential equation

$$\frac{dy}{dx} = \frac{-2xy^2 + 6xy}{2x^2y - 3x^2 - 4y},$$

which can also be written as $(2xy^2 - 6xy)dx + (2x^2y - 3x^2 - 4y)dy = 0$. The resulting ODE is exact, with

$$\frac{\partial H}{\partial x} = 2xy^2 - 6xy \quad \text{and} \quad \frac{\partial H}{\partial y} = 2x^2y - 3x^2 - 4y.$$

Integrating the first equation, we find that $H(x,y) = x^2y^2 - 3x^2y + f(y)$. It follows that

$$\frac{\partial H}{\partial y} = 2x^2y - 3x^2 + f'(y).$$

Comparing the two partial derivatives, we obtain $f(y) = -2y^2 + c$. Hence

$$H(x,y) = x^2y^2 - 3x^2y - 2y^2.$$

(b) The associated direction field shows the direction of motion along the trajectories.

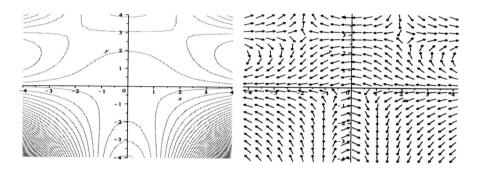

24.(a) The trajectories are solutions of the differential equation

$$\frac{dy}{dx} = \frac{-6x + x^3}{6y},$$

which can also be written as $(6x - x^3)dx + 6y\,dy = 0$. The resulting ODE is exact, with

$$\frac{\partial H}{\partial x} = 6x - x^3 \text{ and } \frac{\partial H}{\partial y} = 6y.$$

Integrating the first equation, we have $H(x,y) = 3x^2 - x^4/4 + f(y)$. It follows that

$$\frac{\partial H}{\partial y} = f'(y).$$

Comparing the two partial derivatives, we conclude that $f(y) = 3y^2 + c$. Hence

$$H(x,y) = 3x^2 - \frac{x^4}{4} + 3y^2.$$

(b)

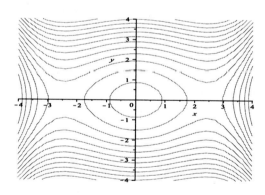

9.3

1. Write the system in the form $\mathbf{x}' = \mathbf{A}\mathbf{x} + \mathbf{g}(\mathbf{x})$. In this case, it is evident that

$$\frac{d}{dt}\begin{pmatrix} x \\ y \end{pmatrix} = \begin{pmatrix} 1 & 0 \\ 1 & -2 \end{pmatrix}\begin{pmatrix} x \\ y \end{pmatrix} + \begin{pmatrix} -y^2 \\ x^2 \end{pmatrix}.$$

That is, $\mathbf{g}(\mathbf{x}) = (-y^2, x^2)^T$. Using polar coordinates, $\|\mathbf{g}(\mathbf{x})\| = r^2\sqrt{\sin^4\theta + \cos^4\theta}$ and $\|\mathbf{x}\| = r$. Hence

$$\lim_{r\to 0}\frac{\|\mathbf{g}(\mathbf{x})\|}{\|\mathbf{x}\|} = \lim_{r\to 0} r\sqrt{\sin^4\theta + \cos^4\theta} = 0,$$

and the system is locally linear. The origin is an isolated critical point of the linear system

$$\frac{d}{dt}\begin{pmatrix} x \\ y \end{pmatrix} = \begin{pmatrix} 1 & 0 \\ 1 & -2 \end{pmatrix}\begin{pmatrix} x \\ y \end{pmatrix}.$$

The characteristic equation of the coefficient matrix is $r^2 + r - 2 = 0$, with roots $r_1 = 1$ and $r_2 = -2$. Hence the critical point is a saddle, which is unstable.

2. The system can be written as

$$\frac{d}{dt}\begin{pmatrix} x \\ y \end{pmatrix} = \begin{pmatrix} -1 & 1 \\ -4 & -1 \end{pmatrix}\begin{pmatrix} x \\ y \end{pmatrix} + \begin{pmatrix} 2xy \\ x^2 - y^2 \end{pmatrix}.$$

Following the discussion in Example 3, we note that $F(x,y) = -x + y + 2xy$ and $G(x,y) = -4x - y + x^2 - y^2$. Both of the functions F and G are twice differentiable, hence the system is locally linear. Furthermore,

$$F_x = -1 + 2y, \quad F_y = 1 + 2x, \quad G_x = -4 + 2x, \quad G_y = -1 - 2y.$$

The origin is an isolated critical point, with

$$\begin{pmatrix} F_x(0,0) & F_y(0,0) \\ G_x(0,0) & G_y(0,0) \end{pmatrix} = \begin{pmatrix} -1 & 1 \\ -4 & -1 \end{pmatrix}.$$

The characteristic equation of the associated linear system is $r^2 + 2r + 5 = 0$, with complex conjugate roots $r_{1,2} = -1 \pm 2i$. The origin is a stable spiral, which is asymptotically stable.

5.(a) The critical points consist of the solution set of the equations

$$(2+x)(y-x) = 0$$
$$(4-x)(y+x) = 0.$$

As shown in Problem 13 of Section 9.2, the only critical points are at $(0,0)$, $(4,4)$ and $(-2,2)$.

(b,c) First note that $F(x,y) = (2+x)(y-x)$ and $G(x,y) = (4-x)(y+x)$. The Jacobian matrix of the vector field is

$$\mathbf{J} = \begin{pmatrix} F_x(x,y) & F_y(x,y) \\ G_x(x,y) & G_y(x,y) \end{pmatrix} = \begin{pmatrix} -2 - 2x + y & 2 + x \\ 4 - y - 2x & 4 - x \end{pmatrix}.$$

At the origin, the coefficient matrix of the linearized system is

$$\mathbf{J}(0,0) = \begin{pmatrix} -2 & 2 \\ 4 & 4 \end{pmatrix},$$

with eigenvalues $r_1 = 1 - \sqrt{17}$ and $r_2 = 1 + \sqrt{17}$. The eigenvalues are real, with opposite sign. Hence the critical point is a saddle, which is unstable. At the point $(-2, 2)$, the coefficient matrix of the linearized system is

$$\mathbf{J}(-2, 2) = \begin{pmatrix} 4 & 0 \\ 6 & 6 \end{pmatrix},$$

with eigenvalues $r_1 = 4$ and $r_2 = 6$. The eigenvalues are real, unequal and positive, hence the critical point is an unstable node. At the point $(4, 4)$, the coefficient matrix of the linearized system is

$$\mathbf{J}(4, 4) = \begin{pmatrix} -6 & 6 \\ -8 & 0 \end{pmatrix},$$

with complex conjugate eigenvalues $r_{1,2} = -3 \pm i\sqrt{39}$. The critical point is a stable spiral, which is asymptotically stable. Based on Table 9.3.1, the nonlinear terms do not affect the stability and type of each critical point.

(d)

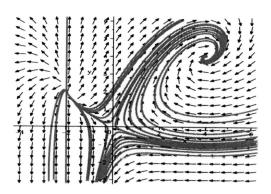

7.(a) The critical points are solutions of the equations

$$1 - y = 0$$
$$(x - y)(x + y) = 0.$$

The first equation requires that $y = 1$. Based on the second equation, $x = \pm 1$. Hence the critical points are $(-1, 1)$ and $(1, 1)$.

(b,c) $F(x, y) = 1 - y$ and $G(x, y) = x^2 - y^2$. The Jacobian matrix of the vector field is

$$\mathbf{J} = \begin{pmatrix} F_x(x, y) & F_y(x, y) \\ G_x(x, y) & G_y(x, y) \end{pmatrix} = \begin{pmatrix} 0 & -1 \\ 2x & -2y \end{pmatrix}.$$

At the critical point $(-1,1)$, the coefficient matrix of the linearized system is

$$\mathbf{J}(-1,1) = \begin{pmatrix} 0 & -1 \\ -2 & -2 \end{pmatrix},$$

with eigenvalues $r_1 = -1 - \sqrt{3}$ and $r_2 = -1 + \sqrt{3}$. The eigenvalues are real, with opposite sign. Hence the critical point is a saddle, which is unstable. At the point $(1,1)$, the coefficient matrix of the linearized system is

$$\mathbf{J}(1,1) = \begin{pmatrix} 0 & -1 \\ 2 & -2 \end{pmatrix},$$

with complex conjugate eigenvalues $r_{1,2} = -1 \pm i$. The critical point is a stable spiral, which is asymptotically stable.

(d)

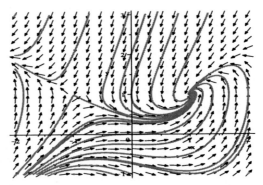

Based on Table 9.3.1, the nonlinear terms do not affect the stability and type of each critical point.

8.(a) The critical points are given by the solution set of the equations

$$x(1 - x - y) = 0$$
$$y(2 - y - 3x) = 0.$$

If $x = 0$, then either $y = 0$ or $y = 2$. If $y = 0$, then $x = 0$ or $x = 1$. If $y = 1 - x$, then either $x = 1/2$ or $x = 1$. If $y = 2 - 3x$, then $x = 0$ or $x = 1/2$. Hence the critical points are at $(0,0)$, $(0,2)$, $(1,0)$ and $(1/2, 1/2)$.

(b,c) Note that $F(x,y) = x - x^2 - xy$ and $G(x,y) = (2y - y^2 - 3xy)/4$. The Jacobian matrix of the vector field is

$$\mathbf{J} = \begin{pmatrix} F_x(x,y) & F_y(x,y) \\ G_x(x,y) & G_y(x,y) \end{pmatrix} = \begin{pmatrix} 1 - 2x - y & -x \\ -3y/4 & 1/2 - y/2 - 3x/4 \end{pmatrix}.$$

At the origin, the coefficient matrix of the linearized system is

$$\mathbf{J}(0,0) = \begin{pmatrix} 1 & 0 \\ 0 & \frac{1}{2} \end{pmatrix},$$

with eigenvalues $r_1 = 1$ and $r_2 = 1/2$. The eigenvalues are real and both positive. Hence the critical point is an unstable node. At the point $(0,2)$, the coefficient matrix of the linearized system is

$$\mathbf{J}(0,2) = \begin{pmatrix} -1 & 0 \\ -\frac{3}{2} & -\frac{1}{2} \end{pmatrix},$$

with eigenvalues $r_1 = -1$ and $r_2 = -1/2$. The eigenvalues are both negative, hence the critical point is a stable node. At the point $(1,0)$, the coefficient matrix of the linearized system is

$$\mathbf{J}(1,0) = \begin{pmatrix} -1 & -1 \\ 0 & -\frac{1}{4} \end{pmatrix},$$

with eigenvalues $r_1 = -1$ and $r_2 = -1/4$. Both of the eigenvalues are negative, and hence the critical point is a stable node. At the critical point $(1/2, 1/2)$, the coefficient matrix of the linearized system is

$$\mathbf{J}(1/2, 1/2) = \begin{pmatrix} -1/2 & -1/2 \\ -3/8 & -1/8 \end{pmatrix},$$

with eigenvalues $r_1 = -5/16 - \sqrt{57}/16$ and $r_2 = -5/16 + \sqrt{57}/16$. The eigenvalues are real, with opposite sign. Hence the critical point is a saddle, which is unstable.

(d)

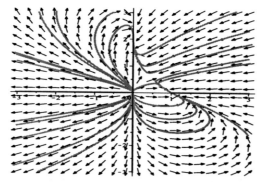

Based on Table 9.3.1, the nonlinear terms do not affect the stability and type of each critical point.

9.(a) The critical points are given by the solution set of the equations

$$(2+y)(y - x/2) = 0$$
$$(2-x)(y + x/2) = 0.$$

If $y = -2$, then either $x = 2$ or $x = 4$. If $x = 2$, then $y = -2$ or $y = 1$. Also, $x = y = 0$ is a solution. Hence the critical points are at $(0,0)$, $(2,-2)$, $(4,-2)$ and $(2,1)$.

(b,c) Note that $F(x,y) = 2y + y^2 - x - xy/2$ and $G(x,y) = 2y - xy + x - x^2/2$. The Jacobian matrix of the vector field is

$$\mathbf{J} = \begin{pmatrix} F_x(x,y) & F_y(x,y) \\ G_x(x,y) & G_y(x,y) \end{pmatrix} = \begin{pmatrix} -1 - y/2 & 2 + 2y - x/2 \\ -y + 1 - x & 2 - x \end{pmatrix}.$$

At the origin, the coefficient matrix of the linearized system is

$$\mathbf{J}(0,0) = \begin{pmatrix} -1 & 2 \\ 1 & 2 \end{pmatrix},$$

with eigenvalues $r_1 = (1 + \sqrt{17})/2$ and $r_2 = (1 - \sqrt{17})/2$. The eigenvalues are real, with opposite sign. Hence the critical point is a saddle, which is unstable. At the point $(2, -2)$, the coefficient matrix of the linearized system is

$$\mathbf{J}(2, -2) = \begin{pmatrix} 0 & -3 \\ 1 & 0 \end{pmatrix},$$

with eigenvalues $r_1 = \sqrt{3}i$ and $r_2 = -\sqrt{3}i$. The eigenvalues are purely imaginary, hence the critical point is either a center or a spiral point. At the point $(4, -2)$, the coefficient matrix of the linearized system is

$$\mathbf{J}(4, -2) = \begin{pmatrix} 0 & -4 \\ -1 & -2 \end{pmatrix},$$

with eigenvalues $r_1 = -1 + \sqrt{5}$ and $r_2 = -1 - \sqrt{5}$. The eigenvalues are real, with opposite sign. Hence the critical point is a saddle, which is unstable. At the critical point $(2, 1)$, the coefficient matrix of the linearized system is

$$\mathbf{J}(2, 1) = \begin{pmatrix} -3/2 & 3 \\ -2 & 0 \end{pmatrix},$$

with eigenvalues $r_1 = -3/4 + \sqrt{87}i/4$ and $r_2 = -3/4 - \sqrt{87}i/4$. The critical point is a stable spiral, which is asymptotically stable.

(d)

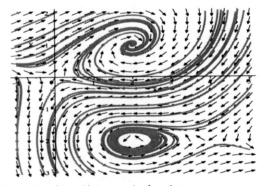

We observe that the point $(2, -2)$ is a spiral point.

11.(a) The critical points are solutions of the equations
$$2x + y + xy^3 = 0$$
$$x - 2y - xy = 0.$$

Substitution of $y = x/(x+2)$ into the first equation results in
$$3x^4 + 13x^3 + 28x^2 + 20x = 0.$$

One root of the resulting equation is $x = 0$. The only other real root of the equation is
$$x = \frac{1}{9}\left[(287 + 18\sqrt{2019})^{1/3} - 83(287 + 18\sqrt{2019})^{-1/3} - 13\right].$$

Hence the critical points are $(0,0)$ and $(-1.19345\ldots, -1.4797\ldots)$.

(b,c) $F(x,y) = 2x + y + xy^3$ and $G(x,y) = x - 2y - xy$. The Jacobian matrix of the vector field is
$$\mathbf{J} = \begin{pmatrix} F_x(x,y) & F_y(x,y) \\ G_x(x,y) & G_y(x,y) \end{pmatrix} = \begin{pmatrix} 2 + y^3 & 1 + 3xy^2 \\ 1 - y & -2 - x \end{pmatrix}.$$

At the origin, the coefficient matrix of the linearized system is
$$\mathbf{J}(0,0) = \begin{pmatrix} 2 & 1 \\ 1 & -2 \end{pmatrix},$$

with eigenvalues $r_1 = \sqrt{5}$ and $r_2 = -\sqrt{5}$. The eigenvalues are real and of opposite sign. Hence the critical point is a saddle, which is unstable. At the point $(-1.19345\ldots, -1.4797\ldots)$, the coefficient matrix of the linearized system is
$$\mathbf{J}(-1.19345, -1.4797) = \begin{pmatrix} -1.2399 & -6.8393 \\ 2.4797 & -0.8065 \end{pmatrix},$$

with complex conjugate eigenvalues $r_{1,2} = -1.0232 \pm 4.1125\,i$. The critical point is a stable spiral, which is asymptotically stable.

(d)

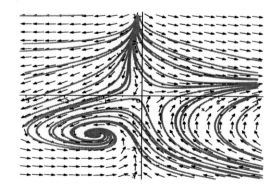

In both cases, the nonlinear terms do not affect the stability and type of the critical point.

12.(a) The critical points are given by the solution set of the equations

$$(1+x)\sin y = 0$$
$$1 - x - \cos y = 0.$$

If $x = -1$, then we must have $\cos y = 2$, which is impossible. Therefore $\sin y = 0$, which implies that $y = n\pi$, $n = 0, \pm 1, 2, \ldots$. Based on the second equation,

$$x = 1 - \cos n\pi.$$

It follows that the critical points are located at $(0, 2k\pi)$ and $(2, (2k+1)\pi)$, where $k = 0, \pm 1, \pm 2, \ldots$.

(b,c) Given that $F(x,y) = (1+x)\sin y$ and $G(x,y) = 1 - x - \cos y$, the Jacobian matrix of the vector field is

$$\mathbf{J} = \begin{pmatrix} \sin y & (1+x)\cos y \\ -1 & \sin y \end{pmatrix}.$$

At the critical points $(0, 2k\pi)$, the coefficient matrix of the linearized system is

$$\mathbf{J}(0, 2k\pi) = \begin{pmatrix} 0 & 1 \\ -1 & 0 \end{pmatrix},$$

with purely complex eigenvalues $r_{1,2} = \pm i$. The critical points of the associated linear systems are centers, which are stable. Note that Theorem 9.3.2 does not provide a definite conclusion regarding the relation between the nature of the critical points of the nonlinear systems and their corresponding linearizations. At the points $(2, (2k+1)\pi)$, the coefficient matrix of the linearized system is

$$\mathbf{J}[2, (2k+1)\pi] = \begin{pmatrix} 0 & -3 \\ -1 & 0 \end{pmatrix},$$

with eigenvalues $r_1 = \sqrt{3}$ and $r_2 = -\sqrt{3}$. The eigenvalues are real, with opposite sign. Hence the critical points of the associated linear systems are saddles, which are unstable.

(d)

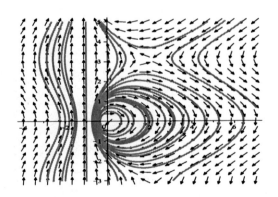

As asserted in Theorem 9.3.2, the trajectories near the critical points $(2,(2k+1)\pi)$ resemble those near a saddle. Upon closer examination, the critical points $(0,2k\pi)$ are indeed centers.

13.(a) The critical points are solutions of the equations

$$x - y^2 = 0$$
$$y - x^2 = 0.$$

Substitution of $y = x^2$ into the first equation results in

$$x - x^4 = 0,$$

with real roots $x = 0, 1$. Hence the critical points are at $(0,0)$ and $(1,1)$.

(b,c) In this problem, $F(x,y) = x - y^2$ and $G(x,y) = y - x^2$. The Jacobian matrix of the vector field is

$$\mathbf{J} = \begin{pmatrix} 1 & -2y \\ -2x & 1 \end{pmatrix}.$$

At the origin, the coefficient matrix of the linearized system is

$$\mathbf{J}(0,0) = \begin{pmatrix} 1 & 0 \\ 0 & 1 \end{pmatrix},$$

with repeated eigenvalues $r_1 = 1$ and $r_2 = 1$. It is easy to see that the corresponding eigenvectors are linearly independent. Hence the critical point is an unstable proper node. Theorem 9.3.2 does not provide a definite conclusion regarding the relation between the nature of the critical point of the nonlinear system and the corresponding linearization. At the critical point $(1,1)$, the coefficient matrix of the linearized system is

$$\mathbf{J}(1,1) = \begin{pmatrix} 1 & -2 \\ -2 & 1 \end{pmatrix},$$

with eigenvalues $r_1 = 3$ and $r_2 = -1$. The eigenvalues are real, with opposite sign. Hence the critical point is a saddle, which is unstable.

(d)

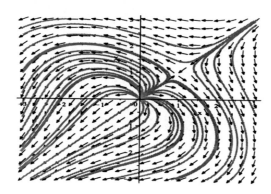

Closer examination reveals that the critical point at the origin is indeed a proper node.

14.(a) The critical points are given by the solution set of the equations
$$1 - xy = 0$$
$$x - y^3 = 0.$$

After multiplying the second equation by y, it follows that $y = \pm 1$. Hence the critical points of the system are at $(1,1)$ and $(-1,-1)$.

(b,c) Note that $F(x,y) = 1 - xy$ and $G(x,y) = x - y^3$. The Jacobian matrix of the vector field is
$$\mathbf{J} = \begin{pmatrix} -y & -x \\ 1 & -3y^2 \end{pmatrix}.$$

At the critical point $(1,1)$, the coefficient matrix of the linearized system is
$$\mathbf{J}(1,1) = \begin{pmatrix} -1 & -1 \\ 1 & -3 \end{pmatrix},$$

with eigenvalues $r_1 = -2$ and $r_2 = -2$. The eigenvalues are real and equal. It is easy to show that there is only one linearly independent eigenvector. Hence the critical point is a stable improper node. Theorem 9.3.2 does not provide a definite conclusion regarding the relation between the nature of the critical point of the nonlinear system and the corresponding linearization. At the point $(-1,-1)$, the coefficient matrix of the linearized system is
$$\mathbf{J}(-1,-1) = \begin{pmatrix} 1 & 1 \\ 1 & -3 \end{pmatrix},$$

with eigenvalues $r_1 = -1 + \sqrt{5}$ and $r_2 = -1 - \sqrt{5}$. The eigenvalues are real, with opposite sign. Hence the critical point of the associated linear system is a saddle, which is unstable.

(d)

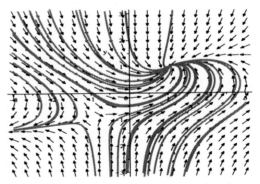

Closer examination reveals that the critical point at $(1,1)$ is indeed a stable improper node, which is asymptotically stable.

15.(a) The critical points are given by the solution set of the equations
$$-2x - y - x(x^2 + y^2) = 0$$
$$x - y + y(x^2 + y^2) = 0.$$
It is clear that the origin is a critical point. Solving the first equation for y, we find that
$$y = \frac{-1 \pm \sqrt{1 - 8x^2 - 4x^4}}{2x}.$$
Substitution of these relations into the second equation results in two equations of the form $f_1(x) = 0$ and $f_2(x) = 0$. Plotting these functions, we note that only $f_1(x) = 0$ has real roots given by $x \approx \pm 0.33076$. It follows that the additional critical points are at $(-0.33076, 1.0924)$ and $(0.33076, -1.0924)$.

(b,c) Given that
$$F(x, y) = -2x - y - x(x^2 + y^2)$$
$$G(x, y) = x - y + y(x^2 + y^2),$$
the Jacobian matrix of the vector field is
$$\mathbf{J} = \begin{pmatrix} -2 - 3x^2 - y^2 & -1 - 2xy \\ 1 + 2xy & -1 + x^2 + 3y^2 \end{pmatrix}.$$
At the critical point $(0, 0)$, the coefficient matrix of the linearized system is
$$\mathbf{J}(0, 0) = \begin{pmatrix} -2 & -1 \\ 1 & -1 \end{pmatrix},$$
with complex conjugate eigenvalues $r_{1,2} = (-3 \pm i\sqrt{3})/2$. Hence the critical point is a stable spiral, which is asymptotically stable. At the point $(-0.33076, 1.0924)$, the coefficient matrix of the linearized system is
$$\mathbf{J}(-0.33076, 1.0924) = \begin{pmatrix} -3.5216 & -0.27735 \\ 0.27735 & 2.6895 \end{pmatrix},$$
with eigenvalues $r_1 = -3.5092$ and $r_2 = 2.6771$. The eigenvalues are real, with opposite sign. Hence the critical point of the associated linear system is a saddle, which is unstable. Identical results hold for the point at $(0.33076, -1.0924)$.

(d)

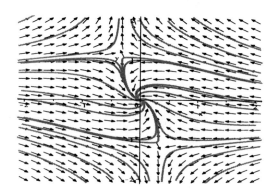

16.(a) The critical points are solutions of the equations

$$y + x(1 - x^2 - y^2) = 0$$
$$-x + y(1 - x^2 - y^2) = 0.$$

Multiply the first equation by y and the second equation by x. The difference of the two equations gives $x^2 + y^2 = 0$. Hence the only critical point is at the origin.

(b,c) With $F(x,y) = y + x(1 - x^2 - y^2)$ and $G(x,y) = -x + y(1 - x^2 - y^2)$, the Jacobian matrix of the vector field is

$$\mathbf{J} = \begin{pmatrix} 1 - 3x^2 - y^2 & 1 - 2xy \\ -1 - 2xy & 1 - x^2 - 3y^2 \end{pmatrix}.$$

At the origin, the coefficient matrix of the linearized system is

$$\mathbf{J}(0,0) = \begin{pmatrix} 1 & 1 \\ -1 & 1 \end{pmatrix},$$

with complex conjugate eigenvalues $r_{1,2} = 1 \pm i$. Hence the origin is an unstable spiral.

(d)

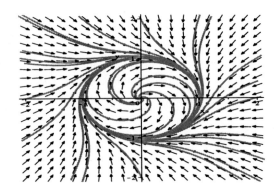

17.(a) The critical points are given by the solution set of the equations

$$4 - y^2 = 0$$
$$(x + 3/2)(y - x) = 0.$$

Clearly, $y = \pm 2$. The second equation tells us that $x = -3/2$ or $x = y$. Hence the critical points are at $(-3/2, 2)$, $(-3/2, -2)$, $(2, 2)$ and $(-2, -2)$.

(b,c) Note that $F(x,y) = 4 - y^2$ and $G(x,y) = xy + 3y/2 - x^2 - 3x/2$. The Jacobian matrix of the vector field is

$$\mathbf{J} = \begin{pmatrix} F_x(x,y) & F_y(x,y) \\ G_x(x,y) & G_y(x,y) \end{pmatrix} = \begin{pmatrix} 0 & -2y \\ y - 2x - 3/2 & x + 3/2 \end{pmatrix}.$$

At $(-3/2, 2)$, the coefficient matrix of the linearized system is

$$\mathbf{J}(-3/2, 2) = \begin{pmatrix} 0 & -4 \\ 7/2 & 0 \end{pmatrix},$$

with eigenvalues $r_1 = \sqrt{14}i$ and $r_2 = -\sqrt{14}i$. The eigenvalues are purely imaginary, hence the critical point is either a center or a spiral point. At the point $(-3/2, -2)$, the coefficient matrix of the linearized system is

$$\mathbf{J}(-3/2, -2) = \begin{pmatrix} 0 & 4 \\ -1/2 & 0 \end{pmatrix},$$

with eigenvalues $r_1 = \sqrt{2}i$ and $r_2 = -\sqrt{2}i$. The eigenvalues are purely imaginary, hence the critical point is either a center or a spiral point. At the point $(2, 2)$, the coefficient matrix of the linearized system is

$$\mathbf{J}(2, 2) = \begin{pmatrix} 0 & -4 \\ -7/2 & 7/2 \end{pmatrix},$$

with eigenvalues $r_1 = (7 + \sqrt{273})/4$ and $r_2 = (7 - \sqrt{273})/4$. The eigenvalues are real, with opposite sign. Hence the critical point is a saddle, which is unstable. At the critical point $(-2, -2)$, the coefficient matrix of the linearized system is

$$\mathbf{J}(-2, -2) = \begin{pmatrix} 0 & 4 \\ 1/2 & -1/2 \end{pmatrix},$$

with eigenvalues $r_1 = (-1 + \sqrt{33})/4$ and $r_2 = -1 - \sqrt{33})/4$. Hence the critical point is a saddle, which is unstable.

(d)

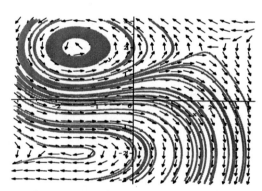

Further observation indicates that $(-3/2, 2)$ and $(-3/2, -2)$ are unstable spiral points.

19.(a) The Jacobian matrix of the vector field is

$$\mathbf{J} = \begin{pmatrix} 0 & 1 \\ 1 + 6x^2 & 0 \end{pmatrix}.$$

At the origin, the coefficient matrix of the linearized system is

$$\mathbf{J}(0, 0) = \begin{pmatrix} 0 & 1 \\ 1 & 0 \end{pmatrix},$$

with eigenvalues $r_1 = 1$ and $r_2 = -1$. The eigenvalues are real, with opposite sign. Hence the critical point is a saddle point.

(b) The trajectories of the linearized system are solutions of the differential equation

$$\frac{dy}{dx} = \frac{x}{y},$$

which is separable. Integrating both sides of the equation $x\,dx - y\,dy = 0$, the solution is $x^2 - y^2 = C$. The trajectories consist of a family of hyperbolas.

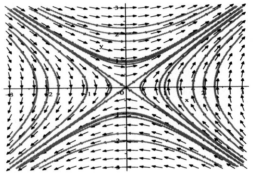

It is easy to show that the general solution is given by $x(t) = c_1 e^t + c_2 e^{-t}$ and $y(t) = c_1 e^t - c_2 e^{-t}$. The only bounded solutions consist of those for which $c_1 = 0$. In that case, $x(t) = c_2 e^{-t} = -y(t)$.

(c) The trajectories of the given system are solutions of the differential equation

$$\frac{dy}{dx} = \frac{x + 2x^3}{y},$$

which can also be written as $(x + 2x^3)dx - y\,dy = 0$. The resulting ODE is exact, with

$$\frac{\partial H}{\partial x} = x + 2x^3 \quad \text{and} \quad \frac{\partial H}{\partial y} = -y.$$

Integrating the first equation, we find that $H(x,y) = x^2/2 + x^4/2 + f(y)$. It follows that

$$\frac{\partial H}{\partial y} = f'(y).$$

Comparing the partial derivatives, we obtain $f(y) = -y^2/2 + c$. Hence the solutions are level curves of the function

$$H(x,y) = x^2/2 + x^4/2 - y^2/2.$$

(The equation is also separable. Separation of variables yields the same $H(x,y)$.) The trajectories approaching to, or diverging from, the origin are no longer straight lines.

21.(a) The solutions of the system of equations

$$y = 0$$
$$-\omega^2 \sin x = 0$$

consist of the points $(\pm n\pi, 0)$, $n = 0, 1, 2, \ldots$. The functions $F(x, y) = y$ and $G(x, y) = -\omega^2 \sin x$ are analytic on the entire plane. It follows that the system is locally linear near each of the critical points.

(b) The Jacobian matrix of the vector field is

$$\mathbf{J} = \begin{pmatrix} 0 & 1 \\ -\omega^2 \cos x & 0 \end{pmatrix}.$$

At the origin, the coefficient matrix of the linearized system is

$$\mathbf{J}(0, 0) = \begin{pmatrix} 0 & 1 \\ -\omega^2 & 0 \end{pmatrix},$$

with purely complex eigenvalues $r_{1,2} = \pm i\omega$. Hence the origin is a center. Since the eigenvalues are purely complex, Theorem 9.3.2 gives no definite conclusion about the critical point of the nonlinear system. Physically, the critical point corresponds to the state $\theta = 0$, $\theta' = 0$. That is, the rest configuration of the pendulum.

(c) At the critical point $(\pi, 0)$, the coefficient matrix of the linearized system is

$$\mathbf{J}(\pi, 0) = \begin{pmatrix} 0 & 1 \\ \omega^2 & 0 \end{pmatrix},$$

with eigenvalues $r_{1,2} = \pm \omega$. The eigenvalues are real and of opposite sign. Hence the critical point is a saddle. Theorem 9.3.2 asserts that the critical point for the nonlinear system is also a saddle, which is unstable. This critical point corresponds to the state $\theta = \pi$, $\theta' = 0$. That is, the upright rest configuration.

(d) Let $\omega^2 = 1$. The following is a plot of the phase curves near $(0, 0)$.

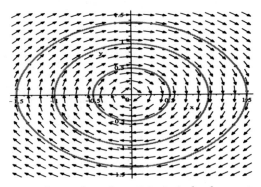

The local phase portrait shows that the origin is indeed a center.

(e)

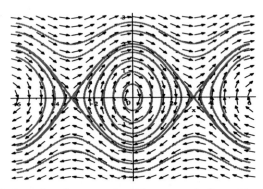

It should be noted that the phase portrait has a periodic pattern, since $\theta = x \bmod 2\pi$.

22.(a) The trajectories of the system in Problem 21 are solutions of the differential equation

$$\frac{dy}{dx} = \frac{-\omega^2 \sin x}{y},$$

which can also be written as $\omega^2 \sin x \, dx + y \, dy = 0$. The resulting ODE is exact, with

$$\frac{\partial H}{\partial x} = \omega^2 \sin x \quad \text{and} \quad \frac{\partial H}{\partial y} = y.$$

Integrating the first equation, we find that $H(x,y) = -\omega^2 \cos x + f(y)$. It follows that

$$\frac{\partial H}{\partial y} = f'(y).$$

Comparing the partial derivatives, we obtain $f(y) = y^2/2 + C$. Hence the solutions are level curves of the function

$$H(x,y) = -\omega^2 \cos x + y^2/2.$$

Adding an arbitrary constant, say ω^2, to the function $H(x,y)$ does not change the nature of the level curves. Hence the trajectories are can be written as
$$\frac{1}{2}y^2 + \omega^2(1 - \cos x) = c,$$
in which c is an arbitrary constant.

(b) Multiplying by mL^2 and reverting to the original physical variables, we obtain
$$\frac{1}{2}mL^2(\frac{d\theta}{dt})^2 + mL^2\omega^2(1 - \cos\theta) = mL^2 c.$$
Since $\omega^2 = g/L$, the equation can be written as
$$\frac{1}{2}mL^2(\frac{d\theta}{dt})^2 + mgL(1 - \cos\theta) = E,$$
in which $E = mL^2 c$.

(c) The absolute velocity of the point mass is given by $v = L\, d\theta/dt$. The kinetic energy of the mass is $T = mv^2/2$. Choosing the rest position as the datum, that is, the level of zero potential energy, the gravitational potential energy of the point mass is
$$V = mgL(1 - \cos\theta).$$
It follows that the total energy, $T + V$, is constant along the trajectories.

23.(a) $A = 0.25$

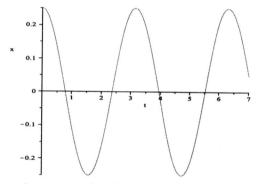

Since the system is undamped, and $y(0) = 0$, the amplitude is 0.25. The period is estimated at $\tau \approx 3.16$.

(b)

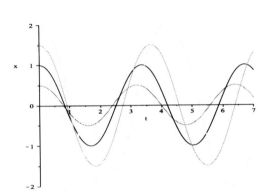

	R	τ
$A = 0.5$	0.5	3.20
$A = 1.0$	1.0	3.35
$A = 1.5$	1.5	3.63
$A = 2.0$	2.0	4.17

(c) Since the system is conservative, the amplitude is equal to the initial amplitude. On the other hand, the period of the pendulum is a monotone increasing function of the initial position A.

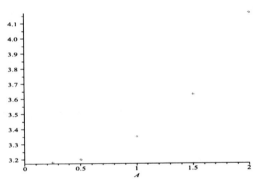

It appears that as $A \to 0$, the period approaches π, the period of the corresponding linear pendulum $(2\pi/\omega)$.

(d)

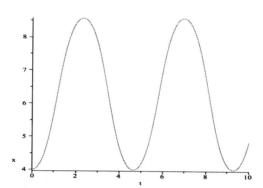

The pendulum is released from rest, at an inclination of $4 - \pi$ radians from the vertical. Based on conservation of energy, the pendulum will swing past the lower equilibrium position ($\theta = 2\pi$) and come to rest, momentarily, at a maximum rotational displacement of $\theta_{max} = 3\pi - (4 - \pi) = 4\pi - 4$. The transition between the two dynamics occurs at $A = \pi$, that is, once the pendulum is released beyond the upright configuration.

26.(a) It is evident that the origin is a critical point of each system. Furthermore, it is easy to see that the corresponding linear system, in each case, is given by

$$\frac{dx}{dt} = y$$
$$\frac{dy}{dt} = -x.$$

The eigenvalues of the coefficient matrix are $r_{1,2} = \pm i$. Hence the critical point of the linearized system is a center.

(b) Using polar coordinates, it is also easy to show that

$$\lim_{r \to 0} \frac{\|\mathbf{g}(\mathbf{x})\|}{\|\mathbf{x}\|} = 0.$$

Alternatively, the nonlinear terms are analytic in the entire plane. Hence both systems are locally linear near the origin.

(c) For system (ii), note that

$$x\frac{dx}{dt} + y\frac{dy}{dt} = xy - x^2(x^2 + y^2) - xy - y^2(x^2 + y^2).$$

Converting to polar coordinates, and differentiating the equation $r^2 = x^2 + y^2$ with respect to t, we find that

$$r\frac{dr}{dt} = x\frac{dx}{dt} + y\frac{dy}{dt} = -(x^2 + y^2)^2 = -r^4.$$

That is, $r' = -r^3$. It follows that $r^2 = 1/(2t + c)$, where $c = 1/r_0^2$. Since $r \to 0$ as $t \to \infty$ and $dr/dt < 0$, regardless of the value of r_0, the origin is an asymptotically stable critical point.

On the other hand, for system (i),

$$r\frac{dr}{dt} = x\frac{dx}{dt} + y\frac{dy}{dt} = (x^2+y^2)^2 = r^4.$$

That is, $r' = r^3$. Solving the differential equation results in

$$r^2 = \frac{c-2t}{(2t-c)^2}.$$

Imposing the initial condition $r(0) = r_0$, we obtain a specific solution

$$r^2 = -\frac{r_0^2}{2r_0^2 t - 1}.$$

Since the solution becomes unbounded as $t \to 1/2r_0^2$, the critical point is unstable.

27. The characteristic equation of the coefficient matrix is $r^2 + 1 = 0$, with complex roots $r_{1,2} = \pm i$. Hence the critical point at the origin is a center. The characteristic equation of the perturbed matrix is $r^2 - 2\epsilon r + 1 + \epsilon^2 = 0$, with complex conjugate roots $r_{1,2} = \epsilon \pm i$. As long as $\epsilon \neq 0$, the critical point of the perturbed system is a spiral point. Its stability depends on the sign of ϵ.

28. The characteristic equation of the coefficient matrix is $(r+1)^2 = 0$, with roots $r_1 = r_2 = -1$. Hence the critical point is an asymptotically stable node. On the other hand, the characteristic equation of the perturbed system is $r^2 + 2r + 1 + \epsilon = 0$, with roots $r_{1,2} = -1 \pm \sqrt{-\epsilon}$. If $\epsilon > 0$, then $r_{1,2} = -1 \pm i\sqrt{\epsilon}$ are complex roots. The critical point is a stable spiral. If $\epsilon < 0$, then $r_{1,2} = -1 \pm \sqrt{|\epsilon|}$ are real and both negative ($|\epsilon| < 1$). The critical point remains a stable node.

9.4

1.(a)

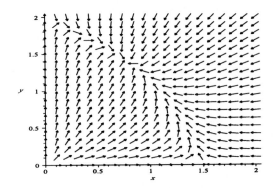

(b) The critical points are solutions of the system of equations
$$x(1.5 - x - 0.5y) = 0$$
$$y(2 - y - 0.75x) = 0.$$
The four critical points are $(0,0)$, $(0,2)$, $(1.5,0)$ and $(0.8,1.4)$.

(c) The Jacobian matrix of the vector field is
$$\mathbf{J} = \begin{pmatrix} 3/2 - 2x - y/2 & -x/2 \\ -3y/4 & 2 - 3x/4 - 2y \end{pmatrix}.$$
At the critical point $(0,0)$, the coefficient matrix of the linearized system is
$$\mathbf{J}(0,0) = \begin{pmatrix} 3/2 & 0 \\ 0 & 2 \end{pmatrix}.$$
The eigenvalues and eigenvectors are
$$r_1 = 3/2 \,, \boldsymbol{\xi}^{(1)} = \begin{pmatrix} 1 \\ 0 \end{pmatrix} \,;\, r_2 = 2 \,, \boldsymbol{\xi}^{(2)} = \begin{pmatrix} 0 \\ 1 \end{pmatrix}.$$
The eigenvalues are positive, hence the origin is an unstable node.
At the critical point $(0,2)$, the coefficient matrix of the linearized system is
$$\mathbf{J}(0,2) = \begin{pmatrix} 1/2 & 0 \\ -3/2 & -2 \end{pmatrix}.$$
The eigenvalues and eigenvectors are
$$r_1 = 1/2 \,, \boldsymbol{\xi}^{(1)} = \begin{pmatrix} 1 \\ -0.6 \end{pmatrix} \,;\, r_2 = -2 \,, \boldsymbol{\xi}^{(2)} = \begin{pmatrix} 0 \\ 1 \end{pmatrix}.$$
The eigenvalues are of opposite sign. Hence the critical point is a saddle, which is unstable.
At the critical point $(1.5,0)$, the coefficient matrix of the linearized system is
$$\mathbf{J}(1.5,0) = \begin{pmatrix} -1.5 & -0.75 \\ 0 & 0.875 \end{pmatrix}.$$
The eigenvalues and eigenvectors are
$$r_1 = -1.5 \,, \boldsymbol{\xi}^{(1)} = \begin{pmatrix} 1 \\ 0 \end{pmatrix} \,;\, r_2 = 0.875 \,, \boldsymbol{\xi}^{(2)} = \begin{pmatrix} -0.31579 \\ 1 \end{pmatrix}.$$
The eigenvalues are of opposite sign. Hence the critical point is also a saddle, which is unstable.
At the critical point $(0.8, 1.4)$, the coefficient matrix of the linearized system is
$$\mathbf{J}(0.8, 1.4) = \begin{pmatrix} -0.8 & -0.4 \\ -1.05 & -1.4 \end{pmatrix}.$$
The eigenvalues and eigenvectors are
$$r_1 = -\frac{11}{10} + \frac{\sqrt{51}}{10} \,, \boldsymbol{\xi}^{(1)} = \begin{pmatrix} 1 \\ \frac{3-\sqrt{51}}{4} \end{pmatrix} \,;\, r_2 = -\frac{11}{10} - \frac{\sqrt{51}}{10} \,, \boldsymbol{\xi}^{(2)} = \begin{pmatrix} 1 \\ \frac{3+\sqrt{51}}{4} \end{pmatrix}.$$

The eigenvalues are both negative. Hence the critical point is a stable node, which is asymptotically stable.

(d,e)

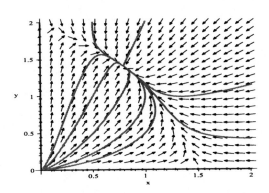

(f) Except for initial conditions lying on the coordinate axes, almost all trajectories converge to the stable node at $(0.8, 1.4)$. Thus the species can coexist.

2.(a)

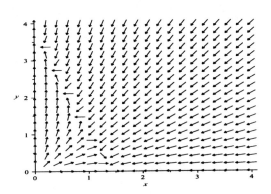

(b) The critical points are the solution set of the system of equations
$$x(1.5 - x - 0.5\,y) = 0$$
$$y(2 - 0.5\,y - 1.5\,x) = 0\,.$$

The four critical points are $(0,0)$, $(0,4)$, $(1.5,0)$ and $(1,1)$.

(c) The Jacobian matrix of the vector field is
$$\mathbf{J} = \begin{pmatrix} 3/2 - 2x - y/2 & -x/2 \\ -3y/2 & 2 - 3x/2 - y \end{pmatrix}.$$

At the origin, the coefficient matrix of the linearized system is
$$\mathbf{J}(0,0) = \begin{pmatrix} 3/2 & 0 \\ 0 & 2 \end{pmatrix}.$$

The eigenvalues and eigenvectors are
$$r_1 = 3/2, \; \boldsymbol{\xi}^{(1)} = \begin{pmatrix} 1 \\ 0 \end{pmatrix} ; r_2 = 2, \; \boldsymbol{\xi}^{(2)} = \begin{pmatrix} 0 \\ 1 \end{pmatrix}.$$

The eigenvalues are positive, hence the origin is an unstable node.
At the critical point $(0, 4)$, the coefficient matrix of the linearized system is
$$\mathbf{J}(0, 4) = \begin{pmatrix} -1/2 & 0 \\ -6 & -2 \end{pmatrix}.$$

The eigenvalues and eigenvectors are
$$r_1 = -1/2, \; \boldsymbol{\xi}^{(1)} = \begin{pmatrix} 1 \\ -4 \end{pmatrix} ; r_2 = -2, \; \boldsymbol{\xi}^{(2)} = \begin{pmatrix} 0 \\ 1 \end{pmatrix}.$$

The eigenvalues are both negative, hence the critical point $(0, 4)$ is a stable node, which is asymptotically stable.
At the critical point $(3/2, 0)$, the coefficient matrix of the linearized system is
$$\mathbf{J}(3/2, 0) = \begin{pmatrix} -3/2 & -3/4 \\ 0 & -1/4 \end{pmatrix}.$$

The eigenvalues and eigenvectors are
$$r_1 = -3/2, \; \boldsymbol{\xi}^{(1)} = \begin{pmatrix} 1 \\ 0 \end{pmatrix} ; r_2 = -1/4, \; \boldsymbol{\xi}^{(2)} = \begin{pmatrix} 3 \\ -5 \end{pmatrix}.$$

The eigenvalues are both negative, hence the critical point is a stable node, which is asymptotically stable.
At the critical point $(1, 1)$, the coefficient matrix of the linearized system is
$$\mathbf{J}(1, 1) = \begin{pmatrix} -1 & -1/2 \\ -3/2 & -1/2 \end{pmatrix}.$$

The eigenvalues and eigenvectors are
$$r_1 = \frac{-3 + \sqrt{13}}{4}, \; \boldsymbol{\xi}^{(1)} = \begin{pmatrix} 1 \\ -\frac{1+\sqrt{13}}{2} \end{pmatrix} ; r_2 = -\frac{3 + \sqrt{13}}{4}, \; \boldsymbol{\xi}^{(2)} = \begin{pmatrix} 0 \\ \frac{-1+\sqrt{13}}{2} \end{pmatrix}.$$

The eigenvalues are of opposite sign, hence $(1, 1)$ is a saddle, which is unstable.

(d,e)

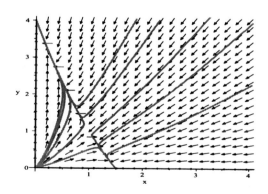

(f) Trajectories approaching the critical point $(1,1)$ form a separatrix. Solutions on either side of the separatrix approach either $(0,4)$ or $(1.5,0)$. Thus depending on the initial conditions, one species will drive the other to extinction.

4.(a)

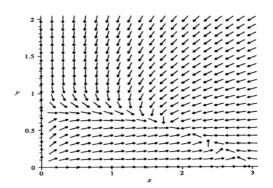

(b) The critical points are solutions of the system of equations
$$x(1.5 - 0.5\,x - y) = 0$$
$$y(0.75 - y - 0.125\,x) = 0.$$

The four critical points are $(0,0)$, $(0,3/4)$, $(3,0)$ and $(2,1/2)$.

(c) The Jacobian matrix of the vector field is
$$\mathbf{J} = \begin{pmatrix} 3/2 - x - y & -x \\ -y/8 & 3/4 - x/8 - 2y \end{pmatrix}.$$

At the origin, the coefficient matrix of the linearized system is
$$\mathbf{J}(0,0) = \begin{pmatrix} 3/2 & 0 \\ 0 & 3/4 \end{pmatrix}.$$

The eigenvalues and eigenvectors are
$$r_1 = 3/2\,,\ \boldsymbol{\xi}^{(1)} = \begin{pmatrix} 1 \\ 0 \end{pmatrix}\,;\ r_2 = 3/4\,,\ \boldsymbol{\xi}^{(2)} = \begin{pmatrix} 0 \\ 1 \end{pmatrix}.$$

The eigenvalues are positive, hence the origin is an unstable node.
At the critical point $(0,3/4)$, the coefficient matrix of the linearized system is
$$\mathbf{J}(0,3/4) = \begin{pmatrix} 3/4 & 0 \\ -3/32 & -3/4 \end{pmatrix}.$$

The eigenvalues and eigenvectors are
$$r_1 = 3/4\,,\ \boldsymbol{\xi}^{(1)} = \begin{pmatrix} -16 \\ 1 \end{pmatrix}\,;\ r_2 = -3/4\,,\ \boldsymbol{\xi}^{(2)} = \begin{pmatrix} 0 \\ 1 \end{pmatrix}.$$

The eigenvalues are of opposite sign, hence the critical point $(0,3/4)$ is a saddle, which is unstable.

At the critical point $(3,0)$, the coefficient matrix of the linearized system is
$$\mathbf{J}(3,0) = \begin{pmatrix} -3/2 & -3 \\ 0 & 3/8 \end{pmatrix}.$$
The eigenvalues and eigenvectors are
$$r_1 = -3/2\,,\; \boldsymbol{\xi}^{(1)} = \begin{pmatrix} 1 \\ 0 \end{pmatrix}\,;\; r_2 = 3/8\,,\; \boldsymbol{\xi}^{(2)} = \begin{pmatrix} -8 \\ 5 \end{pmatrix}.$$
The eigenvalues are of opposite sign, hence the critical point $(0,3/4)$ is a saddle, which is unstable.

At the critical point $(2,1/2)$, the coefficient matrix of the linearized system is
$$\mathbf{J}(2,1/2) = \begin{pmatrix} -1 & -2 \\ -1/16 & -1/2 \end{pmatrix}.$$
The eigenvalues and eigenvectors are
$$r_1 = \frac{-3+\sqrt{3}}{4}\,,\; \boldsymbol{\xi}^{(1)} = \begin{pmatrix} 1 \\ -\frac{1+\sqrt{3}}{8} \end{pmatrix}\,;\; r_2 = -\frac{3+\sqrt{3}}{4}\,,\; \boldsymbol{\xi}^{(2)} = \begin{pmatrix} 0 \\ \frac{-1+\sqrt{3}}{8} \end{pmatrix}.$$
The eigenvalues are negative, hence the critical point $(2,1/2)$ is a stable node, which is asymptotically stable.

(d,e)

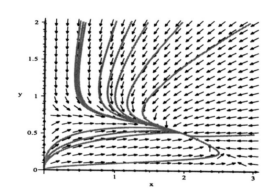

(f) Except for initial conditions along the coordinate axes, almost all solutions converge to the stable node $(2,1/2)$. Thus the species can coexist.

7. It follows immediately that
$$(\sigma_1 X + \sigma_2 Y)^2 - 4\sigma_1\sigma_2 XY = \sigma_1^2 X^2 + 2\sigma_1\sigma_2 XY + \sigma_2^2 Y^2 - 4\sigma_1\sigma_2 XY = (\sigma_1 X - \sigma_2 Y)^2,$$
so
$$(\sigma_1 X + \sigma_2 Y)^2 - 4(\sigma_1\sigma_2 - \alpha_1\alpha_2)XY = (\sigma_1 X - \sigma_2 Y)^2 + 4\alpha_1\alpha_2 XY.$$
Since all parameters and variables are positive, it follows that
$$(\sigma_1 X + \sigma_2 Y)^2 - 4(\sigma_1\sigma_2 - \alpha_1\alpha_2)XY \geq 0\,.$$
Hence the radicand in Eq.(39) is nonnegative.

10.(a) The critical points consist of the solution set of the equations

$$x(\epsilon_1 - \sigma_1 x - \alpha_1 y) = 0$$
$$y(\epsilon_2 - \sigma_2 y - \alpha_2 x) = 0.$$

If $x = 0$, then either $y = 0$ or $y = \epsilon_2/\sigma_2$. If $\epsilon_1 - \sigma_1 x - \alpha_1 y = 0$, then solving for x results in $x = (\epsilon_1 - \alpha_1 y)/\sigma_1$. Substitution into the second equation yields

$$(\sigma_1\sigma_2 - \alpha_1\alpha_2)y^2 - (\sigma_1\epsilon_2 - \epsilon_1\alpha_2)y = 0.$$

Based on the hypothesis, it follows that $(\sigma_1\epsilon_2 - \epsilon_1\alpha_2)y = 0$. So if $\sigma_1\epsilon_2 - \epsilon_1\alpha_2 \neq 0$, then $y = 0$, and the critical points are located at $(0,0)$, $(0, \epsilon_2/\sigma_2)$ and $(\epsilon_1/\sigma_1, 0)$. For the case $\sigma_1\epsilon_2 - \epsilon_1\alpha_2 = 0$, y can be arbitrary. From the relation

$$x = (\epsilon_1 - \alpha_1 y)/\sigma_1,$$

we conclude that all points on the line $\sigma_1 x + \alpha_1 y = \epsilon_1$ are critical points, in addition to the point $(0,0)$.

(b) The Jacobian matrix of the vector field is

$$\mathbf{J} = \begin{pmatrix} \epsilon_1 - 2\sigma_1 x - \alpha_1 y & -\alpha_1 x \\ -\alpha_2 y & \epsilon_2 - 2\sigma_2 y - \alpha_2 x \end{pmatrix}.$$

At the origin, the coefficient matrix of the linearized system is

$$\mathbf{J}(0,0) = \begin{pmatrix} \epsilon_1 & 0 \\ 0 & \epsilon_2 \end{pmatrix},$$

with eigenvalues $r_1 = \epsilon_1$ and $r_2 = \epsilon_2$. Since both eigenvalues are positive, the origin is an unstable node.
At the point $(0, \epsilon_2/\sigma_2)$, the coefficient matrix of the linearized system is

$$\mathbf{J}(0, \epsilon_2/\sigma_2) = \begin{pmatrix} (\epsilon_1\sigma_2 - \alpha_1\epsilon_2)/\sigma_2 & 0 \\ \epsilon_2\alpha_2/\sigma_2 & -\epsilon_2 \end{pmatrix} = \begin{pmatrix} (\epsilon_1\alpha_2 - \sigma_1\epsilon_2)/\alpha_2 & 0 \\ \epsilon_2\alpha_2/\sigma_2 & -\epsilon_2 \end{pmatrix},$$

since $\sigma_1\sigma_2 - \alpha_1\alpha_2 = 0$ implies that $\alpha_1/\sigma_2 = \sigma_1/\alpha_2$. Thus the matrix has eigenvalues $r_1 = (\epsilon_1\alpha_2 - \sigma_1\epsilon_2)/\alpha_2$ and $r_2 = -\epsilon_2$. If $\sigma_1\epsilon_2 - \epsilon_1\alpha_2 > 0$, then both eigenvalues are negative. Hence the point $(0, \epsilon_2/\sigma_2)$ is a stable node, which is asymptotically stable. If $\sigma_1\epsilon_2 - \epsilon_1\alpha_2 < 0$, then the eigenvalues are of opposite sign. Hence the point $(0, \epsilon_2/\sigma_2)$ is a saddle, which is unstable.
At the point $(\epsilon_1/\sigma_1, 0)$, the coefficient matrix of the linearized system is

$$\mathbf{J}(\epsilon_1/\sigma_1, 0) = \begin{pmatrix} -\epsilon_1 & -\epsilon_1\alpha_1/\sigma_1 \\ 0 & (\sigma_1\epsilon_2 - \epsilon_1\alpha_2)/\sigma_1 \end{pmatrix},$$

with eigenvalues $r_1 = (\sigma_1\epsilon_2 - \epsilon_1\alpha_2)/\sigma_1$ and $r_2 = -\epsilon_1$. If $\sigma_1\epsilon_2 - \epsilon_1\alpha_2 > 0$, then the eigenvalues are of opposite sign. Hence the point $(\epsilon_1/\sigma_1, 0)$ is a saddle, which is unstable. If $\sigma_1\epsilon_2 - \epsilon_1\alpha_2 < 0$, then both eigenvalues are negative. In that case the point $(\epsilon_1/\sigma_1, 0)$ is a stable node, which is asymptotically stable.

(c) As shown in part (a), when $\sigma_1\epsilon_2 - \epsilon_1\alpha_2 = 0$, the set of critical points consists of $(0,0)$ and all the points on the straight line $\sigma_1 x + \alpha_1 y = \epsilon_1$. Based on part (b), the origin is still an unstable node. Setting $y = (\epsilon_1 - \sigma_1 x)/\alpha_1$, the Jacobian matrix of the vector field, along the given straight line, is

$$\mathbf{J}(\epsilon_1/\sigma_1, 0) = \begin{pmatrix} -\sigma_1 x & -\alpha_1 x \\ -\alpha_2(\epsilon_1 - \sigma_1 x)/\alpha_1 & \epsilon_2 - 2\sigma_2(\epsilon_1 - \sigma_1 x)/\alpha_1 - \alpha_2 x \end{pmatrix}$$

$$= \begin{pmatrix} -\sigma_1 x & -\alpha_1 x \\ -\alpha_2(\epsilon_1 - \sigma_1 x)/\alpha_1 & \alpha_2 x - \epsilon_1 \alpha_2/\sigma_1 \end{pmatrix},$$

since $\sigma_1 \sigma_2 - \alpha_1 \alpha_2 = 0$ implies that $\sigma_2/\alpha_1 = \alpha_2/\sigma_1$ and since $\sigma_1 \epsilon_2 - \epsilon_1 \alpha_2 = 0$ implies that $\epsilon_2 = \epsilon_1 \alpha_2/\sigma_1$. The characteristic equation of the matrix is

$$r^2 + \left[\frac{\epsilon_1 \alpha_2 - \alpha_2 \sigma_1 x + \sigma_1^2 x}{\sigma_1}\right] r = 0.$$

Since $\epsilon_2 = \epsilon_1 \alpha_2/\sigma_1$, we have that $(\epsilon_1 \alpha_2 - \alpha_2 \sigma_1 x + \sigma_1^2 x)/\sigma_1 = \epsilon_2 - \alpha_2 x + \sigma_1 x$. Hence the characteristic equation can be written as

$$r^2 + [\epsilon_2 - \alpha_2 x + \sigma_1 x] r = 0.$$

First note that $0 \leq x \leq \epsilon_1/\sigma_1$. Since the coefficient in the quadratic equation is linear, and

$$\epsilon_2 - \alpha_2 x + \sigma_1 x = \begin{cases} \epsilon_2 \text{ at } x = 0 \\ \epsilon_1 \text{ at } x = \epsilon_1/\sigma_1, \end{cases}$$

it follows that the coefficient is positive for $0 \leq x \leq \epsilon_1/\sigma_1$. Therefore, along the straight line $\sigma_1 x + \alpha_1 y = \epsilon_1$, one eigenvalue is zero and the other one is negative. Hence the continuum of critical points consists of stable nodes, which are asymptotically stable.

11.(a) The critical points are solutions of the system of equations

$$x(1 - x - y) + \delta a = 0$$
$$y(0.75 - y - 0.5 x) + \delta b = 0.$$

Assume solutions of the form

$$x = r_0 + r_1 \delta + r_2 \delta^2 + \ldots$$
$$y = y_0 + y_1 \delta + y_2 \delta^2 + \ldots.$$

Substitution of the series expansions results in

$$x_0(1 - x_0 - y_0) + (x_1 - 2x_1 x_0 - x_0 y_1 - x_1 y_0 + a)\delta + \ldots = 0$$
$$y_0(0.75 - y_0 - 0.5 x_0) + (0.75 y_1 - 2y_0 y_1 - x_1 y_0/2 - x_0 y_1/2 + b)\delta + \ldots = 0.$$

(b) Taking a limit as $\delta \to 0$, the equations reduce to the original system of equations. It follows that $x_0 = y_0 = 0.5$.

(c) Setting the coefficients of the linear terms equal to zero, we find that

$$-y_1/2 - x_1/2 + a = 0$$
$$-x_1/4 - y_1/2 + b = 0,$$

with solution $x_1 = 4a - 4b$ and $y_1 = -2a + 4b$.

(d) Consider the ab-parameter space. The collection of points for which $b < a$ represents an increase in the level of species 1. At points where $b > a$, $x_1\delta < 0$. Likewise, the collection of points for which $2b > a$ represents an increase in the level of species 2. At points where $2b < a$, $y_1\delta < 0$.

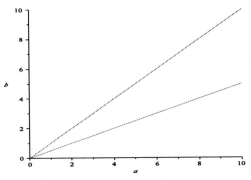

It follows that if $b < a < 2b$, the level of both species will increase. This condition is represented by the wedge-shaped region on the graph. Otherwise, the level of one species will increase, whereas the level of the other species will simultaneously decrease. Only for $a = b = 0$ will both populations remain the same.

12.(a) The critical points consist of the solution set of the equations
$$-y = 0$$
$$-\gamma y - x(x - 0.15)(x - 2) = 0.$$

Setting $y = 0$, the second equation becomes $x(x - 0.15)(x - 2) = 0$, with roots $x = 0$, 0.15 and 2. Hence the critical points are located at $(0,0)$, $(0.15,0)$ and $(2,0)$. The Jacobian matrix of the vector field is
$$\mathbf{J} = \begin{pmatrix} 0 & -1 \\ -3x^2 + 4.3\,x - 0.3 & -\gamma \end{pmatrix}.$$

At the origin, the coefficient matrix of the linearized system is
$$\mathbf{J}(0,0) = \begin{pmatrix} 0 & -1 \\ -0.3 & -\gamma \end{pmatrix},$$

with eigenvalues
$$r_{1,2} = -\frac{\gamma}{2} \pm \frac{1}{10}\sqrt{25\gamma^2 + 30}.$$

Regardless of the value of γ, the eigenvalues are real and of opposite sign. Hence $(0,0)$ is a saddle, which is unstable.

At the critical point $(0.15, 0)$, the coefficient matrix of the linearized system is
$$\mathbf{J}(0.15,0) = \begin{pmatrix} 0 & -1 \\ 0.2775 & -\gamma \end{pmatrix},$$

with eigenvalues
$$r_{1,2} = -\frac{\gamma}{2} \pm \frac{1}{20}\sqrt{100\gamma^2 - 111}.$$

If $100\gamma^2 - 111 \geq 0$, then the eigenvalues are real. Furthermore, since $r_1 r_2 = 0.2775$, both eigenvalues will have the same sign. Therefore the critical point is a node, with its stability dependent on the sign of γ. If $100\gamma^2 - 111 < 0$, the eigenvalues are complex conjugates. In that case the critical point $(0.15, 0)$ is a spiral, with its stability dependent on the sign of γ.

At the critical point $(2, 0)$, the coefficient matrix of the linearized system is

$$\mathbf{J}(2, 0) = \begin{pmatrix} 0 & -1 \\ -3.7 & -\gamma \end{pmatrix},$$

with eigenvalues

$$r_{1,2} = -\frac{\gamma}{2} \pm \frac{1}{10}\sqrt{25\gamma^2 + 370}.$$

Regardless of the value of γ, the eigenvalues are real and of opposite sign. Hence $(2, 0)$ is a saddle, which is unstable.

(b)

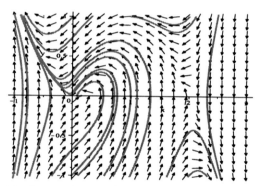

For $\gamma = 0.8$, the critical point $(0.15, 0)$ is a stable spiral.

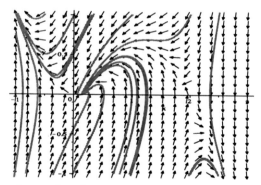

Closer examination shows that for $\gamma = 1.5$, the critical point $(0.15, 0)$ is a stable node.

(c) Based on the phase portraits in part (b), it is apparent that the required value of γ satisfies $0.8 < \gamma < 1.5$. Using the initial condition $x(0) = 2$ and $y(0) = 0.01$, it is possible to solve the initial value problem for various values of γ. A reasonable first guess is $\gamma = \sqrt{1.11}$. This value marks the change in qualitative behavior of the critical point $(0.15, 0)$. Numerical experiments show that the solution remains positive for $\gamma \approx 1.20$.

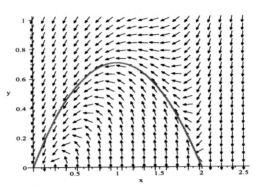

14.(a) Nullclines:

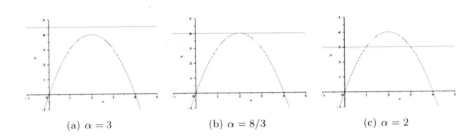

(a) $\alpha = 3$ (b) $\alpha = 8/3$ (c) $\alpha = 2$

(b) The critical points are solutions of the algebraic system

$$\frac{3}{2}\alpha - y = 0$$
$$-4x + y + x^2 = 0$$

which are

$$(2 \pm \sqrt{4 - \frac{3}{2}\alpha}, \frac{3}{2}\alpha),$$

and exist for $\alpha \leq 8/3$.

(c) For $\alpha = 2$, the critical points are at $(1, 3)$ and $(3, 3)$. The Jacobian matrix of the vector field is

$$\mathbf{J} = \begin{pmatrix} 0 & -1 \\ -4 + 2x & 1 \end{pmatrix}.$$

At the critical point $(1, 3)$, the coefficient matrix of the linearized system is

$$\mathbf{J}(1, 3) = \begin{pmatrix} 0 & -1 \\ -2 & 1 \end{pmatrix}.$$

The eigenvalues of the Jacobian, $r = -1$ and 2, are real and opposite in sign; hence the critical point is a saddle point.

At the critical point $(3,3)$, the coefficient matrix of the linearized system is

$$\mathbf{J}(3,3) = \begin{pmatrix} 0 & -1 \\ 2 & 1 \end{pmatrix}.$$

The eigenvalues of the Jacobian are

$$r = \frac{1 \pm i\sqrt{7}}{2},$$

hence the critical point is an unstable spiral.

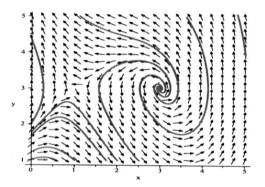

(d) The bifurcation value is $\alpha_0 = 8/3$. The coincident critical points are at $(2,4)$. The coefficient matrix of the linearized system is

$$\mathbf{J}(2,4) = \begin{pmatrix} 0 & -1 \\ 0 & 1 \end{pmatrix}.$$

The eigenvalues of the Jacobian are $r = 0$ and 1.

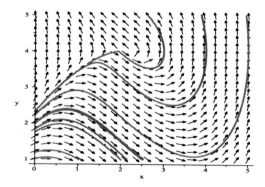

(e)

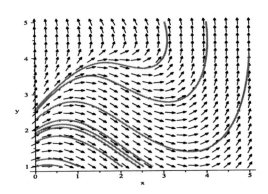

15.(a) Nullclines:

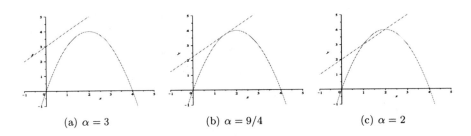

(a) $\alpha = 3$ (b) $\alpha = 9/4$ (c) $\alpha = 2$

(b) The critical points are solutions of the algebraic system
$$-4x + y + x^2 = 0$$
$$-\alpha - x + y = 0$$
which are
$$x_0 = \frac{3 + \sqrt{9 - 4\alpha}}{2} \ , \ y_0 = \alpha + \frac{3 + \sqrt{9 - 4\alpha}}{2}$$
and
$$x_0 = \frac{3 - \sqrt{9 - 4\alpha}}{2} \ , \ y_0 = \alpha + \frac{3 - \sqrt{9 - 4\alpha}}{2} \ .$$
These critical points exist for $\alpha \leq 9/4$.

(c) For $\alpha = 2$, the critical points are at $(1,3)$ and $(2,4)$. The Jacobian matrix of the vector field is
$$\mathbf{J} = \begin{pmatrix} 2x - 4 & 1 \\ -1 & 1 \end{pmatrix}.$$

At the critical point $(1,3)$, the coefficient matrix of the linearized system is
$$\mathbf{J}(1,3) = \begin{pmatrix} -2 & 1 \\ -1 & 1 \end{pmatrix}.$$

The eigenvalues of the Jacobian are
$$r = \frac{-1 \pm \sqrt{5}}{2},$$
hence the critical point is a saddle point.

At the critical point $(2, 4)$, the coefficient matrix of the linearized system is
$$\mathbf{J}(2, 4) = \begin{pmatrix} 0 & 1 \\ -1 & 1 \end{pmatrix}.$$

The eigenvalues of the Jacobian are
$$r = \frac{1 \pm i\sqrt{3}}{2},$$
hence the critical point is an unstable spiral.

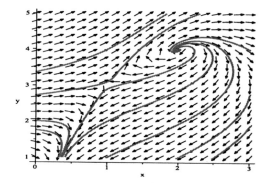

(d) A bifurcation occurs for $\alpha_0 = 9/4$. The coincident critical points are at $(3/2, 15/4)$. The coefficient matrix of the linearized system is
$$\mathbf{J}(3/2, 15/4) = \begin{pmatrix} -1 & 1 \\ -1 & 1 \end{pmatrix},$$
with eigenvalues both equal to zero.

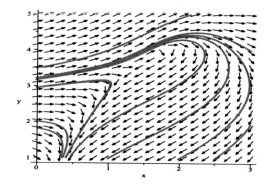

(e)

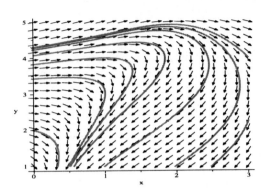

16.(a) Nullclines:

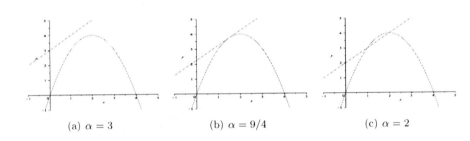

(a) $\alpha = 3$ (b) $\alpha = 9/4$ (c) $\alpha = 2$

(b) The critical points are solutions of the algebraic system
$$-\alpha - x + y = 0$$
$$-4x + y + x^2 = 0$$

which are
$$x_0 = \frac{3 + \sqrt{9 - 4\alpha}}{2}, \ y_0 = \alpha + \frac{3 + \sqrt{9 - 4\alpha}}{2}$$

and
$$x_0 = \frac{3 - \sqrt{9 - 4\alpha}}{2}, \ y_0 = \alpha + \frac{3 - \sqrt{9 - 4\alpha}}{2}.$$

These critical points exist for $\alpha \leq 9/4$.

(c) For $\alpha = 2$, the critical points are at $(1, 3)$ and $(2, 4)$. The Jacobian matrix of the vector field is
$$\mathbf{J} = \begin{pmatrix} -1 & 1 \\ 2x - 4 & 1 \end{pmatrix}.$$

At the critical point $(1, 3)$, the coefficient matrix of the linearized system is
$$\mathbf{J}(1, 3) = \begin{pmatrix} -1 & 1 \\ -2 & 1 \end{pmatrix}.$$

The eigenvalues of the Jacobian are
$$r = \pm i,$$
hence the critical point is a center.

At the critical point $(2, 4)$, the coefficient matrix of the linearized system is
$$\mathbf{J}(2, 4) = \begin{pmatrix} -1 & 1 \\ 0 & 1 \end{pmatrix}.$$

The eigenvalues of the Jacobian, $r = -1$ and 1, are real and opposite in sign; hence the critical point is a saddle point.

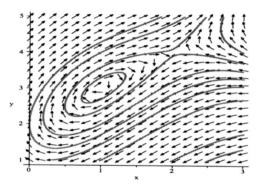

(d) The bifurcation value is $\alpha_0 = 8/3$. The coincident critical point is at $(3/2, 14/4)$. The coefficient matrix of the linearized system is
$$\mathbf{J}(3/2, 15/4) = \begin{pmatrix} -1 & 1 \\ -1 & 1 \end{pmatrix}.$$

The eigenvalues are both equal to zero.

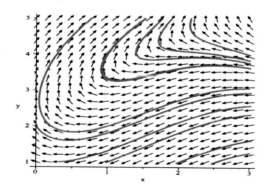

(e)

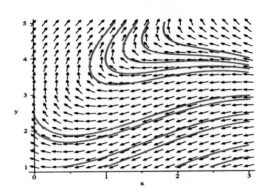

18.(a) Nullclines:

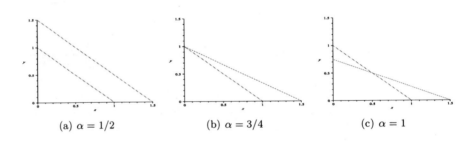

(a) $\alpha = 1/2$ (b) $\alpha = 3/4$ (c) $\alpha = 1$

(b) The critical points are $(0,0)$, $(1,0)$, $((4\alpha - 3)/(4\alpha - 2), 1/(4\alpha - 2))$, and $(0, 3/(4\alpha))$. The third critical point is in the first quadrant as long as $\alpha \geq 3/4$.

(c) The third and fourth critical points will coincide (see part (b)) when $\alpha = 3/4$.

(d,e) The Jacobian is

$$\mathbf{J} = \begin{pmatrix} 1 - 2x - y & -x \\ -y/2 & 3/4 - 2\alpha y - x/2 \end{pmatrix}.$$

This means that at the origin

$$\mathbf{J}(0,0) = \begin{pmatrix} 1 & 0 \\ 0 & 3/4 \end{pmatrix},$$

so the origin is an unstable node. (The eigenvalues are clearly 1 and $3/4$.) At the critical point $(1,0)$ the Jacobian is

$$\mathbf{J}(1,0) = \begin{pmatrix} -1 & -1 \\ 0 & 1/4 \end{pmatrix},$$

which means that this critical point is a saddle.

At the critical point $(0, 3/(4\alpha))$ the Jacobian is

$$\mathbf{J}(0, 3/(4\alpha)) = \begin{pmatrix} 1 - 3/(4\alpha) & 0 \\ -3/(8\alpha) & -3/4 \end{pmatrix},$$

which implies that this critical point is a saddle when $\alpha > 3/4$ and an asymptotically stable node when $0 < \alpha < 3/4$.

At the critical point $(\frac{4\alpha-3}{4\alpha-2}, \frac{1}{4\alpha-2})$ the Jacobian is given by

$$\mathbf{J}(\frac{4\alpha-3}{4\alpha-2}, \frac{1}{4\alpha-2}) = \begin{pmatrix} \frac{3-4\alpha}{4\alpha-2} & \frac{3-4\alpha}{4\alpha-2} \\ \frac{-1/2}{4\alpha-2} & \frac{-\alpha}{4\alpha-2} \end{pmatrix}.$$

It can be shown that this is an asymptotically stable node when $\alpha > 3/4$.

(f) Phase portraits:

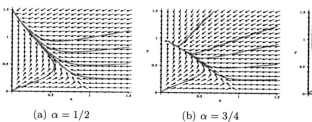
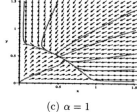

(a) $\alpha = 1/2$ (b) $\alpha = 3/4$ (c) $\alpha = 1$

19.(a) Nullclines:

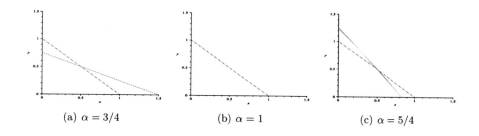

(a) $\alpha = 3/4$ (b) $\alpha = 1$ (c) $\alpha = 5/4$

(b) The critical points are $(0,0)$, $(1,0)$, $(0,\alpha)$ and $(1/2, 1/2)$. Also, when $\alpha = 1$, then all points on the line $x + y = 1$ are critical points.

(c) Clearly, $\alpha = 1$ is the bifurcation value.

(d,e) The Jacobian is
$$\mathbf{J} = \begin{pmatrix} 1 - 2x - y & -x \\ -(2\alpha - 1)y & \alpha - 2y - (2\alpha - 1)x \end{pmatrix}.$$

This means that at the origin
$$\mathbf{J}(0,0) = \begin{pmatrix} 1 & 0 \\ 0 & \alpha \end{pmatrix},$$

so the origin is an unstable node when $\alpha > 0$.

At the critical point $(1, 0)$ the Jacobian is
$$\mathbf{J}(1,0) = \begin{pmatrix} -1 & -1 \\ 0 & 1 - \alpha \end{pmatrix},$$

which means that this critical point is a saddle when $0 < \alpha < 1$ and an asymptotically stable node when $\alpha > 1$.

At the critical point $(0, \alpha)$ the Jacobian is
$$\mathbf{J}(0,\alpha)) = \begin{pmatrix} 1 - \alpha & 0 \\ -\alpha(2\alpha - 1) & -\alpha \end{pmatrix},$$

which implies that this critical point is a saddle when $0 < \alpha < 1$ and an asymptotically stable node when $\alpha > 1$.

At the critical point $(1/2, 1/2)$ the Jacobian is given by
$$\mathbf{J}(1/2, 1/2) = \begin{pmatrix} -1/2 & -1/2 \\ 1/2 - \alpha & -1/2 \end{pmatrix}.$$

The eigenvalues of $\mathbf{J}(1/2, 1/2)$ are $(-1 \pm \sqrt{-1 + 2\alpha})/2$. Thus the critical point is an asymptotically stable spiral for $0 < \alpha < 1/2$, an asymptotically stable node for $1/2 \leq \alpha < 1$, and a saddle for $\alpha > 1$.

(f) Phase portraits:

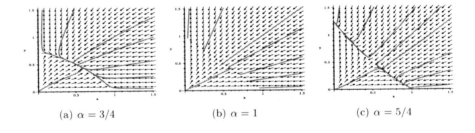

(a) $\alpha = 3/4$ (b) $\alpha = 1$ (c) $\alpha = 5/4$

9.5

1.(a)

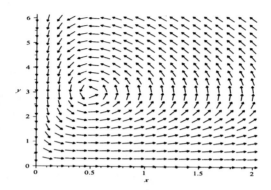

(b) The critical points are solutions of the system of equations
$$x(1.5 - 0.5\,y) = 0$$
$$y(-0.5 + x) = 0.$$
The two critical points are $(0\,,0)$ and $(0.5\,,3)$.

(c) The Jacobian matrix of the vector field is
$$\mathbf{J} = \begin{pmatrix} 3/2 - y/2 & -x/2 \\ y & -1/2 + x \end{pmatrix}.$$
At the critical point $(0\,,0)$, the coefficient matrix of the linearized system is
$$\mathbf{J}(0\,,0) = \begin{pmatrix} 3/2 & 0 \\ 0 & -1/2 \end{pmatrix}.$$
The eigenvalues and eigenvectors are
$$r_1 = 3/2\,,\;\boldsymbol{\xi}^{(1)} = \begin{pmatrix} 1 \\ 0 \end{pmatrix}\,;\; r_2 = -1/2\,,\;\boldsymbol{\xi}^{(2)} = \begin{pmatrix} 0 \\ 1 \end{pmatrix}.$$
The eigenvalues are of opposite sign, hence the origin is a saddle, which is unstable. At the critical point $(0.5\,,3)$, the coefficient matrix of the linearized system is
$$\mathbf{J}(0.5\,,3) = \begin{pmatrix} 0 & -1/4 \\ 3 & 0 \end{pmatrix}.$$
The eigenvalues and eigenvectors are
$$r_1 = i\,\frac{\sqrt{3}}{2}\,,\;\boldsymbol{\xi}^{(1)} = \begin{pmatrix} 1 \\ -2\,i\sqrt{3} \end{pmatrix}\,;\; r_2 = -i\,\frac{\sqrt{3}}{2}\,,\;\boldsymbol{\xi}^{(2)} = \begin{pmatrix} 1 \\ 2\,i\sqrt{3} \end{pmatrix}.$$
The eigenvalues are purely imaginary. Hence the critical point is a center, which is stable.

(d,e)

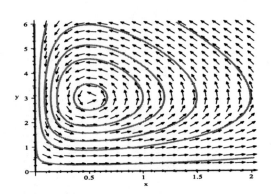

(f) Except for solutions along the coordinate axes, almost all trajectories are closed curves about the critical point $(0.5, 3)$.

2.(a)

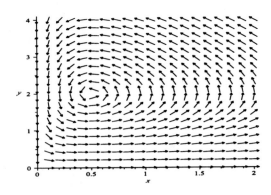

(b) The critical points are the solution set of the system of equations
$$x(1 - 0.5\,y) = 0$$
$$y(-0.25 + 0.5\,x) = 0.$$

The two critical points are $(0,0)$ and $(0.5, 2)$.

(c) The Jacobian matrix of the vector field is
$$\mathbf{J} = \begin{pmatrix} 1 - y/2 & -x/2 \\ y/2 & -1/4 + x/2 \end{pmatrix}.$$

At the critical point $(0,0)$, the coefficient matrix of the linearized system is
$$\mathbf{J}(0,0) = \begin{pmatrix} 1 & 0 \\ 0 & -1/4 \end{pmatrix}.$$

The eigenvalues and eigenvectors are
$$r_1 = 1,\ \boldsymbol{\xi}^{(1)} = \begin{pmatrix} 1 \\ 0 \end{pmatrix};\ r_2 = -1/4,\ \boldsymbol{\xi}^{(2)} = \begin{pmatrix} 0 \\ 1 \end{pmatrix}.$$

The eigenvalues are of opposite sign, hence the origin is a saddle, which is unstable. At the critical point $(0.5, 2)$, the coefficient matrix of the linearized system is

$$\mathbf{J}(0.5, 2) = \begin{pmatrix} 0 & -1/4 \\ 1 & 0 \end{pmatrix}.$$

The eigenvalues and eigenvectors are

$$r_1 = i/2 \, , \, \boldsymbol{\xi}^{(1)} = \begin{pmatrix} 1 \\ -2i \end{pmatrix} \, ; \, r_2 = -i/2 \, , \, \boldsymbol{\xi}^{(2)} = \begin{pmatrix} 1 \\ 2i \end{pmatrix}.$$

The eigenvalues are purely imaginary. Hence the critical point is a center, which is stable.

(d,e)

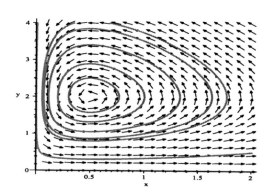

(f) Except for solutions along the coordinate axes, almost all trajectories are closed curves about the critical point $(0.5, 2)$.

4.(a)

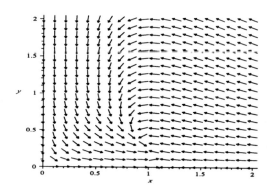

(b) The critical points are the solution set of the system of equations

$$x(9/8 - x - y/2) = 0$$
$$y(-1 + x) = 0.$$

The three critical points are $(0, 0)$, $(9/8, 0)$ and $(1, 1/4)$.

(c) The Jacobian matrix of the vector field is

$$\mathbf{J} = \begin{pmatrix} 9/8 - 2x - y/2 & -x/2 \\ y & -1 + x \end{pmatrix}.$$

At the critical point $(0,0)$, the coefficient matrix of the linearized system is

$$\mathbf{J}(0,0) = \begin{pmatrix} 9/8 & 0 \\ 0 & -1 \end{pmatrix}.$$

The eigenvalues and eigenvectors are

$$r_1 = 9/8, \ \boldsymbol{\xi}^{(1)} = \begin{pmatrix} 1 \\ 0 \end{pmatrix}; \ r_2 = -1, \ \boldsymbol{\xi}^{(2)} = \begin{pmatrix} 0 \\ 1 \end{pmatrix}.$$

The eigenvalues are of opposite sign, hence the origin is a saddle, which is unstable.
At the critical point $(9/8, 0)$, the coefficient matrix of the linearized system is

$$\mathbf{J}(9/8, 0) = \begin{pmatrix} -9/8 & -9/16 \\ 0 & 1/8 \end{pmatrix}.$$

The eigenvalues and eigenvectors are

$$r_1 = -\frac{9}{8}, \ \boldsymbol{\xi}^{(1)} = \begin{pmatrix} 1 \\ 0 \end{pmatrix}; \ r_2 = \frac{1}{8}, \ \boldsymbol{\xi}^{(2)} = \begin{pmatrix} 9 \\ -20 \end{pmatrix}.$$

The eigenvalues are of opposite sign, hence the critical point $(9/8, 0)$ is a saddle, which is unstable.
At the critical point $(1, 1/4)$, the coefficient matrix of the linearized system is

$$\mathbf{J}(1, 1/4) = \begin{pmatrix} -1 & -1/2 \\ 1/4 & 0 \end{pmatrix}.$$

The eigenvalues and eigenvectors are

$$r_1 = \frac{-2 + \sqrt{2}}{4}, \ \boldsymbol{\xi}^{(1)} = \begin{pmatrix} -2 + \sqrt{2} \\ 1 \end{pmatrix}; \ r_2 = \frac{-2 - \sqrt{2}}{4}, \ \boldsymbol{\xi}^{(2)} = \begin{pmatrix} -2 - \sqrt{2} \\ 1 \end{pmatrix}.$$

The eigenvalues are both negative. Hence the critical point is a stable node, which is asymptotically stable.

(d,e)

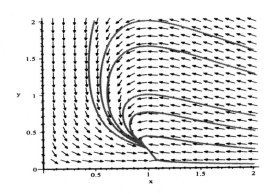

(f) Except for solutions along the coordinate axes, all solutions converge to the critical point $(1, 1/4)$.

5.(a)

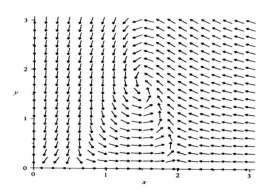

(b) The critical points are solutions of the system of equations
$$x(-1 + 2.5x - 0.3y - x^2) = 0$$
$$y(-1.5 + x) = 0.$$
The four critical points are $(0,0)$, $(1/2, 0)$, $(2, 0)$ and $(3/2, 5/3)$.

(c) The Jacobian matrix of the vector field is
$$\mathbf{J} = \begin{pmatrix} -1 + 5x - 3x^2 - 3y/10 & -3x/10 \\ y & -3/2 + x \end{pmatrix}.$$

At the critical point $(0,0)$, the coefficient matrix of the linearized system is
$$\mathbf{J}(0,0) = \begin{pmatrix} -1 & 0 \\ 0 & -3/2 \end{pmatrix}.$$

The eigenvalues and eigenvectors are
$$r_1 = -1, \; \boldsymbol{\xi}^{(1)} = \begin{pmatrix} 1 \\ 0 \end{pmatrix} ; \; r_2 = -3/2, \; \boldsymbol{\xi}^{(2)} = \begin{pmatrix} 0 \\ 1 \end{pmatrix}.$$

The eigenvalues are both negative, hence the critical point $(0,0)$ is a stable node, which is asymptotically stable.

At the critical point $(1/2, 0)$, the coefficient matrix of the linearized system is
$$\mathbf{J}(1/2, 0) = \begin{pmatrix} 3/4 & -3/20 \\ 0 & -1 \end{pmatrix}.$$

The eigenvalues and eigenvectors are
$$r_1 = \frac{3}{4}, \; \boldsymbol{\xi}^{(1)} = \begin{pmatrix} 1 \\ 0 \end{pmatrix} ; \; r_2 = -1, \; \boldsymbol{\xi}^{(2)} = \begin{pmatrix} 3 \\ 35 \end{pmatrix}.$$

The eigenvalues are of opposite sign, hence the critical point $(1/2, 0)$ is a saddle, which is unstable.

At the critical point $(2,0)$, the coefficient matrix of the linearized system is

$$\mathbf{J}(2,0) = \begin{pmatrix} -3 & -3/5 \\ 0 & 1/2 \end{pmatrix}.$$

The eigenvalues and eigenvectors are

$$r_1 = -3, \ \boldsymbol{\xi}^{(1)} = \begin{pmatrix} 1 \\ 0 \end{pmatrix} ; r_2 = 1/2, \ \boldsymbol{\xi}^{(2)} = \begin{pmatrix} 6 \\ -35 \end{pmatrix}.$$

The eigenvalues are of opposite sign, hence the critical point $(2,0)$ is a saddle, which is unstable.

At the critical point $(3/2, 5/3)$, the coefficient matrix of the linearized system is

$$\mathbf{J}(3/2, 5/3) = \begin{pmatrix} -3/4 & -9/20 \\ 5/3 & 0 \end{pmatrix}.$$

The eigenvalues and eigenvectors are

$$r_1 = \frac{-3 + i\sqrt{39}}{8}, \ \boldsymbol{\xi}^{(1)} = \begin{pmatrix} \frac{-9+i3\sqrt{39}}{40} \\ 1 \end{pmatrix} ; r_2 = \frac{-3 - i\sqrt{39}}{8}, \ \boldsymbol{\xi}^{(2)} = \begin{pmatrix} \frac{-9-i3\sqrt{39}}{40} \\ 1 \end{pmatrix}.$$

The eigenvalues are complex conjugates. Hence the critical point $(3/2, 5/3)$ is a stable spiral, which is asymptotically stable.

(d,e)

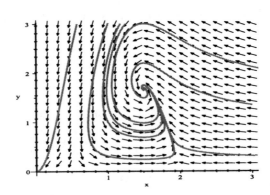

(f) The single solution curve that converges to the node at $(1/2, 0)$ is a separatrix. Except for initial conditions on the coordinate axes, trajectories on either side of the separatrix converge to the node at $(0,0)$ or the stable spiral at $(3/2, 5/3)$.

6. Given that t is measured from the time that x is a maximum, we have

$$x = \frac{c}{\gamma} + \frac{cK}{\gamma} \cos(\sqrt{ac}\, t)$$

$$y = \frac{a}{\alpha} + K\frac{a}{\alpha}\sqrt{\frac{c}{\alpha}} \sin(\sqrt{ac}\, t).$$

(a) Since $\sin\theta$ reaches a maximum at $\theta = \pi/2$, y reaches a maximum at $\sqrt{ac}\, t = \pi/2$, or $t = \pi/(2\sqrt{ac}) = T/4$, where $T = 2\pi/\sqrt{ac}$ is the period of oscillation.

(b) Note that
$$\frac{dx}{dt} = -\frac{cK\sqrt{ac}}{\gamma}\sin(\sqrt{ac}\,t).$$

Since $\sin\theta$ reaches a minimum of -1 at $\theta = 3\pi/2, 7\pi/2, \ldots$, x is increasing most rapidly at $\sqrt{ac}\,t = 3\pi/2, 7\pi/2, \ldots$, or $t = 3T/4, 7T/4, \ldots$. Since $\sin\theta$ reaches a maximum of 1 at $\theta = \pi/2, 5\pi/2, \ldots$, x is decreasing most rapidly at $\sqrt{ac}\,t = \pi/2, 5\pi/2, \ldots$, or $t = T/4, 5T/4, \ldots$. Since $\cos\theta$ reaches a minimum at $\theta = \pi, 3\pi, \ldots$, x reaches a minimum at $\sqrt{ac}\,t = \pi, 3\pi, \ldots$, or $t = T/2, 3T/2, \ldots$.

(c) Note that
$$\frac{dy}{dt} = Kc\left(\frac{a}{\alpha}\right)^{3/2}\cos(\sqrt{ac}\,t).$$

Since $\cos\theta$ reaches a minimum of -1 at $\theta = \pi, 3\pi, \ldots$, y is decreasing most rapidly at $\sqrt{ac}\,t = \pi, 3\pi, \ldots$, or $t = T/2, 3T/2, \ldots$. Since $\cos\theta$ reaches a maximum of 1 at $\theta = 0, 2\pi, \ldots$, y is increasing most rapidly at $\sqrt{ac}\,t = 0, 2\pi, \ldots$, or $t = 0, T, \ldots$. Since $\sin\theta$ reaches a minimum at $\theta = 3\pi/2, 7\pi/2, \ldots$, y reaches a minimum at $\sqrt{ac}\,t = 3\pi/2, 7\pi/2, \ldots$, or $t = 3T/4, 7T/4, \ldots$.

(d) In the following example, the system in Problem 2 is solved numerically with the initial conditions $x(0) = 0.7$ and $y(0) = 2$. The critical point of interest is at $(0.5, 2)$. Since $a = 1$ and $c = 1/4$, it follows that the period of oscillation is $T = 4\pi$.

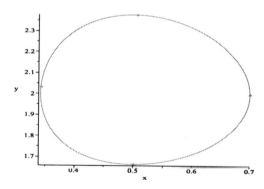

8.(a) The period of oscillation for the linear system is $T = 2\pi/\sqrt{ac}$. In system (2), $a = 1$ and $c = 0.75$. Hence the period is estimated as $T = 2\pi/\sqrt{0.75} \approx 7.2552$.

(b) The estimated period appears to agree with the graphic in Figure 9.5.3.

(c) The critical point of interest is at $(3, 2)$. The system is solved numerically, with $y(0) = 2$ and $x(0) = 3.5, 4.0, 4.5, 5.0$. The resulting periods are shown in the table:

	$x(0) = 3.5$	$x(0) = 4.0$	$x(0) = 4.5$	$x(0) = 5.0$
T	7.26	7.29	7.34	7.42

The actual period steadily increases as $x(0)$ increases.

9. The system
$$\frac{dx}{dt} = a\,x(1 - \frac{y}{2})$$
$$\frac{dy}{dt} = b\,y(-1 + \frac{x}{3})$$
is solved numerically for various values of the parameters. The initial conditions are $x(0) = 5$, $y(0) = 2$.

(a) $a = 1$ and $b = 1$:

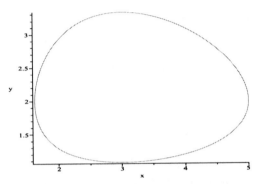

The period is estimated by observing when the trajectory becomes a closed curve. In this case, $T \approx 6.45$.

(b) $a = 3$ and $a = 1/3$, with $b = 1$:

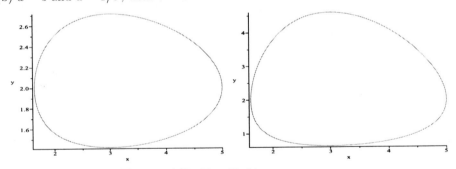

For $a = 3$, $T \approx 3.69$. For $a = 1/3$, $T \approx 11.44$.

(c) $b = 3$ and $b = 1/3$, with $a = 1$:

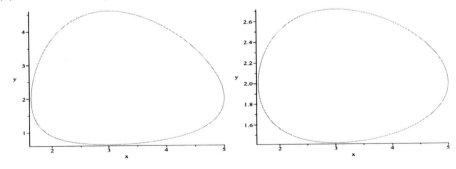

For $b = 3$, $T \approx 3.82$. For $b = 1/3$, $T \approx 11.06$.

(d) It appears that if one of the parameters is fixed, the period varies inversely with the other parameter. Hence one might postulate the relation

$$T = \frac{k}{f(a,b)}.$$

10.(a) Since $T = 2\pi/\sqrt{ac}$, we first note that

$$\int_A^{A+T} \cos(\sqrt{ac}\, t + \phi)dt = \int_A^{A+T} \sin(\sqrt{ac}\, t + \phi)dt = 0.$$

Hence

$$\overline{x} = \frac{1}{T}\int_A^{A+T} \frac{c}{\gamma} dt = \frac{c}{\gamma} \quad \text{and} \quad \overline{y} = \frac{1}{T}\int_A^{A+T} \frac{a}{\alpha} dt = \frac{a}{\alpha}.$$

(b) One way to estimate the mean values is to find a horizontal line such that the area above the line is approximately equal to the area under the line. From Figure 9.5.3, it appears that $\overline{x} \approx 3.25$ and $\overline{y} \approx 2.0$. In Example 1, $a = 1$, $c = 0.75$, $\alpha = 0.5$ and $\gamma = 0.25$. Using the result in part (a), $\overline{x} = 3$ and $\overline{y} = 2$.

(c) The system

$$\frac{dx}{dt} = x(1 - \frac{y}{2})$$
$$\frac{dy}{dt} = y(-\frac{3}{4} + \frac{x}{4})$$

is solved numerically for various initial conditions.

$x(0) = 3$ and $y(0) = 2.5$:

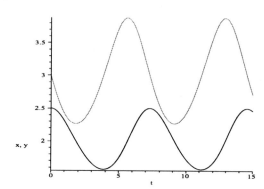

$x(0) = 3$ and $y(0) = 3.0$:

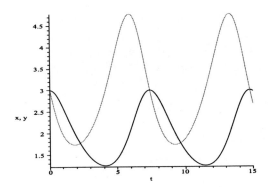

$x(0) = 3$ and $y(0) = 3.5$:

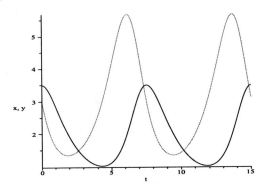

$x(0) = 3$ and $y(0) = 4.0$:

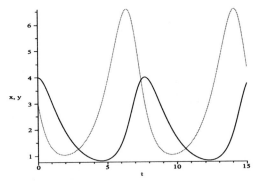

It is evident that the mean values increase as the amplitude increases. That is, the mean values increase as the initial conditions move farther from the critical point.

12.(a) The critical points are the solutions of the system

$$x(a - \sigma x - \alpha y) = 0$$
$$y(-c + \gamma x) = 0.$$

If $x = 0$, then $y = 0$. If $y = 0$, then $x = a/\sigma$. The third solution is found by substituting $x = c/\gamma$ into the first equation. This implies that $y = a/\alpha - \sigma c/(\gamma \alpha)$.

So the critical points are $(0,0)$, $(\frac{a}{\sigma},0)$ and $(\frac{c}{\gamma}, \frac{a}{\alpha} - \frac{\sigma c}{\gamma \alpha})$. When σ is increasing, the critical point $(\frac{a}{\sigma}, 0)$ moves to the left and the critical point $(\frac{c}{\gamma}, \frac{a}{\alpha} - \frac{\sigma c}{\gamma \alpha})$ moves down. The assumption $a > \frac{\sigma c}{\gamma}$ is necessary for the third critical point to be in the first quadrant. (When $a = \frac{\sigma c}{\gamma}$, then the two nonzero critical points coincide.)

(b,c) The Jacobian of the system is

$$\mathbf{J} = \begin{pmatrix} a - 2\sigma x - \alpha y & -\alpha x \\ \gamma y & -c + \gamma x \end{pmatrix}.$$

This implies that at the origin

$$\mathbf{J}(0,0) = \begin{pmatrix} a & 0 \\ 0 & -c \end{pmatrix},$$

which implies that the origin is a saddle point. ($a > 0$ and $c > 0$ by our assumption.) At the critical point $(\frac{a}{\sigma}, 0)$

$$\mathbf{J}(\frac{a}{\sigma}, 0) = \begin{pmatrix} -a & -\alpha a/\sigma \\ 0 & -c + \gamma a/\sigma \end{pmatrix},$$

which implies that this critical point is also a saddle as long as our assumption $a > \frac{\sigma c}{\gamma}$ is valid.
At the critical point $(\frac{c}{\gamma}, \frac{a}{\alpha} - \frac{\sigma c}{\gamma \alpha})$

$$\mathbf{J}(\frac{c}{\gamma}, \frac{a}{\alpha} - \frac{\sigma c}{\gamma \alpha}) = \begin{pmatrix} -\sigma c/\gamma & -\alpha c/\gamma \\ \gamma a/\alpha - \sigma c/\alpha & 0 \end{pmatrix}.$$

The eigenvalues of the matrix are

$$\frac{-c\sigma \pm \sqrt{c^2 \sigma^2 + 4c^2 \gamma \sigma - 4ac\gamma^2}}{2\gamma}.$$

We set the discriminant equal to zero and find that the greater solution is

$$\sigma_1 = -2\gamma + \frac{2\gamma \sqrt{ac + c^2}}{c}.$$

First note that $\sigma_1 > 0$, since $\sqrt{ac + c^2} > c$. Next we note that $\sigma_1 < a\gamma/c$. Since

$$\sqrt{ac + c^2} < \sqrt{\frac{a^2}{4} + ac + c^2} = \frac{a}{2} + c,$$

we see that

$$\sigma_1 = -2\gamma + \frac{2\gamma \sqrt{ac + c^2}}{c} < -2\gamma + \frac{2\gamma}{c}\left(\frac{a}{2} + c\right) = -2\gamma + \frac{a\gamma}{c} + 2\gamma = \frac{a\gamma}{c}.$$

For $0 < \sigma < \sigma_1$, the eigenvalues will be complex conjugates with negative real part, so the critical point will be an asymptotically stable spiral point. For $\sigma = \sigma_1$, the eigenvalues will be repeated and negative, so the critical point will be an asymptotically stable spiral point or node. For $\sigma_1 < \sigma < ac/\gamma$, the eigenvalues will be distinct and negative, so the critical point will be an asymptotically stable node.

(d) Since the third critical point is asymptotically stable for $0 < \sigma < ac/\gamma$, and the other critical points are saddle points, the populations will coexist for all such values of σ.

13.(a) The critical points are the solutions of the system
$$x(1 - \frac{x}{5} - \frac{2y}{x+6}) = 0$$
$$y(-\frac{1}{4} + \frac{x}{x+6}) = 0.$$

If $x = 0$, then $y = 0$. If $y = 0$, then $x = 5$. The third critical point can be found by setting $1/4 = x/(x+6)$, which gives $x = 2$ and then $y = 2.4$. So the critical points are $(0,0)$, $(5,0)$ and $(2, 2.4)$.

(b) The Jacobian of the system is
$$\mathbf{J} = \begin{pmatrix} 1 - \frac{2x}{5} - \frac{12y}{(x+6)^2} & -\frac{2x}{x+6} \\ \frac{6y}{(x+6)^2} & -\frac{1}{4} + \frac{x}{x+6} \end{pmatrix}.$$

This implies that at the origin
$$\mathbf{J}(0,0) = \begin{pmatrix} 1 & 0 \\ 0 & -1/4 \end{pmatrix},$$
which implies that the origin is a saddle point.
At the critical point $(5, 0)$
$$\mathbf{J}(5,0) = \begin{pmatrix} -1 & -10/11 \\ 0 & 9/44 \end{pmatrix},$$
which implies that this critical point is also a saddle point.
At the critical point $(2, 2.4)$
$$\mathbf{J}(2, 2.4) = \begin{pmatrix} -1/4 & -1/2 \\ 9/40 & 0 \end{pmatrix},$$
whose eigenvalues are complex with negative real part, which implies that this critical point is an asymptotically stable spiral.

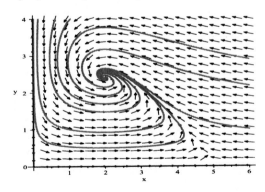

15.(a) Solving for the equilibrium of interest we obtain
$$x = \frac{E_2 + c}{\gamma} \qquad y = \frac{a}{\alpha} - \frac{\sigma}{\alpha} \cdot \frac{E_2 + c}{\gamma} - \frac{E_1}{\alpha}.$$

So if $E_1 > 0$ and $E_2 = 0$, then we have the same amount of prey and fewer predators.

(b) If $E_1 = 0$ and $E_2 > 0$, then we have more prey and fewer predators.

(c) If $E_1 > 0$ and $E_2 > 0$, then we have more prey and even fewer predators.

16.(b) The equilibrium solutions are given by the solutions of the system
$$x(1 - \frac{y}{2}) = H_1$$
$$y(-\frac{3}{4} + \frac{x}{4}) = H_2.$$

Now if $H_2 = 0$, then $x = 3$ and the first equation gives $y = 2 - 2H_1/3$. This means we have the same amount of prey and fewer predators.

(c) If $H_1 = 0$, then $y = 2$ and the second equation gives $x = 3 + 2H_2$. This means we have the same amount of predators and more prey.

(d) If $H_1 > 0$ and $H_2 > 0$, then the second equation gives $(x-3)y = 4H_2$ and using this we obtain from the first equation that $x(1 - \frac{2H_2}{x-3}) = H_1$. This gives the quadratic equation $x^2 - (3 + 2H_2 + H_1)x + 3H_1 = 0$. Now at the old value $x = 3$ this expression is $-6H_2$, so there is a root which is bigger than $x = 3$. The other root of the quadratic equation is closer to 0, so the equilibrium increases here: we have more prey. (Check this with e.g. $H_1 = H_2 = 1$: the roots are $3 \pm \sqrt{6}$, so both of the original roots get bigger.) A similar analysis shows that we will have fewer predators in this case.

9.6

2. We consider the function $V(x, y) = ax^2 + cy^2$. The rate of change of V along any trajectory is
$$\dot{V} = V_x \frac{dx}{dt} + V_y \frac{dy}{dt} = 2ax(-\frac{1}{2}x^3 + 2xy^2) + 2cy(-y^3) = -ax^4 + 4ax^2y^2 - 2cy^4.$$

Let us complete the square now the following way:
$$\dot{V} = -ax^4 + 4ax^2y^2 - 2cy^4 = -a(x^4 - 4x^2y^2) - 2cy^4 =$$
$$= -a(x^2 - 2y^2)^2 + 4ay^4 - 2cy^4 = -a(x^2 - 2y^2)^2 + (4a - 2c)y^4.$$

If $a > 0$ and $c > 0$, then $V(x, y)$ is positive definite. Clearly, if $4a - 2c < 0$, i.e. when $c > 2a$, then $\dot{V}(x,y)$ is negative definite. One such example is $V(x,y) =$

$x^2 + 3y^2$. It follows from Theorem 9.6.1 that the origin is an asymptotically stable critical point.

4. Given $V(x,y) = a x^2 + c y^2$, the rate of change of V along any trajectory is

$$\dot{V} = V_x \frac{dx}{dt} + V_y \frac{dy}{dt} = 2ax(x^3 - y^3) + 2cy(2xy^2 + 4x^2y + 2y^3) =$$
$$= 2a x^4 + (4c - 2a)xy^3 + 8c x^2 y^2 + 4c y^4.$$

Setting $a = 2c$,

$$\dot{V} = 4c x^4 + 8c x^2 y^2 + 4c y^4 \geq 4c x^4 + 4c y^4.$$

As long as $a = 2c > 0$, the function $V(x,y)$ is positive definite and $\dot{V}(x,y)$ is also positive definite. It follows from Theorem 9.6.2 that $(0,0)$ is an unstable critical point.

5. Given $V(x,y) = c(x^2 + y^2)$, the rate of change of V along any trajectory is

$$\dot{V} = V_x \frac{dx}{dt} + V_y \frac{dy}{dt} = 2cx\,[y - xf(x,y)] + 2cy\,[-x - yf(x,y)] = -2c(x^2 + y^2)f(x,y).$$

If $c > 0$, then $V(x,y)$ is positive definite. Furthermore, if $f(x,y)$ is positive in some neighborhood of the origin, then $\dot{V}(x,y)$ is negative definite. Theorem 9.6.1 asserts that the origin is an asymptotically stable critical point. On the other hand, if $f(x,y)$ is negative in some neighborhood of the origin, then $V(x,y)$ and $\dot{V}(x,y)$ are both positive definite. It follows from Theorem 9.6.2 that the origin is an unstable critical point.

9.(a) Letting $x = u$ and $y = u'$, we obtain the system of equations

$$\frac{dx}{dt} = y$$
$$\frac{dy}{dt} = -g(x) - y.$$

Since $g(0) = 0$, it is evident that $(0,0)$ is a critical point of the system. Consider the function

$$V(x,y) = \frac{1}{2}y^2 + \int_0^x g(s)ds.$$

It is clear that $V(0,0) = 0$. Since $g(u)$ is an odd function in a neighborhood of $u = 0$,

$$\int_0^x g(s)ds > 0 \text{ for } x > 0,$$

and

$$\int_0^x g(s)ds = -\int_x^0 g(s)ds > 0 \text{ for } x < 0.$$

Therefore $V(x,y)$ is positive definite. The rate of change of V along any trajectory is

$$\dot{V} = V_x \frac{dx}{dt} + V_y \frac{dy}{dt} = g(x) \cdot (y) + y\,[-g(x) - y] = -y^2.$$

It follows that $\dot{V}(x,y)$ is only negative semidefinite. Hence the origin is a stable critical point.

(b) Given
$$V(x,y) = \frac{1}{2}y^2 + \frac{1}{2}y \sin(x) + \int_0^x \sin(s)ds,$$

It is easy to see that $V(0,0) = 0$. The rate of change of V along any trajectory is
$$\dot{V} = V_x \frac{dx}{dt} + V_y \frac{dy}{dt} = \left[\sin x + \frac{y}{2}\cos x\right](y) + \left[y + \frac{1}{2}\sin x\right][-\sin x - y] =$$
$$= \frac{1}{2}y^2 \cos x - \frac{1}{2}\sin^2 x - \frac{y}{2}\sin x - y^2.$$

For $-\pi/2 < x < \pi/2$, we can write $\sin x = x - \alpha x^3/6$ and $\cos x = 1 - \beta x^2/2$, in which $\alpha = \alpha(x)$, $\beta = \beta(x)$. Note that $0 < \alpha, \beta < 1$. Then
$$\dot{V}(x,y) = \frac{y^2}{2}(1 - \frac{\beta x^2}{2}) - \frac{1}{2}(x - \frac{\alpha x^3}{6})^2 - \frac{y}{2}(x - \frac{\alpha x^3}{6}) - y^2.$$

Using polar coordinates,
$$\dot{V}(r,\theta) = -\frac{r^2}{2}[1 + \sin\theta\cos\theta + h(r,\theta)] = -\frac{r^2}{2}\left[1 + \frac{1}{2}\sin 2\theta + h(r,\theta)\right].$$

It is easy to show that
$$|h(r,\theta)| \leq \frac{1}{2}r^2 + \frac{1}{72}r^4.$$

So if r is sufficiently small, then $|h(r,\theta)| < 1/2$ and $\left|\frac{1}{2}\sin 2\theta + h(r,\theta)\right| < 1$. Hence $\dot{V}(x,y)$ is negative definite. Now we show that $V(x,y)$ is positive definite. Since $g(u) = \sin u$,
$$V(x,y) = \frac{1}{2}y^2 + \frac{1}{2}y \sin(x) + 1 - \cos x.$$

This time we set
$$\cos x = 1 - \frac{x^2}{2} + \gamma \frac{x^4}{24}.$$

Note that $0 < \gamma < 1$ for $-\pi/2 < x < \pi/2$. Converting to polar coordinates,
$$V(r,\theta) = \frac{r^2}{2}\left[1 + \sin\theta\cos\theta - \frac{r^2}{12}\sin\theta\cos^3\theta - \gamma\frac{r^2}{24}\cos^4\theta\right]$$
$$= \frac{r^2}{2}\left[1 + \frac{1}{2}\sin 2\theta - \frac{r^2}{12}\sin\theta\cos^3\theta - \gamma\frac{r^2}{24}\cos^4\theta\right].$$

Now
$$-\frac{r^2}{12}\sin\theta\cos^3\theta - \gamma\frac{r^2}{24}\cos^4\theta > -\frac{1}{8} \text{ for } r < 1.$$

It follows that when $r > 0$,
$$V(r,\theta) > \frac{r^2}{2}\left[\frac{7}{8} + \frac{1}{2}\sin 2\theta\right] \geq \frac{3r^2}{16} > 0.$$

Therefore $V(x,y)$ is indeed positive definite, and by Theorem 9.6.1, the origin is an asymptotically stable critical point.

12.(a) We consider the linear system

$$\begin{pmatrix} x \\ y \end{pmatrix}' = \begin{pmatrix} a_{11} & a_{12} \\ a_{21} & a_{22} \end{pmatrix} \begin{pmatrix} x \\ y \end{pmatrix}.$$

Let $V(x,y) = Ax^2 + Bxy + Cy^2$, in which

$$A = -\frac{a_{21}^2 + a_{22}^2 + (a_{11}a_{22} - a_{12}a_{21})}{2\Delta}$$

$$B = \frac{a_{12}a_{22} + a_{11}a_{21}}{\Delta}$$

$$C = -\frac{a_{11}^2 + a_{12}^2 + (a_{11}a_{22} - a_{12}a_{21})}{2\Delta},$$

and $\Delta = (a_{11} + a_{22})(a_{11}a_{22} - a_{12}a_{21})$. Based on the hypothesis, the coefficients A and B are negative. Therefore, except for the origin, $V(x,y)$ is negative on each of the coordinate axes. Along each trajectory,

$$\dot{V} = (2Ax + By)(a_{11}x + a_{12}y) + (2Cy + Bx)(a_{21}x + a_{22}y) = -x^2 - y^2.$$

Hence $\dot{V}(x,y)$ is negative definite. Theorem 9.6.2 asserts that the origin is an unstable critical point.

(b) We now consider the system

$$\begin{pmatrix} x \\ y \end{pmatrix}' = \begin{pmatrix} a_{11} & a_{12} \\ a_{21} & a_{22} \end{pmatrix} \begin{pmatrix} x \\ y \end{pmatrix} + \begin{pmatrix} F_1(x,y) \\ G_1(x,y) \end{pmatrix},$$

in which $F_1(x,y)/r \to 0$ and $G_1(x,y)/r \to 0$ as $r \to 0$. Let

$$V(x,y) = Ax^2 + Bxy + Cy^2,$$

in which

$$A = \frac{a_{21}^2 + a_{22}^2 + (a_{11}a_{22} - a_{12}a_{21})}{2\Delta}$$

$$B = -\frac{a_{12}a_{22} + a_{11}a_{21}}{\Delta}$$

$$C = \frac{a_{11}^2 + a_{12}^2 + (a_{11}a_{22} - a_{12}a_{21})}{2\Delta},$$

and $\Delta = (a_{11} + a_{22})(a_{11}a_{22} - a_{12}a_{21})$. Based on the hypothesis, $A, B > 0$. Except for the origin, $V(x,y)$ is positive on each of the coordinate axes. Along each trajectory,

$$\dot{V} = x^2 + y^2 + (2Ax + By)F_1(x,y) + (2Cy + Bx)G_1(x,y).$$

Converting to polar coordinates, for $r \neq 0$,

$$\dot{V} = = r^2 + r(2A\cos\theta + B\sin\theta)F_1 + r(2C\sin\theta + B\cos\theta)G_1$$

$$= r^2 + r^2\left[(2A\cos\theta + B\sin\theta)\frac{F_1}{r} + (2C\sin\theta + B\cos\theta)\frac{G_1}{r}\right].$$

Since the system is almost linear, there is an R such that

$$\left|(2A\cos\theta + B\sin\theta)\frac{F_1}{r} + (2C\sin\theta + B\cos\theta)\frac{G_1}{r}\right| < \frac{1}{2},$$

and hence

$$(2A\cos\theta + B\sin\theta)\frac{F_1}{r} + (2C\sin\theta + B\cos\theta)\frac{G_1}{r} > -\frac{1}{2}$$

for $r < R$. It follows that

$$\dot{V} > \frac{1}{2}r^2$$

as long as $0 < r < R$. Hence $\dot{V}$ is positive definite on the domain

$$D = \{(x,y) \,|\, x^2 + y^2 < R^2\}.$$

By Theorem 9.6.2, the origin is an unstable critical point.

9.7

3. The critical points of the ODE

$$\frac{dr}{dt} = r(r-1)(r-3)$$

are given by $r_1 = 0$, $r_2 = 1$ and $r_3 = 3$. Note that

$$\frac{dr}{dt} > 0 \text{ for } 0 < r < 1 \text{ and } r > 3 \,; \quad \frac{dr}{dt} < 0 \text{ for } 1 < r < 3.$$

$r = 0$ corresponds to an unstable critical point. The critical point $r_2 = 1$ is asymptotically stable, whereas the critical point $r_3 = 3$ is unstable. Since the critical values are isolated, a limit cycle is given by

$$r = 1, \, \theta = t + t_0$$

which is asymptotically stable. Another periodic solution is found to be

$$r = 3, \, \theta = t + t_0$$

which is unstable.

5. The critical points of the ODE

$$\frac{dr}{dt} = \sin \pi r$$

are given by $r = n$, $n = 0, 1, 2, \ldots$. Based on the sign of r' in the neighborhood of each critical value, the critical points $r = 2k$, $k = 1, 2, \ldots$ correspond to unstable periodic solutions, with $\theta = t + t_0$. The critical points $r = 2k + 1$, $k = 0, 1, 2, \ldots$ correspond to stable limit cycles, with $\theta = t + t_0$. The solution $r = 0$ represents an unstable critical point.

6. The critical points of the ODE
$$\frac{dr}{dt} = r|r-2|(r-3)$$
are given by $r_1 = 0$, $r_2 = 2$ and $r_3 = 3$. Note that
$$\frac{dr}{dt} < 0 \text{ for } 0 < r < 3 \, ; \, \frac{dr}{dt} > 0 \text{ for } r > 3 \, .$$
$r = 0$ corresponds to an asymptotically stable critical point. The critical points $r_2 = 2$ is semistable, whereas the critical point $r_3 = 3$ is unstable. Since the critical values are isolated, a semistable limit cycle is given by
$$r = 2 \, , \, \theta = -t + t_0.$$
Another periodic solution is found to be
$$r = 3 \, , \, \theta = -t + t_0$$
which is unstable.

10. Given $F(x,y) = a_{11}x + a_{12}y$ and $G(x,y) = a_{21}x + a_{22}y$, it follows that
$$F_x + G_y = a_{11} + a_{22} \, .$$
Based on the hypothesis, $F_x + G_y$ is either positive or negative on the entire plane. By Theorem 9.7.2, the system cannot have a nontrivial periodic solution.

12. Given that $F(x,y) = -2x - 3y - xy^2$ and $G(x,y) = y + x^3 - x^2y$,
$$F_x + G_y = -1 - x^2 - y^2.$$
Since $F_x + G_y < 0$ on the entire plane, Theorem 9.7.2 asserts that the system cannot have a nontrivial periodic solution.

14.(a) Based on the given graphs, the following table shows the estimated values:

$\mu = 0.2$	$T \approx 6.29$
$\mu = 1.0$	$T \approx 6.66$
$\mu = 5.0$	$T \approx 11.60$

(b) The initial conditions were chosen as $x(0) = 2$, $y(0) = 0$. For $\mu = 0.5$, $T \approx 6.38$:

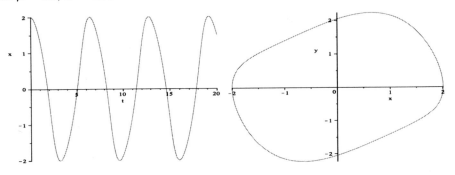

For $\mu = 2$, $T \approx 7.65$:

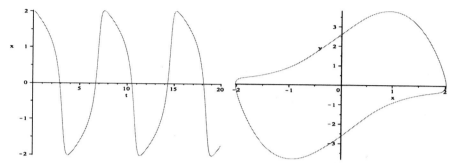

For $\mu = 3$, $T \approx 8.86$:

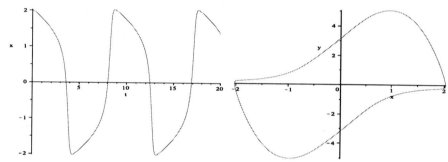

For $\mu = 5$, $T \approx 10.25$:

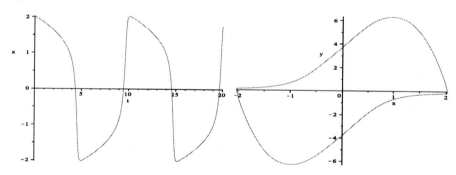

(c) The period, T, appears to be a quadratic function of μ.

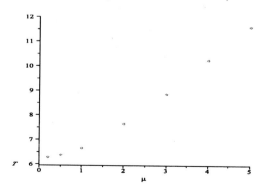

15.(a) Setting $x = u$ and $y = u'$, we obtain the system of equations

$$\frac{dx}{dt} = y$$
$$\frac{dy}{dt} = -x + \mu(1 - \frac{1}{3}y^2)y.$$

(b) Evidently, $y = 0$. It follows that $x = 0$. Hence the only critical point of the system is at $(0,0)$. The components of the vector field are infinitely differentiable everywhere. Therefore the system is almost linear. The Jacobian matrix of the vector field is

$$\mathbf{J} = \begin{pmatrix} 0 & 1 \\ -1 & \mu - \mu y^2 \end{pmatrix}.$$

At the critical point $(0,0)$, the coefficient matrix of the linearized system is

$$\mathbf{J}(0,0) = \begin{pmatrix} 0 & 1 \\ -1 & \mu \end{pmatrix},$$

with eigenvalues

$$r_{1,2} = \frac{\mu}{2} \pm \frac{1}{2}\sqrt{\mu^2 - 4}.$$

If $\mu = 0$, the equation reduces to the ODE for a simple harmonic oscillator. For the case $0 < \mu < 2$, the eigenvalues are complex, and the critical point is an unstable spiral. For $\mu \geq 2$, the eigenvalues are real, and the origin is an unstable node.

(c) The initial conditions were chosen as $x(0) = 2$, $y(0) = 0$.

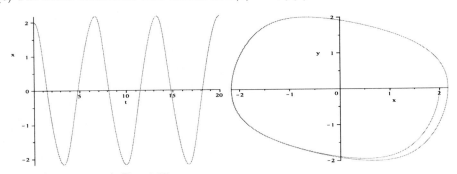

$\mu = 1$: $A \approx 2.16$ and $T \approx 6.65$.

(d)

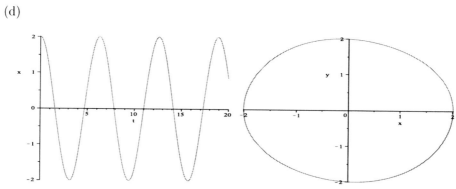

$\mu = 0.2$: $A \approx 2.00$ and $T \approx 6.30$.

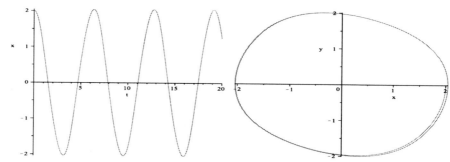

$\mu = 0.5$: $A \approx 2.04$ and $T \approx 6.38$.

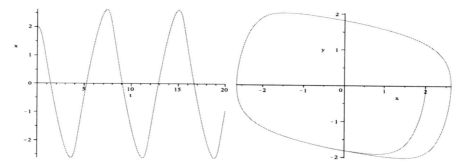

$\mu = 2$: $A \approx 2.6$ and $T \approx 7.62$.

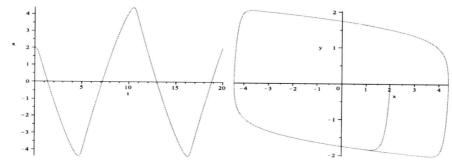

$\mu = 5$: $A \approx 4.37$ and $T \approx 11.61$.

(e)

	A	T
$\mu = 0.2$	2.00	6.30
$\mu = 0.5$	2.04	6.38
$\mu = 1.0$	2.16	6.65
$\mu = 2.0$	2.6	7.62
$\mu = 5.0$	4.37	11.61

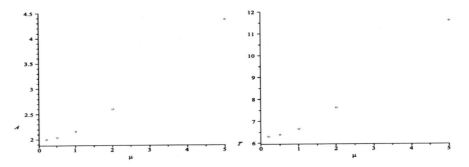

16.(a) The critical points are solutions of the algebraic system

$$\mu x + y - x(x^2 + y^2) = 0$$
$$-x + \mu y - y(x^2 + y^2) = 0.$$

Multiply the first equation by y and the second equation by x to obtain

$$\mu xy + y^2 - xy(x^2 + y^2) = 0$$
$$-x^2 + \mu xy - xy(x^2 + y^2) = 0.$$

Subtraction of the two equations results in

$$x^2 + y^2 = 0,$$

which is satisfied only for $x = y = 0$.

(b) The Jacobian matrix of the vector field is

$$\mathbf{J} = \begin{pmatrix} \mu - 3x^2 - y^2 & 1 - 2xy \\ -1 - 2xy & \mu - x^2 - 3y^2 \end{pmatrix}.$$

At the critical point $(0,0)$, the coefficient matrix of the linearized system is

$$\mathbf{J}(0,0) = \begin{pmatrix} \mu & 1 \\ -1 & \mu \end{pmatrix},$$

resulting in the linear system

$$x' = \mu x + y$$
$$y' = -x + \mu y.$$

The characteristic equation for the coefficient matrix is $\lambda^2 - 2\mu\lambda + \mu^2 + 1 = 0$, with solution
$$\lambda = \mu \pm i.$$
For $\mu < 0$, the origin is a stable spiral. When $\mu = 0$, the origin is a center. For $\mu > 0$, the origin is an unstable spiral.

(c) Introduce polar coordinates r and θ, so that $x = r\cos\theta$ and $y = r\sin\theta$ for $r \geq 0$. Multiply the first of Eqns (i) by x and the second equation by y to obtain
$$x\,x' = \mu x^2 + xy - x^2(x^2 + y^2)$$
$$y\,y' = -xy + \mu y^2 - y^2(x^2 + y^2).$$
Addition of the two equations results in
$$x\,x' + y\,y' = \mu(x^2 + y^2) - (x^2 + y^2)^2.$$
Since $r^2 = x^2 + y^2$ and $rr' = x\,x' + y\,y'$, it follows that $rr' = \mu r^2 - r^4$ and
$$\frac{dr}{dt} = \mu r - r^3$$
for $r > 0$. Multiply the first of Eqns (i) by y and the second equation by x, the difference of the two equations results in
$$y\,x' - x\,y' = x^2 + y^2.$$
Since $y\,x' - x\,y' = -r^2\,\theta'$, the above equation reduces to
$$\frac{d\theta}{dt} = -1.$$

(d) From $r' = r(\mu - r^2)$ and $\theta' = -1$, it follows that one solution of the system is given by
$$r = \sqrt{\mu} \quad \text{and} \quad \theta = -t + t_0$$
valid for $\mu > 0$. This corresponds to a periodic solution with a circular trajectory. Since $r \geq 0$, observe that
$$\frac{dr}{dt} = r(\mu - r^2) < 0 \text{ for } r > \sqrt{\mu}$$
$$= r(\mu - r^2) > 0 \text{ for } \sqrt{\mu} > r > 0.$$
Hence solutions with initial condition $r(0) \neq \sqrt{\mu}$ are attracted to the limit cycle.

17.(a) The critical points are solutions of the algebraic system
$$y = 0$$
$$-x + \mu(1 - x^2)y = 0.$$
Clearly, $y = 0$, and this implies that $x = 0$. So the origin is the only critical point. The Jacobian matrix of the vector field is
$$\mathbf{J} = \begin{pmatrix} 0 & 1 \\ -1 - 2\mu xy & \mu(1 - x^2) \end{pmatrix}.$$

At the critical point $(0,0)$, the coefficient matrix of the linearized system is

$$\mathbf{J}(0,0) = \begin{pmatrix} 0 & 1 \\ -1 & \mu \end{pmatrix}.$$

The characteristic equation is $\lambda^2 - \mu\lambda + 1 = 0$, and the roots are

$$\lambda = \frac{\mu \pm \sqrt{\mu^2 - 4}}{2}.$$

This implies that the origin is an asymptotically stable node for $\mu < -2$, an asymptotically stable spiral point for $-2 < \mu < 0$, an unstable spiral point for $0 < \mu < 2$ and an unstable node for $\mu > 2$.

(b)

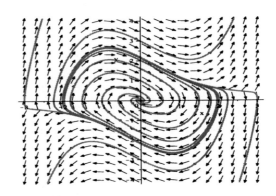

(c)

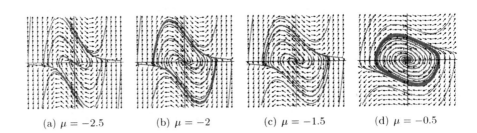

(a) $\mu = -2.5$ (b) $\mu = -2$ (c) $\mu = -1.5$ (d) $\mu = -0.5$

(d)

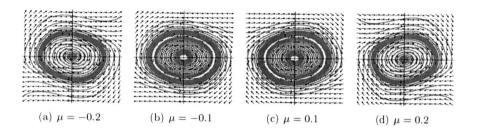

(a) $\mu = -0.2$ (b) $\mu = -0.1$ (c) $\mu = 0.1$ (d) $\mu = 0.2$

19.(a) The critical points are solutions of the algebraic system
$$x(a - \frac{x}{5} - \frac{2y}{x+6}) = 0$$
$$y(-\frac{1}{4} + \frac{x}{x+6}) = 0.$$

If $x = 0$, then $y = 0$. If $y = 0$, then $x = 5a$ from the first equation. The third critical point comes from $x = 2$, and then the first equation implies that $y = 4a - 8/5$. So the critical points are $(0,0)$, $(5a, 0)$ and $(2, 4a - 8/5)$.

(b) The Jacobian matrix of the vector field is
$$\mathbf{J} = \begin{pmatrix} a - \frac{2x}{5} - \frac{12y}{(x+6)^2} & -\frac{2x}{x+6} \\ \frac{6y}{(x+6)^2} & -\frac{1}{4} + \frac{x}{x+6} \end{pmatrix}.$$

At the critical point $(2, 4a - 8/5)$, the coefficient matrix of the linearized system is
$$\mathbf{J}(2, 4a - 8/5) = \begin{pmatrix} a/4 - 1/2 & -1/2 \\ 3a/8 - 3/20 & 0 \end{pmatrix}.$$

The characteristic equation is $\lambda^2 + \lambda(1/2 - a/4) + (3a/8 - 3/20)/2 = 0$, and the roots are
$$\lambda = \frac{a}{8} - \frac{1}{4} \pm \frac{1}{2}\sqrt{\frac{a^2}{16} - a + \frac{11}{20}}.$$

We can conclude that $a_0 = 2$.

(c)

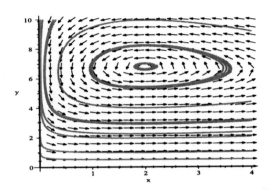

20.(a) The critical points are solutions of the algebraic system
$$1 - (b+1)x + x^2 y/4 = 0$$
$$b\,x - x^2 y/4 = 0.$$

Add the two equations to obtain $1 - x = 0$, with solution $x = 1$. The second of the above equations yields $y = 4\,b$.

(b) The Jacobian matrix of the vector field is
$$\mathbf{J} = \frac{1}{4}\begin{pmatrix} -4(b+1) + 2xy & x^2 \\ 4b - 2xy & -x^2 \end{pmatrix}.$$

At the critical point $(1, 4b)$, the coefficient matrix of the linearized system is
$$\mathbf{J}(1, 4b) = \begin{pmatrix} b-1 & 1/4 \\ -b & -1/4 \end{pmatrix},$$

with characteristic equation
$$\lambda^2 + (\frac{5}{4} - b)\lambda + \frac{1}{4} = 0$$

and eigenvalues
$$\lambda = -\frac{5 - 4b \pm \sqrt{9 - 40\,b + 16\,b^2}}{8} = -\frac{(5/4 - b) \pm \sqrt{(5/4 - b)^2 - 1}}{2}.$$

(c) Let $L = 5/4 - b$. The eigenvalues can be expressed as
$$\lambda = -\frac{L \pm \sqrt{L^2 - 1}}{2}.$$

We find that the eigenvalues are real if $L^2 \geq 1$ and are complex if $L^2 < 1$.
For the complex case, that is, $-1 < L < 1$, the critical point is a stable spiral if $0 < L < 1$; it is an unstable spiral if $-1 < L < 0$. That is, the critical point is an asymptotically stable spiral if $1/4 < b < 5/4$ and is an unstable spiral if $5/4 < b < 9/4$. When $L^2 > 1$, the critical point is a node. The critical point is a stable node if $L > 1$; it is an unstable node if $L < -1$. That is, the critical point is an asymptotically stable node if $0 < b < 1/4$ and is an unstable node if $b > 9/4$.

(d) From part (c), the critical point changes from an asymptotically stable spiral to an unstable spiral when $b_0 = 5/4$.

(e,f)

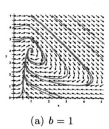
(a) $b = 1$

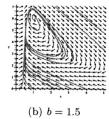
(b) $b = 1.5$

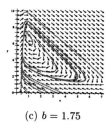
(c) $b = 1.75$

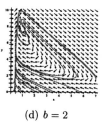
(d) $b = 2$

9.8

6. $r = 28$, with initial point $(5, 5, 5)$:

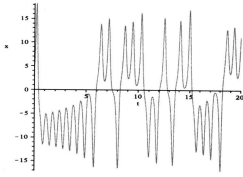

$r = 28$, with initial point $(5.01, 5, 5)$:

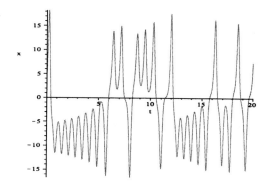

7. $r = 28$:

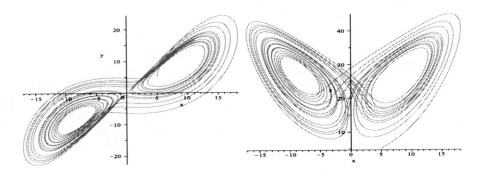

9.(a) $r = 100$, initial point $(-5, -13, 55)$:

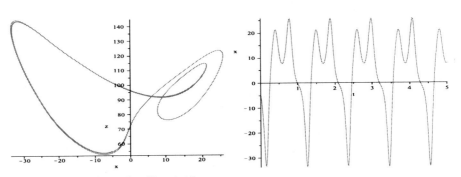

The period appears to be $T \approx 1.12$.

(b) $r = 99.94$, initial point $(-5, -13, 55)$:

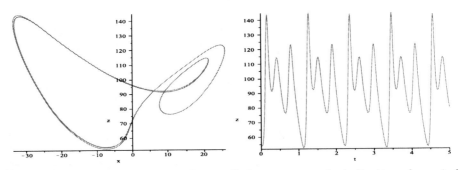

The periodic trajectory appears to have split into two strands, indicative of a period-doubling. Closer examination reveals that the peak values of $z(t)$ are slightly different.

$r = 99.7$, initial point $(-5, -13, 55)$:

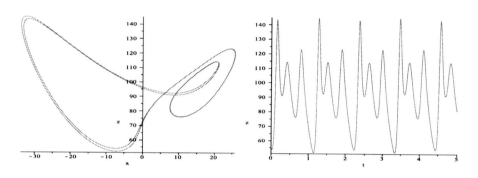

(c) $r = 99.6$, initial point $(-5, -13, 55)$:

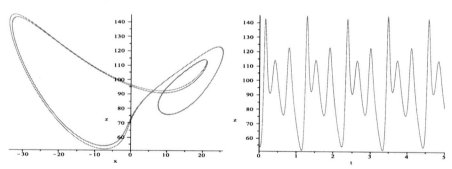

The strands again appear to have split. Closer examination reveals that the peak values of $z(t)$ are different.

10.(a) $r = 100.5$, initial point $(-5, -13, 55)$:

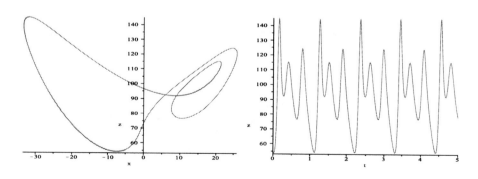

$r = 100.7$, initial point $(-5, -13, 55)$:

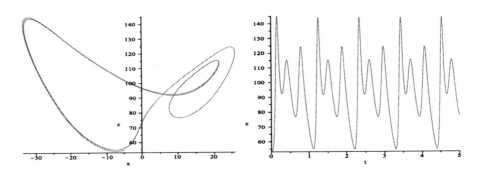

(b) $r = 100.8$, initial point $(-5, -13, 55)$:

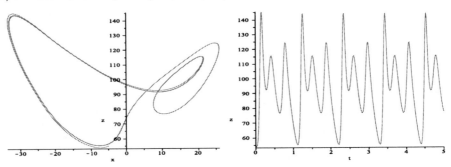

$r = 100.81$, initial point $(-5, -13, 55)$:

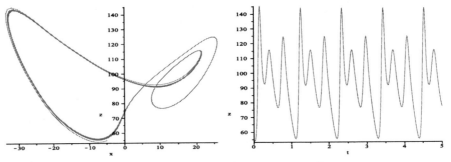

The strands of the periodic trajectory are beginning to split apart.

$r = 100.82$, initial point $(-5, -13, 55)$:

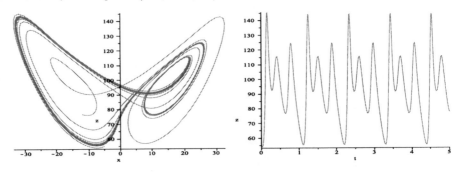

$r = 100.83$, initial point $(-5, -13, 55)$:

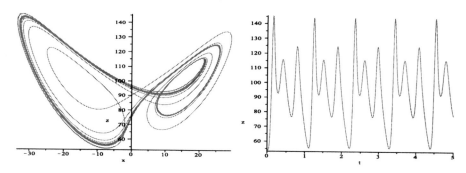

$r = 100.84$, initial point $(-5, -13, 55)$:

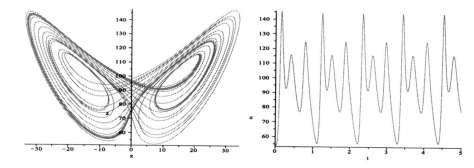

12. The system is given by

$$x' = -y - z$$
$$y' = x + y/4$$
$$z' = 1/2 + z(x - c).$$

(a) We obtain that the critical points (when $c^2 > 1/2$) are

$$(\frac{c}{2} + \frac{1}{4}\sqrt{4c^2 - 2}, -2c - \sqrt{4c^2 - 2}, 2c + \sqrt{4c^2 - 2}),$$
$$(\frac{c}{2} - \frac{1}{4}\sqrt{4c^2 - 2}, -2c + \sqrt{4c^2 - 2}, 2c - \sqrt{4c^2 - 2}).$$

The Jacobian of the system is

$$\mathbf{J} = \begin{pmatrix} 0 & -1 & -1 \\ 1 & 1/4 & 0 \\ z & 0 & x - c \end{pmatrix}.$$

In the following computations, we approximate the values to 4 decimal points. When $c = 1.3$, the two critical points are

$$(1.1954, -4.7817, 4.7817) \quad \text{and} \quad (0.1046, -0.4183, 0.4183).$$

The corresponding eigenvalues of the Jacobian are

$$0.1893, \ -0.0219 \pm 2.4007i \quad \text{and} \ -0.9613, \ 0.0080 \pm 1.0652i.$$

(b)

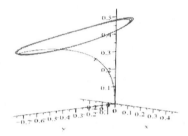

(c)

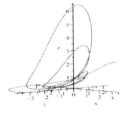 

(d) $T_1 \approx 5.9$.

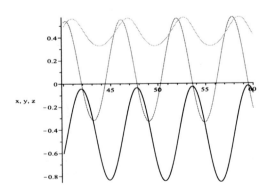

13.(b) Using the eigenvalue idea:

c	Critical Point	Complex Eigenvalue Pair
1.2	$(0.1152, -0.4609, 0.4609)$	$-0.0063 \pm 1.0859i$
1.25	$(0.1096, -0.4384, 0.4384)$	$0.0010 \pm 1.0747i$

Clearly, somewhere in between we have a Hopf bifurcation. Similar computations show that the bifurcation value is $c \approx 1.243$.

14.(a) When $c = 3$, the two critical points are

$$(2.9577, -11.8310, 11.8310) \quad \text{and} \quad (0.0423, -0.1690, 0.1690).$$

The corresponding eigenvalues of the Jacobian are

$$0.2273, \; -0.0098 \pm 3.5812i \quad \text{and} \quad -2.9053, \; 0.0988 \pm 0.9969i.$$

(b)

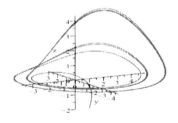

(c) $T_2 \approx 11.8$.

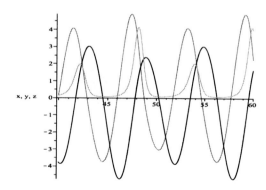

15.(a) When $c = 3.8$, the two critical points are

$$(3.7668, -15.0673, 15.0673) \quad \text{and} \quad (0.0332, -0.1327, 0.1327).$$

The corresponding eigenvalues of the Jacobian are

$$0.2324, \ -0.0078 \pm 4.0078i \quad \text{and} \ -3.7335, \ 0.1083 \pm 0.9941i.$$

(b) $T_4 \approx 23.6$.

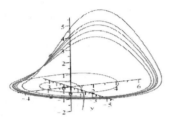

(c)

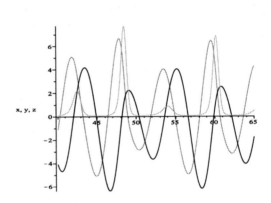

CHAPTER 10

Partial Differential Equations and Fourier Series

10.1

1. The general solution of the ODE is $y(x) = c_1 \cos x + c_2 \sin x$. Imposing the first boundary condition, it is necessary that $c_1 = 0$. Therefore $y(x) = c_2 \sin x$. Taking its derivative, $y'(x) = c_2 \cos x$. Imposing the second boundary condition, we require that $c_2 \cos \pi = 1$. The latter equation is satisfied only if $c_2 = -1$. Hence the solution of the boundary value problem is $y(x) = -\sin x$.

4. The general solution of the differential equation is $y(x) = c_1 \cos x + c_2 \sin x$. It follows that $y'(x) = -c_1 \sin x + c_2 \cos x$. Imposing the first boundary condition, we find that $c_2 = 1$. Therefore $y(x) = c_1 \cos x + \sin x$. Imposing the second boundary condition, we require that $c_1 \cos L + \sin L = 0$. If $\cos L \neq 0$, that is, as long as $L \neq (2k-1)\pi/2$, with k an integer, then $c_1 = -\tan L$. The solution of the boundary value problem is $y(x) = -\tan L \cos x + \sin x$. If $\cos L = 0$, the boundary condition results in $\sin L = 0$. The latter two equations are inconsistent, which implies that the BVP has no solution.

5. The general solution of the homogeneous differential equation is $y(x) = c_1 \cos x + c_2 \sin x$. Using any of a number of methods, including the method of undetermined coefficients, it is easy to show that a particular solution is $Y(x) = x$. Hence the general solution of the given differential equation is $y(x) = c_1 \cos x + c_2 \sin x + x$. The first boundary condition requires that $c_1 = 0$. Imposing the second boundary condition, it is necessary that $c_2 \sin \pi + \pi = 0$. The resulting equation has no solution. We conclude that the boundary value problem has no solution.

6. Using the method of undetermined coefficients, it is easy to show that the general solution of the ODE is $y(x) = c_1 \cos \sqrt{2}\, x + c_2 \sin \sqrt{2}\, x + x/2$. Imposing the first boundary condition, we find that $c_1 = 0$. The second boundary condition requires that $c_2 \sin \sqrt{2}\,\pi + \pi/2 = 0$. It follows that $c_2 = -\pi/(2 \sin \sqrt{2}\,\pi)$. Hence the solution of the boundary value problem is

$$y(x) = -\frac{\pi}{2 \sin \sqrt{2}\,\pi} \sin \sqrt{2}\, x + \frac{x}{2}.$$

8. The general solution of the homogeneous differential equation is $y(x) = c_1 \cos 2x + c_2 \sin 2x$. Using the method of undetermined coefficients, a particular solution is $Y(x) = \sin x/3$. Hence the general solution of the given differential equation is $y(x) = c_1 \cos 2x + c_2 \sin 2x + \sin x/3$. The first boundary condition requires that $c_1 = 0$. The second boundary requires that $c_2 \sin 2\pi + (1/3) \sin \pi = 0$. The latter equation is valid for all values of c_2. Therefore the solution of the boundary value problem is $y(x) = c_2 \sin 2x + \sin x/3$.

9. Using the method of undetermined coefficients, it is easy to show that the general solution of the ODE is $y(x) = c_1 \cos 2x + c_2 \sin 2x + \cos x/3$. It follows that $y'(x) = -2c_1 \sin 2x + 2c_2 \cos 2x - \sin x/3$. Imposing the first boundary condition, we find that $c_2 = 0$. The second boundary condition requires that $-2c_1 \sin 2\pi - (1/3) \sin \pi = 0$. The resulting equation is satisfied for all values of c_1. Hence the solution is the family of functions $y(x) = c_1 \cos 2x + \cos x/3$.

10. The general solution of the differential equation is given by $y(x) = c_1 \cos \sqrt{3}\, x + c_2 \sin \sqrt{3}\, x + \cos x/2$. Its derivative is $y'(x) = -\sqrt{3}\, c_1 \sin \sqrt{3}\, x + \sqrt{3}\, c_2 \cos \sqrt{3}\, x - \sin x/2$. The first boundary condition requires that $c_2 = 0$. Imposing the second boundary condition, we obtain $-\sqrt{3}\, c_1 \sin \sqrt{3}\,\pi = 0$. It follows that $c_1 = 0$. Hence the solution of the BVP is $y(x) = \cos x/2$.

11. The general solution of the differential equation is $y(x) = c_1 x + c_2 x^2$. Imposing the boundary conditions, we find that $c_1 = -5/2$ and $c_2 = 3/2$. That is, $y(x) = -5x/2 + 3x^2/2$.

13. With the change of variable $x = e^t$, the ODE can be written as

$$y'' + 4y' + (4 + \pi^2) y = t,$$

with corresponding initial conditions $y(0) = 0$ and $y(1) = 0$. The general solution of this ODE is

$$y(t) = c_1 e^{-2t} \cos \pi t + c_2 e^{-2t} \sin \pi t + \frac{t\pi^2 + 4t - 4}{(4 + \pi^2)^2}.$$

Imposing the boundary conditions, it is necessary that

$$c_1 - \frac{4}{(4 + \pi^2)^2} = 0, \qquad -e^{-2} c_1 + \frac{\pi^2}{(4 + \pi^2)^2} = 0.$$

Hence no solution exists.

15. Assuming that $\lambda > 0$, we can set $\lambda = \mu^2$. The general solution of the differential equation is $y(x) = c_1 \cos \mu x + c_2 \sin \mu x$, so that $y'(x) = -\mu c_1 \sin \mu x + \mu c_2 \cos \mu x$. Imposing the first boundary condition, it follows that $c_2 = 0$. Therefore $y(x) = c_1 \cos \mu x$. The second boundary condition requires that $c_1 \cos \mu \pi = 0$. For a nontrivial solution, it is necessary that $\cos \mu \pi = 0$, that is, $\mu \pi = (2n-1)\pi/2$, with $n = 1, 2, \ldots$. Therefore the eigenvalues are

$$\lambda_n = \frac{(2n-1)^2}{4}, \ n = 1, 2, \ldots.$$

The corresponding eigenfunctions are given by

$$y_n = \cos \frac{(2n-1)x}{2}, \ n = 1, 2, \ldots.$$

Assuming that $\lambda < 0$, we can set $\lambda = -\mu^2$. The general solution of the differential equation is $y(x) = c_1 \cosh \mu x + c_2 \sinh \mu x$, so that $y'(x) = \mu c_1 \sinh \mu x + \mu c_2 \cosh \mu x$. Imposing the first boundary condition, it follows that $c_2 = 0$. Therefore $y(x) = c_1 \cosh \mu x$. The second boundary condition requires that $c_1 \cosh \mu \pi = 0$, which results in $c_1 = 0$. Hence the only solution is the trivial solution. Finally, with $\lambda = 0$, the general solution of the ODE is $y(x) = c_1 x + c_2$. It is easy to show that the boundary conditions require that $c_1 = c_2 = 0$. Therefore all of the eigenvalues are positive.

16. Assuming that $\lambda > 0$, we can set $\lambda = \mu^2$. The general solution of the differential equation is $y(x) = c_1 \cos \mu x + c_2 \sin \mu x$, so that $y'(x) = -\mu c_1 \sin \mu x + \mu c_2 \cos \mu x$. Imposing the first boundary condition, it follows that $c_2 = 0$. The second boundary condition requires that $c_1 \sin \mu \pi = 0$. For a nontrivial solution, we must have $\mu \pi = n \pi$, $n = 1, 2, \ldots$. It follows that the eigenvalues are

$$\lambda_n = n^2, \ n = 1, 2, \ldots,$$

and the corresponding eigenfunctions are

$$y_n = \cos nx, \ n = 1, 2, \ldots.$$

Assuming that $\lambda < 0$, we can set $\lambda = -\mu^2$. The general solution of the differential equation is $y(x) = c_1 \cosh \mu x + c_2 \sinh \mu x$, so that $y'(x) = \mu c_1 \sinh \mu x + \mu c_2 \cosh \mu x$. Imposing the first boundary condition, it follows that $c_2 = 0$. The second boundary condition requires that $c_1 \sinh \mu \pi = 0$. The latter equation is satisfied only for $c_1 = 0$. Finally, for $\lambda = 0$, the solution is $y(x) = c_1 x + c_2$. Imposing the boundary conditions, we find that $y(x) = c_2$. Therefore $\lambda = 0$ is also an eigenvalue, with corresponding eigenfunction $y_0(x) = 1$.

17. It can be shown, as in Problem 15, that $\lambda > 0$. Setting $\lambda = \mu^2$, the general solution of the resulting ODE is $y(x) = c_1 \cos \mu x + c_2 \sin \mu x$, with $y'(x) = -\mu c_1 \sin \mu x + \mu c_2 \cos \mu x$. Imposing the first boundary condition, we find that $c_2 = 0$. Therefore $y(x) = c_1 \cos \mu x$. The second boundary condition gives $c_1 \cos \mu L = 0$. For a nontrivial solution, it is necessary that $\cos \mu L = 0$, that is, $\mu = (2n-1)\pi/(2L)$, with $n = 1, 2, \ldots$. Therefore the eigenvalues are

$$\lambda_n = \frac{(2n-1)^2 \pi^2}{4L^2}, \ n = 1, 2, \ldots.$$

The corresponding eigenfunctions are given by

$$y_n = \cos \frac{(2n-1)\pi x}{2L}, \quad n = 1, 2, \ldots.$$

19. Assuming that $\lambda > 0$, we can set $\lambda = \mu^2$. The general solution of the differential equation is $y(x) = c_1 \cosh \mu x + c_2 \sinh \mu x$. The first boundary condition requires that $c_1 = 0$. Therefore $y(x) = c_2 \sinh \mu x$ and $y'(x) = \mu c_2 \cosh \mu x$. Imposing the second boundary condition, it is necessary that $\mu c_2 \cosh \mu L = 0$. The latter equation is valid only for $c_2 = 0$. The only solution is the trivial solution. Assuming that $\lambda < 0$, we set $\lambda = -\mu^2$. The general solution of the resulting ODE is $y(x) = c_1 \cos \mu x + c_2 \sin \mu x$. Imposing the first boundary condition, we find that $c_1 = 0$. Hence $y(x) = c_2 \sin \mu x$ and $y'(x) = \mu c_2 \cos \mu x$. In order to satisfy the second boundary condition, it is necessary that $\mu c_2 \cos \mu L = 0$. For a nontrivial solution, $\mu = (2n-1)\pi/(2L)$, with $n = 1, 2, \ldots$. Therefore the eigenvalues are

$$\lambda_n = -\frac{(2n-1)^2 \pi^2}{4L^2}, \quad n = 1, 2, \ldots.$$

The corresponding eigenfunctions are given by

$$y_n = \sin \frac{(2n-1)\pi x}{2L}, \quad n = 1, 2, \ldots.$$

Finally, for $\lambda = 0$, the general solution is linear. Based on the boundary conditions, it follows that $y(x) = 0$. Therefore all of the eigenvalues are negative.

20. With the change of variable $x = e^t$, the ODE can be written as

$$y'' - 2y' + \lambda y = 0,$$

with corresponding initial conditions $y(0) = 0$ and $y(\ln(L)) = 0$. Let $\lambda = \mu^2 + 1$. The general solution of the ODE is $y(t) = c_1 e^t \cos \mu t + c_2 e^t \sin \mu t$. Imposing the first boundary condition, we find that $c_1 = 0$. In order to satisfy the second boundary condition, it is necessary that $c_2 L \sin \mu(\ln(L)) = 0$. For a nontrivial solution, $\mu \ln(L) = n\pi$, or $\mu = n\pi/\ln(L)$, with $n = 1, 2, \ldots$. Therefore the eigenvalues are

$$\lambda_n = 1 + \left[\frac{n\pi}{\ln(L)}\right]^2, \quad n = 1, 2, \ldots.$$

The corresponding eigenfunctions are given by

$$y_n = x \sin \left[\frac{n\pi \ln(x)}{\ln(L)}\right], \quad n = 1, 2, \ldots.$$

22. Integrating the differential equation four times, the general solution is

$$y(x) = \frac{k x^4}{24} + \frac{c_3 x^3}{6} + \frac{c_2 x^2}{2} + c_1 x + c_0.$$

(a) Imposing the boundary conditions $y(0) = y(L) = y''(0) = y''(L) = 0$,

$$y(x) = \frac{k x^4}{24} - \frac{kL x^3}{12} + \frac{kL^3 x}{24}.$$

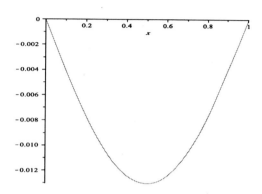

(b) Imposing the boundary conditions $y(0) = y(L) = y'(0) = y'(L) = 0$,

$$y(x) = \frac{k\,x^4}{24} - \frac{kL\,x^3}{12} + \frac{kL^2 x^2}{24}.$$

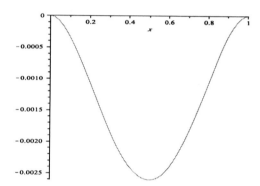

(c) Imposing the boundary conditions $y(0) = y'(0) = y''(L) = y'''(L) = 0$,

$$y(x) = \frac{k\,x^4}{24} - \frac{kL\,x^3}{6} + \frac{kL^2 x^2}{4}.$$

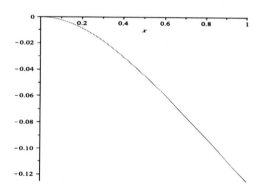

23.(a) Setting $\lambda = \mu^2$, write the general solution of the ODE $y'' + \mu^2 y = 0$ as $y(x) = k_1 e^{i\mu x} + k_2 e^{-i\mu x}$. Imposing the boundary conditions $y(0) = y(\pi) = 0$, we obtain the system of equations $k_1 + k_2 = 0$, $k_1 e^{i\mu\pi} + k_2 e^{-i\mu\pi} = 0$. The system has a nontrivial solution if and only if the coefficient matrix is singular. Set the determinant equal to zero to obtain $e^{-i\mu\pi} - e^{i\mu\pi} = 0$.

(b) Let $\mu = \nu + i\sigma$. Then $i\mu\pi = i\nu\pi - \sigma\pi$, and the previous equation can be written as $e^{\sigma\pi}e^{-i\nu\pi} - e^{-\sigma\pi}e^{i\nu\pi} = 0$. Using Euler's relation, $e^{i\nu\pi} = \cos\nu\pi + i\sin\nu\pi$, we obtain

$$e^{\sigma\pi}(\cos\nu - i\sin\nu) - e^{-\sigma\pi}(\cos\nu + i\sin\nu) = 0.$$

Equating the real and imaginary parts of the equation,

$$(e^{\sigma\pi} - e^{-\sigma\pi})\cos\nu\pi = 0, \qquad (e^{\sigma\pi} + e^{-\sigma\pi})\sin\nu\pi = 0.$$

(c) Based on the second equation, $\nu = n$, $n \in \mathbb{I}$. Since $\cos n\pi \neq 0$, it follows that $e^{\sigma\pi} = e^{-\sigma\pi}$, or $e^{2\sigma\pi} = 1$. Hence $\sigma = 0$, and $\mu = n$, $n \in \mathbb{I}$.

10.2

1. The period of the function $\sin\alpha x$ is $T = 2\pi/\alpha$. Therefore the function $\sin 5x$ has period $T = 2\pi/5$.

2. The period of the function $\cos\alpha x$ is also $T = 2\pi/\alpha$. Therefore the function $\cos 2\pi x$ has period $T = 2\pi/2\pi = 1$.

4. Based on Problem 1, the period of the function $\sin\pi x/L$ is $T = 2\pi/(\pi/L) = 2L$.

6. Let $T > 0$ and consider the equation $(x+T)^2 = x^2$. It follows that $2Tx + T^2 = 0$ and $2x + T = 0$. Since the latter equation is not an identity, the function x^2 cannot be periodic with finite period.

8. The function is defined on intervals of length $(2n+1) - (2n-1) = 2$. On any two consecutive intervals, $f(x)$ is identically equal to 1 on one of the intervals and alternates between 1 and -1 on the other. It follows that the period is $T = 4$.

9. On the interval $L < x < 2L$, a shift to the right results in $f(x) = -(x - 2L) = 2L - x$. On the interval $-3L < x < -2L$, a shift to the left results in $f(x) = -(x + 2L) = -2L - x$.

11. The next fundamental period to the left is on the interval $-2L < x < 0$. Hence the interval $-L < x < 0$ is the second half of a fundamental period. A shift to the left results in $f(x) = L - (x + 2L) = -L - x$.

12. First note that

$$\cos\frac{m\pi x}{L}\cos\frac{n\pi x}{L} = \frac{1}{2}\left[\cos\frac{(m-n)\pi x}{L} + \cos\frac{(m+n)\pi x}{L}\right]$$

and
$$\cos\frac{m\pi x}{L}\sin\frac{n\pi x}{L} = \frac{1}{2}\left[\sin\frac{(n-m)\pi x}{L} + \sin\frac{(m+n)\pi x}{L}\right].$$

It follows that
$$\int_{-L}^{L} \cos\frac{m\pi x}{L}\cos\frac{n\pi x}{L}dx = \frac{1}{2}\int_{-L}^{L}\left[\cos\frac{(m-n)\pi x}{L} + \cos\frac{(m+n)\pi x}{L}\right]dx$$
$$= \frac{1}{2}\frac{L}{\pi}\left\{\frac{\sin[(m-n)\pi x/L]}{m-n} + \frac{\sin[(m+n)\pi x/L]}{m+n}\right\}\bigg|_{-L}^{L} = 0,$$

as long as $m+n$ and $m-n$ are not zero. For the case $m=n$,
$$\int_{-L}^{L}(\cos\frac{n\pi x}{L})^2 dx = \frac{1}{2}\int_{-L}^{L}\left[1 + \cos\frac{2n\pi x}{L}\right]dx = \frac{1}{2}\left\{x + \frac{\sin(2n\pi x/L)}{2n\pi/L}\right\}\bigg|_{-L}^{L} = L.$$

Likewise,
$$\int_{-L}^{L}\cos\frac{m\pi x}{L}\sin\frac{n\pi x}{L}dx = \frac{1}{2}\int_{-L}^{L}\left[\sin\frac{(n-m)\pi x}{L} + \sin\frac{(m+n)\pi x}{L}\right]dx$$
$$= \frac{1}{2}\frac{L}{\pi}\left\{\frac{\cos[(n-m)\pi x/L]}{m-n} - \frac{\cos[(m+n)\pi x/L]}{m+n}\right\}\bigg|_{-L}^{L} = 0,$$

as long as $m+n$ and $m-n$ are not zero. For the case $m=n$,
$$\int_{-L}^{L}\cos\frac{n\pi x}{L}\sin\frac{n\pi x}{L}dx = \frac{1}{2}\int_{-L}^{L}\sin\frac{2n\pi x}{L}dx = -\frac{1}{2}\left\{\frac{\cos(2n\pi x/L)}{2n\pi/L}\right\}\bigg|_{-L}^{L} = 0.$$

14.(a) For $L=1$,

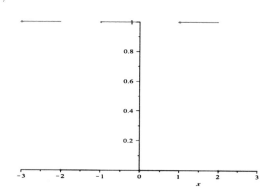

(b) The Fourier coefficients are calculated using the Euler-Fourier formulas:
$$a_0 = \frac{1}{L}\int_{-L}^{L} f(x)dx = \frac{1}{L}\int_{-L}^{0} dx = 1.$$

For $n > 0$,
$$a_n = \frac{1}{L}\int_{-L}^{L} f(x)\cos\frac{n\pi x}{L}dx = \frac{1}{L}\int_{-L}^{0}\cos\frac{n\pi x}{L}dx = 0.$$

Likewise,
$$b_n = \frac{1}{L}\int_{-L}^{L} f(x)\sin\frac{n\pi x}{L}dx = \frac{1}{L}\int_{-L}^{0}\sin\frac{n\pi x}{L}dx = \frac{-1+(-1)^n}{n\pi}.$$

It follows that $b_{2k}=0$ and $b_{2k-1} = -2/[(2k-1)\pi]$, $k=1,2,3,\ldots$. Therefore the Fourier series for the given function is
$$f(x) = \frac{1}{2} - \frac{2}{\pi}\sum_{k=1}^{\infty}\frac{1}{2k-1}\sin\frac{(2k-1)\pi x}{L}.$$

16.(a)

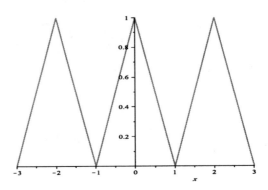

(b) The Fourier coefficients are calculated using the Euler-Fourier formulas:
$$a_0 = \frac{1}{L}\int_{-L}^{L} f(x)dx = \int_{-1}^{0}(x+1)dx + \int_{0}^{1}(1-x)dx = 1.$$

For $n>0$,
$$a_n = \frac{1}{L}\int_{-L}^{L} f(x)\cos\frac{n\pi x}{L}dx =$$
$$= \int_{-1}^{0}(x+1)\cos n\pi x\, dx + \int_{0}^{1}(1-x)\cos n\pi x\, dx = -2\frac{-1+(-1)^n}{n^2\pi^2}.$$

It follows that $a_{2k}=0$ and $a_{2k-1} = 4/[(2k-1)^2\pi^2]$, $k=1,2,3,\ldots$. Likewise,
$$b_n = \frac{1}{L}\int_{-L}^{L} f(x)\sin\frac{n\pi x}{L}dx =$$
$$= \int_{-1}^{0}(x+1)\sin n\pi x\, dx + \int_{0}^{1}(1-x)\sin n\pi x\, dx = 0.$$

Therefore the Fourier series for the given function is
$$f(x) = \frac{1}{2} + \frac{4}{\pi^2}\sum_{k=1}^{\infty}\frac{1}{(2k-1)^2}\cos(2k-1)\pi x.$$

17.(a) For $L = 1$,

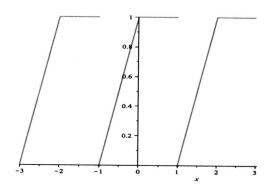

(b) The Fourier coefficients are calculated using the Euler-Fourier formulas:
$$a_0 = \frac{1}{L}\int_{-L}^{L} f(x)dx = \frac{1}{L}\int_{-L}^{0}(x+L)dx + \frac{1}{L}\int_{0}^{L} L\,dx = 3L/2.$$

For $n > 0$,
$$a_n = \frac{1}{L}\int_{-L}^{L} f(x)\cos\frac{n\pi x}{L}dx =$$
$$= \frac{1}{L}\int_{-L}^{0}(x+L)\cos\frac{n\pi x}{L}dx + \frac{1}{L}\int_{0}^{L} L\cos\frac{n\pi x}{L}dx = \frac{L(1-\cos n\pi)}{n^2\pi^2}.$$

Likewise,
$$b_n = \frac{1}{L}\int_{-L}^{L} f(x)\sin\frac{n\pi x}{L}dx =$$
$$= \frac{1}{L}\int_{-L}^{0}(x+L)\sin\frac{n\pi x}{L}dx + \frac{1}{L}\int_{0}^{L} L\sin\frac{n\pi x}{L}dx = -\frac{L\cos n\pi}{n\pi}.$$

Note that $\cos n\pi = (-1)^n$. It follows that the Fourier series for the given function is
$$f(x) = \frac{3L}{4} + \frac{L}{\pi^2}\sum_{n=1}^{\infty}\left[\frac{2}{(2n-1)^2}\cos\frac{(2n-1)\pi x}{L} - \frac{(-1)^n\pi}{n}\sin\frac{n\pi x}{L}\right].$$

18.(a)

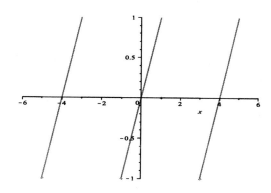

(b) The Fourier coefficients are calculated using the Euler-Fourier formulas:

$$a_0 = \frac{1}{L}\int_{-L}^{L} f(x)dx = \frac{1}{2}\int_{-1}^{1} x\,dx = 0.$$

For $n > 0$,

$$a_n = \frac{1}{L}\int_{-L}^{L} f(x)\cos\frac{n\pi x}{L}dx = \frac{1}{2}\int_{-1}^{1} x\cos\frac{n\pi x}{L}dx = 0.$$

Likewise,

$$b_n = \frac{1}{L}\int_{-L}^{L} f(x)\sin\frac{n\pi x}{L}dx = \frac{1}{2}\int_{-1}^{1} x\sin\frac{n\pi x}{L}dx = \frac{2}{n^2\pi^2}\left(2\sin\frac{n\pi}{2} - n\pi\cos\frac{n\pi}{2}\right).$$

Therefore the Fourier series for the given function is

$$f(x) = \sum_{n=1}^{\infty}\left[\frac{4}{n^2\pi^2}\sin\frac{n\pi}{2} - \frac{2}{n\pi}\cos\frac{n\pi}{2}\right]\sin\frac{n\pi x}{2}.$$

19.(a)

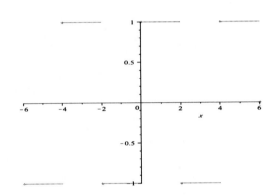

(b) The Fourier cosine coefficients are given by

$$a_n = \frac{1}{L}\int_{-L}^{L} f(x)\cos\frac{n\pi x}{L}dx =$$

$$= \frac{1}{2}\int_{-2}^{0} -\cos\frac{n\pi x}{2}dx + \frac{1}{2}\int_{0}^{2} \cos\frac{n\pi x}{2}dx = 0.$$

The Fourier sine coefficients are given by

$$b_n = \frac{1}{L}\int_{-L}^{L} f(x)\sin\frac{n\pi x}{L}dx =$$

$$= \frac{1}{2}\int_{-2}^{0} -\sin\frac{n\pi x}{2}dx + \frac{1}{2}\int_{0}^{2} \sin\frac{n\pi x}{2}dx = 2\frac{1-\cos n\pi}{n\pi}.$$

Therefore the Fourier series for the given function is
$$f(x) = \frac{4}{\pi} \sum_{n=1}^{\infty} \frac{1}{2n-1} \sin \frac{(2n-1)\pi x}{2}.$$

(c)

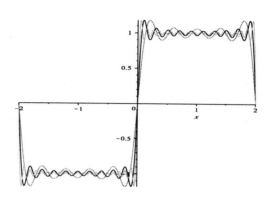

20.(a)

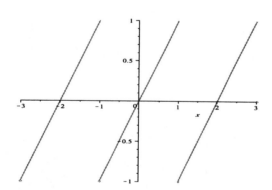

(b) The Fourier cosine coefficients are given by
$$a_n = \frac{1}{L} \int_{-L}^{L} f(x) \cos \frac{n\pi x}{L} dx = \int_{-1}^{1} x \cos n\pi x \, dx = 0.$$
The Fourier sine coefficients are given by
$$b_n = \frac{1}{L} \int_{-L}^{L} f(x) \sin \frac{n\pi x}{L} dx = \int_{-1}^{1} x \sin n\pi x \, dx = -2\frac{\cos n\pi}{n\pi}.$$
Therefore the Fourier series for the given function is
$$f(x) = -\frac{2}{\pi} \sum_{n=1}^{\infty} \frac{(-1)^n}{n} \sin n\pi x.$$

(c)

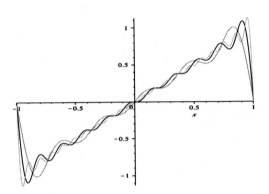

22.(a)

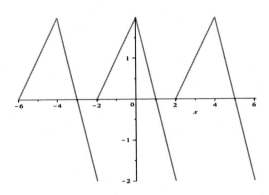

(b) The Fourier cosine coefficients are given by

$$a_0 = \frac{1}{L}\int_{-L}^{L} f(x)dx = \frac{1}{2}\int_{-2}^{0}(x+2)dx + \frac{1}{2}\int_{0}^{2}(2-2x)dx = 1,$$

and for $n > 0$,

$$a_n = \frac{1}{L}\int_{-L}^{L} f(x)\cos\frac{n\pi x}{L}dx =$$

$$= \frac{1}{2}\int_{-2}^{0}(x+2)\cos\frac{n\pi x}{2}dx + \frac{1}{2}\int_{0}^{2}(2-2x)\cos\frac{n\pi x}{2}dx = 6\frac{(1-\cos n\pi)}{n^2\pi^2}.$$

The Fourier sine coefficients are given by

$$b_n = \frac{1}{L}\int_{-L}^{L} f(x)\sin\frac{n\pi x}{L}dx =$$

$$= \frac{1}{2}\int_{-2}^{0}(x+2)\sin\frac{n\pi x}{2}dx + \frac{1}{2}\int_{0}^{2}(2-2x)\sin\frac{n\pi x}{2}dx = 2\frac{\cos n\pi}{n\pi}.$$

Therefore the Fourier series for the given function is

$$f(x) = \frac{1}{2} + \frac{12}{\pi^2}\sum_{n=1}^{\infty}\frac{1}{(2n-1)^2}\cos\frac{(2n-1)\pi x}{2} + \frac{2}{\pi}\sum_{n=1}^{\infty}\frac{(-1)^n}{n}\sin\frac{n\pi x}{2}.$$

(c)

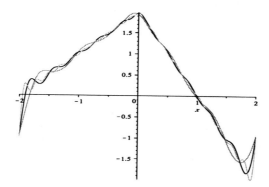

23.(a)

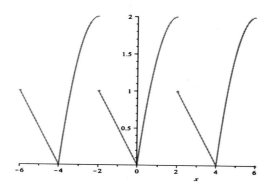

(b) The Fourier cosine coefficients are given by

$$a_0 = \frac{1}{L}\int_{-L}^{L} f(x)dx = \frac{1}{2}\int_{-2}^{0}(-\frac{x}{2})dx + \frac{1}{2}\int_{0}^{2}(2x - \frac{1}{2}x^2)dx = 11/6,$$

and for $n > 0$,

$$a_n = \frac{1}{L}\int_{-L}^{L} f(x)\cos\frac{n\pi x}{L}dx =$$

$$= \frac{1}{2}\int_{-2}^{0}(-\frac{x}{2})\cos\frac{n\pi x}{2}dx + \frac{1}{2}\int_{0}^{2}(2x - \frac{1}{2}x^2)\cos\frac{n\pi x}{2}dx = -\frac{(5 - \cos n\pi)}{n^2\pi^2}.$$

The Fourier sine coefficients are given by

$$b_n = \frac{1}{L}\int_{-L}^{L} f(x)\sin\frac{n\pi x}{L}dx =$$

$$= \frac{1}{2}\int_{-2}^{0}(-\frac{x}{2})\sin\frac{n\pi x}{2}dx + \frac{1}{2}\int_{0}^{2}(2x - \frac{1}{2}x^2)\sin\frac{n\pi x}{2}dx = \frac{4 - (4 + n^2\pi^2)\cos n\pi}{n^3\pi^3}.$$

Therefore the Fourier series for the given function is

$$f(x) = \frac{11}{12} + \frac{1}{\pi^2}\sum_{n=1}^{\infty}\frac{[(-1)^n - 5]}{n^2}\cos\frac{n\pi x}{2} + \frac{1}{\pi^3}\sum_{n=1}^{\infty}\frac{[4 - (4 + n^2\pi^2)(-1)^n]}{n^3}\sin\frac{n\pi x}{2}.$$

(c)

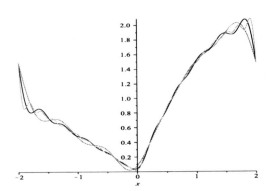

24.(a)

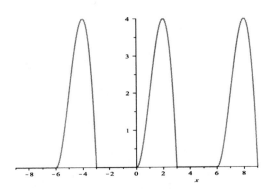

(b) The Fourier cosine coefficients are given by

$$a_0 = \frac{1}{L}\int_{-L}^{L} f(x)dx = \frac{1}{3}\int_0^3 x^2(3-x)dx = 9/4,$$

and for $n > 0$,

$$a_n = \frac{1}{L}\int_{-L}^{L} f(x)\cos\frac{n\pi x}{L}dx =$$

$$= \frac{1}{3}\int_0^3 x^2(3-x)\cos\frac{n\pi x}{3}dx = -27\frac{(6 - 6\cos n\pi + n^2\pi^2\cos n\pi)}{n^4\pi^4}.$$

The Fourier sine coefficients are given by

$$b_n = \frac{1}{L}\int_{-L}^{L} f(x)\sin\frac{n\pi x}{L}dx = \frac{1}{3}\int_0^3 x^2(3-x)\sin\frac{n\pi x}{3}dx = -54\frac{1 + 2\cos n\pi}{n^3\pi^3}.$$

Therefore the Fourier series for the given function is

$$f(x) = \frac{9}{8} - 27 \sum_{n=1}^{\infty} \left[\frac{6[1-(-1)^n]}{n^4\pi^4} + \frac{(-1)^n}{n^2\pi^2} \right] \cos \frac{n\pi x}{3}$$

$$- \frac{54}{\pi^3} \sum_{n=1}^{\infty} \frac{[1+2(-1)^n]}{n^3} \sin \frac{n\pi x}{3}.$$

(c)

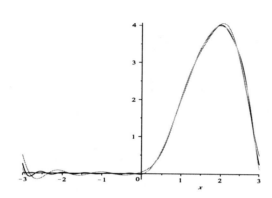

26.(a)

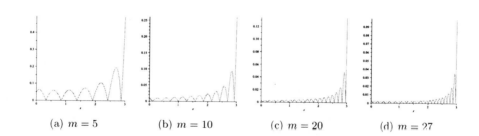

(a) $m = 5$ (b) $m = 10$ (c) $m = 20$ (d) $m = 27$

(b) It is evident that $|e_m(x)|$ is greatest at $x = \pm 3$. Increasing the number of terms in the partial sums, we find that if $m \geq 27$, then $|e_m(x)| \leq 0.1$, for all $x \in [-3, 3]$.

28. Let $x = T + a$, for some $a \in [0, T]$. First note that for any value of h,

$$f(x+h) - f(x) = f(T+a+h) - f(T+a) = f(a+h) - f(a).$$

Since f is differentiable,

$$f'(x) = \lim_{h \to 0} \frac{f(x+h) - f(x)}{h} = \lim_{h \to 0} \frac{f(a+h) - f(a)}{h} = f'(a).$$

That is, $f'(a+T) = f'(a)$. By induction, it follows that $f'(a+T) = f'(a)$ for every value of a. On the other hand, if $f(x) = 1 + \cos x$, then the function

$$F(x) = \int_0^x [1 + \cos t]\, dt = x + \sin x$$

is not periodic.

29.(a) Based on the hypothesis, the vectors $\mathbf{v}_1$, $\mathbf{v}_2$ and $\mathbf{v}_3$ are a basis for $\mathbb{R}^3$. Given any vector $\mathbf{u} \in \mathbb{R}^3$, it can be expressed as a linear combination

$$\mathbf{u} = a_1 \mathbf{v}_1 + a_2 \mathbf{v}_2 + a_3 \mathbf{v}_3.$$

Taking the inner product of both sides of this equation with $\mathbf{v}_i$, we have

$$\mathbf{u} \cdot \mathbf{v}_i = (a_1 \mathbf{v}_1 + a_2 \mathbf{v}_2 + a_3 \mathbf{v}_3) \cdot \mathbf{v}_i = a_i \mathbf{v}_i \cdot \mathbf{v}_i,$$

since the basis vectors are mutually orthogonal. Hence

$$a_i = \frac{\mathbf{u} \cdot \mathbf{v}_i}{\mathbf{v}_i \cdot \mathbf{v}_i}, \ i = 1, 2, 3.$$

Recall that $\mathbf{u} \cdot \mathbf{v}_i = \|\mathbf{u}\|\|\mathbf{v}_i\| \cos \theta$, where θ is the angle between $\mathbf{u}$ and $\mathbf{v}_i$ and $\|\mathbf{b}\| = \sqrt{\mathbf{b} \cdot \mathbf{b}}$ is the length of a vector $\mathbf{b}$. Therefore

$$a_i = \frac{\|\mathbf{u}\|\|\mathbf{v}_i\| \cos \theta}{\|\mathbf{v}_i\|^2} = \frac{\|\mathbf{u}\| \cos \theta}{\|\mathbf{v}_i\|}.$$

Here $\|\mathbf{u}\| \cos \theta$ can be interpreted as the magnitude of the component of $\mathbf{u}$ in the direction of $\mathbf{v}_i$, which is the magnitude of the (vector) projection of $\mathbf{u}$ onto $\mathbf{v}_i$.

(b) Assuming that a Fourier series converges to a periodic function, $f(x)$,

$$f(x) = \frac{a_0}{2} \phi_0(x) + \sum_{m=1}^{\infty} a_m \phi_m(x) + \sum_{m=1}^{\infty} b_m \psi_m(x).$$

Taking the inner product, defined by

$$(u, v) = \int_{-L}^{L} u(x) v(x)\, dx,$$

of both sides of the series expansion with the specified trigonometric functions, we have

$$(f, \phi_n) = \frac{a_0}{2}(\phi_0, \phi_n) + \sum_{m=1}^{\infty} a_m (\phi_m, \phi_n) + \sum_{m=1}^{\infty} b_m (\psi_m, \phi_n)$$

for $n = 0, 1, 2, \ldots$.

(c) It also follows that

$$(f, \psi_n) = \frac{a_0}{2}(\phi_0, \psi_n) + \sum_{m=1}^{\infty} a_m (\phi_m, \psi_n) + \sum_{m=1}^{\infty} b_m (\psi_m, \psi_n)$$

for $n = 1, 2, \ldots$. Based on the orthogonality conditions,

$$(\phi_m, \phi_n) = L\, \delta_{mn}\ , \ (\psi_m, \psi_n) = L\, \delta_{mn}\ ,$$

and $(\psi_m, \phi_n) = 0$. Note that $(\phi_0, \phi_0) = 2L$. Therefore

$$a_0 = \frac{2(f, \phi_0)}{(\phi_0, \phi_0)} = \frac{1}{L}\int_{-L}^{L} f(x)\phi_0(x)dx$$

and

$$a_n = \frac{(f, \phi_n)}{(\phi_n, \phi_n)} = \frac{1}{L}\int_{-L}^{L} f(x)\phi_n(x)dx, \ n = 1, 2, \ldots.$$

Likewise,

$$b_n = \frac{(f, \psi_n)}{(\psi_n, \psi_n)} = \frac{1}{L}\int_{-L}^{L} f(x)\psi_n(x)dx, \ n = 1, 2, \ldots.$$

10.3

1.(a) The given function is assumed to be periodic with $2L = 2$. The Fourier cosine coefficients are given by

$$a_0 = \frac{1}{L}\int_{-L}^{L} f(x)dx = \int_{-1}^{0}(-1)dx + \int_{0}^{1}(1)dx = 0,$$

and for $n > 0$,

$$a_n = \frac{1}{L}\int_{-L}^{L} f(x)\cos\frac{n\pi x}{L}dx = -\int_{-1}^{0}\cos n\pi x\, dx + \int_{0}^{1}\cos n\pi x\, dx = 0.$$

The Fourier sine coefficients are given by

$$b_n = \frac{1}{L}\int_{-L}^{L} f(x)\sin\frac{n\pi x}{L}dx = -\int_{-1}^{0}\sin n\pi x\, dx + \int_{0}^{1}\sin n\pi x\, dx = 2\frac{1 - \cos n\pi}{n\pi}.$$

Therefore the Fourier series for the specified function is

$$f(x) = \frac{4}{\pi}\sum_{n=1}^{\infty}\frac{1}{2n-1}\sin(2n-1)\pi x.$$

(b)

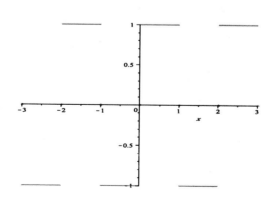

The function is piecewise continuous on each finite interval. The points of discontinuity are at integer values of x. At these points, the series converges to

$$|f(x-) + f(x+)| = 0.$$

3.(a) The given function is assumed to be periodic with $T = 2L$. The Fourier cosine coefficients are given by

$$a_0 = \frac{1}{L}\int_{-L}^{L} f(x)dx = \frac{1}{L}\int_{-L}^{0}(L+x)dx + \frac{1}{L}\int_{0}^{L}(L-x)dx = L,$$

and for $n > 0$,

$$a_n = \frac{1}{L}\int_{-L}^{L} f(x)\cos\frac{n\pi x}{L}dx =$$

$$= \frac{1}{L}\int_{-L}^{0}(L+x)\cos\frac{n\pi x}{L}dx + \frac{1}{L}\int_{0}^{L}(L-x)\cos\frac{n\pi x}{L}dx = 2L\frac{1-\cos n\pi}{n^2\pi^2}.$$

The Fourier sine coefficients are given by

$$b_n = \frac{1}{L}\int_{-L}^{L} f(x)\sin\frac{n\pi x}{L}dx =$$

$$= \frac{1}{L}\int_{-L}^{0}(L+x)\sin\frac{n\pi x}{L}dx + \frac{1}{L}\int_{0}^{L}(L-x)\sin\frac{n\pi x}{L}dx = 0.$$

Therefore the Fourier series of the specified function is

$$f(x) = \frac{L}{2} + \frac{4L}{\pi^2}\sum_{n=1}^{\infty}\frac{1}{(2n-1)^2}\cos\frac{(2n-1)\pi x}{L}.$$

(b) For $L = 1$,

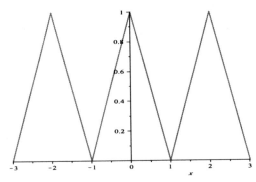

Note that $f(x)$ is continuous. Based on Theorem 10.3.1, the series converges to the continuous function $f(x)$.

5.(a) The given function is assumed to be periodic with $2L = 2\pi$. The Fourier cosine coefficients are given by

$$a_0 = \frac{1}{L}\int_{-L}^{L} f(x)dx = \frac{1}{\pi}\int_{-\pi/2}^{\pi/2}(1)dx = 1,$$

and for $n > 0$,
$$a_n = \frac{1}{L}\int_{-L}^{L} f(x)\cos\frac{n\pi x}{L}dx = \frac{1}{\pi}\int_{-\pi/2}^{\pi/2}(1)\cos nx\, dx = \frac{2}{n\pi}\sin(\frac{n\pi}{2}).$$

The Fourier sine coefficients are given by
$$b_n = \frac{1}{L}\int_{-L}^{L} f(x)\sin\frac{n\pi x}{L}dx = \frac{1}{\pi}\int_{-\pi/2}^{\pi/2}(1)\sin nx\, dx = 0.$$

Observe that
$$\sin(\frac{n\pi}{2}) = \begin{cases} 0, & n = 2k \\ (-1)^{k+1}, & n = 2k-1 \end{cases}, \quad k = 1, 2, \ldots.$$

Therefore the Fourier series of the specified function is
$$f(x) = \frac{1}{2} - \frac{2}{\pi}\sum_{n=1}^{\infty}\frac{(-1)^n}{2n-1}\cos(2n-1)x.$$

(b)

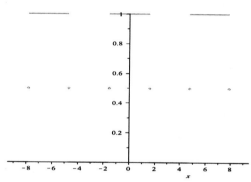

The given function is piecewise continuous, with discontinuities at odd multiples of $\pi/2$. At $x_d = (2k-1)\pi/2$, $k = 0, 1, 2, \ldots$, the series converges to
$$|f(x_d-) + f(x_d+)| = 1/2.$$

6.(a) The given function is assumed to be periodic with $2L = 2$. The Fourier cosine coefficients are given by
$$a_0 = \frac{1}{L}\int_{-L}^{L} f(x)dx = \int_0^1 x^2 dx = 1/3,$$
and for $n > 0$,
$$a_n = \frac{1}{L}\int_{-L}^{L} f(x)\cos\frac{n\pi x}{L}dx = \int_0^1 x^2\cos n\pi x\, dx = \frac{2\cos n\pi}{n^2\pi^2}.$$

The Fourier sine coefficients are given by
$$b_n = \frac{1}{L}\int_{-L}^{L} f(x)\sin\frac{n\pi x}{L}dx = \int_0^1 x^2\sin n\pi x\, dx = -\frac{2 - 2\cos n\pi + n^2\pi^2\cos n\pi}{n^3\pi^3}.$$

Therefore the Fourier series for the specified function is

$$f(x) = \frac{1}{6} + \frac{2}{\pi^2} \sum_{n=1}^{\infty} \frac{(-1)^n}{n^2} \cos n\pi x - \sum_{n=1}^{\infty} \left[\frac{2[1-(-1)^n]}{n^3\pi^3} + \frac{(-1)^n}{n\pi} \right] \sin n\pi x.$$

(b)

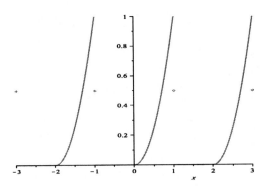

The given function is piecewise continuous, with discontinuities at the odd integers. At $x_d = 2k - 1$, $k = 0, 1, 2, \ldots$, the series converges to

$$|f(x_d-) + f(x_d+)| = 1/2.$$

8.(a) As shown in Problem 16 of Section 10.2,

$$f(x) = \frac{1}{2} + \frac{4}{\pi^2} \sum_{n=1}^{\infty} \frac{1}{(2n-1)^2} \cos(2n-1)\pi x.$$

(b)

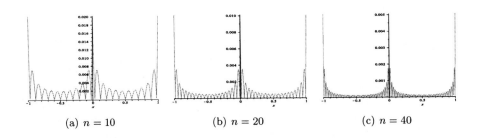

(a) $n = 10$ (b) $n = 20$ (c) $n = 40$

The maximum value of $|e_n(x)|$ is approximately 0.02 for $n = 10$, 0.01 for $n = 20$, and 0.005 for $n = 40$.

(c)

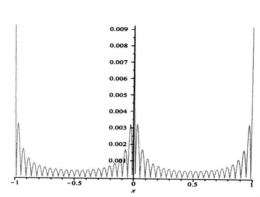

The smallest value of n for which $|e_n(x)| \leq 0.01$ for all x is $n = 21$ as shown above.

9.(a) As shown in Problem 20 of Section 10.2,

$$f(x) = -\frac{2}{\pi} \sum_{n=1}^{\infty} \frac{(-1)^n}{n} \sin n\pi x.$$

(b)

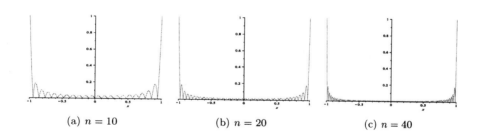

(a) $n = 10$ (b) $n = 20$ (c) $n = 40$

For all n, the maximum value of $|e_n(x)|$ is 1.

(c) The given function is discontinuous at $x = \pm 1$. At these points, the series will converge to a value of zero. The error can never be made arbitrarily small.

10.(a) As shown in Problem 22 of Section 10.2,

$$f(x) = \frac{1}{2} + \frac{12}{\pi^2} \sum_{n=1}^{\infty} \frac{1}{(2n-1)^2} \cos \frac{(2n-1)\pi x}{2} + \frac{2}{\pi} \sum_{n=1}^{\infty} \frac{(-1)^n}{n} \sin \frac{n\pi x}{2}.$$

(b)

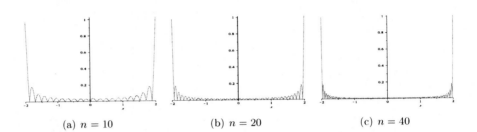

(a) $n = 10$ (b) $n = 20$ (c) $n = 40$

The maximum value of $|e_n(x)|$ is approximately 1.06 for $n = 10$, 1.03 for $n = 20$, and 1.015 for $n = 40$.

(c) The given function is discontinuous at $x = \pm 2$. At these points, the series will converge to a value of -1. The error can never be made arbitrarily small.

11.(a) As shown in Problem 6, above,

$$f(x) = \frac{1}{6} + \frac{2}{\pi^2} \sum_{n=1}^{\infty} \frac{(-1)^n}{n^2} \cos n\pi x - \sum_{n=1}^{\infty} \left[\frac{2[1-(-1)^n]}{n^3\pi^3} + \frac{(-1)^n}{n\pi} \right] \sin n\pi x.$$

(b)

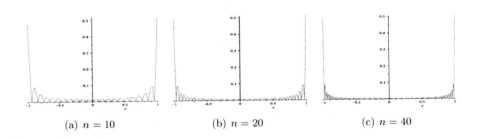

(a) $n = 10$ (b) $n = 20$ (c) $n = 40$

The maximum value of $|e_n(x)|$ is approximately 0.52 for $n = 10$, 0.51 for $n = 20$, and 0.505 for $n = 40$.

(c) The given function is piecewise continuous, with discontinuities at the odd integers. At $x_d = 2k - 1$, $k = 0, 1, 2, \ldots$, the series converges to

$$|f(x_d-) + f(x_d+)| = 1/2.$$

At these points the error can never be made arbitrarily small.

13. The solution of the homogenous differential equation is $y_c(t) = c_1 \cos \omega t + c_2 \sin \omega t$. Given that $\omega^2 \neq n^2$, we can use the method of undetermined coefficients to find a particular solution

$$Y(t) = \frac{1}{\omega^2 - n^2} \sin nt.$$

Hence the general solution of the ODE is

$$y(t) = c_1 \cos \omega t + c_2 \sin \omega t + \frac{1}{\omega^2 - n^2} \sin nt.$$

Imposing the initial conditions, we obtain the equations

$$c_1 = 0, \quad \omega c_2 + \frac{n}{\omega^2 - n^2} = 0.$$

It follows that $c_2 = -n/\left[\omega(\omega^2 - n^2)\right]$. The solution of the IVP is

$$y(t) = \frac{1}{\omega^2 - n^2} \sin nt - \frac{n}{\omega(\omega^2 - n^2)} \sin \omega t.$$

If $\omega^2 = n^2$, then the forcing function is also one of the fundamental solutions of the ODE. The method of undetermined coefficients may still be used, with a more elaborate trial solution. Using the method of variation of parameters, we obtain

$$Y(t) = -\cos nt \int \frac{\sin^2 nt}{n} dt + \sin nt \int \frac{\cos nt \sin nt}{n} dt = \frac{\sin nt - nt \cos nt}{2n^2}.$$

In this case, the general solution is

$$y(t) = c_1 \cos nt + c_2 \sin nt - \frac{t}{2n} \cos nt.$$

Invoking the initial conditions, we obtain $c_1 = 0$ and $c_2 = 1/2n^2$. Therefore the solution of the IVP is

$$y(t) = \frac{1}{2n^2} \sin nt - \frac{t}{2n} \cos nt.$$

16. Note that the function $f(t)$ and the function given in Problem 8 have the same Fourier series. Therefore

$$f(t) = \frac{1}{2} + \frac{4}{\pi^2} \sum_{n=1}^{\infty} \frac{1}{(2n-1)^2} \cos(2n-1)\pi t.$$

The solution of the homogeneous problem is $y_c(t) = c_1 \cos \omega t + c_2 \sin \omega t$. Using the method of undetermined coefficients, we assume a particular solution of the form

$$Y(t) = A_0 + \sum_{n=1}^{\infty} A_n \cos n\pi t.$$

Substitution into the ODE and equating like terms results in $A_0 = 1/2\omega^2$ and

$$A_n = \frac{a_n}{\omega^2 - n^2 \pi^2} \text{ so } A_n = 0 \text{ for } n \text{ even; also, } a_n = \frac{4}{\pi^2(2n-1)^2}.$$

It follows that the general solution is

$$y(t) = c_1 \cos \omega t + c_2 \sin \omega t + \frac{1}{2\omega^2} + \frac{4}{\pi^2} \sum_{n=1}^{\infty} \frac{\cos(2n-1)\pi t}{(2n-1)^2 \left[\omega^2 - (2n-1)^2\pi^2\right]}.$$

Setting $y(0) = 1$, we find that

$$c_1 = 1 - \frac{1}{2\omega^2} - \frac{4}{\pi^2} \sum_{n=1}^{\infty} \frac{1}{(2n-1)^2 \left[\omega^2 - (2n-1)^2\pi^2\right]}.$$

Invoking the initial condition $y'(0) = 0$, we obtain $c_2 = 0$. Hence the solution of the initial value problem is

$$y(t) = \cos \omega t - \frac{1}{2\omega^2} \cos \omega t + \frac{1}{2\omega^2} + \frac{4}{\pi^2} \sum_{n=1}^{\infty} \frac{\cos(2n-1)\pi t - \cos \omega t}{(2n-1)^2 \left[\omega^2 - (2n-1)^2\pi^2\right]}.$$

17. Let

$$f(x) = \frac{a_0}{2} + \sum_{n=1}^{\infty} \left[a_n \cos \frac{n\pi x}{L} + b_n \sin \frac{n\pi x}{L}\right].$$

Squaring both sides of the equation, we formally have

$$f(x)^2 = \frac{a_0^2}{4} + a_0 \sum_{n=1}^{\infty} \left[a_n \cos \frac{n\pi x}{L} + b_n \sin \frac{n\pi x}{L}\right]$$

$$+ \left(\sum_{m=1}^{\infty} \left[a_m \cos \frac{m\pi x}{L} + b_m \sin \frac{m\pi x}{L}\right]\right) \left(\sum_{n=1}^{\infty} \left[a_n \cos \frac{n\pi x}{L} + b_n \sin \frac{n\pi x}{L}\right]\right)$$

$$= \frac{a_0^2}{4} + a_0 \sum_{n=1}^{\infty} \left[a_n \cos \frac{n\pi x}{L} + b_n \sin \frac{n\pi x}{L}\right]$$

$$+ \sum_{n=1}^{\infty} \left[a_n^2 \cos^2 \frac{n\pi x}{L} + b_n^2 \sin^2 \frac{n\pi x}{L}\right] + 2 \sum_{m=1}^{\infty} \sum_{n=m+1}^{\infty} a_m a_n \cos \frac{m\pi x}{L} \cos \frac{n\pi x}{L}$$

$$+ 2 \sum_{m=1}^{\infty} \sum_{n=m+1}^{\infty} a_m b_n \cos \frac{m\pi x}{L} \sin \frac{n\pi x}{L} + 2 \sum_{m=1}^{\infty} \sum_{n=m+1}^{\infty} b_m b_n \sin \frac{m\pi x}{L} \sin \frac{n\pi x}{L}.$$

Integrating both sides of the last equation, and using the orthogonality conditions,

$$\int_{-L}^{L} f(x)^2 dx = \int_{-L}^{L} \frac{a_0^2}{4} dx + \sum_{n=1}^{\infty} \left[\int_{-L}^{L} a_n^2 \cos^2 \frac{n\pi x}{L} dx + \int_{-L}^{L} b_n^2 \sin^2 \frac{n\pi x}{L} dx\right]$$

$$= \frac{a_0^2}{2} L + \sum_{n=1}^{\infty} \left[a_n^2 L + b_n^2 L\right].$$

Therefore,

$$\frac{1}{L} \int_{-L}^{L} f(x)^2 dx = \frac{a_0^2}{2} + \sum_{n=1}^{\infty} (a_n^2 + b_n^2).$$

19.(a) As shown in the Example, the Fourier series of the function
$$f(x) = \begin{cases} 0, & -L < x < 0 \\ L, & 0 < x < L, \end{cases}$$
is given by
$$f(x) = \frac{L}{2} + \frac{2L}{\pi} \sum_{n=1}^{\infty} \frac{1}{2n-1} \sin\frac{(2n-1)\pi x}{L}.$$
Setting $L = 1$,
$$f(x) = \frac{1}{2} + \frac{2}{\pi} \sum_{n=1}^{\infty} \frac{1}{2n-1} \sin(2n-1)\pi x.$$
It follows that
$$\sum_{n=1}^{\infty} \frac{1}{2n-1} \sin(2n-1)\pi x = \frac{\pi}{2}\left[f(x) - \frac{1}{2}\right]. \quad \text{(ii)}$$

(b) Given that
$$g(x) = \sum_{n=1}^{\infty} \frac{2n-1}{1+(2n-1)^2} \sin(2n-1)\pi x, \quad \text{(i)}$$
and subtracting Eq.(ii) from Eq.(i), we find that
$$g(x) - \frac{\pi}{2}\left[f(x) - \frac{1}{2}\right] = \sum_{n=1}^{\infty} \frac{2n-1}{1+(2n-1)^2} \sin(2n-1)\pi x - \sum_{n=1}^{\infty} \frac{1}{2n-1} \sin(2n-1)\pi x.$$
Based on the fact that
$$\frac{2n-1}{1+(2n-1)^2} - \frac{1}{2n-1} = -\frac{1}{(2n-1)[1+(2n-1)^2]},$$
and the fact that we can combine the two series, it follows that
$$g(x) = \frac{\pi}{2}\left[f(x) - \frac{1}{2}\right] - \sum_{n=1}^{\infty} \frac{\sin(2n-1)\pi x}{(2n-1)[1+(2n-1)^2]}.$$

10.4

1. Since the function contains only odd powers of x, the function is odd.

2. Since the function contains both odd and even powers of x, the function is neither even nor odd.

4. We have $\sec x = 1/\cos x$. Since the quotient of two even functions is even, the function is even.

492 Chapter 10. Partial Differential Equations and Fourier Series

5. We can write $|x|^3 = |x| \cdot |x|^2 = |x| \cdot x^2$. Since both factors are even, it follows that the function is even.

8. $L = 2$.

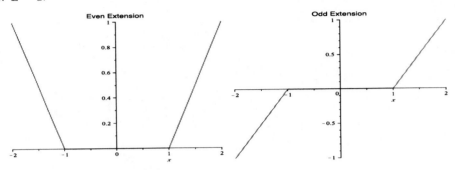

9. $L = 2$.

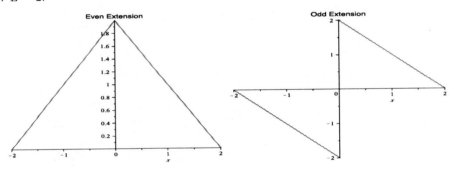

11. $L = 2$.

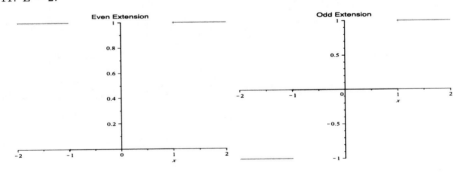

12. $L = 1$.

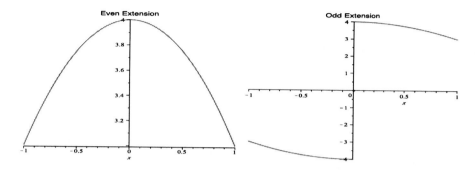

16.(a) $L = 2$. For an odd extension of the function, the cosine coefficients are zero. The sine coefficients are given by

$$b_n = \frac{2}{L}\int_0^L f(x)\sin\frac{n\pi x}{L}dx = \int_0^1 x\sin\frac{n\pi x}{2}dx + \int_1^2 \sin\frac{n\pi x}{2}dx =$$

$$= 2\frac{2\sin\frac{n\pi}{2} - n\pi\cos n\pi}{n^2\pi^2}.$$

Observe that

$$\sin(\frac{n\pi}{2}) = \begin{cases} 0, & n = 2k \\ (-1)^{k+1}, & n = 2k-1 \end{cases}, \ k = 1, 2, \ldots.$$

Likewise,

$$\cos n\pi = \begin{cases} 1, & n = 2k \\ -1, & n = 2k-1 \end{cases}, \ k = 1, 2, \ldots.$$

Therefore the Fourier sine series of the specified function is

$$f(x) = -\frac{1}{\pi}\sum_{n=1}^{\infty}\frac{1}{n}\sin n\pi x + \frac{2}{\pi^2}\sum_{n=1}^{\infty}\frac{2(-1)^{n+1} + (2n-1)\pi}{(2n-1)^2}\sin\frac{(2n-1)\pi x}{2}.$$

(b)

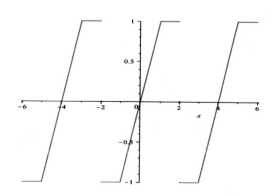

17.(a) $L = \pi$. For an even extension of the function, the sine coefficients are zero. The cosine coefficients are given by

$$a_0 = \frac{2}{L}\int_0^L f(x)dx = \frac{2}{\pi}\int_0^\pi (1)dx = 2,$$

and for $n > 0$,

$$a_n = \frac{2}{L}\int_0^L f(x)\cos\frac{n\pi x}{L}dx = \frac{2}{\pi}\int_0^\pi (1)\cos nx\, dx = 0.$$

The even extension of the given function is a constant function. As expected, the Fourier cosine series is

$$f(x) = \frac{a_0}{2} = 1.$$

(b)

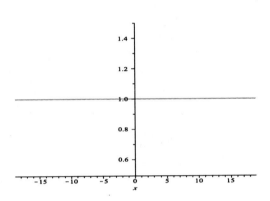

19.(a) $L = 3\pi$. For an odd extension of the function, the cosine coefficients are zero. The sine coefficients are given by

$$b_n = \frac{2}{L}\int_0^L f(x)\sin\frac{n\pi x}{L}dx = \frac{2}{3\pi}\int_\pi^{2\pi}\sin\frac{nx}{3}dx + \frac{2}{3\pi}\int_{2\pi}^{3\pi} 2\sin\frac{nx}{3}dx =$$

$$= -2\frac{2\cos n\pi - \cos\frac{n\pi}{3} - \cos\frac{2n\pi}{3}}{n\pi}.$$

Therefore the Fourier sine series of the specified function is

$$f(x) = \frac{2}{\pi}\sum_{n=1}^\infty \frac{1}{n}\left[\cos\frac{n\pi}{3} + \cos\frac{2n\pi}{3} - 2\cos n\pi\right]\sin\frac{nx}{3}.$$

(b)

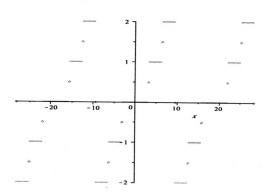

21.(a) Extend the function over the interval $[-L, L]$ as

$$f(x) = \begin{cases} x + L, & -L \leq x < 0 \\ L - x, & 0 \leq x \leq L. \end{cases}$$

Since the extended function is even, the sine coefficients are zero. The cosine coefficients are given by

$$a_0 = \frac{2}{L} \int_0^L f(x)dx = \frac{2}{L} \int_0^L (L - x)dx = L,$$

and for $n > 0$,

$$a_n = \frac{2}{L} \int_0^L f(x) \cos \frac{n\pi x}{L} dx = \frac{2}{L} \int_0^L (L - x) \cos \frac{n\pi x}{L} dx = 2L \frac{1 - \cos n\pi}{n^2 \pi^2}.$$

Therefore the Fourier cosine series of the extended function is

$$f(x) = \frac{L}{2} + \frac{4L}{\pi^2} \sum_{n=1}^{\infty} \frac{1}{(2n-1)^2} \cos \frac{(2n-1)\pi x}{L}.$$

(b) In order to compare the result with Example 1 of Section 10.2, set $L = 2$. The cosine series converges to the function graphed below:

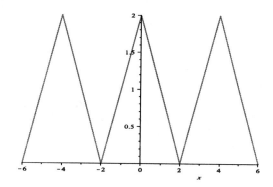

This function is a shift of the function in Example 1 of Section 10.2.

22.(a) Extend the function over the interval $[-L, L]$ as

$$f(x) = \begin{cases} -x - L, & -L \leq x < 0 \\ L - x, & 0 < x \leq L, \end{cases}$$

with $f(0) = 0$. Since the extended function is odd, the cosine coefficients are zero. The sine coefficients are given by

$$b_n = \frac{2}{L} \int_0^L f(x) \sin \frac{n\pi x}{L} dx = \frac{2}{L} \int_0^L (L - x) \sin \frac{n\pi x}{L} dx = \frac{2L}{n\pi}.$$

Therefore the Fourier cosine series of the extended function is

$$f(x) = \frac{2L}{\pi} \sum_{n=1}^{\infty} \frac{1}{n} \sin \frac{n\pi x}{L}.$$

(b) Setting $L = 2$, for example, the series converges to the function graphed below:

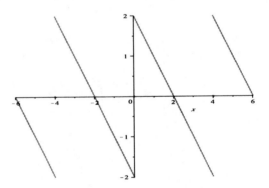

23.(a) $L = 2\pi$. For an even extension of the function, the sine coefficients are zero. The cosine coefficients are given by

$$a_0 = \frac{2}{L} \int_0^L f(x) dx = \frac{1}{\pi} \int_0^\pi x \, dx = \pi/2,$$

and for $n > 0$,

$$a_n = \frac{2}{L} \int_0^L f(x) \cos \frac{n\pi x}{L} dx = \frac{1}{\pi} \int_0^\pi x \cos \frac{nx}{2} dx =$$

$$= 2 \frac{2 \cos(\frac{n\pi}{2}) + n\pi \sin(\frac{n\pi}{2}) - 2}{n^2 \pi}.$$

Therefore the Fourier cosine series of the given function is

$$f(x) = \frac{\pi}{4} + \frac{2}{\pi} \sum_{n=1}^{\infty} \left[\frac{\pi}{n} \sin \frac{n\pi}{2} + \frac{2}{n^2} (\cos \frac{n\pi}{2} - 1) \right] \cos \frac{nx}{2}.$$

(b)

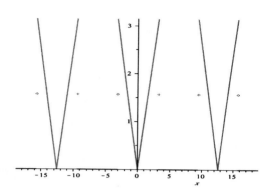

(c) $n = 10, 20, 30$:

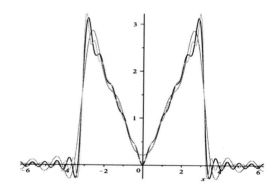

24.(a) $L = \pi$. For an odd extension of the function, the cosine coefficients are zero. Note that $f(x) = -x$ on $0 \leq x < \pi$. The sine coefficients are given by

$$b_n = \frac{2}{L}\int_0^L f(x)\sin\frac{n\pi x}{L}dx = -\frac{2}{\pi}\int_0^\pi x\sin nx\,dx = \frac{2\cos n\pi}{n}.$$

Therefore the Fourier sine series of the given function is

$$f(x) = 2\sum_{n=1}^\infty \frac{(-1)^n}{n}\sin nx.$$

(b)

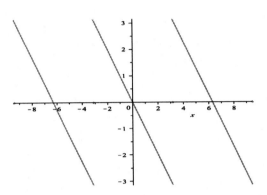

(c) $n = 10, 20, 30$:

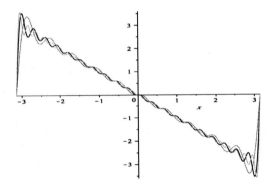

26. (a) $L = 4$. For an even extension of the function, the sine coefficients are zero. The cosine coefficients are given by

$$a_0 = \frac{2}{L}\int_0^L f(x)dx = \frac{1}{2}\int_0^4 (x^2 - 2x)dx = 8/3,$$

and for $n > 0$,

$$a_n = \frac{2}{L}\int_0^L f(x)\cos\frac{n\pi x}{L}dx = \frac{1}{2}\int_0^4 (x^2 - 2x)\cos\frac{n\pi x}{4}dx = 16\frac{1 + 3\cos n\pi}{n^2\pi^2}.$$

Therefore the Fourier cosine series of the given function is

$$f(x) = \frac{4}{3} + \frac{16}{\pi^2}\sum_{n=1}^{\infty}\frac{1 + 3(-1)^n}{n^2}\cos\frac{n\pi x}{4}.$$

(b)

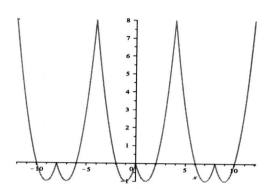

(c) $n = 10, 20, 30$:

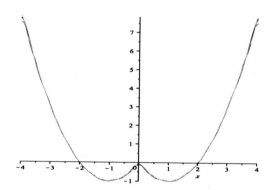

27.(a)

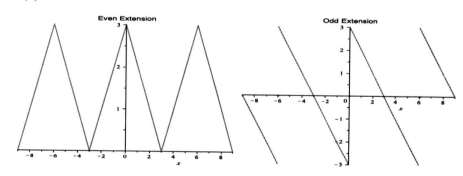

(b) $L = 3$. For an even extension of the function, the cosine coefficients are given by

$$a_0 = \frac{2}{L}\int_0^L f(x)dx = \frac{2}{3}\int_0^3 (3-x)dx = 3,$$

and for $n > 0$,

$$a_n = \frac{2}{L}\int_0^L f(x)\cos\frac{n\pi x}{L}dx = \frac{2}{3}\int_0^3 (3-x)\cos\frac{n\pi x}{3}dx = 6\frac{1-\cos n\pi}{n^2\pi^2}.$$

Therefore the Fourier cosine series of the given function is

$$g(x) = \frac{3}{2} + \frac{6}{\pi^2} \sum_{n=1}^{\infty} \frac{1-(-1)^n}{n^2} \cos \frac{n\pi x}{3}.$$

For an odd extension of the function, the sine coefficients are given by

$$b_n = \frac{2}{L} \int_0^L f(x) \sin \frac{n\pi x}{L} dx = \frac{2}{3} \int_0^3 (3-x) \sin \frac{n\pi x}{3} dx = \frac{6}{n\pi}.$$

Therefore the Fourier sine series of the given function is

$$h(x) = \frac{6}{\pi} \sum_{n=1}^{\infty} \frac{1}{n} \sin \frac{n\pi x}{3}.$$

(c) For the even extension, $n = 10, 20, 30$:

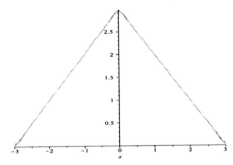

For the odd extension, $n = 10, 20, 30$:

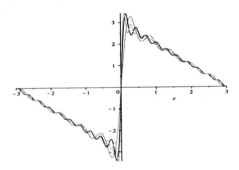

(d) Since the even extension is continuous, the series converges uniformly. On the other hand, the odd extension is discontinuous. Gibbs' phenomenon results in a finite error for all values of n.

29.(a)

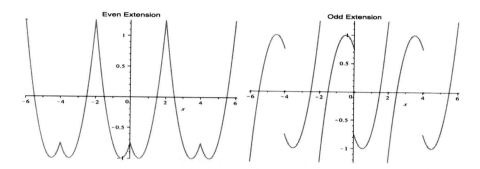

(b) $L = 2$. For an even extension of the function, the cosine coefficients are given by

$$a_0 = \frac{2}{L} \int_0^L f(x)dx = \int_0^2 \left[\frac{4x^2 - 4x - 3}{4}\right] dx = -5/6,$$

and for $n > 0$,

$$a_n = \frac{2}{L} \int_0^L f(x) \cos \frac{n\pi x}{L} dx = \int_0^2 \left[\frac{4x^2 - 4x - 3}{4}\right] \cos \frac{n\pi x}{2} dx = 4 \frac{1 + 3 \cos n\pi}{n^2 \pi^2}.$$

Therefore the Fourier cosine series of the given function is

$$g(x) = -\frac{5}{12} + \frac{4}{\pi^2} \sum_{n=1}^{\infty} \frac{1 + 3(-1)^n}{n^2} \cos \frac{n\pi x}{2}.$$

For an odd extension of the function, the sine coefficients are given by

$$b_n = \frac{2}{L} \int_0^L f(x) \sin \frac{n\pi x}{L} dx = \int_0^2 \left[\frac{4x^2 - 4x - 3}{4}\right] \sin \frac{n\pi x}{2} dx =$$

$$= -\frac{32 + 3n^2\pi^2 + 5n^2\pi^2 \cos n\pi - 32 \cos n\pi}{2 n^3 \pi^3}.$$

Therefore the Fourier sine series of the given function is

$$h(x) = -\frac{1}{2\pi^3} \sum_{n=1}^{\infty} \frac{32(1 - \cos n\pi) + n^2\pi^2(3 + 5 \cos n\pi)}{n^3} \sin \frac{n\pi x}{2}.$$

(c) For the even extension, $n = 10, 20, 30$:

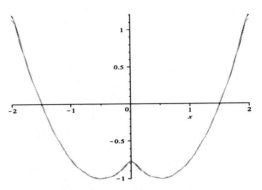

For the odd extension, $n = 10, 20, 30$:

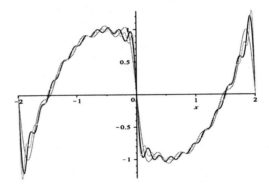

(d) Since the even extension is continuous, the series converges uniformly. On the other hand, the odd extension is discontinuous. Gibbs' phenomenon results in a finite error for all values of n.

30.(a)

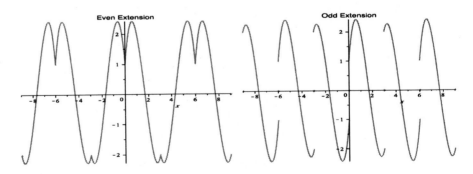

(b) $L = 3$. For an even extension of the function, the cosine coefficients are given by

$$a_0 = \frac{2}{L}\int_0^L f(x)dx = \frac{2}{3}\int_0^3 (x^3 - 5x^2 + 5x + 1)dx = 1/2,$$

and for $n > 0$,

$$a_n = \frac{2}{L}\int_0^L f(x)\cos\frac{n\pi x}{L}dx = \frac{2}{3}\int_0^3 (x^3 - 5x^2 + 5x + 1)\cos\frac{n\pi x}{3}dx =$$

$$= 2\frac{162 - 15n^2\pi^2 + 6n^2\pi^2\cos n\pi - 162\cos n\pi}{n^4\pi^4}.$$

Therefore the Fourier cosine series of the given function is

$$g(x) = \frac{1}{4} + \frac{2}{\pi^4}\sum_{n=1}^\infty \frac{162(1 - \cos n\pi) - 3n^2\pi^2(5 - 2\cos n\pi)}{n^4}\cos\frac{n\pi x}{3}.$$

For an odd extension of the function, the sine coefficients are given by

$$b_n = \frac{2}{L}\int_0^L f(x)\sin\frac{n\pi x}{L}dx = \frac{2}{3}\int_0^3 (x^3 - 5x^2 + 5x + 1)\sin\frac{n\pi x}{3}dx =$$

$$= 2\frac{90 + n^2\pi^2 + 2n^2\pi^2\cos n\pi + 72\cos n\pi}{n^3\pi^3}.$$

Therefore the Fourier sine series of the given function is

$$h(x) = \frac{2}{\pi^3}\sum_{n=1}^\infty \frac{18(5 + 4\cos n\pi) + n^2\pi^2(1 + 2\cos n\pi)}{n^3}\sin\frac{n\pi x}{3}.$$

(c) For the even extension, $n = 10, 20, 30$:

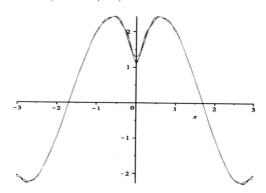

For the odd extension, $n = 10, 20, 30$:

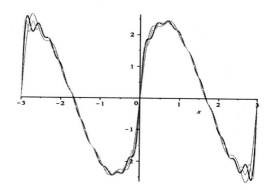

(d) Since the even extension is continuous, the series converges uniformly. On the other hand, the odd extension is discontinuous. Gibbs' phenomenon results in a finite error for all values of n; particularly at $x = \pm 3$.

33. Let $f(x)$ be a differentiable even function. For any x in its domain,
$$f(-x + h) - f(-x) = f(x - h) - f(x).$$

It follows that
$$f'(-x) = \lim_{h \to 0} \frac{f(-x + h) - f(-x)}{h} = \lim_{h \to 0} \frac{f(x - h) - f(x)}{h} =$$
$$= -\lim_{h \to 0} \frac{f(x - h) - f(x)}{(-h)}.$$

Setting $h = -\delta$, we have
$$f'(-x) = -\lim_{-\delta \to 0} \frac{f(x + \delta) - f(x)}{\delta} = -\lim_{\delta \to 0} \frac{f(x + \delta) - f(x)}{\delta} = -f'(x).$$

Therefore $f'(-x) = -f'(x)$. If $f(x)$ is a differentiable odd function, for any x in its domain,
$$f(-x + h) - f(-x) = -f(x - h) + f(x).$$

It follows that
$$f'(-x) = \lim_{h \to 0} \frac{f(-x + h) - f(-x)}{h} = \lim_{h \to 0} \frac{-f(x - h) + f(x)}{h} =$$
$$= \lim_{h \to 0} \frac{f(x - h) - f(x)}{(-h)}.$$

Setting $h = -\delta$, we have
$$f'(-x) = \lim_{-\delta \to 0} \frac{f(x + \delta) - f(x)}{\delta} = \lim_{\delta \to 0} \frac{f(x + \delta) - f(x)}{\delta} = f'(x).$$

Therefore $f'(-x) = f'(x)$.

36. From Example 1 of Section 10.2, the function
$$f(x) = \begin{cases} -x, & -2 \leq x < 0 \\ x, & 0 \leq x < 2, \end{cases}$$

has a convergent Fourier series ($L = 2$):
$$f(x) = 1 - \frac{8}{\pi^2} \sum_{n=1}^{\infty} \frac{1}{(2n-1)^2} \cos \frac{(2n-1)\pi x}{2}.$$

Since $f(x)$ is continuous, the series converges everywhere. In particular, at $x = 0$, we have
$$0 = f(0) = 1 - \frac{8}{\pi^2} \sum_{n=1}^{\infty} \frac{1}{(2n-1)^2}.$$

It follows immediately that

$$\frac{\pi^2}{8} = \sum_{n=1}^{\infty} \frac{1}{(2n-1)^2} = 1 + \frac{1}{3^2} + \frac{1}{5^2} + \frac{1}{7^2} + \ldots.$$

40.(a) Since one objective is to obtain a Fourier series containing only cosine terms, any extension of $f(x)$ should be an even function. Another objective is to derive a series containing only the terms

$$\cos\frac{(2n-1)\pi x}{2L}, \, n = 1, 2, \ldots.$$

First note that the functions

$$\cos\frac{n\pi x}{L}, \, n = 1, 2, \ldots$$

are symmetric about $x = L$. Indeed,

$$\cos\frac{n\pi(2L-x)}{L} = \cos(2n\pi - \frac{n\pi x}{L}) = \cos(-\frac{n\pi x}{L}) = \cos\frac{n\pi x}{L}.$$

It follows that if $f(x)$ is extended into $(L, 2L)$ as an antisymmetric function about $x = L$, that is, $f(2L - x) = -f(x)$ for $0 \leq x \leq 2L$, then

$$\int_0^{2L} f(x) \cos\frac{n\pi x}{L} dx = 0.$$

This follows from the fact that the integrand is an antisymmetric function about $x = L$. Now extend the function $f(x)$ to obtain

$$\tilde{f}(x) = \begin{cases} f(x), & 0 \leq x < L \\ -f(2L-x), & L < x < 2L. \end{cases}$$

Finally, extend the resulting function into $(-2L, 0)$ as an even function, and then as a periodic function of period $4L$. By construction, the Fourier series will contain only cosine terms. We first note that

$$a_0 = \frac{2}{2L}\int_0^{2L} \tilde{f}(x)\, dx = \frac{1}{L}\int_0^L f(x)dx - \frac{1}{L}\int_L^{2L} f(2L-x)dx =$$

$$= \frac{1}{L}\int_0^L f(x)dx - \frac{1}{L}\int_0^L f(u)du = 0.$$

For $n > 0$,

$$a_n = \frac{2}{2L}\int_0^{2L} \tilde{f}(x) \cos\frac{n\pi x}{2L} dx =$$

$$= \frac{1}{L}\int_0^L f(x)\cos\frac{n\pi x}{2L}dx - \frac{1}{L}\int_L^{2L} f(2L-x)\cos\frac{n\pi x}{2L}dx.$$

For the second integral, let $u = 2L - x$. Then

$$\cos\frac{n\pi x}{2L} = \cos\frac{n\pi(2L+u)}{2L} = (-1)^n \cos\frac{n\pi u}{2L}$$

and therefore

$$\int_L^{2L} f(2L-x)\cos\frac{n\pi x}{2L}dx = (-1)^n \int_0^L f(u)\cos\frac{n\pi u}{2L}du.$$

Hence

$$a_n = \frac{1-(-1)^n}{L}\int_0^L f(x)\cos\frac{n\pi x}{2L}dx.$$

It immediately follows that $a_n = 0$ for $n = 2k$, $k = 0, 1, 2, \ldots$, and

$$a_{2k-1} = \frac{2}{L}\int_0^L f(x)\cos\frac{(2k-1)\pi x}{2L}dx, \text{ for } k = 1, 2, \ldots.$$

The associated Fourier series representation

$$f(x) = \sum_{n=0}^{\infty} a_{2n-1}\cos\frac{(2n-1)\pi x}{2L}$$

converges almost everywhere on $(-2L, 2L)$ and hence on $(0, L)$.

(b) If $f(x) = x$ for $0 \leq x \leq L = 1$, the graph of the extended function is:

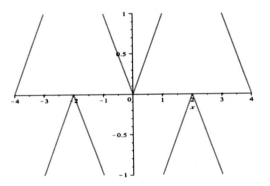

10.5

1. We consider solutions of the form $u(x,t) = X(x)T(t)$. Substitution into the partial differential equation results in

$$xX''T + XT' = 0.$$

Divide both sides of the differential equation by the product XT to obtain

$$x\frac{X''}{X} + \frac{T'}{T} = 0,$$

so that

$$x\frac{X''}{X} = -\frac{T'}{T}.$$

Since both sides of the resulting equation are functions of different variables, each must be equal to a constant, say λ. We obtain the ordinary differential equations

$$xX'' - \lambda X = 0 \text{ and } T' + \lambda T = 0.$$

2. In order to apply the method of separation of variables, we consider solutions of the form $u(x,t) = X(x)T(t)$. Substituting the assumed form of the solution into the partial differential equation, we obtain

$$t X''T + x XT' = 0.$$

Divide both sides of the differential equation by the product $xtXT$ to obtain

$$\frac{X''}{xX} + \frac{T'}{tT} = 0,$$

so that

$$\frac{X''}{xX} = -\frac{T'}{tT}.$$

Since both sides of the resulting equation are functions of different variables, it follows that

$$\frac{X''}{xX} = -\frac{T'}{tT} = \lambda.$$

Therefore $X(x)$ and $T(t)$ are solutions of the ordinary differential equations

$$X'' - \lambda x X = 0 \text{ and } T' + \lambda t T = 0.$$

4. Assume that the solution of the PDE has the form $u(x,t) = X(x)T(t)$. Substitution into the partial differential equation results in

$$[p(x)X']' T - r(x)X T'' = 0.$$

Divide both sides of the differential equation by the product $r(x)XT$ to obtain

$$\frac{[p(x)X']'}{r(x)X} - \frac{T''}{T} = 0,$$

that is,

$$\frac{[p(x)X']'}{r(x)X} = \frac{T''}{T}.$$

Since both sides of the resulting equation are functions of different variables, each must be equal to a constant, say $-\lambda$. We obtain the ordinary differential equations

$$[p(x)X']' + \lambda r(x)X = 0 \text{ and } T'' + \lambda T = 0.$$

6. We consider solutions of the form $u(x,y) = X(x)Y(y)$. Substitution into the partial differential equation results in

$$X''Y + XY'' + xXY = 0.$$

Divide both sides of the differential equation by the product XY to obtain

$$\frac{X''}{X} + \frac{Y''}{Y} + x = 0,$$

that is,

$$\frac{X''}{X} + x = -\frac{Y''}{Y}.$$

Since both sides of the resulting equation are functions of different variables, it follows that

$$\frac{X''}{X} + x = -\frac{Y''}{Y} = -\lambda.$$

We obtain the ordinary differential equations

$$X'' + (x + \lambda)X = 0 \text{ and } Y'' - \lambda Y = 0.$$

7. The heat conduction equation, $100\, u_{xx} = u_t$, and the given boundary conditions are homogeneous. We consider solutions of the form $u(x,t) = X(x)T(t)$. Substitution into the partial differential equation results in

$$100\, X''T = XT'.$$

Divide both sides of the differential equation by the product XT to obtain

$$\frac{X''}{X} = \frac{T'}{100\, T}.$$

Since both sides of the resulting equation are functions of different variables, it follows that

$$\frac{X''}{X} = \frac{T'}{100\, T} = -\lambda.$$

Therefore $X(x)$ and $T(t)$ are solutions of the ordinary differential equations

$$X'' + \lambda X = 0 \text{ and } T' + 100\lambda T = 0.$$

The general solution of the spatial equation is $X = c_1 \cos \lambda^{1/2} x + c_2 \sin \lambda^{1/2} x$. In order to satisfy the homogeneous boundary conditions, we require that $c_1 = 0$, and

$$\lambda^{1/2} = n\pi.$$

Hence the eigenfunctions are $X_n = \sin n\pi x$, with associated eigenvalues $\lambda_n = n^2\pi^2$. We thus obtain the family of equations $T' + 100\lambda_n T = 0$. The solutions are given by

$$T_n = e^{-100\lambda_n t}.$$

Hence the fundamental solutions of the PDE are

$$u_n(x,t) = e^{-100n^2\pi^2 t} \sin n\pi x,$$

which yield the general solution

$$u(x,t) = \sum_{n=1}^{\infty} c_n e^{-100n^2\pi^2 t} \sin n\pi x.$$

Finally, the initial condition $u(x,0) = \sin 2\pi x - \sin 5\pi x$ must be satisfied. Therefore is it necessary that

$$\sum_{n=1}^{\infty} c_n \sin n\pi x = \sin 2\pi x - \sin 5\pi x.$$

From the orthogonality conditions we obtain $c_2 = -c_5 = 1$, with all other $c_n = 0$. Therefore the solution of the given heat conduction problem is

$$u(x,t) = e^{-400\pi^2 t} \sin 2\pi x - e^{-2500\pi^2 t} \sin 5\pi x.$$

9. The heat conduction problem is formulated as

$$u_{xx} = u_t, \qquad 0 < x < 40, \; t > 0;$$
$$u(0,t) = 0, \qquad u(40,t) = 0, \; t > 0;$$
$$u(x,0) = 50, \qquad 0 < x < 40.$$

Assume a solution of the form $u(x,t) = X(x)T(t)$. Following the procedure in this section, we obtain the eigenfunctions $X_n = \sin n\pi x/40$, with associated eigenvalues $\lambda_n = n^2\pi^2/1600$. The solutions of the temporal equations are $T_n = e^{-\lambda_n t}$ Hence the general solution of the given problem is

$$u(x,t) = \sum_{n=1}^{\infty} c_n e^{-n^2\pi^2 t/1600} \sin \frac{n\pi x}{40}.$$

The coefficients c_n are the Fourier sine coefficients of $u(x,0) = 50$. That is,

$$c_n = \frac{2}{L} \int_0^L f(x) \sin \frac{n\pi x}{L} dx = \frac{5}{2} \int_0^{40} \sin \frac{n\pi x}{40} dx = 100 \frac{1 - \cos n\pi}{n\pi}.$$

The sine series of the initial condition is

$$50 = \frac{100}{\pi} \sum_{n=1}^{\infty} \frac{1 - \cos n\pi}{n} \sin \frac{n\pi x}{40}.$$

Therefore the solution of the given heat conduction problem is

$$u(x,t) = \frac{100}{\pi} \sum_{n=1}^{\infty} \frac{1 - \cos n\pi}{n} e^{-n^2\pi^2 t/1600} \sin \frac{n\pi x}{40}.$$

11. Refer to Problem 9 for the formulation of the problem. In this case, the initial condition is given by

$$u(x,0) = \begin{cases} 0, & 0 \leq x < 10, \\ 50, & 10 \leq x \leq 30, \\ 0, & 30 < x \leq 40. \end{cases}$$

All other data being the same, the solution of the given problem is

$$u(x,t) = \sum_{n=1}^{\infty} c_n e^{-n^2\pi^2 t/1600} \sin \frac{n\pi x}{40}.$$

The coefficients c_n are the Fourier sine coefficients of $u(x,0)$. That is,

$$c_n = \frac{2}{L}\int_0^L f(x)\sin\frac{n\pi x}{L}dx = \frac{5}{2}\int_{10}^{30}\sin\frac{n\pi x}{40}dx = 100\,\frac{\cos\frac{n\pi}{4} - \cos\frac{3n\pi}{4}}{n\pi}.$$

Therefore the solution of the given heat conduction problem is

$$u(x,t) = \frac{100}{\pi}\sum_{n=1}^{\infty}\frac{\cos\frac{n\pi}{4} - \cos\frac{3n\pi}{4}}{n}e^{-n^2\pi^2 t/1600}\sin\frac{n\pi x}{40}.$$

12. Refer to Problem 9 for the formulation of the problem. In this case, the initial condition is given by

$$u(x,0) = x,\ 0 < x < 40.$$

All other data being the same, the solution of the given problem is

$$u(x,t) = \sum_{n=1}^{\infty} c_n e^{-n^2\pi^2 t/1600}\sin\frac{n\pi x}{40}.$$

The coefficients c_n are the Fourier sine coefficients of $u(x,0) = x$. That is,

$$c_n = \frac{2}{L}\int_0^L f(x)\sin\frac{n\pi x}{L}dx = \frac{1}{20}\int_0^{40} x\sin\frac{n\pi x}{40}dx = -80\,\frac{\cos n\pi}{n\pi}.$$

Therefore the solution of the given heat conduction problem is

$$u(x,t) = \frac{80}{\pi}\sum_{n=1}^{\infty}\frac{(-1)^{n+1}}{n}e^{-n^2\pi^2 t/1600}\sin\frac{n\pi x}{40}.$$

13. Substituting $x = 20$, into the solution, we have

$$u(20,t) = \frac{100}{\pi}\sum_{n=1}^{\infty}\frac{1 - \cos n\pi}{n}e^{-n^2\pi^2 t/1600}\sin\frac{n\pi}{2}.$$

We can also write

$$u(20,t) = \frac{200}{\pi}\sum_{k=1}^{\infty}\frac{(-1)^{k+1}}{2k-1}e^{-(2k-1)^2\pi^2 t/1600}.$$

Therefore,

$$u(20,5) = \frac{200}{\pi}\sum_{k=1}^{\infty}\frac{(-1)^{k+1}}{2k-1}e^{-(2k-1)^2\pi^2/320}.$$

Let

$$A_k = \frac{(-1)^{k+1}200}{\pi(2k-1)}e^{-(2k-1)^2\pi^2/320}.$$

It follows that $|A_k| < 0.005$ for $k \geq 8$. So for $n = 2k - 1 \geq 15$, the summation is correct to three decimal places.
For $t = 20$,

$$u(20,20) = \frac{200}{\pi}\sum_{k=1}^{\infty}\frac{(-1)^{k+1}}{2k-1}e^{-(2k-1)^2\pi^2/80}.$$

Let
$$A_k = \frac{(-1)^{k+1}200}{\pi(2k-1)}e^{-(2k-1)^2\pi^2/80}.$$

It follows that $|A_k| < 0.005$ for $k \geq 5$. So for $n = 2k-1 \geq 9$, the summation is correct to three decimal places.

For $t = 80$,
$$u(20,80) = \frac{200}{\pi}\sum_{k=1}^{\infty}\frac{(-1)^{k+1}}{2k-1}e^{-(2k-1)^2\pi^2/20}.$$

Let
$$A_k = \frac{(-1)^{k+1}200}{\pi(2k-1)}e^{-(2k-1)^2\pi^2/20}.$$

It follows that $|A_k| < 0.00005$ for $k \geq 3$. So for $n = 2k-1 \geq 5$, the summation is correct to three decimal places. The series solution converges faster as t increases.

14.(a) The solution of the given heat conduction problem is
$$u(x,t) = \frac{100}{\pi}\sum_{n=1}^{\infty}\frac{1-\cos n\pi}{n}e^{-n^2\pi^2 t/1600}\sin\frac{n\pi x}{40}.$$

Setting $t = 5, 10, 20, 40, 100, 200$:

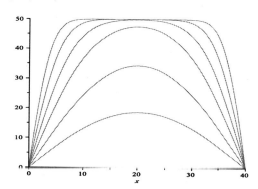

(b) Setting $x = 5, 10, 15, 20$:

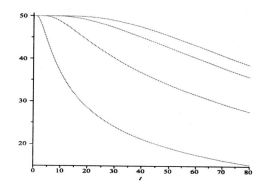

(c) Surface plot of $u(x,t)$:

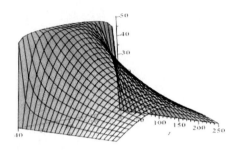

(d) $0 \leq u(x,t) \leq 1$ for $t \geq 673.35$ sec.

16.(a) The solution of the given heat conduction problem is

$$u(x,t) = \frac{100}{\pi} \sum_{n=1}^{\infty} \frac{\cos \frac{n\pi}{4} - \cos \frac{3n\pi}{4}}{n} e^{-n^2\pi^2 t/1600} \sin \frac{n\pi x}{40}.$$

Setting $t = 5, 10, 20, 40, 100, 200$:

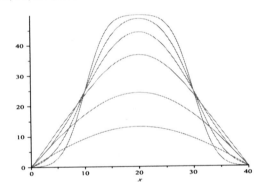

(b) Setting $x = 5, 10, 15, 20$:

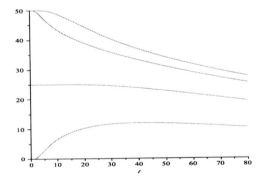

(c) Surface plot of $u(x,t)$:

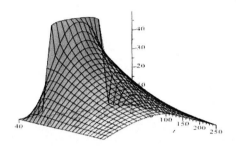

(d) $0 \leq u(x,t) \leq 1$ for $t \geq 617.17$ sec.

17.(a) The solution of the given heat conduction problem is

$$u(x,t) = \frac{80}{\pi} \sum_{n=1}^{\infty} \frac{(-1)^{n+1}}{n} e^{-n^2\pi^2 t/1600} \sin\frac{n\pi x}{40}.$$

Setting $t = 5, 10, 20, 40, 100, 200$:

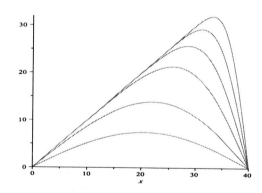

(b) Analyzing the individual plots, we find that the 'hot spot' varies with time:

t	5	10	20	40	100	200
x_h	33	31	29	26	22	21

Location of the 'hot spot', x_h, versus time:

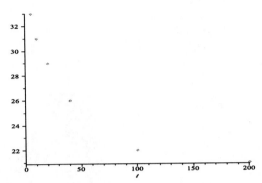

Evidently, the location of the greatest temperature migrates to the center of the rod.

(c) Setting $x = 10, 20, 30$:

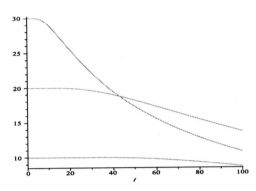

(d) Surface plot of $u(x,t)$:

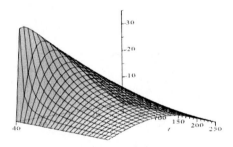

(e) $0 \leq u(x,t) \leq 1$ for $t \geq 525$ sec.

21.(a) Given the partial differential equation
$$a\, u_{xx} - b\, u_t + c\, u = 0,$$
in which a, b, and c are constants, set $u(x,t) = e^{\delta t} w(x,t)$. Substitution into the

PDE results in
$$a\,e^{\delta t}w_{xx} - b(\delta\,e^{\delta t}w + e^{\delta t}w_t) + c\,e^{\delta t}w = 0.$$

Dividing both sides of the equation by $e^{\delta t}$, we obtain
$$a\,w_{xx} - b\,w_t + (c - b\delta)\,w = 0.$$

(b) As long as $b \neq 0$, choosing $\delta = c/b$ yields
$$\frac{a}{b}w_{xx} - w_t = 0,$$

which is the heat conduction equation with dependent variable w.

23. The heat conduction equation in polar coordinates is given by
$$\alpha^2\left[u_{rr} + \frac{1}{r}u_r + \frac{1}{r^2}u_{\theta\theta}\right] = u_t.$$

We consider solutions of the form $u(r,\theta,t) = R(r)\Theta(\theta)T(t)$. Substitution into the PDE results in
$$\alpha^2\left[R''\Theta T + \frac{1}{r}R'\Theta T + \frac{1}{r^2}R\Theta''T\right] = R\Theta T'.$$

Dividing both sides of the equation by the factor $R\Theta T$, we obtain
$$\frac{R''}{R} + \frac{1}{r}\frac{R'}{R} + \frac{1}{r^2}\frac{\Theta''}{\Theta} = \frac{T'}{\alpha^2 T}.$$

Since both sides of the resulting differential equation depend on different variables, each side must be equal to a constant, say $-\lambda^2$. That is,
$$\frac{R''}{R} + \frac{1}{r}\frac{R'}{R} + \frac{1}{r^2}\frac{\Theta''}{\Theta} = \frac{T'}{\alpha^2 T} = -\lambda^2.$$

It follows that $T' + \alpha^2\lambda^2 T = 0$, and
$$\frac{R''}{R} + \frac{1}{r}\frac{R'}{R} + \frac{1}{r^2}\frac{\Theta''}{\Theta} = -\lambda^2.$$

Multiplying both sides of this differential equation by r^2, we find that
$$r^2\frac{R''}{R} + r\frac{R'}{R} + \frac{\Theta''}{\Theta} = -\lambda^2 r^2,$$

which can be written as
$$r^2\frac{R''}{R} + r\frac{R'}{R} + \lambda^2 r^2 = -\frac{\Theta''}{\Theta}.$$

Once again, since both sides of the resulting differential equation depend on different variables, each side must be equal to a constant. Hence
$$r^2\frac{R''}{R} + r\frac{R'}{R} + \lambda^2 r^2 = \mu^2 \quad \text{and} \quad -\frac{\Theta''}{\Theta} = \mu^2.$$

The resulting ordinary equations are

$$r^2 R'' + rR' + (\lambda^2 r^2 - \mu^2)R = 0$$
$$\Theta'' + \mu^2 \Theta = 0$$
$$T' + \alpha^2 \lambda^2 T = 0.$$

10.6

1. The steady-state solution, $v(x)$, satisfies the boundary value problem

$$v''(x) = 0, \ 0 < x < 50, \ v(0) = 10, \ v(50) = 40.$$

The general solution of the ODE is $v(x) = Ax + B$. Imposing the boundary conditions, we have

$$v(x) = \frac{40 - 10}{50}x + 10 = \frac{3x}{5} + 10.$$

2. The steady-state solution, $v(x)$, satisfies the boundary value problem

$$v''(x) = 0, \ 0 < x < 40, \ v(0) = 30, \ v(40) = -20.$$

The solution of the ODE is linear. Imposing the boundary conditions, we have

$$v(x) = \frac{-20 - 30}{40}x + 30 = -\frac{5x}{4} + 30.$$

4. The steady-state solution is also a solution of the boundary value problem given by $v''(x) = 0$, $0 < x < L$, and the conditions $v'(0) = 0$, $v(L) = T$. The solution of the ODE is $v(x) = Ax + B$. The boundary condition $v'(0) = 0$ requires that $A = 0$. The other condition requires that $B = T$. Hence $v(x) = T$.

5. As in Problem 4, the steady-state solution has the form $v(x) = Ax + B$. The boundary condition $v(0) = 0$ requires that $B = 0$. The boundary condition $v'(L) = 0$ requires that $A = 0$. Hence $v(x) = 0$.

6. The steady-state solution has the form $v(x) = Ax + B$. The first boundary condition, $v(0) = T$, requires that $B = T$. The other boundary condition, $v'(L) = 0$, requires that $A = 0$. Hence $v(x) = T$.

8. The steady-state solution, $v(x)$, satisfies the differential equation $v''(x) = 0$, along with the boundary conditions $v(0) = T$, $v'(L) + v(L) = 0$. The general solution of the ODE is $v(x) = Ax + B$. The boundary condition $v'(0) = 0$ requires that $B = T$. It follows that $v(x) = Ax + T$, and $v'(L) + v(L) = A + AL + T$. The second boundary condition requires that $A = -T/(1 + L)$. Therefore

$$v(x) = -\frac{Tx}{1 + L} + T.$$

10.(a) Based on the symmetry of the problem, consider only the left half of the bar. The steady-state solution satisfies the ODE $v''(x) = 0$, along with the boundary conditions $v(0) = 0$ and $v(50) = 100$. The solution of this boundary value problem is $v(x) = 2x$. It follows that the steady-state temperature in the entire rod is given by

$$f(x) = \begin{cases} 2x, & 0 \le x \le 50 \\ 200 - 2x, & 50 \le x \le 100. \end{cases}$$

(b) The heat conduction problem is formulated as

$$\alpha^2 u_{xx} = u_t, \qquad 0 < x < 100, \ t > 0;$$
$$u(0,t) = 20, \qquad u(100,t) = 0, \ t > 0;$$
$$u(x,0) = f(x), \qquad 0 < x < 100.$$

First express the solution as $u(x,t) = g(x) + w(x,t)$, where $g(x) = -x/5 + 20$ and w satisfies the heat conduction problem

$$\alpha^2 w_{xx} = w_t, \qquad 0 < x < 100, \ t > 0;$$
$$w(0,t) = 0, \qquad w(100,t) = 0, \ t > 0;$$
$$w(x,0) = f(x) - g(x), \qquad 0 < x < 100.$$

Based on the results in Section 10.5,

$$w(x,t) = \sum_{n=1}^{\infty} c_n e^{-n^2 \pi^2 \alpha^2 t/10000} \sin \frac{n\pi x}{100},$$

in which the coefficients c_n are the Fourier sine coefficients of $f(x) - g(x)$. That is,

$$c_n = \frac{2}{L} \int_0^L [f(x) - g(x)] \sin \frac{n\pi x}{L} dx = \frac{1}{50} \int_0^{100} [f(x) - g(x)] \sin \frac{n\pi x}{100} dx =$$

$$= 40 \, \frac{20 \sin \frac{n\pi}{2} - n\pi}{n^2 \pi^2}.$$

Finally, the thermal diffusivity of copper is $1.14 \text{ cm}^2/\text{sec}$. Therefore the temperature distribution in the rod is

$$u(x,t) = 20 - \frac{x}{5} + \frac{40}{\pi^2} \sum_{n=1}^{\infty} \frac{20 \sin \frac{n\pi}{2} - n\pi}{n^2} e^{-1.14 n^2 \pi^2 t/10000} \sin \frac{n\pi x}{100}.$$

(c) $t = 5, 10, 20, 40$ sec, then $t = 100, 200, 300, 500$ sec:

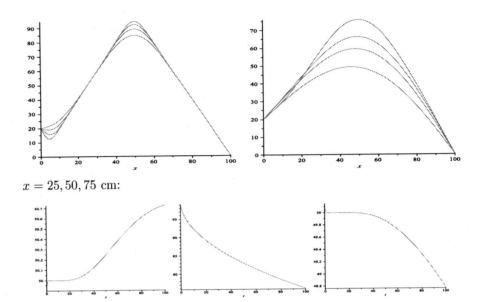

$x = 25, 50, 75$ cm:

(d) The steady-state temperature of the center of the rod will be $g(50) = 10°C$. Using a one-term approximation,

$$u(x,t) \approx 10 + \frac{800 - 40\pi}{\pi^2} e^{-1.14\pi^2 t/10000}.$$

Numerical investigation shows that $10 < u(50,t) < 11$ for $t \geq 3755$ sec.

11.(a) The heat conduction problem is formulated as

$$\begin{aligned} u_{xx} &= u_t, & 0 < x < 30, \ t > 0; \\ u(0,t) &= 30, & u(30,t) = 0, \ t > 0; \\ u(x,0) &= f(x), & 0 < x < 30, \end{aligned}$$

in which the initial condition is given by $f(x) = x(60-x)/30$. Express the solution as $u(x,t) = v(x) + w(x,t)$, where $v(x) = 30 - x$ and w satisfies the heat conduction problem

$$\begin{aligned} w_{xx} &= w_t, & 0 < x < 30, \ t > 0; \\ w(0,t) &= 0, & w(30,t) = 0, \ t > 0; \\ w(x,0) &= f(x) - v(x), & 0 < x < 30. \end{aligned}$$

As shown in Section 10.5,

$$w(x,t) = \sum_{n=1}^{\infty} c_n e^{-n^2\pi^2 t/900} \sin\frac{n\pi x}{30},$$

in which the coefficients c_n are the Fourier sine coefficients of $f(x) - v(x)$. That is,

$$c_n = \frac{2}{L}\int_0^L [f(x) - g(x)] \sin\frac{n\pi x}{L} dx = \frac{1}{15}\int_0^{30} [f(x) - g(x)] \sin\frac{n\pi x}{30} dx =$$

$$= 60\,\frac{2(1-\cos n\pi)-n^2\pi^2(1+\cos n\pi)}{n^3\pi^3}\,.$$

Therefore the temperature distribution in the rod is

$$u(x,t) = 30 - x + \frac{60}{\pi^3}\sum_{n=1}^{\infty}\frac{2(1-\cos n\pi)-n^2\pi^2(1+\cos n\pi)}{n^3}\,e^{-n^2\pi^2 t/900}\sin\frac{n\pi x}{30}\,.$$

(b) $t = 5, 10, 20, 40$ sec, then $t = 50, 75, 100, 200$ sec:

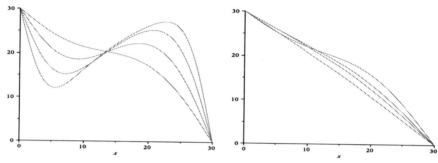

$x = 7.5, 15, 22.5$ cm:

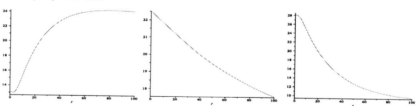

(c)

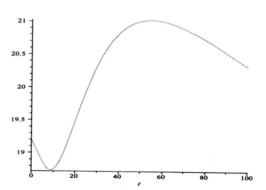

Based on the heat conduction equation, the rate of change of the temperature at any given point is proportional to the concavity of the graph of u versus x, that is, u_{xx}. Evidently, near $t = 60$, the concavity of $u(x,t)$ changes.

13.(a) The heat conduction problem is formulated as

$$\begin{aligned}u_{xx} &= 4\,u_t, & 0 &< x < 40,\ t > 0;\\ u_x(0,t) &= 0, & u_x(40,t) &= 0,\ t > 0;\\ u(x,0) &= f(x), & 0 &< x < 40,\end{aligned}$$

in which the initial condition is given by $f(x) = x(60-x)/30$. As shown in the discussion on rods with insulated ends, the solution is given by

$$u(x,t) = \frac{c_0}{2} + \sum_{n=1}^{\infty} c_n e^{-n^2\pi^2\alpha^2 t/1600} \cos\frac{n\pi x}{40},$$

where c_n are the Fourier cosine coefficients. In this problem,

$$c_0 = \frac{2}{L}\int_0^L f(x)dx = \frac{1}{20}\int_0^{40} \frac{x(60-x)}{30}dx = 400/9,$$

and for $n \geq 1$,

$$c_n = \frac{2}{L}\int_0^L f(x)\cos\frac{n\pi x}{L}dx = \frac{1}{20}\int_0^{40} \frac{x(60-x)}{30}\cos\frac{n\pi x}{40}dx = -\frac{160(3+\cos n\pi)}{3n^2\pi^2}.$$

Therefore the temperature distribution in the rod is

$$u(x,t) = \frac{200}{9} - \frac{160}{3\pi^2}\sum_{n=1}^{\infty}\frac{(3+\cos n\pi)}{n^2}e^{-n^2\pi^2 t/6400}\cos\frac{n\pi x}{40}.$$

(b) $t = 50, 100, 150, 200$ sec, then $t = 400, 600, 800, 1000$ sec:

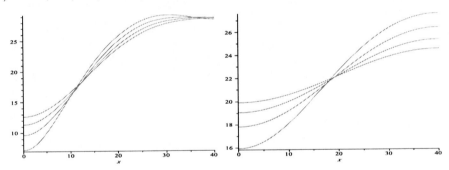

$x = 0, 10, 20, 40$ cm:

(c) Since

$$\lim_{t \to \infty} e^{-n^2\pi^2 t/6400}\cos\frac{n\pi x}{40} = 0$$

for each x, it follows that the steady-state temperature is $u_\infty = 200/9$.

(d) We first note that

$$u(40,t) = \frac{200}{9} - \frac{160}{3\pi^2}\sum_{n=1}^{\infty}\frac{(-1)^n(3+\cos n\pi)}{n^2}e^{-n^2\pi^2 t/6400}.$$

For large values of t, an approximation is given by
$$u(40,t) \approx \frac{200}{9} + \frac{320}{3\pi^2} e^{-\pi^2 t/6400}.$$
Numerical investigation shows that $22.22 < u(40,t) < 23.22$ for $t \geq 1543$ sec.

16.(a) The heat conduction problem is formulated as
$$u_{xx} = u_t, \qquad 0 < x < 30, \ t > 0;$$
$$u(0,t) = 0, \qquad u_x(30,t) = 0, \ t > 0;$$
$$u(x,0) = f(x), \qquad 0 < x < 30,$$
in which the initial condition is given by $f(x) = 30 - x$. Based on the results of Problem 15, the solution is given by
$$u(x,t) = \sum_{n=1}^{\infty} c_n e^{-(2n-1)^2 \pi^2 t/3600} \sin \frac{(2n-1)\pi x}{60},$$
in which
$$c_n = \frac{2}{L} \int_0^L f(x) \sin \frac{(2n-1)\pi x}{60} dx = \frac{1}{15} \int_0^{30} (30-x) \sin \frac{(2n-1)\pi x}{60} dx =$$
$$= 120 \frac{2 \cos n\pi + (2n-1)\pi}{(2n-1)^2 \pi^2}.$$

Therefore the solution of the heat conduction problem is
$$u(x,t) = 120 \sum_{n=1}^{\infty} \frac{2 \cos n\pi + (2n-1)\pi}{(2n-1)^2 \pi^2} e^{-(2n-1)^2 \pi^2 t/3600} \sin \frac{(2n-1)\pi x}{60}.$$

(b) $t = 10, 20, 30, 40, 50, 60$ sec, then $t = 70, 80, 90, 100, 110, 120$ sec:

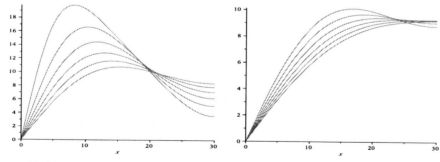

$x = 10, 20, 30$ cm:

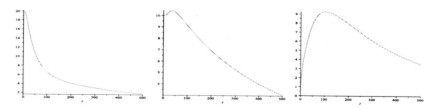

(c)

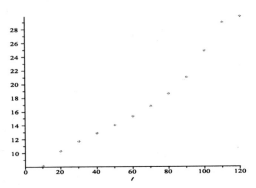

The location of x_h moves from $x = 0$ to $x = 30$.

(d)

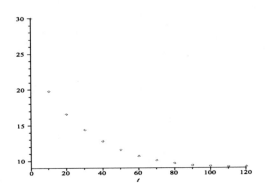

17.(a) The heat conduction problem is formulated as

$$u_{xx} = u_t, \quad 0 < x < 30, \ t > 0;$$
$$u(0,t) = 40, \quad u_x(30,t) = 0, \ t > 0;$$
$$u(x,0) = 30 - x, \quad 0 < x < 30,$$

The steady-state temperature satisfies the boundary value problem

$$v'' = 0, \ v(0) = 40 \ \text{and} \ v'(30) = 0.$$

It easy to see we must have $v(x) = 40$. Express the solution as

$$u(x,t) = 40 + w(x,t),$$

in which w satisfies the heat conduction problem

$$w_{xx} = w_t, \quad 0 < x < 30, \ t > 0;$$
$$w(0,t) = 0, \quad w_x(30,t) = 0, \ t > 0;$$
$$w(x,0) = -10 - x, \quad 0 < x < 30.$$

As shown in Problem 15, the solution is given by

$$w(x,t) = \sum_{n=1}^{\infty} c_n e^{-(2n-1)^2 \pi^2 t/3600} \sin \frac{(2n-1)\pi x}{60},$$

in which

$$c_n = \frac{2}{L} \int_0^L f(x) \sin \frac{(2n-1)\pi x}{60} dx = \frac{1}{15} \int_0^{30} (-10 - x) \sin \frac{(2n-1)\pi x}{60} dx =$$

$$= 40 \frac{6 \cos n\pi - (2n-1)\pi}{(2n-1)^2 \pi^2}.$$

Therefore the solution of the original heat conduction problem is

$$u(x,t) = 40 + 40 \sum_{n=1}^{\infty} \frac{6 \cos n\pi - (2n-1)\pi}{(2n-1)^2 \pi^2} e^{-(2n-1)^2 \pi^2 t/3600} \sin \frac{(2n-1)\pi x}{60}.$$

(b) $t = 20, 40, 60, 80, 100, 120$ sec, then $t = 140, 160, 180, 200, 220, 240$ sec:

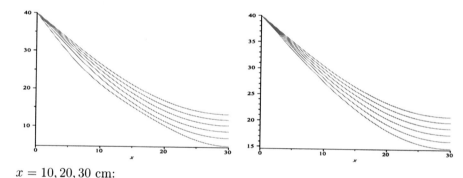

$x = 10, 20, 30$ cm:

(c) Observe the concavity of the curves. Note also that the temperature at the insulated end tends to the value of the fixed temperature at the boundary $x = 0$.

18. Setting $\lambda = \mu^2$, the general solution of the ODE $X'' + \mu^2 X = 0$ is

$$X(x) = k_1 e^{i\mu x} + k_2 e^{-i\mu x}.$$

The boundary conditions $y'(0) = y'(L) = 0$ lead to the system of equations

$$\mu k_1 - \mu k_2 = 0, \qquad \mu k_1 e^{i\mu L} - \mu k_2 e^{-i\mu L} = 0. \tag{$*$}$$

If $\mu = 0$, then the solution of the ODE is $X = Ax + B$. The boundary conditions require that $X = B$. If $\mu \neq 0$, then the system of algebraic equations has a nontrivial solution if and only if the coefficient matrix is singular. Set the determinant equal to zero to obtain

$$e^{-i\mu L} - e^{i\mu L} = 0.$$

Let $\mu = \nu + i\sigma$. Then $i\mu L = i\nu L - \sigma L$, and the previous equation can be written as

$$e^{\sigma L} e^{-i\nu L} - e^{-\sigma L} e^{i\nu L} = 0.$$

Using Euler's relation, $e^{i\nu L} = \cos \nu L + i \sin \nu L$, we obtain

$$e^{\sigma L}(\cos \nu L - i \sin \nu L) - e^{-\sigma L}(\cos \nu L + i \sin \nu L) = 0.$$

Equating the real and imaginary parts of the equation,

$$(e^{\sigma L} - e^{-\sigma L}) \cos \nu L = 0, \qquad (e^{\sigma L} + e^{-\sigma L}) \sin \nu L = 0.$$

Based on the second equation, $\nu L = n\pi$, $n \in \mathbb{I}$. Since $\cos nL \neq 0$, it follows that $e^{\sigma L} = e^{-\sigma L}$, or $e^{2\sigma L} = 1$. Hence $\sigma = 0$, and $\mu = n\pi/L$, $n \in \mathbb{I}$. Note that if $\sigma \neq 0$, then the last two equations have no solution. It follows that the system of equations (∗) has no nontrivial solutions.

20.(a) Consider solutions of the form $u(x,t) = X(x)T(t)$. Substitution into the partial differential equation results in

$$\alpha^2 X'' T = T'.$$

Divide both sides of the differential equation by the product XT to obtain

$$\frac{X''}{X} = \frac{T'}{\alpha^2 T}.$$

Since both sides of the resulting equation are functions of different variables, each must be equal to a constant, say $-\lambda$. We obtain the ordinary differential equations

$$X'' + \lambda X = 0 \quad \text{and} \quad T' + \lambda \alpha^2 T = 0.$$

Invoking the first boundary condition,

$$u(0,t) = X(0)T(t) = 0.$$

At the other boundary,

$$u_x(L,t) + \gamma u(L,t) = [X'(L) + \gamma X(L)] T(t) = 0.$$

Since these conditions are valid for all $t > 0$, it follows that

$$X(0) = 0 \quad \text{and} \quad X'(L) + \gamma X(L) = 0.$$

(b) We consider the boundary value problem

$$X'' + \lambda X = 0, \, 0 < x < L;$$
$$X(0) = 0, \, X'(L) + \gamma X(L) = 0.$$

(∗)

Assume that λ is real, with $\lambda = -\mu^2$. The general solution of the ODE is

$$X(x) = c_1 \cosh(\mu x) + c_2 \sinh(\mu x).$$

The first boundary condition requires that $c_1 = 0$. Imposing the second boundary condition,

$$c_2 \mu \cosh(\mu L) + \gamma c_2 \sinh(\mu L) = 0.$$

If $c_2 \neq 0$, then $\mu \cosh(\mu L) + \gamma \sinh(\mu L) = 0$, which can also be written as

$$(\mu + \gamma)e^{\mu L} - (\mu + \gamma)e^{-\mu L} = 0.$$

If $\gamma = -\mu$, then it follows that $\cosh(\mu L) = \sinh(\mu L)$, and hence $\mu = 0$. If $\gamma \neq -\mu$, then $e^{\mu L} = e^{-\mu L}$ again implies that $\mu = 0$. For the case $\mu = 0$, the general solution is $X(x) = Ax + B$. Imposing the boundary conditions, we have $B = 0$ and

$$A + \gamma AL = 0.$$

If $\gamma = -1/L$, then $X(x) = Ax$ is a solution of $(*)$. Otherwise $A = 0$.

(c) Let $\lambda = \mu^2$, with $\mu > 0$. The general solution of $(*)$ is $X(x) = c_1 \cos(\mu x) + c_2 \sin(\mu x)$ The first boundary condition requires that $c_1 = 0$. From the second boundary condition,

$$c_2 \mu \cos(\mu L) + \gamma c_2 \sin(\mu L) = 0.$$

For a nontrivial solution, we must have

$$\mu \cos(\mu L) + \gamma \sin(\mu L) = 0.$$

(d) The last equation can also be written as

$$\tan \mu L = -\frac{\mu}{\gamma}. \qquad (**)$$

The eigenvalues λ obtained from the solutions of $(**)$, which are infinite in number. In the graph below, we assume $\gamma L = 1$.

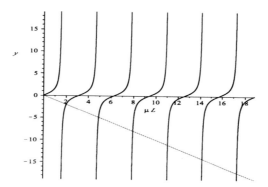

For $\gamma L = -1$:

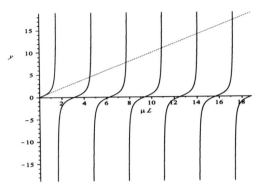

Denote the nonzero solutions of (**) by μ_1, μ_2, μ_3, ...

(e) We can in principle calculate the eigenvalues $\lambda_n = \mu_n^2$. Hence the associated eigenfunctions are $X_n = \sin \mu_n x$. Furthermore, the solutions of the temporal equations are $T_n = e^{-\alpha^2 \mu_n^2 t}$. The fundamental solutions of the heat conduction problem are given as

$$u_n(x,t) = e^{-\alpha^2 \mu_n^2 t} \sin \mu_n x,$$

which lead to the general solution

$$u(x,t) = \sum_{n=1}^{\infty} c_n e^{-\alpha^2 \mu_n^2 t} \sin \mu_n x.$$

23.(a) The heat conduction problem is formulated as

$$u_t = u_{xx} + s(x), \qquad 0 < x < L, \; t > 0;$$
$$u(0,t) = T_1, \qquad u(L,t) = T_2, \; t > 0;$$
$$u(x,0) = f(x), \qquad 0 < x < L.$$

Express the solution as $u(x,t) = v(x) + w(x,t)$, where $v(x)$ is the solution of the boundary value problem

$$v'' + s(x) = 0, \; v(0) = T_1, \; v(L) = T_2.$$

and w satisfies the heat conduction problem

$$w_t = w_{xx}, \qquad 0 < x < L, \; t > 0;$$
$$w(0,t) = 0, \qquad w(L,t) = 0, \; t > 0;$$
$$w(x,0) = f(x) - v(x), \qquad 0 < x < L.$$

Given that $s(x) = kx/L$, the general solution of the BVP is

$$v(x) = -\frac{k x^3}{6 L} + c_1 x + c_0.$$

Imposing the boundary conditions,

$$v(x) = -\frac{k x^3}{6 L} + \frac{(kL^2 - 6T_1 + 6T_2)x}{6 L} + T_1.$$

(b) As shown in Section 10.5,

$$w(x,t) = \sum_{n=1}^{\infty} c_n e^{-n^2\pi^2 t/400} \sin\frac{n\pi x}{20},$$

in which the coefficients c_n are the Fourier sine coefficients of $f(x) - v(x)$. That is,

$$c_n = \frac{2}{L}\int_0^L [f(x)-v(x)]\sin\frac{n\pi x}{L}dx = \frac{1}{10}\int_0^{20}\left[\frac{x^3}{240} - \frac{8x}{3} - 10\right]\sin\frac{n\pi x}{20}dx =$$

$$= 20\,\frac{3n^3\pi^3 \cos(n\pi) + 20 n\pi \cos(n\pi) - n^3\pi^3}{n^4\pi^4}.$$

Therefore the temperature distribution in the rod is

$$u(x,t) = v(x) + \frac{20}{\pi^4}\sum_{n=1}^{\infty}\frac{3n^3\pi^3 \cos(n\pi) + 20 n\pi \cos(n\pi) - n^3\pi^3}{n^4}e^{-n^2\pi^2 t/400}\sin\frac{n\pi x}{20}.$$

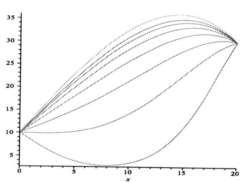

10.7

2.(a) The initial velocity is zero. Therefore the solution, as given by Eq.(20), is

$$u(x,t) = \sum_{n=1}^{\infty} c_n \sin\frac{n\pi x}{L}\cos\frac{n\pi a t}{L},$$

in which the coefficients are the Fourier sine coefficients of $f(x)$. That is,

$$c_n = \frac{2}{L}\int_0^L f(x)\sin\frac{n\pi x}{L}dx =$$

$$= \frac{2}{L}\left[\int_0^{L/4}\frac{4x}{L}\sin\frac{n\pi x}{L}dx + \int_{L/4}^{3L/4}\sin\frac{n\pi x}{L}dx + \int_{3L/4}^{L}\frac{4L-4x}{L}\sin\frac{n\pi x}{L}dx\right] =$$

$$= 8\,\frac{\sin n\pi/4 + \sin 3n\pi/4}{n^2\pi^2}.$$

Therefore the displacement of the string is given by
$$u(x,t) = \frac{8}{\pi^2} \sum_{n=1}^{\infty} \frac{\sin\frac{n\pi}{4} + \sin\frac{3n\pi}{4}}{n^2} \sin\frac{n\pi x}{L} \cos\frac{n\pi a t}{L}.$$

(b) With $a = 1$ and $L = 10$,
$$u(x,t) = \frac{8}{\pi^2} \sum_{n=1}^{\infty} \frac{\sin\frac{n\pi}{4} + \sin\frac{3n\pi}{4}}{n^2} \sin\frac{n\pi x}{10} \cos\frac{n\pi t}{10}.$$

For $t = 5, 10, 15, 20$:

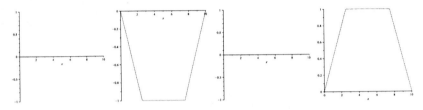

(c) For $x = 2.5, 5, 7.5$:

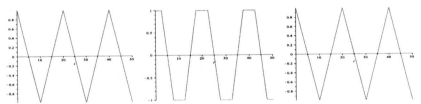

(d)

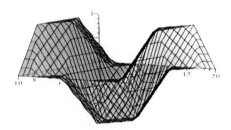

3.(a) The initial velocity is zero. As given by Eq.(20), the solution is
$$u(x,t) = \sum_{n=1}^{\infty} c_n \sin\frac{n\pi x}{L} \cos\frac{n\pi a t}{L},$$
in which the coefficients are the Fourier sine coefficients of $f(x)$. That is,
$$c_n = \frac{2}{L} \int_0^L f(x) \sin\frac{n\pi x}{L} dx = \frac{2}{L} \int_0^L \frac{8x(L-x)^2}{L^3} \sin\frac{n\pi x}{L} dx =$$

$$= 32 \frac{2 + \cos n\pi}{n^3 \pi^3}.$$

Therefore the displacement of the string is given by

$$u(x,t) = \frac{32}{\pi^3} \sum_{n=1}^{\infty} \frac{2 + \cos n\pi}{n^3} \sin \frac{n\pi x}{L} \cos \frac{n\pi a t}{L}.$$

(b) With $a = 1$ and $L = 10$,

$$u(x,t) = \frac{32}{\pi^3} \sum_{n=1}^{\infty} \frac{2 + \cos n\pi}{n^3} \sin \frac{n\pi x}{10} \cos \frac{n\pi t}{10}.$$

For $t = 5, 10, 15, 20$:

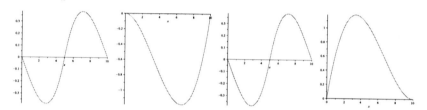

(c) For $x = 2.5, 5, 7.5$:

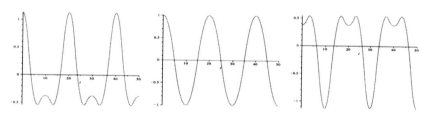

(d)

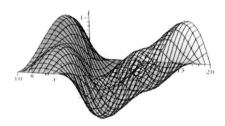

4.(a) As given by Eq.(20), the solution is

$$u(x,t) = \sum_{n=1}^{\infty} c_n \sin \frac{n\pi x}{L} \cos \frac{n\pi a t}{L},$$

in which the coefficients are the Fourier sine coefficients of $f(x)$. That is,

$$c_n = \frac{2}{L}\int_0^L f(x)\sin\frac{n\pi x}{L}dx = \frac{2}{L}\int_{L/2-1}^{L/2+1} \sin\frac{n\pi x}{L}dx = 4\,\frac{\sin\frac{n\pi}{2}\sin\frac{n\pi}{L}}{n\pi}.$$

Therefore the displacement of the string is given by

$$u(x,t) = \frac{4}{\pi}\sum_{n=1}^{\infty}\frac{1}{n}\left[\sin\frac{n\pi}{2}\sin\frac{n\pi}{L}\right]\sin\frac{n\pi x}{L}\cos\frac{n\pi a t}{L}.$$

(b) With $a = 1$ and $L = 10$,

$$u(x,t) = \frac{4}{\pi}\sum_{n=1}^{\infty}\frac{1}{n}\left[\sin\frac{n\pi}{2}\sin\frac{n\pi}{10}\right]\sin\frac{n\pi x}{10}\cos\frac{n\pi t}{10}.$$

For $t = 0, 2.5, 5, 7.5, 10, 12.5$:

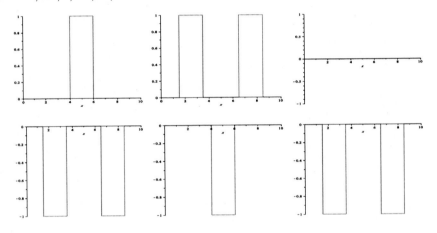

(c) For $x = 2.5, 5, 7.5$:

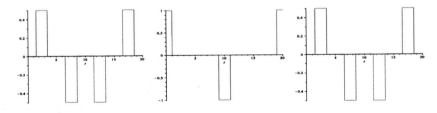

(d)

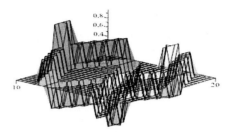

5.(a) The initial displacement is zero. Therefore the solution, as given by Eq.(34), is

$$u(x,t) = \sum_{n=1}^{\infty} k_n \sin\frac{n\pi x}{L} \sin\frac{n\pi a t}{L},$$

in which the coefficients are the Fourier sine coefficients of $u_t(x,0) = g(x)$. It follows that

$$k_n = \frac{2}{n\pi a}\int_0^L g(x)\sin\frac{n\pi x}{L}dx =$$

$$= \frac{2}{n\pi a}\left[\int_0^{L/2}\frac{2x}{L}\sin\frac{n\pi x}{L}dx + \int_{L/2}^L\frac{2(L-x)}{L}\sin\frac{n\pi x}{L}dx\right] =$$

$$= 8L\frac{\sin n\pi/2}{n^3\pi^3 a}.$$

Therefore the displacement of the string is given by

$$u(x,t) = \frac{8L}{a\pi^3}\sum_{n=1}^{\infty}\frac{1}{n^3}\sin\frac{n\pi}{2}\sin\frac{n\pi x}{L}\sin\frac{n\pi a t}{L}.$$

(b) With $a=1$ and $L=10$,

$$u(x,t) = \frac{80}{\pi^3}\sum_{n=1}^{\infty}\frac{1}{n^3}\sin\frac{n\pi}{2}\sin\frac{n\pi x}{10}\sin\frac{n\pi t}{10}.$$

For $t = 5, 10, 15, 20$:

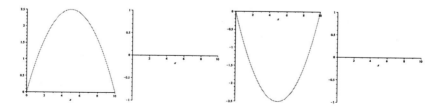

(c) For $x = 2.5, 5, 7.5$:

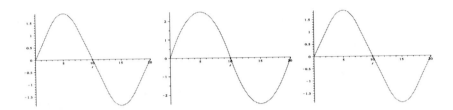

(d)

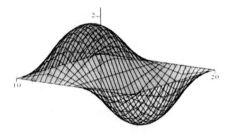

7.(a) The initial displacement is zero. As given by Eq.(34), the solution is

$$u(x,t) = \sum_{n=1}^{\infty} k_n \sin\frac{n\pi x}{L} \sin\frac{n\pi a t}{L},$$

in which the coefficients are the Fourier sine coefficients of $u_t(x,0) = g(x)$. It follows that

$$k_n = \frac{2}{n\pi a}\int_0^L g(x)\sin\frac{n\pi x}{L}dx = \frac{2}{n\pi a}\int_0^L \frac{8x(L-x)^2}{L^3}\sin\frac{n\pi x}{L}dx =$$

$$= 32L\,\frac{2+\cos n\pi}{n^4\pi^4 a}.$$

Therefore the displacement of the string is given by

$$u(x,t) = \frac{32L}{a\pi^4}\sum_{n=1}^{\infty}\frac{2+\cos n\pi}{n^4}\sin\frac{n\pi x}{L}\sin\frac{n\pi a t}{L}.$$

(b) With $a = 1$ and $L = 10$,

$$u(x,t) = \frac{320}{\pi^4}\sum_{n=1}^{\infty}\frac{2+\cos n\pi}{n^4}\sin\frac{n\pi x}{10}\sin\frac{n\pi t}{10}.$$

For $t = 5, 10, 15, 20$:

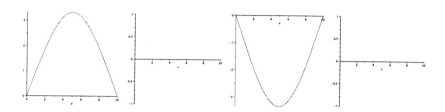

(c) For $x = 2.5, 5, 7.5$:

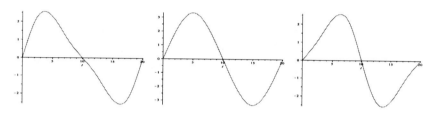

(d)

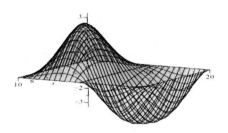

8.(a) As given by Eq.(34), the solution is

$$u(x,t) = \sum_{n=1}^{\infty} k_n \sin \frac{n\pi x}{L} \sin \frac{n\pi a t}{L},$$

in which the coefficients are the Fourier sine coefficients of $u_t(x,0) = g(x)$. It follows that

$$k_n = \frac{2}{n\pi a} \int_0^L g(x) \sin \frac{n\pi x}{L} dx = \frac{2}{n\pi a} \int_{L/2-1}^{L/2+1} \sin \frac{n\pi x}{L} dx = 4L \frac{\sin \frac{n\pi}{2} \sin \frac{n\pi}{L}}{n^2 \pi^2 a}.$$

Therefore the displacement of the string is given by

$$u(x,t) = \frac{4L}{a\pi^2} \sum_{n=1}^{\infty} \frac{1}{n^2} \left[\sin \frac{n\pi}{2} \sin \frac{n\pi}{L} \right] \sin \frac{n\pi x}{L} \sin \frac{n\pi a t}{L}.$$

(b) With $a = 1$ and $L = 10$,

$$u(x,t) = \frac{40}{\pi^2} \sum_{n=1}^{\infty} \frac{1}{n^2} \left[\sin \frac{n\pi}{2} \sin \frac{n\pi}{10} \right] \sin \frac{n\pi x}{10} \sin \frac{n\pi t}{10}.$$

For $t = 5, 10, 15, 20$:

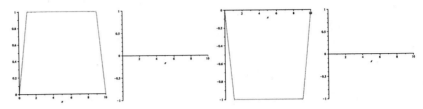

(c) For $x = 2.5, 5, 7.5$:

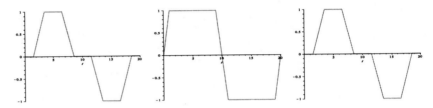

(d)

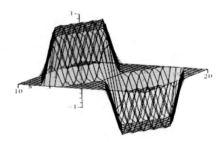

11.(a) As shown in Problem 9, the solution is

$$u(x,t) = \sum_{n=1}^{\infty} c_n \sin \frac{(2n-1)\pi x}{2L} \cos \frac{(2n-1)\pi a t}{2L},$$

in which the coefficients are the Fourier sine coefficients of $f(x)$. It follows that

$$c_n = \frac{2}{L}\int_0^L f(x) \sin \frac{(2n-1)\pi x}{2L} dx = \frac{2}{L}\int_0^L \frac{8x(L-x)^2}{L^3} \sin \frac{(2n-1)\pi x}{2L} dx =$$

$$= 512 \frac{3\cos n\pi + (2n-1)\pi}{(2n-1)^4 \pi^4}.$$

Therefore the displacement of the string is given by

$$u(x,t) = \frac{512}{\pi^4} \sum_{n=1}^{\infty} \frac{3\cos n\pi + (2n-1)\pi}{(2n-1)^4} \sin \frac{(2n-1)\pi x}{2L} \cos \frac{(2n-1)\pi a t}{2L}.$$

Note that the period is $T = 4L/a$.

(b) With $a = 1$ and $L = 10$,

$$u(x,t) = \frac{512}{\pi^4} \sum_{n=1}^{\infty} \frac{3\cos n\pi + (2n-1)\pi}{(2n-1)^4} \sin \frac{(2n-1)\pi x}{20} \cos \frac{(2n-1)\pi t}{20}.$$

For $t = 5, 10, 15, 20$:

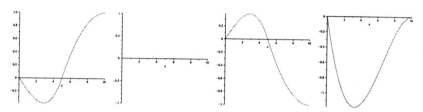

(c) For $x = 2.5, 5, 7.5$:

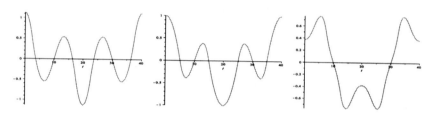

(d)

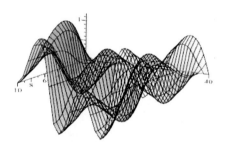

12.(a) The wave equation is given by

$$a^2 \frac{\partial^2 u}{\partial x^2} = \frac{\partial^2 u}{\partial t^2}.$$

Setting $s = x/L$, we have

$$\frac{\partial u}{\partial x} = \frac{\partial u}{\partial s}\frac{ds}{dx} = \frac{1}{L}\frac{\partial u}{\partial s}.$$

It follows that

$$\frac{\partial^2 u}{\partial x^2} = \frac{1}{L^2}\frac{\partial^2 u}{\partial s^2}.$$

(b) Likewise, with $\tau = at/L$,
$$\frac{\partial u}{\partial t} = \frac{a}{L}\frac{\partial u}{\partial \tau} \quad \text{and} \quad \frac{\partial^2 u}{\partial t^2} = \frac{a^2}{L^2}\frac{\partial^2 u}{\partial \tau^2}.$$
Substitution into the original equation results in
$$\frac{\partial^2 u}{\partial s^2} = \frac{\partial^2 u}{\partial \tau^2}.$$

15. The given specifications are $L = 5\,\text{ft}$, $T = 50\,\text{lb}$, and weight per unit length $\gamma = 0.026\,\text{lb/ft}$. It follows that $\rho = \gamma/32.2 = 80.75 \times 10^{-5}\,\text{slugs/ft}$.

(a) The transverse waves propagate with a speed of $a = \sqrt{T/\rho} = 248\,\text{ft/sec}$.

(b) The natural frequencies are $\omega_n = n\pi a/L = 49.8\,\pi n\,\text{rad/sec}$.

(c) The new wave speed is $a = \sqrt{(T+\Delta T)/\rho}$. For a string with fixed ends, the natural modes are proportional to the functions
$$M_n(x) = \sin\frac{n\pi x}{L},$$
which are independent of a.

18. The solution of the wave equation
$$a^2 v_{xx} = v_{tt}$$
in an infinite one-dimensional medium subject to the initial conditions
$$v(x,0) = f(x),\ v_t(x,0) = 0,\ -\infty < x < \infty$$
is given by
$$v(x,t) = \frac{1}{2}\left[f(x-at) + f(x+at)\right].$$
The solution of the wave equation
$$a^2 w_{xx} = w_{tt},$$
on the same domain, subject to the initial conditions
$$w(x,0) = 0,\quad w_t(x,0) = g(x),\quad -\infty < x < \infty$$
is given by
$$w(x,t) = \frac{1}{2a}\int_{x-at}^{x+at} g(\xi)d\xi.$$
Let $u(x,t) = v(x,t) + w(x,t)$. Since the PDE is linear, it is easy to see that $u(x,t)$ is a solution of the wave equation $a^2 u_{xx} = u_{tt}$. Furthermore, we have
$$u(x,0) = v(x,0) + w(x,0) = f(x)$$
and
$$u_t(x,0) = v_t(x,0) + w_t(x,0) = g(x).$$
Hence $u(x,t)$ is a solution of the general wave propagation problem.

19. The solution of the specified wave propagation problem is
$$u(x,t) = \sum_{n=1}^{\infty} c_n \sin\frac{n\pi x}{L} \cos\frac{n\pi a t}{L}.$$

Using the given trigonometric identity,
$$\sin\frac{n\pi x}{L}\cos\frac{n\pi a t}{L} = \frac{1}{2}\left[\sin(\frac{n\pi x}{L}+\frac{n\pi a t}{L})+\sin(\frac{n\pi x}{L}-\frac{n\pi a t}{L})\right] =$$
$$= \frac{1}{2}\left[\sin\frac{n\pi}{L}(x+at)+\sin\frac{n\pi}{L}(x-at)\right].$$

We can therefore also write the solution as
$$u(x,t) = \frac{1}{2}\sum_{n=1}^{\infty} c_n \left[\sin\frac{n\pi}{L}(x+at)+\sin\frac{n\pi}{L}(x-at)\right].$$

Assuming that the series can be split up,
$$u(x,t) = \frac{1}{2}\left[\sum_{n=1}^{\infty} c_n \sin\frac{n\pi}{L}(x-at)+\sum_{n=1}^{\infty} c_n \sin\frac{n\pi}{L}(x+at)\right].$$

Comparing the solution to the one given by Eq.(28), we can infer that
$$h(x) = \sum_{n=1}^{\infty} c_n \sin\frac{n\pi x}{L}.$$

20. Let $h(\xi)$ be a $2L$-periodic function defined by
$$h(\xi) = \begin{cases} f(\xi), & 0 \le \xi \le L; \\ -f(-\xi), & -L \le \xi \le 0. \end{cases}$$

Set $u(x,t) = \frac{1}{2}[h(x-at)+h(x+at)]$. Assuming the appropriate differentiability conditions on h,
$$\frac{\partial u}{\partial x} = \frac{1}{2}[h'(x-at)+h'(x+at)]$$

and
$$\frac{\partial^2 u}{\partial x^2} = \frac{1}{2}[h''(x-at)+h''(x+at)].$$

Likewise,
$$\frac{\partial^2 u}{\partial t^2} = \frac{a^2}{2}[h''(x-at)+h''(x+at)].$$

It follows immediately that
$$a^2\frac{\partial^2 u}{\partial x^2} = \frac{\partial^2 u}{\partial t^2}.$$

Let $t \ge 0$. Checking the first boundary condition,
$$u(0,t) = \frac{1}{2}[h(-at)+h(at)] = \frac{1}{2}[-h(at)+h(at)] = 0.$$

Checking the other boundary condition,
$$u(L,t) = \frac{1}{2}[h(L-at) + h(L+at)] = \frac{1}{2}[-h(at-L) + h(at+L)].$$

Since h is $2L$-periodic, $h(at-L) = h(at-L+2L)$. Therefore $u(L,t) = 0$. Furthermore, for $0 \le x \le L$,
$$u(x,0) = \frac{1}{2}[h(x) + h(x)] = h(x) = f(x).$$

Hence $u(x,t)$ is a solution of the problem.

22. Assuming that we can differentiate term-by-term,
$$\frac{\partial u}{\partial t} = -\pi a \sum_{n=1}^{\infty} \frac{c_n n}{L} \sin \frac{n\pi x}{L} \sin \frac{n\pi a t}{L}$$

and
$$\frac{\partial u}{\partial x} = \pi \sum_{n=1}^{\infty} \frac{c_n n}{L} \cos \frac{n\pi x}{L} \cos \frac{n\pi a t}{L}.$$

Formally,
$$(\frac{\partial u}{\partial t})^2 = \pi^2 a^2 \sum_{n=1}^{\infty} (\frac{c_n n}{L})^2 \sin^2 \frac{n\pi x}{L} \sin^2 \frac{n\pi a t}{L} + \pi^2 a^2 \sum_{n \ne m} F_{nm}(x,t)$$

and
$$(\frac{\partial u}{\partial x})^2 = \pi^2 \sum_{n=1}^{\infty} (\frac{c_n n}{L})^2 \cos^2 \frac{n\pi x}{L} \cos^2 \frac{n\pi a t}{L} + \pi^2 \sum_{n \ne m} G_{nm}(x,t),$$

in which $F_{nm}(x,t)$ and $G_{nm}(x,t)$ contain products of the natural modes and their derivatives. Based on the orthogonality of the natural modes,
$$\int_0^L (\frac{\partial u}{\partial t})^2 dx = \pi^2 a^2 \frac{L}{2} \sum_{n=1}^{\infty} (\frac{c_n n}{L})^2 \sin^2 \frac{n\pi a t}{L}$$

and
$$\int_0^L (\frac{\partial u}{\partial x})^2 dx = \pi^2 \frac{L}{2} \sum_{n=1}^{\infty} (\frac{c_n n}{L})^2 \cos^2 \frac{n\pi a t}{L}.$$

Recall that $a^2 = T/\rho$. It follows that
$$\int_0^L \left[\rho(\frac{\partial u}{\partial t})^2 + T(\frac{\partial u}{\partial x})^2\right] dx = \pi^2 \frac{TL}{2} \sum_{n=1}^{\infty} (\frac{c_n n}{L})^2 \sin^2 \frac{n\pi a t}{L} +$$
$$+ \pi^2 \frac{TL}{2} \sum_{n=1}^{\infty} (\frac{c_n n}{L})^2 \cos^2 \frac{n\pi a t}{L}.$$

Therefore,
$$\int_0^L \left[\frac{1}{2}\rho(\frac{\partial u}{\partial t})^2 + \frac{1}{2}T(\frac{\partial u}{\partial x})^2\right] dx = \pi^2 \frac{T}{4L} \sum_{n=1}^{\infty} n^2 c_n^2.$$

24.(a) Based on Problem 23, the coefficients c_n are the Fourier sine coefficients of $f(x)$. That is,

$$c_n = \frac{2}{L} \int_0^L f(x) \sin \frac{n\pi x}{L} dx = \frac{1}{5} \left[\int_4^5 (x-4) \sin \frac{n\pi x}{10} dx + \int_5^6 (6-x) \sin \frac{n\pi x}{10} dx \right] =$$

$$= 20 \frac{2\sin(n\pi/2) - \sin(2n\pi/5) - \sin(3n\pi/5)}{n^2 \pi^2}.$$

(b)

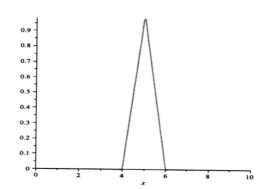

(c) $\gamma = 0$, $t = 60$:

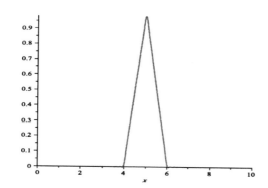

(d) $\gamma = 1/8$, $t = 20, 40, 60$:

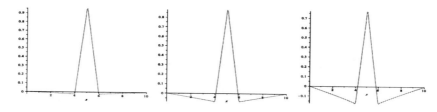

(e) $\gamma = 1/4$, $t = 20, 40, 60$:

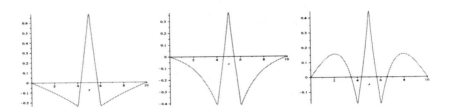

10.8

2. Using the method of separation of variables, write $u(x,y) = X(x)Y(y)$. We arrive at the ordinary differential equations

$$X'' + \lambda X = 0, \text{ with } X(0) = X(a) = 0;$$

and

$$Y'' - \lambda Y = 0, \text{ with } Y(b) = 0.$$

It follows that $\lambda_n = (n\pi/a)^2$ and $X_n(x) = \sin(n\pi x/a)$. Write the solution of the other ODE as

$$Y(y) = d_1 \cosh \lambda(b-y) + d_2 \sinh \lambda(b-y).$$

Imposing the boundary condition, we have $d_1 = 0$. Therefore the fundamental solutions are given by

$$u_n(x,y) = \sin \frac{n\pi x}{a} \sinh \lambda_n(b-y),$$

and the general solution is

$$u(x,y) = \sum_{n=1}^{\infty} c_n \sin \frac{n\pi x}{a} \sinh \frac{n\pi(b-y)}{a}.$$

Based on the boundary condition,

$$h(x) = \sum_{n=1}^{\infty} c_n \sin \frac{n\pi x}{a} \sinh \frac{n\pi b}{a}.$$

The coefficients are calculated using the equation

$$c_n \sinh \frac{n\pi b}{a} = \frac{2}{a} \int_0^a h(x) \sin \frac{n\pi x}{a} dx.$$

6.(a) The method of separation of variables considers solutions of the form

$$u(r,\theta) = R(r)\Theta(\theta).$$

We arrive at the ordinary differential equations

$$r^2 R'' + rR' - \lambda R = 0 \text{ and } \Theta'' + \lambda\Theta = 0.$$

The general solution of the second equation is $\Theta(\theta) = c_1 \cos \lambda^{1/2}\theta + c_2 \sin \lambda^{1/2}\theta$. The associated boundary conditions are $\Theta(0) = \Theta(\pi) = 0$. It follows that $c_1 = 0$

and $\lambda^{1/2} = n$, $n = 1, 2, \ldots$. Hence the eigenvalues are $\lambda_n = n^2$, with corresponding eigenfunctions $\Theta_n(\theta) = \sin n\theta$. Substituting for λ in the radial equation, $r^2 R'' + rR' - n^2 R = 0$. The general solution is $R(r) = k_1 r^{-n} + k_2 r^n$. Using the boundedness condition, we require that $k_1 = 0$. We find that the fundamental solutions are given by $u_n(r, \theta) = r^n \sin n\theta$, and therefore

$$u(r, \theta) = \sum_{n=1}^{\infty} A_n r^n \sin n\theta.$$

From the boundary condition on the semicircle $r = a$, $0 \leq \theta \leq \pi$,

$$f(\theta) = \sum_{n=1}^{\infty} A_n a^n \sin n\theta.$$

The series representation is a Fourier sine series for $f(\theta)$. The coefficients are given by

$$A_n a^n = \frac{2}{\pi} \int_0^{\pi} f(\theta) \sin n\theta \, d\theta.$$

(b) The Fourier sine coefficients are

$$A_n = \frac{2}{a^n \pi} \int_0^{\pi} f(\theta) \sin n\theta \, d\theta = \frac{2}{a^n \pi} \int_0^{\pi} \theta(\pi - \theta) \sin n\theta \, d\theta = 4 \frac{1 - \cos n\pi}{a^n n^3 \pi}.$$

Hence the solution is

$$u(r, \theta) = \frac{4}{\pi} \sum_{n=1}^{\infty} \frac{1 - \cos n\pi}{a^n n^3} r^n \sin n\theta.$$

(c) Set $a = 2$.
$\theta = \pi/4, \pi/3, \pi/2$:

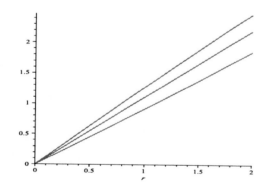

$r = 1/2,\ 1,\ 3/2,\ 2$:

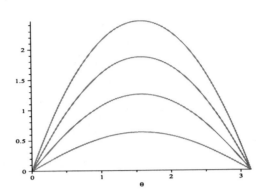

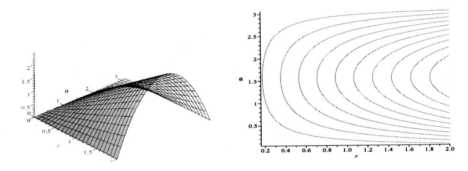

9. Consider the equation $\Theta'' + \lambda \Theta = 0$. Let $\lambda = -\mu^2$, where $\mu = \nu + i\sigma$. The ODE can then be written as

$$\Theta'' - (\nu + i\sigma)^2 \Theta = 0.$$

The general solution is

$$\Theta(\theta) = c_1 e^{(\nu + i\sigma)t} + c_2 e^{-(\nu + i\sigma)t} = c_1 e^{\nu t}[\cos \sigma t + i \sin \sigma t] + c_2 e^{-\nu t}[\cos \sigma t - i \sin \sigma t].$$

Collecting the real and imaginary parts,

$$\Theta(\theta) = (c_1 e^{\nu t} + c_2 e^{-\nu t}) \cos \sigma t + i(c_1 e^{\nu t} - c_2 e^{-\nu t}) \sin \sigma t.$$

For $\nu \neq 0$, the functions $e^{\nu t}$ and $e^{-\nu t}$ are linearly independent. If the coefficients are nonzero, then the real and imaginary parts of $\Theta(\theta)$ are not periodic. Hence Eq. (23) has periodic solutions only if $\lambda = -(i\sigma)^2 = \sigma^2$, with $\sigma > 0$.

10.(a) Applying the method of separation of variables, set $u(x, y) = X(x)Y(y)$. The resulting ordinary differential equations are

$$X'' - \lambda X = 0 \text{ and } Y'' + \lambda Y = 0.$$

Based on the given boundary conditions, we also have

$$X'(0) = 0 \text{ and } Y'(0) = Y'(b) = 0.$$

The general solution of the equation for Y is

$$Y(y) = c_1 \cos \lambda^{1/2} y + c_2 \sin \lambda^{1/2} y,$$

with $Y'(y) = -c_1 \lambda^{1/2} \sin \lambda^{1/2} y + c_2 \lambda^{1/2} \cos \lambda^{1/2} y$. Imposing the boundary conditions, it is necessary that $c_2 = 0$ and $\lambda^{1/2} = 0$ or $\lambda^{1/2} b = n\pi$, $n = 1, 2, \ldots$. Therefore the eigenvalues are $\lambda_n = n^2 \pi^2 / b^2$, $n = 0, 1, 2, \ldots$ and the corresponding eigenfunctions are $Y_n = \cos(n\pi y / b)$, $n = 0, 1, 2, \ldots$. The solution of the equation for X becomes $X(x) = d_1 \cosh(n\pi x / b) + d_2 \sinh(n\pi x / b)$, with

$$X'(x) = d_1 \frac{n\pi}{b} \sinh \frac{n\pi x}{b} + d_2 \frac{n\pi}{b} \cosh \frac{n\pi x}{b}.$$

Invoking the boundary condition, it follows that $X(x) = d_1 \cosh(n\pi x / b)$. Therefore the fundamental solutions of the Neumann problem are

$$u_n(x, y) = \cosh \frac{n\pi x}{b} \cos \frac{n\pi y}{b}, \quad n = 0, 1, 2, \ldots.$$

The general solution is given by

$$u(x, y) = \frac{a_0}{2} + \sum_{n=1}^{\infty} a_n \cosh \frac{n\pi x}{b} \cos \frac{n\pi y}{b}.$$

(b) Differentiating the series term-by-term,

$$u_x(x, y) = \sum_{n=1}^{\infty} a_n \frac{n\pi}{b} \sinh \frac{n\pi x}{b} \cos \frac{n\pi y}{b}.$$

In order to satisfy the nonhomogeneous boundary condition, we must have

$$f(y) = \sum_{n=1}^{\infty} a_n \frac{n\pi}{b} \sinh \frac{n\pi a}{b} \cos \frac{n\pi y}{b}. \tag{$*$}$$

For the series representation to be valid, it is necessary that $f(y)$ be expressible as a Fourier series of period $2b$, and

$$\int_0^b f(y) dy = 0,$$

since there is no constant term in the series $(*)$. Under these conditions, the coefficients in $(*)$ are

$$a_n = \frac{2}{n\pi} \left[\sinh \frac{n\pi a}{b} \right]^{-1} \int_0^b f(y) \cos \frac{n\pi y}{b} dy.$$

The constant a_0 remains arbitrary.

11. The method of separation of variables considers solutions of the form

$$u(r, \theta) = R(r) \Theta(\theta).$$

The resulting ordinary differential equations are

$$r^2 R'' + r R' - \lambda R = 0 \quad \text{and} \quad \Theta'' + \lambda \Theta = 0.$$

The general solution of the second equation is $\Theta(\theta) = c_1 \cos \lambda^{1/2}\theta + c_2 \sin \lambda^{1/2}\theta$. The domain of the given problem is the region bounded by $r = a$. A well-defined solution requires that $\Theta(\theta)$ be periodic with period 2π. It follows that $\lambda^{1/2} = n$, $n = 0, 1, 2, \ldots$. Hence the eigenvalues are $\lambda_n = n^2$, with corresponding eigenfunctions $\Theta_0^{(1)}(\theta) = 1$,

$$\Theta_n^{(1)}(\theta) = \cos n\theta \text{ and } \Theta_n^{(2)}(\theta) = \sin n\theta.$$

Substituting for λ_n in the R equation, $r^2 R'' + rR' - n^2 R = 0$, with general solution

$$R_n(r) = k_1 r^{-n} + k_2 r^n.$$

For a bounded solution, it is necessary that $k_1 = 0$. Therefore the fundamental solutions are $u_0(r, \theta) = 1$, along with

$$u_n(r, \theta) = r^n \cos n\theta \text{ and } v_n(r, \theta) = r^n \sin n\theta, \; n = 1, 2, \ldots.$$

The general solution of the Neumann problem has the form

$$u(r, \theta) = \frac{A_0}{2} + \sum_{n=1}^{\infty} r^n (A_n \cos n\theta + B_n \sin n\theta).$$

Differentiating the series term-by-term,

$$u_r(r, \theta) = \sum_{n=1}^{\infty} n r^{n-1} (A_n \cos n\theta + B_n \sin n\theta).$$

In order to satisfy the nonhomogeneous boundary condition, we must have

$$g(y) = \sum_{n=1}^{\infty} n a^{n-1} (A_n \cos n\theta + B_n \sin n\theta), \; 0 \leq \theta \leq 2\pi. \tag{*}$$

For the series representation to be valid, it is necessary that $g(y)$ be expressible as a Fourier series of period 2π, and

$$\int_0^{2\pi} g(y) dy = 0,$$

since there is no constant term in the series $(*)$. Under these conditions, the coefficients in $(*)$ are

$$A_n = \frac{1}{n\pi a^{n-1}} \int_0^{2\pi} g(y) \cos n\theta \, d\theta,$$

and

$$B_n = \frac{1}{n\pi a^{n-1}} \int_0^{2\pi} g(y) \sin n\theta \, d\theta.$$

The constant A_0 remains arbitrary.

12.(a) Applying the method of separation of variables, set $u(x, y) = X(x)Y(y)$. The resulting ordinary differential equations are

$$X'' + \lambda X = 0 \text{ and } Y'' - \lambda Y = 0.$$

Based on the given boundary conditions, we also have

$$X(0) = X(a) = 0 \text{ and } Y'(0) = 0.$$

The general solution of the equation for X is

$$X(x) = c_1 \cos \lambda^{1/2} x + c_2 \sin \lambda^{1/2} x,$$

Imposing the boundary conditions, we find that $c_1 = 0$ and $\lambda^{1/2} a = n\pi$, $n = 1, 2, \ldots$. Therefore the eigenvalues are $\lambda_n = n^2 \pi^2 / a^2$, and the corresponding eigenfunctions are $X_n = \sin(n\pi x/a)$, $n = 1, 2, \ldots$. The general solution of the equation for Y is

$$Y(y) = d_1 \cosh \lambda^{1/2} y + d_2 \sinh \lambda^{1/2} y,$$

with $Y'(y) = d_1 \lambda^{1/2} \sinh \lambda^{1/2} y + d_2 \lambda^{1/2} \cosh \lambda^{1/2} y$. Invoking the boundary condition, it is necessary that $d_2 = 0$. Therefore

$$Y_n(y) = \cosh \frac{n\pi y}{a}, \quad n = 1, 2, \ldots.$$

The system of fundamental solutions is given by

$$u_n(x, y) = \sin \frac{n\pi x}{a} \cosh \frac{n\pi y}{a},$$

so that the general solution may be written as

$$u(x, y) = \sum_{n=1}^{\infty} B_n \sin \frac{n\pi x}{a} \cosh \frac{n\pi y}{a}.$$

In order to satisfy the nonhomogeneous boundary condition, we must have

$$g(x) = \sum_{n=1}^{\infty} B_n \cosh \frac{n\pi b}{a} \sin \frac{n\pi x}{a}.$$

The coefficients in the general solution are the Fourier sine coefficients of $g(x)$. That is,

$$B_n = \frac{2}{a} \left[\cosh \frac{n\pi b}{a} \right]^{-1} \int_0^a g(x) \sin \frac{n\pi x}{a} dx.$$

(b) From part (a),

$$B_n = \frac{2}{a} \left[\cosh \frac{n\pi b}{a} \right]^{-1} \left[\int_0^{a/2} x \sin \frac{n\pi x}{a} dx + \int_{a/2}^a (a - x) \sin \frac{n\pi x}{a} dx \right] =$$

$$= 4a \frac{\sin \frac{n\pi}{2}}{n^2 \pi^2 \cosh(n\pi b/a)}.$$

Therefore the solution of the given problem is

$$u(x, y) = \frac{4a}{\pi^2} \sum_{n=1}^{\infty} \frac{\sin \frac{n\pi}{2}}{n^2 \cosh(n\pi b/a)} \sin \frac{n\pi x}{a} \cosh \frac{n\pi y}{a}.$$

(c) Setting $a = 3$ and $b = 1$,

$$u(x,y) = \frac{12}{\pi^2} \sum_{n=1}^{\infty} \frac{\sin \frac{n\pi}{2}}{n^2 \cosh(n\pi/3)} \sin \frac{n\pi x}{3} \cosh \frac{n\pi y}{3}.$$

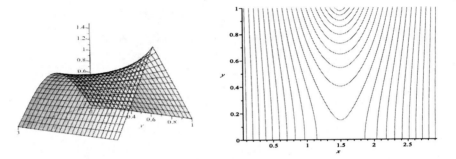

14.(a) Applying the method of separation of variables, set $u(x,y) = X(x)Y(y)$. The resulting ordinary differential equations are

$$X'' + \lambda X = 0 \text{ and } Y'' - \lambda Y = 0.$$

Based on the given boundary conditions, we also have

$$X'(0) = X'(a) = 0 \text{ and } Y(0) = 0.$$

The general solution of the equation for X is

$$X(x) = c_1 \cos \lambda^{1/2} x + c_2 \sin \lambda^{1/2} x,$$

with $X'(x) = -c_1 \lambda^{1/2} \sin \lambda^{1/2} x + c_2 \lambda^{1/2} \cos \lambda^{1/2} x$. Imposing the boundary conditions, it is necessary that $c_2 = 0$ and $\lambda^{1/2} = 0$ or $\lambda^{1/2} a = n\pi$, $n = 1, 2, \ldots$. Therefore the eigenvalues are $\lambda_n = n^2 \pi^2 / a^2$, $n = 0, 1, 2, \ldots$ and the corresponding eigenfunctions are $X_n = \cos(n\pi x/a)$, $n = 0, 1, 2, \ldots$. Substituting for λ_n in the equation for Y, the general solution is

$$Y(y) = d_1 \cosh(n\pi y/a) + d_2 \sinh(n\pi y/a).$$

Invoking the boundary condition, it follows that $Y(y) = d_2 \sinh(n\pi y/a)$. For the case $\lambda_0 = 0$, the solution for the Y equation is $Y(y) = Ay + B$. Based on the same boundary condition, $B = 0$, and $Y_0(y) = Ay$. Therefore the fundamental solutions are given by $u_0(x,y) = y$, along with

$$u_n(x,y) = \cos \frac{n\pi x}{a} \sinh \frac{n\pi y}{a}, \quad n = 1, 2, \ldots.$$

The general solution of the given problem can be written as

$$u(x,y) = \frac{A_0 y}{2} + \sum_{n=1}^{\infty} A_n \cos \frac{n\pi x}{a} \sinh \frac{n\pi y}{a}.$$

In order to satisfy the nonhomogeneous boundary condition, we must have

$$g(x) = \frac{A_0 b}{2} + \sum_{n=1}^{\infty} A_n \sinh \frac{n\pi b}{a} \cos \frac{n\pi x}{a}, \quad 0 \le x \le a.$$

The coefficients in the general solution are the Fourier cosine coefficients of $g(x)$. Hence

$$A_0 = \frac{2}{ab}\int_0^a g(x)dx,$$

and for $n \geq 1$,

$$A_n = \frac{2}{a}\left[\sinh\frac{n\pi b}{a}\right]^{-1}\int_0^a g(x)\cos\frac{n\pi x}{a}dx.$$

(b) For $g(x) = 1 + x^2(x-a)^2$,

$$A_0 = \frac{30a + a^5}{15ab},$$

and for $n \geq 1$,

$$A_n = -24a^4\frac{1+\cos n\pi}{n^4\pi^4\sinh(n\pi b/a)}.$$

The solution of the Laplace's equation is

$$u(x,y) = \frac{30a+a^5}{30ab}y - \frac{24a^4}{\pi^4}\sum_{n=1}^{\infty}\frac{1+\cos n\pi}{n^4\sinh(n\pi b/a)}\cos\frac{n\pi x}{a}\sinh\frac{n\pi y}{a}.$$

(c) Setting $a = 3$ and $b = 2$,

$$u(x,y) = \frac{37}{20}y - \frac{1944}{\pi^4}\sum_{n=1}^{\infty}\frac{1+\cos n\pi}{n^4\sinh(2n\pi/3)}\cos\frac{n\pi x}{3}\sinh\frac{n\pi y}{3}.$$

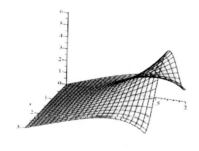
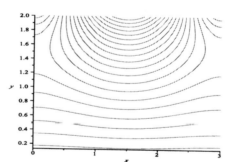

15. For an axially symmetric problem, Laplace's equation in cylindrical coordinates is

$$\frac{\partial^2 u}{\partial r^2} + \frac{1}{r}\frac{\partial u}{\partial r} + \frac{\partial^2 u}{\partial z^2} = 0.$$

Consider solutions of the form $u(r,z) = R(r)Z(z)$. Substitution into the PDE results in

$$R''Z + \frac{1}{r}R'Z + RZ'' = 0.$$

Now divide both sides of the equation by RZ to obtain

$$\frac{R''}{R} + \frac{1}{r}\frac{R'}{R} = -\frac{Z''}{Z}.$$

Since both sides of the resulting equation depend on different variables, each combination of expressions must be equal to a constant, say $-\lambda^2$. That is,

$$\frac{R''}{R} + \frac{1}{r}\frac{R'}{R} = -\frac{Z''}{Z} = -\lambda^2$$

Therefore we obtain two ordinary differential equations for R and Z:

$$rR'' + R' + \lambda^2 rR = 0 \quad \text{and} \quad Z'' - \lambda^2 Z = 0.$$

The first ODE is related to a Bessel's equation of order zero.

CHAPTER 11

Boundary Value Problems and Sturm-Liouville Theory

11.1

1. Since the right hand sides of the ODE and the boundary conditions are all zero, the boundary value problem is homogeneous.

3. The right hand side of the ODE is nonzero. Therefore the boundary value problem is nonhomogeneous.

6. The ODE can also be written as $y'' + \lambda(1 + x^2)y = 0$. Although the second boundary condition has a more general form, the boundary value problem is homogeneous.

7.(a) Suppose that $\lambda = -\mu^2$. In this case, the general solution of the ODE is

$$y(x) = c_1 \cosh \mu x + c_2 \sinh \mu x.$$

The first boundary condition requires that $c_1 = 0$. Imposing the second condition, $c_1(\cosh \mu\pi + \mu \sinh \mu\pi) + c_2(\sinh \mu\pi + \mu \cosh \mu\pi) = 0$. The two boundary conditions result in $c_2(\tanh \mu\pi + \mu) = 0$. Since the only solution of the equation $\tanh \mu\pi + \mu = 0$ is $\mu = 0$, we have $c_2 = 0$. Hence there are no nontrivial solutions. Let $\lambda = \mu^2$, with $\mu > 0$. Then the general solution of the ODE is $y(x) = c_1 \cos \mu x + c_2 \sin \mu x$. Imposing the boundary conditions, we obtain $c_1 = 0$ and $c_1(\cos \mu\pi - \mu \sin \mu\pi) + c_2(\sin \mu\pi + \mu \cos \mu\pi) = 0$. For a nontrivial solution of the ODE, we require that $\sin \mu\pi + \mu \cos \mu\pi = 0$. Note that $\cos \mu\pi = 0$ would imply that $\sin \mu\pi = 0$, which is false. It follows that $\tan \mu\pi = -\mu$, or $\tan \sqrt{\lambda}\,\pi =$

$-\sqrt{\lambda}$. Thus the eigenfunctions have the form $\phi_n(x) = \sin\sqrt{\lambda_n}\,x$, where λ_n satisfies $\tan\sqrt{\lambda}\,\pi = -\sqrt{\lambda}$.

(b) Assume that $\lambda = 0$. The general solution of the ODE is $y(x) = c_1 x + c_2$. The boundary condition at $x = 0$ requires that $c_2 = 0$. Imposing the second condition, $c_1(\pi + 1) + c_2 = 0$. It follows that $c_1 = c_2 = 0$. Hence there are no nontrivial solutions.

(c) From a plot of $f(\sqrt{\lambda}) = -\sqrt{\lambda}$ and $g(\sqrt{\lambda}) = \tan\sqrt{\lambda}\,\pi$,

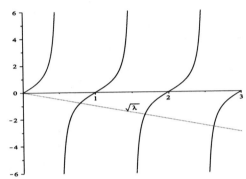

we find that there is a sequence of solutions beginning with $\sqrt{\lambda_1} \approx 0.7876$ and $\sqrt{\lambda_2} \approx 1.6716$, so $\lambda_1 \approx 0.6204$ and $\lambda_2 \approx 2.7943$.

(d) For large values of n, the graph of $g(x)$ may be approximated by its vertical asymptotes, which have the form $x = (2n-1)/2$. Solving $\sqrt{\lambda_n} = (2n-1)/2$ yields the estimate

$$\lambda_n \approx \frac{(2n-1)^2}{4}$$

for large values of n.

8.(a) Setting $\lambda = -\mu^2$, the general solution of the ODE is

$$y(x) = c_1 \cosh\mu x + c_2 \sinh\mu x.$$

The first boundary condition requires that $c_2 = 0$. Imposing the second condition,

$$c_1(\cosh\mu + \mu\sinh\mu) + c_2(\sinh\mu + \mu\cosh\mu) = 0.$$

The two boundary conditions result in $c_1(1 + \mu\tanh\mu) = 0$. Since $\mu\tanh\mu \geq 0$, it follows that $c_1 = 0$, and there are no nontrivial solutions. Let $\lambda = \mu^2$, with $\mu > 0$. Then the general solution of the ODE is

$$y(x) = c_1\cos\mu x + c_2\sin\mu x.$$

Imposing the boundary conditions, we obtain $c_2 = 0$ and

$$c_1(\cos\mu - \mu\sin\mu) + c_2(\sin\mu + \mu\cos\mu) = 0.$$

For a nontrivial solution of the ODE, we require that $\cos\mu - \mu\sin\mu = 0$. First note that $\cos\mu = 0$ would imply that either $\mu = 0$ or $\sin\mu = 0$. It follows that

$\cot\mu = \mu$, or $\cot\sqrt{\lambda} = \sqrt{\lambda}$. Thus the eigenfunctions have the form $\phi_n(x) = \cos\sqrt{\lambda_n}\,x$, where λ_n satisfies $\cot\sqrt{\lambda} = \sqrt{\lambda}$.

(b) With $\lambda = 0$, the general solution of the ODE is $y(x) = c_1 x + c_2$. Imposing the two boundary conditions, $c_1 = 0$ and $2c_1 + c_2 = 0$. It follows that $c_1 = c_2 = 0$. Hence there are no nontrivial solutions.

(c) From a plot of $f(\sqrt{\lambda}) = \sqrt{\lambda}$ and $g(\sqrt{\lambda}) = \cot\sqrt{\lambda}$,

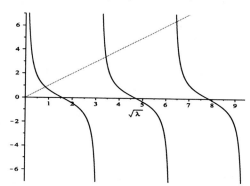

we find that there is a sequence of solutions beginning with $\sqrt{\lambda_1} \approx 0.8603$ and $\sqrt{\lambda_2} \approx 3.4526$, so $\lambda_1 \approx 0.7402$ and $\lambda_2 \approx 11.7349$.

(d) For large values of n, the graph of $g(x)$ may be approximated by its vertical asymptotes, which have the form $x = (n-1)\pi$. Solving $\sqrt{\lambda_n} = (n-1)\pi$ yields the estimate
$$\lambda_n \approx (n-1)^2 \pi^2$$
for large values of n.

12. First note that $P(x) = 1$, $Q(x) = -2x$ and $R(x) = \lambda$. Based on Problem 11, the integrating factor is a solution of the ODE
$$\mu'(x) = -2x\,\mu(x).$$
The differential equation is first order linear, with solution $\mu(x) = e^{-x^2}$. It then follows that the Hermite equation can be written as
$$\left[e^{-x^2} y'\right]' + \lambda e^{-x^2} y = 0.$$

14. For the Laguerre equation, $P(x) = x$, $Q(x) = 1-x$ and $R(x) = \lambda$. Using the result of Problem 11, the integrating factor is a solution of the ODE
$$x\,\mu'(x) = -x\,\mu(x).$$
A solution of $\mu'(x) = -\mu(x)$ is $\mu(x) = e^{-x}$. Therefore the Laguerre equation can be written as
$$\left[x e^{-x} y'\right]' + \lambda e^{-x} y = 0.$$

15. For the Chebyshev equation, $P(x) = 1 - x^2$, $Q(x) = -x$ and $R(x) = \alpha^2$. The integrating factor is a solution of the ODE

$$(1 - x^2)\mu'(x) = x\mu(x).$$

The differential equation is separable, with

$$\frac{d\mu}{\mu} = \frac{x}{1 - x^2}.$$

A solution of the resulting ODE is

$$\mu(x) = \frac{1}{\sqrt{|1 - x^2|}}.$$

Recall that the Chebyshev equation is typically defined for $|x| \leq 1$. Therefore it can also be written as

$$\left[\sqrt{1 - x^2}\, y'\right]' + \frac{\alpha^2}{\sqrt{1 - x^2}}\, y = 0.$$

16. We consider solutions of the form $u(x, t) = X(x)T(t)$. Substitution into the PDE results in

$$XT'' + c\, XT' + k\, XT = a^2 X''T.$$

Dividing both sides of the equation by XT, we obtain

$$\frac{XT''}{XT} + c\frac{XT'}{XT} + k = a^2 \frac{X''T}{XT},$$

that is,

$$\frac{T''}{T} + c\frac{T'}{T} = a^2 \frac{X''}{X} - k.$$

Since both sides of the resulting equation are functions of different variables, each must be equal to a constant, say $-\tilde{\lambda}$. Therefore we obtain two ordinary differential equations

$$a^2 X'' + (\tilde{\lambda} - k)X = 0 \quad \text{and} \quad T'' + cT' + \tilde{\lambda}T = 0.$$

We can also set $\lambda = (\tilde{\lambda} - k)/a^2$. The two ODEs can be written as

$$X'' + \lambda X = 0 \quad \text{and} \quad T'' + cT' + (k + \lambda a^2)T = 0.$$

17.(a) Setting $y = s(x)u$, we have $y' = s'u + su'$ and $y'' = s''u + 2s'u' + su''$. Substitution into the given ODE results in

$$s''u + 2s'u' + su'' - 2(s'u + su') + (1 + \lambda)su = 0.$$

Collecting the various terms,

$$su'' + (2s' - 2s)u' + [s'' - 2s' + (1 + \lambda)s]\, u = 0.$$

The second term on the left vanishes as long as $s' = s$, so we may take $s(x) = e^x$.

(b) With $s(x) = e^x$, the transformed differential equation can be written as

$$u'' + \lambda u = 0.$$

Since the boundary conditions are homogeneous, we also have $u(0) = u(1) = 0$. It now follows that the eigenfunctions are $u_n = \sin\sqrt{\lambda_n}\,x$, with corresponding eigenvalues $\lambda_n = n^2\pi^2$. Therefore the eigenfunctions for the original problem are $\phi_n(x) = e^x \sin n\pi x$.

(c) The given equation is a second order constant coefficient differential equation. The characteristic equation is

$$r^2 - 2r + (1+\lambda) = 0,$$

with roots $r_{1,2} = 1 \pm \sqrt{-\lambda}$. If $\lambda = 0$, then the general solution is $y = c_1 e^x + c_2 x e^x$. Imposing the two boundary conditions, we find that $c_1 = c_2 = 0$, and hence there are no nontrivial solutions. If $\lambda < 0$, then the general solution is

$$y = c_1 e^{1+\sqrt{-\lambda}\,x} + c_2 e^{1-\sqrt{-\lambda}\,x}.$$

It again follows that $c_1 = c_2 = 0$, and hence there are no nontrivial solutions. Therefore $\lambda > 0$, and the general solution is

$$y = c_1 e^x \cos\sqrt{\lambda}\,x + c_2 e^x \sin\sqrt{\lambda}\,x.$$

Invoking the boundary conditions, we have $c_1 = 0$ and $c_2 e \sin\sqrt{\lambda} = 0$. For a nontrivial solution, $\sqrt{\lambda} = n\pi$.

19. First write the differential equation as $y'' + (1+\lambda)y' + \lambda y = 0$, which is a second order constant coefficient differential equation. The characteristic equation is $r^2 + (1+\lambda)r + \lambda = 0$, with roots $r_1 = -1$ and $r_2 = -\lambda$. For $\lambda \neq 1$, the general solution is $y = c_1 e^{-x} + c_2 e^{-\lambda x}$. Imposing the boundary conditions, we need that $c_1 + c_2 = 0$ and $c_1 e^{-1} + c_2 e^{-\lambda} = 0$. For a nontrivial solution, it follows that $e^{-1} = e^{-\lambda}$, and hence $\lambda = 1$, which is contrary to the assumption. If $\lambda = 1$, then the general solution is $y = c_1 e^{-x} + c_2 x e^{-x}$. The boundary conditions require that $c_1 = 0$ and $c_1 e^{-1} + c_2 e^{-1} = 0$. Hence there are no nontrivial solutions.

21. Suppose that $\lambda = 0$. In that case the general solution is $y = c_1 x + c_2$. The boundary conditions require that $c_1 + 2c_2 = 0$ and $c_1 + c_2 = 0$. We find that $c_1 = c_2 = 0$, and hence there are no nontrivial solutions.

(a) Let $\lambda = \mu^2$, with $\mu > 0$. Then the general solution of the ODE is

$$y(x) = c_1 \cos\mu x + c_2 \sin\mu x.$$

The boundary conditions require that $2c_1 + \mu c_2 = 0$ and $c_1 \cos\mu + c_2 \sin\mu = 0$. These equations have a nonzero solution only if $2\sin\mu - \mu\cos\mu = 0$, which can also be written as $2\tan\mu - \mu = 0$.

(b)

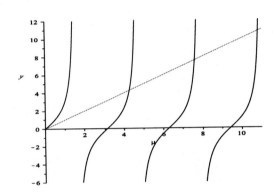

(c) Based on the graph, the positive roots of the determinantal equation are

$$\mu_1 \approx 4.2748, \ \mu_2 \approx 7.5965, \ldots \ ; \text{ for large } n, \ \mu_n \approx (2n+1)\frac{\pi}{2}.$$

Therefore the eigenvalues are

$$\lambda_1 \approx 18.2738, \ \lambda_2 \approx 57.7075, \ldots \ ; \text{ for large } n, \ \lambda_n \approx (2n+1)^2 \frac{\pi^2}{4}.$$

(d) Setting $\lambda = -\mu^2 < 0$, the general solution of the ODE is

$$y(x) = c_1 \cosh \mu x + c_2 \sinh \mu x.$$

Imposing the boundary conditions, we obtain the equations $2c_1 + \mu c_2 = 0$ and $c_1 \cosh \mu + c_2 \sinh \mu = 0$. These equations have a nonzero solution only if $2\sinh \mu - \mu \cosh \mu = 0$.

(e) The latter equation is satisfied only for $\mu = 0$ and $\mu = \pm 1.9150$. Hence the only negative eigenvalue is $\lambda_{-1} = -3.6673$.

24. Based on the physical problem, $\lambda = m\omega^2/EI > 0$. Let $\lambda = \mu^4$. The characteristic equation is $r^4 - \mu^4 = 0$, with roots $r_{1,2} = \pm \mu i$, $r_3 = -\mu$ and $r_4 = \mu$. Hence the general solution is

$$y(x) = c_1 \cosh \mu x + c_2 \sinh \mu x + c_3 \cos \mu x + c_4 \sin \mu x.$$

(a) Simply supported on both ends: $y(0) = y''(0) = 0$; $y(L) = y''(L) = 0$. Invoking the boundary conditions, we obtain the system of equations

$$c_1 + c_3 = 0$$
$$c_1 - c_3 = 0$$
$$c_1 \cosh \mu L + c_2 \sinh \mu L + c_3 \cos \mu L + c_4 \sin \mu L = 0$$
$$c_1 \mu^2 \cosh \mu L + c_2 \mu^2 \sinh \mu L - c_3 \mu^2 \cos \mu L - c_4 \mu^2 \sin \mu L = 0.$$

The determinantal equation is

$$\mu^4 \sinh \mu L \sin \mu L = 0.$$

The nonzero roots are $\mu_n = n\pi/L$, $n = 1, 2, \ldots$. The first two equations result in $c_1 = c_3 = 0$. The last two equations,

$$c_2 \sinh n\pi + c_4 \sin n\pi = 0$$
$$c_2 \sinh n\pi - c_4 \sin n\pi = 0,$$

imply that $c_2 = 0$. Therefore the eigenfunctions are $\phi_n = \sin \mu_n x$, with corresponding eigenvalues $\lambda_n = n^4\pi^4/L^4$. Thus $\lambda_1 \approx 97.409/L^4$ and $\lambda_2 \approx 1558.5/L^4$.

(b) Simply supported : $y(0) = y''(0) = 0$; clamped : $y(L) = y'(L) = 0$. Invoking the boundary conditions, we obtain the system of equations

$$c_1 + c_3 = 0$$
$$c_1 - c_3 = 0$$
$$c_1 \cosh \mu L + c_2 \sinh \mu L + c_3 \cos \mu L + c_4 \sin \mu L = 0$$
$$c_1 \mu \sinh \mu L + c_2 \mu \cosh \mu L - c_3 \mu \sin \mu L + c_4 \mu \cos \mu L = 0.$$

The determinantal equation is

$$2\mu^3 \sinh \mu L \cos \mu L - 2\mu^3 \cosh \mu L \sin \mu L = 0.$$

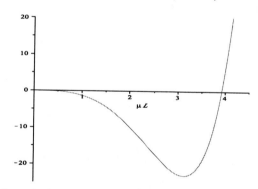

Based on numerical analysis, $\mu_1 \approx 3.9266/L$ and $\mu_2 \approx 7.0686/L$. The first two equations result in $c_1 = c_3 = 0$. The last two equations,

$$c_2 \sinh \mu_n L + c_4 \sin \mu_n L = 0$$
$$c_2 \cosh \mu_n L + c_4 \cos \mu_n L = 0,$$

imply that

$$c_2 = -\frac{\sin \mu_n L}{\sinh \mu_n L} c_4.$$

Therefore the eigenfunctions are

$$\phi_n = -\frac{\sin \mu_n L}{\sinh \mu_n L} \sinh \mu_n x + \sin \mu_n x,$$

with corresponding eigenvalues $\lambda_n = \mu_n^4$. Thus $\lambda_1 \approx 237.72/L^4$ and $\lambda_2 \approx 2496.5/L^4$.

(c) Clamped: $y(0) = y'(0) = 0$; free: $y''(L) = y'''(L) = 0$. Invoking the boundary conditions, we obtain the system of equations

$$c_1 + c_3 = 0$$
$$\mu c_2 + \mu c_4 = 0$$
$$c_1\mu^2 \cosh \mu L + c_2\mu^2 \sinh \mu L - c_3\mu^2 \cos \mu L - c_4\mu^2 \sin \mu L = 0$$
$$c_1\mu^3 \sinh \mu L + c_2\mu^3 \cosh \mu L + c_3\mu^3 \sin \mu L - c_4\mu^3 \cos \mu L = 0.$$

The determinantal equation is

$$1 + \cosh \mu L \cos \mu L = 0.$$

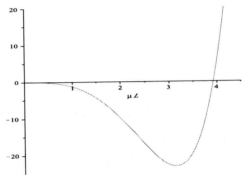

The first two nonzero roots are $\mu_1 \approx 1.8751/L$ and $\mu_2 \approx 4.6941/L$. With $c_3 = -c_1$ and $c_4 = -c_2$, the system of equations reduce to

$$c_1(\cosh \mu_n L + \cos \mu_n L) + c_2(\sinh \mu_n L + \sin \mu_n L) = 0$$
$$c_1(\sinh \mu_n L - \sin \mu_n L) + c_2(\cosh \mu_n L + \cos \mu_n L) = 0.$$

Let $A_n = (\cosh \mu_n L + \cos \mu_n L)/(\sinh \mu_n L + \sin \mu_n L)$. The eigenfunctions are given by

$$\phi_n(x) = \cosh \mu_n x - \cos \mu_n x + A_n(\sin \mu_n x - \sinh \mu_n x),$$

with corresponding eigenvalues $\lambda_n = \mu_n^4$. Thus $\lambda_1 \approx 12.362/L^4$ and $\lambda_2 \approx 485.52/L^4$.

25.(a) Assume that the solution has the form $u(x,t) = X(x)T(t)$. Substitution into the PDE results in

$$\frac{E}{\rho}X''T = XT''.$$

Dividing both sides of the equation by XT, we obtain

$$\frac{E}{\rho}\frac{X''T}{XT} = \frac{XT''}{XT},$$

that is,

$$\frac{X''}{X} = \frac{\rho}{E}\frac{T''}{T}.$$

Since both sides of the resulting equation are functions of different variables, each must be equal to a constant, say $-\lambda$. Therefore we obtain two ordinary differential equations

$$X'' + \lambda X = 0 \quad \text{and} \quad T'' + \lambda \frac{E}{\rho} T = 0.$$

(b) Given that $u(0,t) = X(0)T(t)$ for $t > 0$, it follows that $X(0) = 0$. The second boundary condition can be expressed as

$$EAX'(L)T(t) + mX(L)T''(t) = 0, \quad t > 0.$$

From the result in part (a),

$$EAX'(L)T(t) - \lambda m \frac{E}{\rho} X(L)T(t) = 0, \quad t > 0.$$

Since the condition is to be satisfied for all $t > 0$, we arrive at the boundary condition

$$X'(L) - \lambda \frac{m}{\rho A} X(L) = 0.$$

(c) If $\lambda = 0$, the general solution of the spatial equation is

$$X(x) = c_1 x + c_2.$$

The boundary condition require that $c_1 = c_2 = 0$. Hence there are no nontrivial solutions. If $\lambda = -\mu^2 < 0$, then the general solution is

$$X(x) = c_1 \cosh \mu x + c_2 \sinh \mu x.$$

The first boundary condition implies that $c_1 = 0$. The second boundary condition requires that

$$c_2 \cosh \mu L + c_2 \mu \frac{m}{\rho A} \sinh \mu L = 0.$$

The solution is nontrivial only if $\mu \tanh \mu L = -\frac{\rho A}{m}$. Since $\mu \tanh \mu L \geq 0$, there are no nontrivial solutions. Let $\lambda = \mu^2 > 0$. The general solution of the spatial equation is

$$X(x) = c_1 \cos \mu x + c_2 \sin \mu x.$$

The first boundary condition implies that $c_1 = 0$. The second boundary condition requires that

$$c_2 \cos \mu L - c_2 \mu \frac{m}{\rho A} \sin \mu L = 0.$$

For a nontrivial solution, it is necessary that

$$\cos \mu L - \mu \frac{m}{\rho A} \sin \mu L = 0,$$

or

$$\tan \mu L = \frac{\rho A}{m \mu}.$$

For the case $(m/\rho A L) = 0.5$,

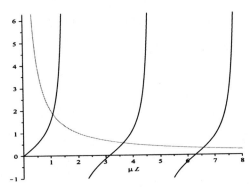

we find that $\mu_1 L \approx 1.0769$ and $\mu_2 L \approx 3.6436$. Therefore the eigenfunctions are given by $\phi_n(x) = \sin \mu_n x$. The corresponding eigenvalues are solutions of

$$\cos \sqrt{\lambda_n} L - \frac{L}{2} \sqrt{\lambda_n} \sin \sqrt{\lambda_n} L = 0.$$

(d) The first two eigenvalues are approximated as

$$\lambda_1 \approx 1.1597/L^2 \quad \text{and} \quad \lambda_2 \approx 13.276/L^2.$$

11.2

2. Based on the boundary conditions, $\lambda > 0$. The general solution of the ODE is

$$y(x) = c_1 \cos \sqrt{\lambda} x + c_2 \sin \sqrt{\lambda} x.$$

The boundary condition $y'(0) = 0$ requires that $c_2 = 0$. Imposing the second boundary condition, we find that $c_1 \cos \sqrt{\lambda} = 0$. So for a nontrivial solution, $\sqrt{\lambda} = (2n-1)\pi/2$, $n = 1, 2, \ldots$. Therefore the eigenfunctions are given by

$$\phi_n(x) = k_n \cos \frac{(2n-1)\pi x}{2}.$$

In this problem, $r(x) = 1$, and the normalization condition is

$$k_n^2 \int_0^1 \left[\cos \frac{(2n-1)\pi x}{2} \right]^2 dx = 1.$$

It follows that $k_n^2 = 2$. Therefore the normalized eigenfunctions are

$$\phi_n(x) = \sqrt{2} \cos \frac{(2n-1)\pi x}{2}, \quad n = 1, 2, \ldots.$$

3. Based on the boundary conditions, $\lambda \geq 0$. For $\lambda = 0$, the eigenfunction is $\phi_0(x) = k_0$. Set $k_0 = 1$. With $\lambda > 0$, the general solution of the ODE is

$$y(x) = c_1 \cos \sqrt{\lambda} x + c_2 \sin \sqrt{\lambda} x.$$

Invoking the boundary conditions, we require that $c_2 = 0$ and $c_1\sqrt{\lambda}\sin\sqrt{\lambda} = 0$. Since $\lambda > 0$, the eigenvalues are $\lambda_n = n^2\pi^2$, $n = 1, 2, \ldots$, with corresponding eigenfunctions

$$\phi_n(x) = k_n \cos n\pi x.$$

The normalization condition is

$$k_n^2 \int_0^1 \cos^2 n\pi x \, dx = 1.$$

It follows that $k_n^2 = 2$. Therefore the normalized eigenfunctions are

$$\phi_0(x) = 1, \text{ and } \phi_n(x) = \sqrt{2}\cos n\pi x, \quad n = 1, 2, \ldots.$$

4. From Problem 8 in Section 11.1, the eigenfunctions are $\phi_n(x) = k_n \cos \sqrt{\lambda_n}\, x$, in which $\cos\sqrt{\lambda_n} - \sqrt{\lambda_n}\sin\sqrt{\lambda_n} = 0$. The normalization condition is

$$k_n^2 \int_0^1 \cos^2 \sqrt{\lambda_n}\, x \, dx = 1.$$

First note that

$$\int_0^1 \cos^2 \sqrt{\lambda_n}\, x\, dx = \frac{\cos\sqrt{\lambda_n}\sin\sqrt{\lambda_n} + \sqrt{\lambda_n}}{2\sqrt{\lambda_n}}.$$

Based on the determinantal equation,

$$\frac{\cos\sqrt{\lambda_n}\sin\sqrt{\lambda_n} + \sqrt{\lambda_n}}{2\sqrt{\lambda_n}} = \frac{1 + \sin^2\sqrt{\lambda_n}}{2} = \frac{3 - \cos 2\sqrt{\lambda_n}}{4}.$$

Therefore

$$k_n^2 = \frac{4}{3 - \cos 2\sqrt{\lambda_n}}$$

and the normalized eigenfunctions are given by

$$\phi_n(x) = \frac{2\cos\sqrt{\lambda_n}\, x}{\sqrt{3 - \cos 2\sqrt{\lambda_n}}}.$$

6. As shown in Problem 1, the normalized eigenfunctions are

$$\phi_n(x) = \sqrt{2}\sin\frac{(2n-1)\pi x}{2}, \quad n = 1, 2, \ldots.$$

Based on Eq.(34), with $r(x) = 1$, the coefficients in the eigenfunction expansion are given by

$$c_m = \int_0^1 f(x)\phi_m(x)dx = \sqrt{2}\int_0^1 \sin\frac{(2m-1)\pi x}{2}dx = \frac{2\sqrt{2}}{(2m-1)\pi}.$$

Therefore we obtain the formal expansion

$$1 = \frac{2\sqrt{2}}{\pi}\sum_{n=1}^{\infty}\frac{1}{2n-1}\sin\frac{(2n-1)\pi x}{2}.$$

8. We consider the normalized eigenfunctions
$$\phi_n(x) = \sqrt{2} \sin \frac{(2n-1)\pi x}{2}, \quad n = 1, 2, \ldots.$$

Based on Eq.(34), with $r(x) = 1$, the coefficients in the eigenfunction expansion are given by
$$c_m = \int_0^1 f(x)\phi_m(x)dx = \sqrt{2} \int_0^{1/2} \sin \frac{(2m-1)\pi x}{2} dx$$
$$= \frac{2\sqrt{2}}{(2m-1)\pi} \left[1 - \cos \frac{(2m-1)\pi}{4} \right].$$

Therefore we obtain the formal expansion
$$f(x) = \frac{2\sqrt{2}}{\pi} \sum_{n=1}^{\infty} \frac{1}{2n-1} \left[1 - \cos \frac{(2n-1)\pi}{4} \right] \sin \frac{(2n-1)\pi x}{2}.$$

9. The normalized eigenfunctions are
$$\phi_n(x) = \sqrt{2} \sin \frac{(2n-1)\pi x}{2}, \quad n = 1, 2, \ldots.$$

Based on Eq.(34), with $r(x) = 1$, the coefficients in the eigenfunction expansion are given by
$$c_m = \int_0^1 f(x)\phi_m(x)dx$$
$$= \sqrt{2} \int_0^{1/2} 2x \sin \frac{(2m-1)\pi x}{2} dx + \sqrt{2} \int_{1/2}^1 \sin \frac{(2m-1)\pi x}{2} dx$$
$$= \frac{8}{(2m-1)^2 \pi^2} \left[\sin \frac{m\pi}{2} - \cos \frac{m\pi}{2} \right].$$

Therefore the formal expansion of the given function is
$$f(x) = \frac{8}{\pi^2} \sum_{n=1}^{\infty} \frac{\sin \frac{n\pi}{2} - \cos \frac{n\pi}{2}}{(2n-1)^2} \sin \frac{(2n-1)\pi x}{2}.$$

11. From Problem 4, the normalized eigenfunctions are given by
$$\phi_n(x) = \frac{2 \cos \sqrt{\lambda_n}\, x}{\sqrt{3 - \cos 2\sqrt{\lambda_n}}},$$

in which the eigenvalues satisfy $\cos \sqrt{\lambda_n} - \sqrt{\lambda_n} \sin \sqrt{\lambda_n} = 0$. Based on Eq.(34), the coefficients in the eigenfunction expansion are given by
$$c_m = \int_0^1 f(x)\phi_m(x)dx = \frac{2}{\sqrt{3 - \cos 2\sqrt{\lambda_m}}} \int_0^1 x \cos \sqrt{\lambda_m}\, x\, dx$$
$$= \frac{\sqrt{2}(2 \cos \sqrt{\lambda_m} - 1)}{\lambda_m \alpha_m},$$

in which $\alpha_m = \sqrt{1 + \sin^2 \sqrt{\lambda_m}}$.

12. The normalized eigenfunctions are given by
$$\phi_n(x) = \frac{2\cos\sqrt{\lambda_n}\,x}{\sqrt{3 - \cos 2\sqrt{\lambda_n}}},$$
in which the eigenvalues satisfy $\cos\sqrt{\lambda_n} - \sqrt{\lambda_n}\sin\sqrt{\lambda_n} = 0$. Based on Eq.(34), the coefficients in the eigenfunction expansion are given by
$$c_m = \int_0^1 f(x)\phi_m(x)dx = \frac{2}{\sqrt{3 - \cos 2\sqrt{\lambda_m}}} \int_0^1 (1-x)\cos\sqrt{\lambda_m}\,x\,dx =$$
$$= \frac{\sqrt{2}(1 - \cos\sqrt{\lambda_m})}{\lambda_m \alpha_m},$$
in which $\alpha_m = \sqrt{1 + \sin^2\sqrt{\lambda_m}}$.

13. We consider the normalized eigenfunctions
$$\phi_n(x) = \frac{2\cos\sqrt{\lambda_n}\,x}{\sqrt{3 - \cos 2\sqrt{\lambda_n}}},$$
in which the eigenvalues satisfy $\cos\sqrt{\lambda_n} - \sqrt{\lambda_n}\sin\sqrt{\lambda_n} = 0$. The coefficients in the eigenfunction expansion are given by
$$c_n = \int_0^1 f(x)\phi_n(x)dx = \frac{2}{\sqrt{3 - \cos 2\sqrt{\lambda_n}}} \int_0^{1/2} \cos\sqrt{\lambda_n}\,x\,dx = \frac{\sqrt{2}\sin(\sqrt{\lambda_n}/2)}{\sqrt{\lambda_n}\,\alpha_n},$$
in which $\alpha_n = \sqrt{1 + \sin^2\sqrt{\lambda_n}}$.

15. The differential equation can be written as
$$[(1+x^2)y']' + y = 0,$$
with $p(x) = -1 - x^2$ and $q(x) = 1$. The boundary conditions are homogeneous and separated. Hence the BVP is self-adjoint.

16. Since the boundary conditions are not separated, the inner product is computed: Given u and v, sufficiently smooth and satisfying the boundary conditions,
$$(L[u], v) = \int_0^1 [u''v + uv]\,dx = u'v\Big|_0^1 - \int_0^1 [u'v' + uv]\,dx =$$
$$= [u'v - uv']\Big|_0^1 + (u, L[v]).$$
Based on the given boundary conditions,
$$u'(1)v(1) - u'(0)v(0) = u(0)v(1) + 2u(1)v(0)$$
$$- u(1)v'(1) + u(0)v'(0) = -u(1)v(0) - 2u(0)v(1).$$
Since
$$[u'v - uv']\Big|_0^1 = u(1)v(0) - u(0)v(1),$$

the BVP is not self-adjoint.

18. The differential equation can be written as $-[y']' = \lambda y$, with $p(x) = 1$, $q(x) = 0$, and $r(x) = 1$. The boundary conditions are homogeneous and separated. Hence the BVP is self-adjoint.

19. If $\alpha_2 = 0$, then $u(0) = v(0) = 0$ and $u'(0)v(0) - u(0)v'(0) = 0$. Furthermore,

$$u'(1)v(1) - u(1)v'(1) = -\frac{\beta_2}{\beta_1}u'(1)v'(1) + \frac{\beta_2}{\beta_1}u'(1)v'(1) = 0.$$

If $\beta_2 = 0$, then $u(1) = v(1) = 0$ and $u'(1)v(1) - u(1)v'(1) = 0$. Furthermore,

$$u'(0)v(0) - u(0)v'(0) = -\frac{\alpha_2}{\alpha_1}u'(0)v'(0) + \frac{\alpha_2}{\alpha_1}u'(0)v'(0) = 0.$$

Clearly, the results are also true if $\alpha_2 = \beta_2 = 0$.

20.(a,b) Suppose that $\phi_1(x)$ and $\phi_2(x)$ are linearly independent eigenfunctions associated with an eigenvalue λ. The Wronskian is given by

$$W(\phi_1, \phi_2)(x) = \phi_1(x)\phi_2'(x) - \phi_2(x)\phi_1'(x).$$

Each of the eigenfunctions satisfies the boundary condition $\alpha_1 y(0) + \alpha_2 y'(0) = 0$. If either $\alpha_1 = 0$ or $\alpha_2 = 0$, then clearly $W(\phi_1, \phi_2)(0) = 0$. On the other hand, if α_2 is not equal to zero, then

$$W(\phi_1, \phi_2)(0) = \phi_1(0)\phi_2'(0) - \phi_2(0)\phi_1'(0) = -\frac{\alpha_1}{\alpha_2}\phi_1(0)\phi_2(0) + \frac{\alpha_1}{\alpha_2}\phi_2(0)\phi_1(0) = 0.$$

(c) By Theorem 3.2.6, $W(\phi_1, \phi_2)(x) = 0$ for all $0 \leq x \leq 1$ and then $\phi_1(x)$ and $\phi_2(x)$ must be linearly dependent. Hence λ must be a simple eigenvalue.

22. We consider the operator

$$L[y] = -[p(x)y']' + q(x)y$$

on the interval $0 < x < 1$, together with the boundary conditions

$$\alpha_1 y(0) + \alpha_2 y'(0) = 0, \qquad \beta_1 y(1) + \beta_2 y'(1) = 0.$$

Let $u = \phi + i\psi$ and $v = \xi + i\eta$. If u and v both satisfy the boundary conditions, then the real and imaginary parts also satisfy the same boundary conditions. Using the inner product

$$(u, v) = \int_0^1 u(x)\overline{v}(x)dx,$$

we obtain that

$$(L[u], v) = \int_0^1 [-[p(x)u']'\overline{v} + q(x)u\overline{v}]dx = \int_0^1 \{-[p(x)(\phi' + i\psi')]'\overline{v} + q(x)u\overline{v}\}dx =$$

$$= -p(x)(\phi' + i\psi')\overline{v}\Big|_0^1 + \int_0^1 \{p(x)(\phi' + i\psi')\overline{v}' + q(x)u\overline{v}\} dx.$$

Integrating by parts, again,

$$\int_0^1 \{p(x)(\phi' + i\psi')\overline{v}'\}\,dx = (\phi + i\psi)p(x)\overline{v}'\Big|_0^1 - \int_0^1 \{[p(x)\overline{v}']' u\}\,dx.$$

Collecting the boundary terms,

$$p(x)\left[(\phi' + i\psi')\overline{v} - (\phi + i\psi)\overline{v}'\right]\Big|_0^1 = p(x)\left[(\phi' + i\psi')(\xi - i\eta) - (\phi + i\psi)(\xi' - i\eta')\right]\Big|_0^1.$$

The real part is given by

$$p(x)\left[(\phi'\xi + \psi'\eta) - (\phi\xi' + \psi\eta')\right]\Big|_0^1 = p(x)\left[(\phi'\xi - \phi\xi') + (\psi'\eta - \psi\eta')\right]\Big|_0^1$$

$$= p(x)\left[\phi'\xi - \phi\xi'\right]\Big|_0^1 + p(x)\left[\psi'\eta - \psi\eta'\right]\Big|_0^1.$$

Since ϕ, ψ, ξ and η satisfy the boundary conditions, it follows that

$$p(x)\left[(\phi'\xi + \psi'\eta) - (\phi\xi' + \psi\eta')\right]\Big|_0^1 = 0.$$

Similarly, the imaginary part also vanishes. That is,

$$p(x)\left[(\psi'\xi - \psi\xi') - (\phi'\eta - \phi\eta')\right]\Big|_0^1 = 0.$$

Therefore

$$(L[u], v) = \int_0^1 \left\{-[p(x)\overline{v}']' u + q(x)u\overline{v}\right\}dx = (L[\overline{v}], \overline{u}) = \overline{(\overline{u}, L[\overline{v}])}.$$

The result follows from the fact that $\overline{(\overline{u}, L[\overline{v}])} = (u, L[v])$.

26. As shown is Problem 25, the general solution is

$$y(x) = c_1 + c_2 x + c_3 \cos \mu x + c_4 \sin \mu x.$$

Imposing the boundary conditions, we obtain the system of equations

$$c_2 = 0$$
$$c_1 + c_3 = 0$$
$$c_2 + \mu c_4 = 0$$
$$c_3 \cos \mu L + c_4 \sin \mu L = 0.$$

For a nontrivial solution, it is necessary that $\cos \mu L = 0$. We find that $c_2 = c_4 = 0$, and hence the eigenfunctions are given by

$$\phi_n(x) = 1 - \cos \sqrt{\lambda_n}\, x.$$

The corresponding eigenvalues are $\lambda_n = (2n-1)^2 \pi^2 / 4L^2$, $n = 1, 2, \ldots$. The smallest eigenvalue is $\lambda_1 = \pi^2/4L^2$. Thus $\phi_1(x) = 1 - \cos(\pi x/2L)$.

28.(a) Setting $c(x,t) = c_0 + u(x,t)$, the BVP satisfied by $u(x,t)$ is given as

$$u_t + v u_x = D u_{xx} \qquad 0 < x < L, t > 0;$$
$$u(0,t) = 0,\ u_x(L,t) = 0, \qquad t > 0;$$

$$u(x,0) = -c_0, \quad 0 < x < L.$$

(b) We consider solutions of the form $u(x,t) = X(x)T(t)$. Substitution into the partial differential equation results in

$$XT' + vX'T = DX''T.$$

Divide both sides of the differential equation by the product XT to obtain

$$\frac{T'}{T} + v\frac{X'}{X} = D\frac{X''}{X},$$

so that

$$D\frac{X''}{X} - v\frac{X'}{X} = \frac{T'}{T}.$$

Since both sides of the resulting equation are functions of different variables, each must be equal to a constant, say $\tilde{\lambda} = -\lambda D$. We obtain the ordinary differential equations

$$DX'' - vX' + \lambda D X = 0 \text{ and } T' + \lambda DT = 0.$$

Multiplying both sides of the spatial equation by $e^{-vx/D}/D$, it can be written as

$$\left[e^{-vx/D} X'\right]' + \lambda e^{-vx/D} X = 0,$$

with associated boundary conditions $X(0) = 0$ and $X'(L) = 0$. The general solution of this ODE is

$$y(t) = c_1 e^{vx/2D} \cos \mu x + c_2 e^{vx/2D} \sin \mu x,$$

where

$$\mu^2 = \lambda - \frac{v^2}{4D^2}.$$

Imposing the boundary conditions, it is necessary that

$$c_1 = 0$$
$$c_2 (v/2D) e^{vL/2D} \sin \mu L + c_2 \mu e^{vL/2D} \cos \mu L = 0.$$

In order to satisfy the second boundary condition, we must have

$$(v/2D) \sin \mu L + \mu \cos \mu L = 0,$$

that is, $\tan \mu L = -2\mu D/v$. Hence the eigenfunctions are $X_n = e^{vx/2D} \sin \mu_n x$, with associated eigenvalues

$$\lambda_n = \mu_n^2 + \frac{v^2}{4D^2}.$$

We also obtain the family of equations $T' + \lambda_n DT = 0$. Solutions are given by

$$T_n = e^{-\lambda_n Dt}.$$

Hence the fundamental solutions of the PDE are

$$u_n(x,t) = e^{-\lambda_n Dt} e^{vx/2D} \sin \mu_n x,$$

which yield the general solution

$$u(x,t) = \sum_{n=1}^{\infty} b_n e^{-\lambda_n Dt} e^{vx/2D} \sin \mu_n x, \text{ where } \lambda_n = \mu_n^2 + \frac{v^2}{4D^2}.$$

Setting $t = 0$, it is necessary that

$$-c_0 = \sum_{n=1}^{\infty} b_n e^{vx/2D} \sin \mu_n x.$$

It follows that the normalized eigenfunctions are

$$\phi_n(x) = \frac{e^{vx/2D} \sin \mu_n x}{\sqrt{\frac{L}{2} + \frac{v}{4D\mu_n^2} \sin^2 \mu_n L}}$$

and

$$b_n = \frac{\int_0^L e^{-vx/D}(-c_0) e^{vx/2D} \sin \mu_n x \, dx}{\frac{L}{2} + \frac{v}{4D\mu_n^2} \sin^2 \mu_n L}$$

$$= \frac{8c_0 D^2 \mu_n^2 (2D\mu_n e^{-vL/2D} \cos(\mu_n L) + v e^{-vL/2D} \sin(\mu_n L) - 2D\mu_n)}{(v^2 + 4D^2\mu_n^2)(2LD\mu_n^2 + v\sin^2(\mu_n L))}.$$

(c) $c(x,t)$, $0 \leq x \leq 10$, $t = 5, 10, 15, 20$:

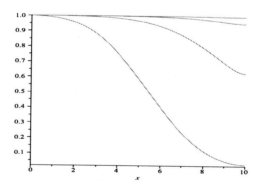

$c(x,t)$, $0 \leq t \leq 10$, $x = 2.5, 5, 7.5, 10$:

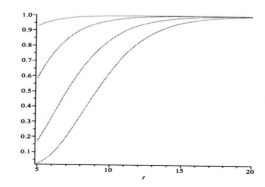

(d) The solution increases monotonically to the steady state solution of c_0 and is essentially attained in approximately 20 s.

11.3

4. The eigensystem of the associated homogeneous problem is given in Problem 11 of Section 11.2. The normalized eigenfunctions are

$$\phi_n(x) = \frac{\sqrt{2}\cos\sqrt{\lambda_n}\,x}{\sqrt{1+\sin^2\sqrt{\lambda_n}}},$$

in which the eigenvalues satisfy $\cos\sqrt{\lambda_n} - \sqrt{\lambda_n}\sin\sqrt{\lambda_n} = 0$. Rewrite the given differential equation as $-y'' = 2y + x$. Since $\mu = 2 \neq \lambda_n$, the formal solution of the nonhomogeneous problem is

$$y(x) = \sum_{n=1}^{\infty} \frac{c_n}{\lambda_n - 2}\phi_n(x),$$

in which

$$c_n = \int_0^1 f(x)\phi_n(x)dx = \frac{\sqrt{2}}{\sqrt{1+\sin^2\sqrt{\lambda_n}}}\int_0^1 x\cos\sqrt{\lambda_n}\,x\,dx =$$

$$= \frac{\sqrt{2}(2\cos\sqrt{\lambda_n} - 1)}{\lambda_n\sqrt{1+\sin^2\sqrt{\lambda_n}}}.$$

Therefore we obtain the formal expansion

$$y(x) = 2\sum_{n=1}^{\infty} \frac{(2\cos\sqrt{\lambda_n} - 1)\cos\sqrt{\lambda_n}\,x}{\lambda_n(\lambda_n - 2)(1+\sin^2\sqrt{\lambda_n})}.$$

5. The solution follows that in Problem 1, except that the coefficients are given by

$$c_n = \int_0^1 f(x)\phi_n(x)dx = \sqrt{2}\int_0^{1/2} 2x\sin n\pi x\,dx + \sqrt{2}\int_{1/2}^1 (2-2x)\sin n\pi x\,dx =$$

$$= 4\frac{\sqrt{2}\sin(n\pi/2)}{n^2\pi^2}.$$

Therefore the formal solution is

$$y(x) = 8\sum_{n=1}^{\infty} \frac{\sin(n\pi/2)\sin n\pi x}{n^2\pi^2(n^2\pi^2 - 2)}.$$

6. The differential equation can be written as $-y'' = \mu y + f(x)$. Note that $q(x) = 0$ and $r(x) = 1$. As shown in Problem 1 in Section 11.2, the normalized eigenfunctions are

$$\phi_n(x) = \sqrt{2}\sin\frac{(2n-1)x}{2},$$

with associated eigenvalues $\lambda_n = (2n-1)^2\pi^2/4$. Based on Theorem 11.3.1, the formal solution is given by

$$y(x) = \sqrt{2} \sum_{n=1}^{\infty} \frac{c_n}{(\lambda_n - \mu)} \sin \frac{(2n-1)x}{2},$$

as long as $\mu \neq \lambda_n$. The coefficients in the series expansion are computed as

$$c_n = \sqrt{2} \int_0^1 f(x) \sin \frac{(2n-1)x}{2} dx.$$

7. As shown in Problem 1 in Section 11.2, the normalized eigenfunctions are

$$\phi_n(x) = \sqrt{2} \cos \frac{(2n-1)x}{2},$$

with associated eigenvalues $\lambda_n = (2n-1)^2\pi^2/4$. Based on Theorem 11.3.1, the formal solution is given by

$$y(x) = \sqrt{2} \sum_{n=1}^{\infty} \frac{c_n}{(\lambda_n - \mu)} \cos \frac{(2n-1)x}{2},$$

as long as $\mu \neq \lambda_n$. The coefficients in the series expansion are computed as

$$c_n = \sqrt{2} \int_0^1 f(x) \cos \frac{(2n-1)x}{2} dx.$$

9. The normalized eigenfunctions are

$$\phi_n(x) = \frac{\sqrt{2} \cos \sqrt{\lambda_n}\, x}{\sqrt{1 + \sin^2 \sqrt{\lambda_n}}}.$$

The eigenvalues satisfy $\cos \sqrt{\lambda_n} - \sqrt{\lambda_n} \sin \sqrt{\lambda_n} = 0$. Based on Theorem 11.3.1, the formal solution is given by

$$y(x) = \sqrt{2} \sum_{n=1}^{\infty} \frac{c_n \cos \sqrt{\lambda_n}\, x}{(\lambda_n - \mu)\sqrt{1 + \sin^2 \sqrt{\lambda_n}}},$$

as long as $\mu \neq \lambda_n$. The coefficients in the series expansion are computed as

$$c_n = \frac{\sqrt{2}}{\sqrt{1 + \sin^2 \sqrt{\lambda_n}}} \int_0^1 f(x) \cos \sqrt{\lambda_n}\, x \, dx.$$

13. The differential equation can be written as $-y'' = \pi^2 y + \cos \pi x - a$. Note that $\mu = \pi^2$ and $f(x) = \cos \pi x - a$. Furthermore, $\mu = \pi^2$ is an eigenvalue corresponding to the eigenfunction $\phi_1(x) = \sqrt{2} \sin \pi x$. A solution exists only if $f(x)$ and $\phi_1(x)$ are orthogonal. Since

$$\int_0^1 (\cos \pi x - a) \sin \pi x \, dx = -2a/\pi,$$

there exists a solution as long as $a = 0$. In that case, the ODE is

$$y'' + \pi^2 y = -\cos \pi x.$$

The complementary solution is $y_c(x) = c_1 \cos \pi x + c_2 \sin \pi x$. A particular solution is $Y(x) = Ax \cos \pi x + Bx \sin \pi x$. Using the method of undetermined coefficients, we find that $A = 0$ and $B = -1/2\pi$. Therefore the general solution is

$$y(x) = c_1 \cos \pi x + c_2 \sin \pi x - \frac{x}{2\pi} \sin \pi x.$$

The boundary conditions require that $c_1 = 0$. Hence the solution of the boundary value problem is

$$y(x) = c_2 \sin \pi x - \frac{x}{2\pi} \sin \pi x.$$

15. Let $y(x) = \phi_1(x) + \phi_2(x)$. It follows that $L[y] = L[\phi_1] + L[\phi_2] = 0 + f(x) = f(x)$. Also,

$$\begin{aligned}\alpha_1 y(0) + \alpha_2 y'(0) &= \alpha_1 \phi_1(0) + \alpha_1 \phi_2(0) + \alpha_2 \phi_1'(0) + \alpha_2 \phi_2'(0) \\ &= \alpha_1 \phi_1(0) + \alpha_2 \phi_1'(0) + \alpha_1 \phi_2(0) + \alpha_2 \phi_2'(0) \\ &= a + 0 = a\end{aligned}$$

and

$$\begin{aligned}\beta_1 y(1) + \beta_2 y'(1) &= \beta_1 \phi_1(1) + \beta_1 \phi_2(1) + \beta_2 \phi_1'(1) + \beta_2 \phi_2'(1) \\ &= \beta_1 \phi_1(1) + \beta_2 \phi_1'(1) + \beta_1 \phi_2(1) + \beta_2 \phi_2'(1) \\ &= b + 0 = b.\end{aligned}$$

16. The complementary solution is $y_c(x) = c_1 \cos \pi x + c_2 \sin \pi x$. A particular solution is $Y(x) = A + Bx$. Using the method of undetermined coefficients, we find that $A = 0$ and $B = 1$. Therefore the general solution is

$$y(x) = c_1 \cos \pi x + c_2 \sin \pi x + x.$$

Imposing the boundary conditions, we find that $c_1 = 1$. Therefore the solution of the BVP is

$$y(x) = \cos \pi x + c_2 \sin \pi x + x.$$

Now attempt to solve the problem as shown in Problem 15. Let BVP-1 be given by

$$u'' + \pi^2 u = \pi^2 x, \quad u(0) = 0, \ u(1) = 0.$$

The general solution of the ODE is

$$u(x) = c_1 \cos \pi x + c_2 \sin \pi x + x.$$

The boundary conditions require that $c_1 = 0$ and $-c_1 + 1 = 0$. We find that BVP-1 has no solution. Let BVP-2 be given by

$$v'' + \pi^2 v = 0, \quad v(0) = 1, \ v(1) = 0.$$

The general solution of the ODE is $v(x) = c_1 \cos \pi x + c_2 \sin \pi x$. Imposing the boundary conditions, we obtain $c_1 = 1$ and $-c_1 = 0$. Thus BVP-2 has no solution.

17. Setting $y(x) = u(x) + v(x)$, substitution results in

$$u'' + v'' + p(x)[u' + v'] + q(x)[u + v] = u'' + p(x)u' + q(x)u + v'' + p(x)v' + q(x)v.$$

Since the left hand side of the equation is zero,
$$u'' + p(x)u' + q(x)u = -[v'' + p(x)v' + q(x)v].$$
Furthermore, $u(0) = y(0) - v(0) = 0$ and $u(1) = y(1) - v(1) = 0$. The simplest function having the assumed properties is $v(x) = (b-a)x + a$. In this case,
$$g(x) = (a-b)p(x) + (a-b)x\,q(x) - a\,q(x).$$

20. The associated homogeneous PDE is $u_t = u_{xx}$, $0 < x < 1$, with
$$u_x(0,t) = 0,\ u_x(1,t) + u(1,t) = 0 \text{ and } u(x,0) = 1 - x.$$
Applying the method of separation of variables, we obtain the eigenvalue problem $X'' + \lambda X = 0$, with boundary conditions $X'(0) = 0$ and $X'(1) + X(1) = 0$. It was shown in Problem 4, in Section 11.2, that the normalized eigenfunctions are
$$\phi_n(x) = \frac{\sqrt{2}\cos\sqrt{\lambda_n}\,x}{\sqrt{1 + \sin^2\sqrt{\lambda_n}}},$$
where $\cos\sqrt{\lambda_n} - \sqrt{\lambda_n}\sin\sqrt{\lambda_n} = 0$. We assume a solution of the form
$$u(x,t) = \sum_{n=1}^{\infty} b_n(t)\phi_n(x).$$
Substitution into the given PDE results in
$$\sum_{n=1}^{\infty} b_n'(t)\phi_n(x) = \sum_{n=1}^{\infty} b_n(t)\phi_n''(x) + e^{-t} = -\sum_{n=1}^{\infty} \lambda_n b_n(t)\phi_n(x) + e^{-t},$$
that is,
$$\sum_{n=1}^{\infty} [b_n'(t) + \lambda_n b_n(t)]\phi_n(x) = e^{-t}.$$
We now note that
$$1 = \sum_{n=1}^{\infty} \frac{\sqrt{2}\sin\sqrt{\lambda_n}}{\sqrt{\lambda_n}\sqrt{1 + \sin^2\sqrt{\lambda_n}}}\phi_n(x).$$
Therefore
$$e^{-t} = \sum_{n=1}^{\infty} \beta_n e^{-t}\phi_n(x),$$
in which $\beta_n = \sqrt{2}\sin\sqrt{\lambda_n}/\left[\sqrt{\lambda_n}\sqrt{1 + \sin^2\sqrt{\lambda_n}}\right]$. Combining these results,
$$\sum_{n=1}^{\infty} [b_n'(t) + \lambda_n b_n(t) - \beta_n e^{-t}]\phi_n(x) = 0.$$
Since the resulting equation is valid for $0 < x < 1$, it follows that
$$b_n'(t) + \lambda_n b_n(t) = \beta_n e^{-t},\ n = 1, 2, \ldots.$$

Prior to solving the sequence of ODEs, we establish the initial conditions. These are obtained from the expansion

$$u(x,0) = 1 - x = \sum_{n=1}^{\infty} \alpha_n \phi_n(x),$$

in which $\alpha_n = \sqrt{2}\,(1 - \cos\sqrt{\lambda_n})/\left[\lambda_n\sqrt{1 + \sin^2\sqrt{\lambda_n}}\right]$. That is, $b_n(0) = \alpha_n$. Therefore the solutions of the first order ODEs are

$$b_n(t) = \frac{\beta_n(e^{-t} - e^{-\lambda_n t})}{(\lambda_n - 1)} + \alpha_n e^{-\lambda_n t}, \quad n = 1, 2, \ldots.$$

Hence the solution of the boundary value problem is

$$u(x,t) = \sum_{n=1}^{\infty} \left[\frac{\beta_n(e^{-t} - e^{-\lambda_n t})}{(\lambda_n - 1)} + \alpha_n e^{-\lambda_n t}\right] \phi_n(x).$$

21. Based on the boundary conditions, the normalized eigenfunctions are given by

$$\phi_n(x) = \sqrt{2}\,\sin n\pi x,$$

with associated eigenvalues $\lambda_n = n^2\pi^2$. We now assume a solution of the form

$$u(x,t) = \sum_{n=1}^{\infty} b_n(t)\phi_n(x).$$

Substitution into the given PDE results in

$$\sum_{n=1}^{\infty} b_n'(t)\phi_n(x) = \sum_{n=1}^{\infty} b_n(t)\phi_n''(x) + 1 - |1 - 2x| = -\sum_{n=1}^{\infty} \lambda_n b_n(t)\phi_n(x) + 1 - |1 - 2x|,$$

that is,

$$\sum_{n=1}^{\infty} [b_n'(t) + \lambda_n b_n(t)]\phi_n(x) = 1 - |1 - 2x|.$$

It was shown in Problem 5 that

$$1 - |1 - 2x| = \sum_{n=1}^{\infty} 4\frac{\sqrt{2}\,\sin(n\pi/2)}{n^2\pi^2}\phi_n(x).$$

Substituting on the right hand side and collecting terms, we obtain

$$\sum_{n=1}^{\infty} \left[b_n'(t) + \lambda_n b_n(t) - 4\frac{\sqrt{2}\,\sin(n\pi/2)}{n^2\pi^2}\right]\phi_n(x) = 0.$$

Since the resulting equation is valid for $0 < x < 1$, it follows that

$$b_n'(t) + n^2\pi^2 b_n(t) = 4\frac{\sqrt{2}\,\sin(n\pi/2)}{n^2\pi^2}, \quad n = 1, 2, \ldots.$$

Based on the given initial condition, we also have $b_n(0) = 0$, for $n = 1, 2, \ldots$. The solutions of the first order ODEs are

$$b_n(t) = 4\frac{\sqrt{2}\,\sin(n\pi/2)}{n^4\pi^4}(1 - e^{-n^2\pi^2 t}), \quad n = 1, 2, \ldots.$$

Hence the solution of the boundary value problem is

$$u(x,t) = \frac{8}{\pi^4} \sum_{n=1}^{\infty} \frac{\sin(n\pi/2)}{n^4}(1 - e^{-n^2\pi^2 t}) \sin n\pi x.$$

23.(a) Let $u(x,t)$ be a solution of the boundary value problem and $v(x)$ be a solution of the related BVP. Substituting for $u(x,t) = w(x,t) + v(x)$, we have

$$r(x)u_t = r(x)w_t$$

and

$$[p(x)u_x]_x - q(x)u + F(x) = [p(x)w_x]_x - q(x)w + [p(x)v']' - q(x)v + F(x) =$$
$$= [p(x)w_x]_x - q(x)w - F(x) + F(x) = [p(x)w_x]_x - q(x)w.$$

Hence $w(x,t)$ is a solution of the homogeneous PDE

$$r(x)w_t = [p(x)w_x]_x - q(x)w.$$

The required boundary conditions are

$$w(0,t) = u(0,t) - v(0) = 0,$$
$$w(1,t) = u(1,t) - v(1) = 0.$$

The associated initial condition is $w(x,0) = u(x,0) - v(x) = f(x) - v(x)$.

(b) Let $v(x)$ be a solution of the ODE

$$[p(x)v']' - q(x)v = -F(x),$$

and satisfying the boundary conditions $v'(0) - h_1 v(0) = T_1$, $v'(1) + h_2 v(1) = T_2$. If $w(x,t) = u(x,t) - v(x)$, then it is easy to show the w satisfies the PDE and initial condition given in part (a). Furthermore,

$$w_x(0,t) - h_1 w(0,t) = u_x(0,t) - v'(0) - h_1 u(0,t) + h_1 v(0)$$
$$= u_x(0,t) - h_1 u(0,t) - v'(0) + h_1 v(0) = 0.$$

Similarly, the other boundary condition is also homogeneous.

25. In this problem, $F(x) = -\pi^2 \cos \pi x$. First find a solution of the boundary value problem

$$v'' = \pi^2 \cos \pi x, \qquad v'(0) = 0, \; v(1) = 1.$$

The general solution is $v(x) = Ax + B - \cos \pi x$. Imposing the initial conditions, the solution of the related BVP is $v(x) = -\cos \pi x$. Now let $w(x,t) = u(x,t) + \cos \pi x$. It follows that $w(x,t)$ satisfies the homogeneous boundary value problem, and the initial condition $w(x,0) = \cos(3\pi x/2) - \cos \pi x - (-\cos \pi x) = \cos(3\pi x/2)$. We now seek solutions of the homogeneous problem of the form

$$w(x,t) = \sum_{n=1}^{\infty} b_n(t)\phi_n(x),$$

in which $\phi_n(x) = \sqrt{2}\cos(2n-1)\pi x/2$ are the normalized eigenfunctions of the homogeneous problem and $\lambda_n = (2n-1)^2\pi^2/4$, with $n = 1, 2, \ldots$. Substitution into the PDE for w, we have

$$\sum_{n=1}^{\infty} b_n'(t)\phi_n(x) = \sum_{n=1}^{\infty} b_n(t)\phi_n''(x) = -\sum_{n=1}^{\infty} \lambda_n b_n(t)\phi_n(x).$$

Since the latter equation is valid for $0 < x < 1$, it follows that

$$b_n'(t) + \lambda_n b_n(t) = 0, \; n = 1, 2, \ldots,$$

with $b_n(t) = b_n(0)e^{-\lambda_n t}$. Hence

$$w(x,t) = \sum_{n=1}^{\infty} b_n(0)e^{-\lambda_n t}\phi_n(x).$$

Imposing the initial condition, we require that

$$\sqrt{2}\sum_{n=1}^{\infty} b_n(0)\cos\frac{(2n-1)\pi x}{2} = \cos\frac{3\pi x}{2}.$$

It is evident that all of the coefficients are zero, except for $b_2(0) = 1/\sqrt{2}$. Therefore

$$w(x,t) = e^{-9\pi^2 t/4}\cos\frac{3\pi x}{2},$$

and the solution of the original BVP is

$$u(x,t) = e^{-9\pi^2 t/4}\cos\frac{3\pi x}{2} - \cos\pi x.$$

26.(a) Let $u(x,t) = X(x)T(t)$. Substituting into the homogeneous form of (i),

$$r(x)XT' = [p(x)X']'T - q(x)XT.$$

Now divide both sides of the resulting equation by XT to obtain

$$\frac{T'}{T} = \frac{[p(x)X']'}{r(x)X} - \frac{q(x)}{r(x)} = -\lambda.$$

It follows that

$$-[p(x)X']' + q(x)X = \lambda r(x)X.$$

Since the boundary conditions (ii) are valid for all $t > 0$, we also have

$$X'(0) - h_1 X(0) = 0, \quad X'(1) + h_2 X(1) = 0.$$

(b) Let λ_n and $\phi_n(x)$ denote the eigenvalues and eigenfunctions of the BVP in the previous part (a). Assume a solution, of the PDE (i), of the form

$$u(x,t) = \sum_{n=1}^{\infty} b_n(t)\phi_n(x).$$

Substituting into (i),

$$r(x)\sum_{n=1}^{\infty} b_n''(t)\phi_n = \sum_{n=1}^{\infty} b_n(t)\left\{[p(x)\phi_n']' - q(x)\phi_n\right\} + F(x,t)$$

$$= \sum_{n=1}^{\infty} b_n(t)[-\lambda_n r(x)\phi_n] + F(x,t).$$

Rearranging the terms,

$$r(x)\sum_{n=1}^{\infty}[b_n''(t) + \lambda_n b_n(t)]\phi_n = F(x,t),$$

or

$$\sum_{n=1}^{\infty}[b_n''(t) + \lambda_n b_n(t)]\phi_n = \frac{F(x,t)}{r(x)}.$$

Now expand the right hand side in terms of the eigenfunctions. That is, write

$$\frac{F(x,t)}{r(x)} = \sum_{n=1}^{\infty} \gamma_n(t)\phi_n(x),$$

in which

$$\gamma_n(t) = \int_0^1 r(x)\frac{F(x,t)}{r(x)}\phi_n(x)dx = \int_0^1 F(x,t)\phi_n(x)dx, \ n = 1, 2, \ldots.$$

Combining these results, we have $\sum_{n=1}^{\infty}[b_n''(t) + \lambda_n b_n(t) - \gamma_n(t)]\phi_n = 0$ It follows that

$$b_n''(t) + \lambda_n b_n(t) = \gamma_n(t), \ n = 1, 2, \ldots.$$

In order to solve this sequence of ODEs, we require initial conditions $b_n(0)$ and $b_n'(0)$. Note that

$$u(x,0) = \sum_{n=1}^{\infty} b_n(0)\phi_n(x) \text{ and } u_t(x,0) = \sum_{n=1}^{\infty} b_n'(0)\phi_n(x).$$

Based on the given initial conditions,

$$f(x) = \sum_{n=1}^{\infty} b_n(0)\phi_n(x) \text{ and } g(x) = \sum_{n=1}^{\infty} b_n'(0)\phi_n(x).$$

Hence $b_n(0) = \alpha_n$ and $b_n'(0) = \beta_n$, the expansion coefficients for $f(x)$ and $g(x)$ in terms of the eigenfunctions, $\phi_n(x)$.

27.(a) Since the eigenvectors are orthogonal, they form a basis. Given any vector $\mathbf{b}$,

$$\mathbf{b} = \sum_{i=1}^{n} b_i \boldsymbol{\xi}^{(i)}.$$

Taking the inner product, with $\boldsymbol{\xi}^{(j)}$, of both sides of the equation, we have

$$(\mathbf{b}, \boldsymbol{\xi}^{(j)}) = b_j(\boldsymbol{\xi}^{(j)}, \boldsymbol{\xi}^{(j)}).$$

(b) Consider solutions of the form
$$\mathbf{x} = \sum_{i=1}^{n} a_i \boldsymbol{\xi}^{(i)}.$$

Substituting into Eq.(i), and using the above form of $\mathbf{b}$,
$$\sum_{i=1}^{n} a_i \mathbf{A}\boldsymbol{\xi}^{(i)} - \sum_{i=1}^{n} \mu a_i \boldsymbol{\xi}^{(i)} = \sum_{i=1}^{n} b_i \boldsymbol{\xi}^{(i)}.$$

It follows that
$$\sum_{i=1}^{n} [a_i \lambda_i - \mu a_i - b_i] \boldsymbol{\xi}^{(i)} = \mathbf{0}.$$

Since the eigenvectors are linearly independent,
$$a_i \lambda_i - \mu a_i - b_i = 0, \text{ for } i = 1, 2, \ldots, n.$$

That is,
$$a_i = b_i / (\lambda_i - \mu), \quad i = 1, 2, \ldots, n.$$

Assuming that the eigenvectors are normalized, the solution is given by
$$\mathbf{x} = \sum_{i=1}^{n} \frac{(\mathbf{b}, \boldsymbol{\xi}^{(i)})}{\lambda_i - \mu} \boldsymbol{\xi}^{(i)},$$

as long as μ is not equal to one of the eigenvalues.

29. First write the ODE as $y'' + y = -f(x)$. A fundamental set of solutions of the homogeneous equation is given by
$$y_1 = \cos x \text{ and } y_2 = \sin x.$$

The Wronskian is equal to $W[\cos x, \sin x] = 1$. Applying the method of variation of parameters, a particular solution is
$$Y(x) = y_1(x)u_1(x) + y_2(x)u_2(x),$$

in which
$$u_1(x) = \int_0^x \sin(s)f(s)ds \text{ and } u_2(x) = -\int_0^x \cos(s)f(s)ds.$$

Therefore the general solution is
$$y = \phi(x) = c_1 \cos x + c_2 \sin x + \cos x \int_0^x \sin(s)f(s)ds - \sin x \int_0^x \cos(s)f(s)ds.$$

Imposing the boundary conditions, we must have $c_1 = 0$ and
$$c_2 \sin 1 + \cos 1 \int_0^1 \sin(s)f(s)ds - \sin 1 \int_0^1 \cos(s)f(s)ds = 0.$$

It follows that
$$c_2 = \frac{1}{\sin 1} \int_0^1 \sin(1-s)f(s)ds,$$

and
$$\phi(x) = \frac{\sin x}{\sin 1} \int_0^1 \sin(1-s) f(s) ds - \int_0^x \sin(x-s) f(s) ds.$$

Using standard identities,
$$\sin x \cdot \sin(1-s) - \sin 1 \cdot \sin(x-s) = \sin s \cdot \sin(1-x).$$

Therefore
$$\frac{\sin x \cdot \sin(1-s)}{\sin 1} - \sin(x-s) = \frac{\sin s \cdot \sin(1-x)}{\sin 1}.$$

Splitting up the first integral, we obtain
$$\phi(x) = \int_0^x \frac{\sin s \cdot \sin(1-x)}{\sin 1} f(s) ds + \int_x^1 \frac{\sin x \cdot \sin(1-s)}{\sin 1} f(s) ds = \int_0^1 G(x,s) f(s) ds,$$

in which
$$G(x,s) = \begin{cases} \frac{\sin s \cdot \sin(1-x)}{\sin 1}, & 0 \le s \le x \\ \frac{\sin x \cdot \sin(1-s)}{\sin 1}, & x \le s \le 1. \end{cases}$$

31. The general solution of the homogeneous problem is $y = c_1 + c_2 x$. By inspection, it is easy to see that $y_1(x) = 1$ satisfies the BC $y'(0) = 0$ and that the function $y_2(x) = 1 - x$ satisfies the BC $y(1) = 0$. The Wronskian of these solutions is $W[y_1, y_2] = -1$. Based on Problem 30, with $p(x) = 1$, the Green's function is given by
$$G(x,s) = \begin{cases} 1-x, & 0 \le s \le x \\ 1-s, & x \le s \le 1. \end{cases}$$

Therefore the solution of the given BVP is
$$\phi(x) = \int_0^x (1-x) f(s) ds + \int_x^1 (1-s) f(s) ds.$$

32. The general solution of the homogeneous problem is $y = c_1 + c_2 x$. We find that $y_1(x) = x$ satisfies the BC $y(0) = 0$. Imposing the boundary condition $y(1) + y'(1) = 0$, we must have $c_1 + 2c_2 = 0$. Hence choose $y_2(x) = -2 + x$. The Wronskian of these solutions is $W[y_1, y_2] = 2$. Based on Problem 30, with $p(x) = 1$, the Green's function is given by
$$G(x,s) = \begin{cases} s(x-2)/2, & 0 \le s \le x \\ x(s-2)/2, & x \le s \le 1. \end{cases}$$

Therefore the solution of the given BVP is
$$\phi(x) = \frac{1}{2} \int_0^x s(x-2) f(s) ds + \frac{1}{2} \int_x^1 x(s-2) f(s) ds.$$

34. The general solution of the homogeneous problem is $y = c_1 + c_2 x$. By inspection, it is easy to see that $y_1(x) = x$ satisfies the BC $y(0) = 0$ and that the function $y_2(x) = 1$ satisfies the BC $y'(1) = 0$. The Wronskian of these solutions

is $W[y_1, y_2] = -1$. Based on Problem 30, with $p(x) = 1$, the Green's function is given by

$$G(x,s) = \begin{cases} s, & 0 \leq s \leq x \\ x, & x \leq s \leq 1. \end{cases}$$

Therefore the solution of the given BVP is

$$\phi(x) = \int_0^x sf(s)ds + \int_x^1 xf(s)ds.$$

35.(a) We proceed to show that if the expression given by Eq.(iv) is substituted into the integral of Eq.(iii), then the result is the solution of the nonhomogeneous problem. As long as we can interchange the summation and integration,

$$y = \phi(x) = \int_0^1 G(x,s,\mu)f(s)ds = \sum_{n=1}^{\infty} \frac{\phi_i(x)}{\lambda_i - \mu} \int_0^1 f(s)\phi_i(s)ds.$$

Note that

$$\int_0^1 f(s)\phi_i(s)ds = c_i.$$

Therefore

$$y = \phi(x) = \sum_{n=1}^{\infty} \frac{c_i \phi_i(x)}{\lambda_i - \mu},$$

as given by Eq.(13) in the text. It is assumed that the eigenfunctions are normalized and $\lambda_i \neq \mu$.

(b) For any fixed value of x, $G(x,s,\mu)$ is a function of s and the parameter μ. With appropriate assumptions on G, we can write the eigenfunction expansion

$$G(x,s,\mu) = \sum_{i=1}^{\infty} a_i(x,\mu)\phi_i(s).$$

Since the eigenfunctions are orthonormal with respect to $r(x)$,

$$\int_0^1 G(x,s,\mu)r(s)\phi_i(s)ds = a_i(x,\mu).$$

Now let

$$y_i(x) = \int_0^1 G(x,s,\mu)r(s)\phi_i(s)ds.$$

Based on the association $f(x) = r(x)\phi_i(x)$, it is evident that

$$L[y_i] = \mu r(x)y_i(x) + r(x)\phi_i(x).$$

In order to evaluate the left hand side, we consider the eigenfunction expansion

$$y_i(x) = \sum_{k=1}^{\infty} b_{ik}\phi_k(x).$$

It follows that
$$L[y_i] = \sum_{k=1}^{\infty} b_{ik} L[\phi_k] = \sum_{k=1}^{\infty} b_{ik} \lambda_k \, r(x) \phi_k(x).$$

Therefore
$$r(x) \sum_{k=1}^{\infty} b_{ik} \lambda_k \phi_k(x) = \mu \, r(x) \sum_{k=1}^{\infty} b_{ik} \phi_k(x) + r(x) \phi_i(x),$$

and since $r(x) \neq 0$,
$$\sum_{k=1}^{\infty} b_{ik} \lambda_k \phi_k(x) = \mu \sum_{k=1}^{\infty} b_{ik} \phi_k(x) + \phi_i(x).$$

Rearranging the terms, we find that
$$\phi_i(x) = \sum_{k=1}^{\infty} b_{ik}(\lambda_k - \mu) \phi_k(x).$$

Since the eigenfunctions are linearly independent, $b_{ik}(\lambda_k - \mu) = \delta_{ik}$, and thus
$$y_i(x) = \sum_{k=1}^{\infty} \frac{\delta_{ik}}{\lambda_k - \mu} \phi_k(x) = \frac{1}{\lambda_i - \mu} \phi_i(x).$$

We conclude that
$$a_i(x, \mu) = \frac{1}{\lambda_i - \mu} \phi_i(x),$$

which verifies that
$$G(x, s, \mu) = \sum_{i=1}^{\infty} \frac{\phi_i(x) \phi_i(s)}{\lambda_i - \mu}.$$

36. First note that $-d^2y/ds^2 = 0$ for $s \neq x$. On the interval $0 < s < x$, the solution of the ODE is $y_1(s) = c_1 + c_2 s$. Given that $y(0) = 0$, we have $y_1(s) = c_2 s$. On the interval $x < s < 1$, the solution is $y_2(s) = d_1 + d_2 s$. Imposing the condition $y(1) = 0$, we have $y_2(s) = d_1(1 - s)$. Assuming continuity of the solution, at $s = x$,
$$c_2 x = d_1(1 - x),$$

which gives $c_2 = d_1(1-x)/x$. Next, integrate both sides of the given ODE over an infinitesimal interval containing $s = x$:
$$-\int_{x^-}^{x^+} \frac{d^2 y}{ds^2} ds = \int_{x^-}^{x^+} \delta(s - x) ds = 1.$$

It follows that
$$y'(x^-) - y'(x^+) = 1,$$

and hence $c_2 - (-d_1) = 1$. Solving for the two coefficients, we obtain $c_2 = 1 - x$ and $d_1 = x$. Therefore the solution of the BVP is given by
$$y(s) = \begin{cases} s(1-x), & 0 \le s \le x \\ x(1-s), & x \le s \le 1, \end{cases}$$

which is identical to the Green's function in Problem 28.

11.4

1. Let $\phi_n(x) = J_0(\sqrt{\lambda_n}\, x)$ be the eigenfunctions of the singular problem
$$-(x\,y')' = \lambda x y\,,\quad 0 < x < 1,$$
$$y\,,y'\text{ bounded as }x \to 0,\ y(1) = 0.$$

Let $\phi(x)$ be a solution of the given BVP, and set
$$\phi(x) = \sum_{n=0}^{\infty} b_n \phi_n(x). \qquad (*)$$

Then
$$-(x\,\phi')' = \mu\, x\phi + f(x) = \mu\, x\phi + x\,\frac{f(x)}{x}.$$

Substituting $(*)$, we obtain
$$\sum_{n=0}^{\infty} b_n \lambda_n x\, \phi_n(x) = \mu x \sum_{n=0}^{\infty} b_n \phi_n(x) + x \sum_{n=0}^{\infty} c_n \phi_n(x),$$

in which the c_n are the expansion coefficients of $f(x)/x$ for $x > 0$. That is,
$$c_n = \frac{1}{\|\phi_n(x)\|^2} \int_0^1 x\,\frac{f(x)}{x}\phi_n(x)\,dx = \frac{1}{\|\phi_n(x)\|^2}\int_0^1 f(x)\phi_n(x)\,dx.$$

It follows that if $x \neq 0$,
$$\sum_{n=0}^{\infty} [c_n - b_n(\lambda_n - \mu)]\,\phi_n(x) = 0.$$

As long as $\mu \neq \lambda_n$, linear independence of the eigenfunctions implies that
$$b_n = \frac{c_n}{\lambda_n - \mu},\quad n = 1, 2, \ldots.$$

Therefore a formal solution is given by
$$\phi(x) = \sum_{n=0}^{\infty} \frac{c_n}{\lambda_n - \mu}\, J_0(\sqrt{\lambda_n}\, x),$$

in which $\sqrt{\lambda_n}$ are the positive roots of $J_0(x) = 0$.

3.(a) Setting $t = \sqrt{\lambda}\, x$, it follows that
$$\frac{dy}{dx} = \sqrt{\lambda}\,\frac{dy}{dt}\quad\text{and}\quad \frac{d^2y}{dx^2} = \lambda\,\frac{d^2y}{dt^2}.$$

The given ODE can be expressed as
$$-\sqrt{\lambda}\,\frac{d}{dt}\Big(\frac{t}{\sqrt{\lambda}}\sqrt{\lambda}\,\frac{dy}{dt}\Big) + \frac{k^2\sqrt{\lambda}}{t} = \sqrt{\lambda}\,t\,y,$$

or
$$-\frac{d}{dt}\left(t\frac{dy}{dt}\right) + \frac{k^2}{t} = ty.$$

An equivalent form is given by
$$t^2\frac{dy}{dt} + t\frac{dy}{dt} + (t^2 - k^2)y = 0,$$

which is known as a Bessel equation of order k. A bounded solution is $J_k(t)$.

(b) $J_k(\sqrt{\lambda}\,x)$ satisfies the boundary condition at $x = 0$. Imposing the other boundary condition, it is necessary that $J_k(\sqrt{\lambda}) = 0$. Therefore the eigenvalues are given by λ_n, $n = 1, 2 \ldots$, where $\sqrt{\lambda_n}$ are the positive zeroes of $J_k(x)$. The eigenfunctions of the BVP are $\phi_n(x) = J_k(\sqrt{\lambda_n}\,x)$.

(c) The BVP is a singular Sturm-Liouville problem with
$$L[y] = -(x\,y')' + \frac{k^2}{x}y \text{ and } r(x) = 1.$$

We note that
$$\lambda_n \int_0^1 x\,\phi_n(x)\phi_m(x)dx = \int_0^1 L[\phi_n]\,\phi_m(x)dx =$$
$$= \int_0^1 \phi_n(x)L[\phi_m]\,dx = \lambda_m \int_0^1 x\,\phi_n(x)\phi_m(x)dx.$$

Therefore
$$(\lambda_n - \lambda_m)\int_0^1 x\,\phi_n(x)\phi_m(x)dx = 0.$$

So for $n \neq m$, we have $\lambda_n \neq \lambda_m$ and
$$\int_0^1 x\,\phi_n(x)\phi_m(x)dx = 0.$$

(d) Consider the expansion
$$f(x) = \sum_{n=0}^{\infty} a_n\phi_n(x).$$

Multiplying both sides of equation by $x\,\phi_j(x)$ and integrating from 0 to 1, and using the orthogonality of the eigenfunction,
$$\int_0^1 x\,f(x)\phi_j(x)dx = \sum_{n=0}^{\infty} a_n \int_0^1 x\,\phi_j(x)\phi_n(x)dx = a_j \int_0^1 x\,\phi_j(x)\phi_j(x)dx.$$

Therefore
$$a_j = \int_0^1 x\,f(x)\phi_j(x)dx \Big/ \int_0^1 x\,[\phi_j(x)]^2\,dx, \quad j = 1, 2, \ldots.$$

(e) Let $\phi(x)$ be a solution of the given BVP, and set

$$\phi(x) = \sum_{n=0}^{\infty} b_n \phi_n(x), \qquad (*)$$

where $\phi_n(x) = J_k(\sqrt{\lambda_n}\, x)$. Then

$$L[\phi] = \mu x \phi + f(x) = \mu x \phi + x \frac{f(x)}{x}.$$

Substituting $(*)$, we obtain

$$\sum_{n=0}^{\infty} b_n \lambda_n x\, \phi_n(x) = \mu x \sum_{n=0}^{\infty} b_n \phi_n(x) + x \sum_{n=0}^{\infty} c_n \phi_n(x),$$

in which the c_n are the expansion coefficients of $f(x)/x$ for $x > 0$. That is,

$$c_n = \frac{1}{\|\phi_n(x)\|^2} \int_0^1 x \frac{f(x)}{x} \phi_n(x)\,dx = \frac{1}{\|J_k(\sqrt{\lambda_n}\,x)\|^2} \int_0^1 f(x) J_k(\sqrt{\lambda_n}\,x)\,dx.$$

It follows that if $x \neq 0$,

$$\sum_{n=0}^{\infty} [c_n - b_n(\lambda_n - \mu)] J_k(\sqrt{\lambda_n}\,x) = 0.$$

As long as $\mu \neq \lambda_n$, linear independence of the eigenfunctions implies that

$$b_n = \frac{c_n}{\lambda_n - \mu}, \quad n = 1, 2, \ldots.$$

Therefore a formal solution is given by

$$\phi(x) = \sum_{n=0}^{\infty} \frac{c_n}{\lambda_n - \mu} J_k(\sqrt{\lambda_n}\,x).$$

5.(a) Setting $\lambda = \alpha^2$ in Problem 15 of Section 11.1, the Chebyshev equation can also be written as

$$-\left[\sqrt{1-x^2}\, y'\right]' = \frac{\lambda}{\sqrt{1-x^2}}\, y.$$

Note that
$$p(x) = \sqrt{1-x^2}\,,\quad q(x) = 0,\ \text{and}\ r(x) = 1/\sqrt{1-x^2}\,,$$

hence both boundary points are singular.

(b) Observe that $p(1-\varepsilon) = \sqrt{2\varepsilon - \varepsilon^2}$ and $p(-1+\varepsilon) = \sqrt{2\varepsilon - \varepsilon^2}$. It follows that if $u(x)$ and $v(x)$ satisfy the boundary conditions (iii), then

$$\lim_{\varepsilon \to 0^+} p(1-\varepsilon)\left[u'(1-\varepsilon)v(1-\varepsilon) - u(1-\varepsilon)v'(1-\varepsilon)\right] = 0$$

and

$$\lim_{\varepsilon \to 0^+} p(-1+\varepsilon)\left[u'(-1+\varepsilon)v(-1+\varepsilon) - u(-1+\varepsilon)v'(-1+\varepsilon)\right] = 0.$$

Therefore Eq.(17) is satisfied and the boundary value problem is self-adjoint.

(c) For $n \neq 0$,
$$n^2 \int_{-1}^{1} \frac{T_0(x) T_n(x)}{\sqrt{1-x^2}} dx = \int_{-1}^{1} T_0(x) L[T_n] dx = \int_{-1}^{1} L[T_0] T_n(x) dx = 0,$$
since $L[T_0] = 0 \cdot T_0 = 0$. Otherwise,
$$n^2 \int_{-1}^{1} \frac{T_n(x) T_m(x)}{\sqrt{1-x^2}} dx = \int_{-1}^{1} L[T_n] T_m(x) dx =$$
$$= \int_{-1}^{1} T_n(x) L[T_m] dx = m^2 \int_{-1}^{1} \frac{T_n(x) T_m(x)}{\sqrt{1-x^2}} dx.$$
Therefore
$$(n^2 - m^2) \int_{-1}^{1} \frac{T_n(x) T_m(x)}{\sqrt{1-x^2}} dx = 0.$$
So for $n \neq m$,
$$\int_{-1}^{1} \frac{T_n(x) T_m(x)}{\sqrt{1-x^2}} dx = 0.$$

11.5

3. The equations relating to this problem are given by Eqs.(2) to (17) in the text. Based on the boundary conditions, the eigenfunctions are $\phi_n(x) = J_0(\lambda_n r)$ and the associated eigenvalues $\lambda_1, \lambda_2, \ldots$ are the positive zeroes of $J_0(\lambda)$. The general solution has the form
$$u(r,t) = \sum_{n=1}^{\infty} [c_n J_0(\lambda_n r) \cos \lambda_n at + k_n J_0(\lambda_n r) \sin \lambda_n at].$$
The initial conditions require that
$$u(r,0) = \sum_{n=1}^{\infty} c_n J_0(\lambda_n r) = f(r)$$
and
$$u_t(r,0) = \sum_{n=1}^{\infty} a\lambda_n k_n J_0(\lambda_n r) = g(r).$$
The coefficients c_n and k_n are obtained from the respective eigenfunction expansions. That is,
$$c_n = \frac{1}{\|J_0(\lambda_n r)\|^2} \int_0^1 r f(r) J_0(\lambda_n r) dr$$
and
$$k_n = \frac{1}{a\lambda_n \|J_0(\lambda_n r)\|^2} \int_0^1 r g(r) J_0(\lambda_n r) dr,$$
in which
$$\|J_0(\lambda_n r)\|^2 = \int_0^1 r [J_0(\lambda_n r)]^2 dr$$

for $n = 1, 2, \ldots$.

8. A more general equation was considered in Problem 23 of Section 10.5. Assuming a solution of the form $u(r,t) = R(r)T(t)$, substitution into the PDE results in

$$\alpha^2 \left[R''T + \frac{1}{r} R'T \right] = RT'.$$

Dividing both sides of the equation by the factor RT, we obtain

$$\frac{R''}{R} + \frac{1}{r} \frac{R'}{R} = \frac{T'}{\alpha^2 T}.$$

Since both sides of the resulting differential equation depend on different variables, each side must be equal to a constant, say $-\lambda^2$. That is,

$$\frac{R''}{R} + \frac{1}{r} \frac{R'}{R} = \frac{T'}{\alpha^2 T} = -\lambda^2.$$

It follows that $T' + \alpha^2 \lambda^2 T = 0$, and

$$\frac{R''}{R} + \frac{1}{r} \frac{R'}{R} = -\lambda^2,$$

which can be written as $r^2 R'' + rR' + \lambda^2 r^2 R = 0$. Introducing the variable $\xi = \lambda r$, the last equation can be expressed as $\xi^2 R'' + \xi R' + \xi^2 R = 0$, which is the Bessel equation of order zero. The temporal equation has solutions which are multiples of $T(t) = e^{-\alpha^2 \lambda^2 t}$. The general solution of the Bessel equation is

$$R(r) = b_1 J_0(\lambda_n r) + b_2 Y_0(\lambda_n r).$$

Since the steady state temperature will be zero, all solutions must be bounded, and hence we set $b_2 = 0$. Furthermore, the boundary condition $u(1,t) = 0$ requires that $R(1) = 0$ and hence $J_0(\lambda) = 0$. It follows that the eigenfunctions are $\phi_n(x) = J_0(\lambda_n r)$, with the associated eigenvalues $\lambda_1, \lambda_2, \ldots$, which are the positive zeroes of $J_0(\lambda)$. Therefore the fundamental solutions of the PDE are $u_n(r,t) = J_0(\lambda_n r) e^{-\alpha^2 \lambda_n^2 t}$, and the general solution has the form

$$u(r,t) = \sum_{n=1}^{\infty} c_n J_0(\lambda_n r) e^{-\alpha^2 \lambda_n^2 t}.$$

The initial condition requires that

$$u(r,0) = \sum_{n=1}^{\infty} c_n J_0(\lambda_n r) = f(r).$$

The coefficients in the general solution are obtained from the eigenfunction expansion of $f(r)$. That is,

$$c_n = \frac{1}{\|J_0(\lambda_n r)\|^2} \int_0^1 r f(r) J_0(\lambda_n r) dr,$$

in which

$$\|J_0(\lambda_n r)\|^2 = \int_0^1 r \left[J_0(\lambda_n r) \right]^2 dr \quad (n = 1, 2, \ldots).$$

11.6

1. The sine expansion of $f(x) = 1$, on $0 < x < 1$, is given by

$$f(x) = 2 \sum_{m=1}^{\infty} \frac{1 - \cos m\pi}{m\pi} \sin m\pi x,$$

with partial sums

$$S_n(x) = 2 \sum_{m=1}^{n} \frac{1 - \cos m\pi}{m\pi} \sin m\pi x.$$

The mean square error in this problem is

$$R_n = \int_0^1 |1 - S_n(x)|^2 \, dx.$$

Several values are shown in the Table:

n	5	10	15	20
R_n	0.067	0.04	0.026	0.02

Further numerical calculation shows that $R_n < 0.02$ for $n \geq 21$.

3.(a) The coefficients b_m are given by

$$b_m = \sqrt{2} \int_0^1 x(1-x) \sin m\pi x \, dx = \frac{2\sqrt{2}(1 - \cos m\pi)}{m^2 \pi^3},$$

so the sine expansion of $f(x) = x(1-x)$, on $0 < x < 1$, is given by

$$f(x) = 2 \sum_{m=1}^{\infty} \frac{1 - \cos m\pi}{m\pi} \sin m\pi x,$$

with partial sums

$$S_n(x) = 4 \sum_{m=1}^{n} \frac{1 - \cos m\pi}{m^3 \pi^3} \sin m\pi x.$$

(b,c) The mean square error in this problem is

$$R_n = \int_0^1 |x(1-x) - S_n(x)|^2 \, dx.$$

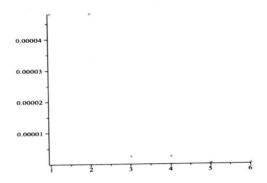

We find that $R_1 = 0.000048$.

6.(a) The function is bounded on intervals not containing $x = 0$, so for $\varepsilon > 0$,

$$\int_\varepsilon^1 f(x)dx = \int_\varepsilon^1 x^{-1/2} dx = 2 - 2\sqrt{\varepsilon}.$$

Hence the improper integral is evaluated as

$$\int_0^1 f(x)dx = \lim_{\varepsilon \to 0^+} \int_\varepsilon^1 x^{-1/2} dx = 2.$$

On the other hand, $f^2(x) = 1/x$ for $x \neq 0$, and

$$\int_\varepsilon^1 f^2(x)dx = \int_\varepsilon^1 x^{-1} dx = -\ln\sqrt{\varepsilon}.$$

Therefore the improper integral does not exist.

(b) Since $f^2(x) \equiv 1$, it is evident that the Riemann integral of $f^2(x)$ exists. Let

$$P_N = \{0 = x_1, x_2, \ldots, x_{N+1} = 1\}$$

be a partition of $[0,1]$ into equal subintervals. We can always choose a rational point, ξ_i, in each of the subintervals so that the Riemann sum

$$R(\xi_1, \xi_2, \ldots, \xi_N) = \sum_{n=1}^N f(\xi_n)\frac{1}{N} = 1.$$

Likewise, can always choose an irrational point, η_i, in each of the subintervals so that the Riemann sum

$$R(\eta_1, \eta_2, \ldots, \eta_N) = \sum_{n=1}^N f(\eta_n)\frac{1}{N} = -1.$$

It follows that $f(x)$ is not Riemann integrable.

8. With $P_0(x) = 1$ and $P_1(x) = x$, the normalization conditions are satisfied. Using the usual inner product on $[-1, 1]$,

$$\int_{-1}^1 P_0(x)P_1(x)dx = 0$$

and hence the polynomials are also orthogonal. Let $P_2(x) = a_2 x^2 + a_1 x + a_0$. The normalization condition requires that $a_2 + a_1 + a_0 = 1$. For orthogonality, we need

$$\int_{-1}^{1}(a_2 x^2 + a_1 x + a_0)dx = 0 \text{ and } \int_{-1}^{1} x(a_2 x^2 + a_1 x + a_0)dx = 0.$$

It follows that $a_2 = 3/2$, $a_1 = 0$ and $a_0 = -1/2$. Hence $P_2(x) = (3x^2 - 1)/2$. Now let $P_3(x) = a_3 x^3 + a_2 x^2 + a_1 x + a_0$. The coefficients must be chosen so that $a_3 + a_2 + a_1 + a_0 = 1$ and the orthogonality conditions

$$\int_{-1}^{1} P_i(x) P_j(x) dx = 0 \ (i \neq j)$$

are satisfied. Solution of the resulting algebraic equations leads to $a_3 = 5/2$, $a_2 = 0$, $a_1 = -3/2$ and $a_0 = 0$. Therefore $P_3(x) = (5x^3 - 3x)/2$.

11. The implied sequence of coefficients is $a_n = 1$, $n \geq 1$. Since the limit of these coefficients is not zero, the series cannot be an eigenfunction expansion.

13. Consider the eigenfunction expansion

$$f(x) = \sum_{i=1}^{\infty} a_i \phi_i(x).$$

Formally,

$$f^2(x) = \sum_{i=1}^{\infty} a_i^2 \phi_i^2(x) + 2 \sum_{i \neq j} a_i a_j \, \phi_i(x) \phi_j(x).$$

Integrating term-by-term,

$$\int_0^1 r(x) f^2(x) dx = \sum_{i=1}^{\infty} \int_0^1 a_i^2 r(x) \phi_i^2(x) dx + 2 \sum_{i \neq j} \int_0^1 a_i a_j \, r(x) \phi_i(x) \phi_j(x) dx$$

$$= \sum_{i=1}^{\infty} a_i^2 \int_0^1 \phi_i^2(x) dx,$$

since the eigenfunctions are orthogonal. Assuming that they are also normalized,

$$\int_0^1 r(x) f^2(x) dx = \sum_{i=1}^{\infty} a_i^2.$$

CPSIA information can be obtained at www.ICGtesting.com
Printed in the USA
LVOW09s1304090813
347032LV00008B/217/P